V. Kunderman

HANDBOOK OF ELECTRIC POWER CALCULATIONS

OTHER McGRAW-HILL HANDBOOKS OF INTEREST

American Institute of Physics · American Institute of Physics Handbook
Baumeister · Marks' Standard Handbook for Mechanical Engineers
Beeman · Industrial Power Systems Handbook
Bovay · Handbook of Mechanical and Electrical Systems for Buildings
Brady and Clauser · Materials Handbook
Condon and Odishaw · Handbook of Physics
Considine · Energy Technology Handbook
Coombs · Basic Electronic Instrument Handbook
Coombs · Printed Circuits Handbook
Croft, Carr, Watt, and Summers · American Electricians' Handbook
Dean · Lange's Handbook of Chemistry
Fink and Beaty · Standard Handbook for Electrical Engineers
Fink and Christiansen · Electronics Engineers' Handbook
Giacoletto · Electronics Designers' Handbook
Harper · Handbook of Components for Electronics
Harper · Handbook of Electronic Packaging
Harper · Handbook of Electronic System Design
Harper · Handbook of Materials and Processes for Electronics
Harper · Handbook of Thick Film Hybrid Microelectronics
Harper · Handbook of Wiring, Cabling, and Interconnecting for Electronics
Hicks · Standard Handbook of Engineering Calculations
Hunter · Handbook of Semiconductor Electronics
Jasik · Antenna Engineering Handbook
Juran · Quality Control Handbook
Kaufman and Seidman · Handbook of Electronics Calculations
Kaufman and Seidman · Handbook for Electronics Engineering Technicians
Kurtz and Shoemaker · The Lineman's and Cableman's Handbook
Machol · System Engineering Handbook
Maissel and Glang · Handbook of Thin Film Technology
Markus · Electronics Dictionary
McPartland · McGraw-Hill's National Electrical Code Handbook
Perry · Engineering Manual
Skolnik · Radar Handbook
Smeaton · Motor Application and Maintenance Handbook
Smeaton · Switchgear and Control Handbook
Stout · Microprocessor Applications
Stout and Kaufman · Handbook of Microcircuit Design and Application
Stout and Kaufman · Handbook of Operational Amplifier Circuit Design
Truxal · Control Engineers' Handbook
Tuma · Engineering Mathematics Handbook
Tuma · Handbook of Physical Calculations
Tuma · Technology Mathematics Handbook
Williams · Electronic Filter Design Handbook

HANDBOOK OF ELECTRIC POWER CALCULATIONS

Editors in Chief
Arthur H. Seidman

Professor of Electrical Engineering
Pratt Institute

and

Haroun Mahrous, D. Sc., P.E.

Professor of Engineering
Pratt Institute

Tyler G. Hicks, P.E., ***Series Editor***

International Engineering Associates

McGRAW-HILL BOOK COMPANY

New York St. Louis San Francisco Auckland
Bogotá Hamburg Johannesburg London Madrid
Mexico Montreal New Delhi Panama Paris
São Paulo Singapore Sydney Tokyo Toronto

Library of Congress Catologing in Publication Data

Main entry under title:

Handbook of electric power calculations.

Includes index.
1. Electric power systems—Handbooks, manuals, etc.
I.Seidman, Arthur H. II.Mahrous, Haroun. III.Hicks, Tyler Gregory, date.
TK1005.H29 1983 621.31 82-24910

ISBN 0-07-056061-7

234567890 KGPKGP 898765

The editors for this book were
Harold B. Crawford and Ruth L. Weine, and the production supervisor was Thomas G. Kowalczyk.
It was set in Baskerville by University Graphics, Inc.

Printed and bound by The Kingsport Press

CONTENTS

CONTRIBUTORS

BARDHAN, SANJIT K. *PSE&G,* Section 5, Three-Phase Induction Motors

BEATY, H. WAYNE *Electric Power Research Institute,* Section 14, System Grounding (Contributing Author: ARMIN BRUNING, *Electric Power Research Institute)*

COX, CYRUS *Department of Electrical Engineering, South Dakota School of Mines and Technology,* Section 16, Power-System Stability (Contributing Author: NORBERT PODWOISKI, *Detroit Edison)*

DeGUILMO, JOSEPH M. *Hudson County Community College,* Section 1, Network Analysis *and* Section 4, Transformers

FRIER, JOHN P. *General Electric Company,* Section 20, Lighting Design

HINMAN, WALTER L. *Gibbs & Hill, Inc.,* Section 15, Power System Grounding (Contributing Authors: W. D. CONNELLY, *Gibbs & Hill, Inc., and* J. GOHARI, *ASEA Inc.)*

HOLLANDER, LAWRENCE J. *Cooper Union,* Section 3, DC Motors and Generators; Section 6, Single-Phase Motors; Section 10, Electric-Power Networks; *and* Section 13, Short-Circuit Computations

JABLONKA, GLEN E. *Wisconsin Power & Light Co.,* Section 8, Generation of Electric Power

KAUPANG, BJORN M. *General Electric Company,* Section 19, Economic Methods

MIGLIARO, MARCO *Ebasco Services, Incorporated,* Section 7, Synchronous Machines (Contributing Author: OMAR S. MAZZONI, *Gibbs & Hill, Inc.) and* Section 18, Batteries

NAHAMKIN, MICHAEL *Gibbs & Hill, Inc.,* Section 17, Cogeneration

TRUNK, EDMUND *Nassau Community College,* Section 2, Instrumentation

WADE, JOHN S., JR. *Pennsylvania State University,* Section 9, Transmission Lines; Section 11, Load Flow Studies; *and* Section 12, Controls

PREFACE

This is the first electric power handbook to provide detailed step-by-step calculation procedures for nearly 300 problems commonly encountered in the electrical field by the engineer and technician. In 20 sections, the topics covered range from network analysis, dc and ac machines, transformers, transmission lines, system stability, and grounding to lighting design, batteries, and economic methods. Because of such a large array and diversity of topics, each section is contributed by an authority on the subject. The treatment throughout the handbook is *practical,* with little emphasis on theory.

Each of the 20 in-depth sections follows the same basic format that will assist the reader in making maximum utilization of the material. The approach to each worked-out problem is:

1. Clear statement of the problem.
2. Step-by-step calculation procedure.
3. Inclusion of suitable graphs and illustrations to clarify the procedure.
4. Use of SI and USCS equivalents.

This relatively simple, yet comprehensive, format adds greatly to the use of the handbook by the engineer and technician. Arithmetic and algebra are employed in the solution of the majority of problems. Each section contains a list of references that is pertinent to the subject matter.

The first section presents a review of dc and ac networks. Voltage, current, power, and power-factor calculations are provided for a variety of circuits. Section 2 is concerned with the methods used to measure voltage, current, power, and so on.

The next five sections cover problems dealing with dc and ac motors and generators, transformers, induction motors, and synchronous machines. Problems range from the evaluation of performance to the method of selecting the right size machine for a particular application. Section 9 concentrates on the analysis of power transmission lines. Sections 8 and 10 cover problems related to power generation and the simulation of power systems, including the per-unit representation. They are followed by Section 11 with problems in load flow studies, including a solution for a digital computer with a FORTRAN program provided.

Section 12 covers problems of power-system controls and Sections 13 through 16 deal with the reliability of power-system operation. Short-circuit problems are presented in Section 13; Section 14 covers grounding, Section 15 is concerned with system protection, and power-system stability is covered in Section 16. The remaining four sections cover the areas of cogeneration, batteries, economic methods, and lighting design.

Although considerable care was exercised to make the handbook as accurate as possible, it is inevitable that in a first edition of this size and scope some errors remain. It will be appreciated if these are brought to our attention.

We wish to gratefully acknowledge the effort and cooperation of all the contributors who helped to make this handbook a reality. They have covered their fields in an up-to-date, comprehensive, and practical manner.

Arthur H. Seidman
Haroun Mahrous

Section 1 NETWORK ANALYSIS

Joseph M. De Guilmo, P.E.
Program Coordinator, Electronics Technology, Hudson County Community College

REFERENCES Bell—*Fundamentals of Electric Circuits,* Reston; Boylestad—*Introductory Circuit Analysis,* Merrill; De Guilmo—*Electricity/Electronics: Principles and Applications,* Delmar; Floyd—*Principles of Electric Circuits,* Merrill; Hayt and Kemmerly—*Engineering Circuit Analysis,* McGraw-Hill; Hicks—*Standard Handbook of Engineering Calculations,* McGraw-Hill; Jackson—*Introduction to Electric Circuits,* Prentice-Hall; Kaufman and Seidman—*Handbook of Electronics Calculations for Engineers and Technicians,* McGraw-Hill; Oppenheimer and Borchers—*Direct and Alternating Currents,* McGraw-Hill; Tocci—*Introduction to Electric Circuit Analysis,* Merrill.

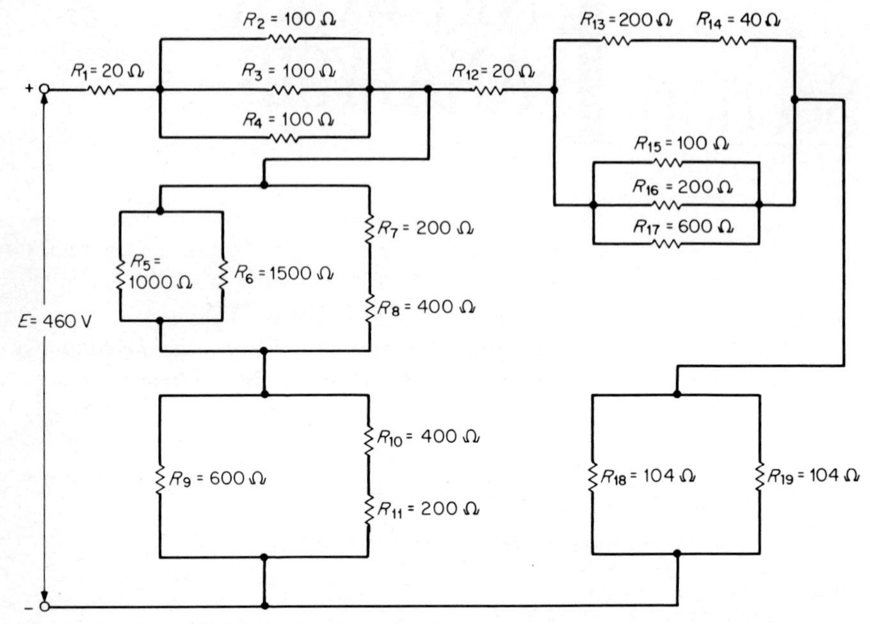

Fig. 1 A series-parallel dc circuit to be analyzed.

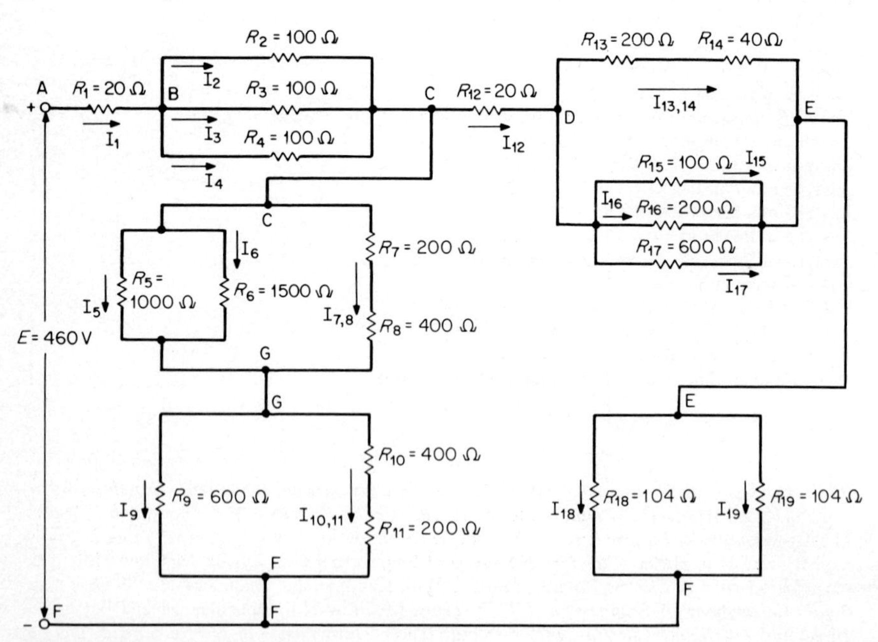

Fig. 2 Labeling the circuit of Fig. 1.

SERIES-PARALLEL DC NETWORK ANALYSIS

A direct-current circuit (network) contains 19 resistors arranged as shown in Fig. 1. Compute the current through and the voltage drop across each resistor in this circuit.

Calculation Procedure:

1. *Label the Circuit*

Label all the sections. Mark on the circuit diagram the direction of current through each resistor (Fig. 2). The equivalent resistance of the series-parallel combination of resistors can be found by successive applications of the rules for combining series resistors and parallel resistors.

2. *Combine All Series Resistors*

In a series circuit, the total or equivalent resistance R_{EQS} seen by the source is equal to the sum of the values of the individual resistors: $R_{EQS} = R_1 + R_2 + R_3 + \cdots + R_N$.

Calculate the series equivalent of the elements connected in series in sections DE, CG, and GF: R_{EQS} (section DE) $= R_{13} + R_{14} = 200 + 40 = 240\ \Omega$, R_{EQS} (section CG) $= R_7 + R_8 = 200 + 400 = 600\ \Omega$, R_{EQS} (section GF) $= R_{10} + R_{11} = 400 + 200 = 600\ \Omega$. Replace the series elements included in sections DE, CG, and GF by their equivalent values (Fig. 3).

3. *Combine All Parallel Resistors*

In the case of a parallel circuit of two unequal resistors in parallel, the total or equivalent resistance R_{EQP} can be found from the product-over-sum equation: $R_{EQP} =$

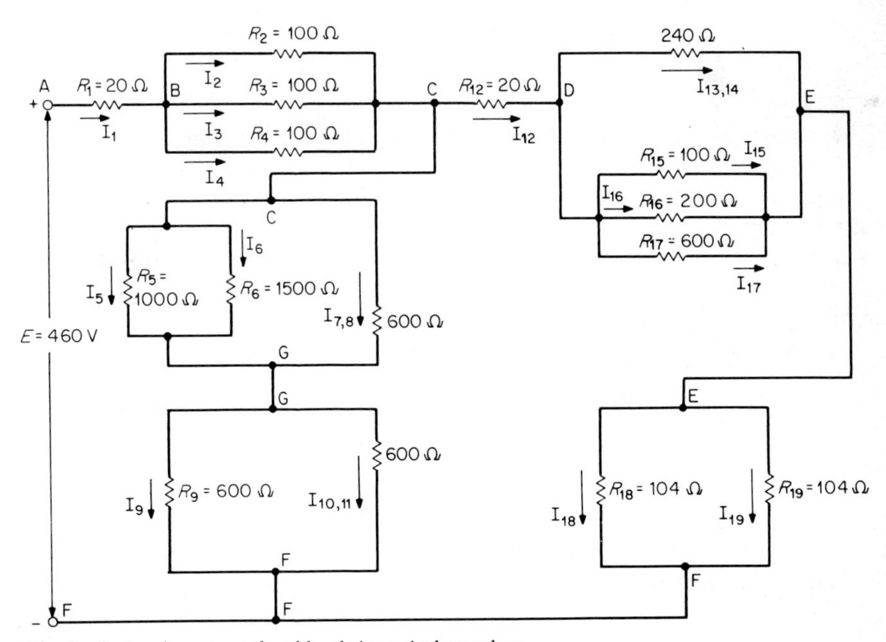

Fig. 3 Series elements replaced by their equivalent values.

$R_1 \| R_2 = R_1R_2/(R_1 + R_2)$, where $\|$ stands for *in parallel with*. The equivalent parallel resistance is always less than the smaller of the two resistors.

In section CG, $R_5 \| R_6 = (1000 \times 1500)/(1000 + 1500) = 600\ \Omega$. Section CG now consists of two 600-Ω resistors in parallel. In a case of a circuit of N equal resistors in parallel, the total, or equivalent, resistance R_{EQP} can be determined from the following equation: $R_{\text{EQP}} = R/N$, where R is the resistance of each of the parallel resistors and N is the number of resistors connected in parallel. For section CG, $R_{CG} = 600/2 = 300\ \Omega$; for section BC, $R_{BC} = 100/3 = 33\tfrac{1}{3}\ \Omega$; for section EF, $R_{EF} = 104/2 = 52\ \Omega$; for section GF, $R_{GF} = 600/2 = 300\ \Omega$

In a circuit of three or more unequal resistors in parallel, the total, or equivalent resistance R_{EQP} is equal to the inverse of the sum of the reciprocals of the individual resistance values: $R_{\text{EQP}} = 1/(1/R_1 + 1/R_2 + 1/R_3 + \cdots + 1/R_N)$. The equivalent parallel resistance is always less than the smallest-value resistor in the parallel combination.

Calculate the equivalent resistance of the elements connected in parallel in section DE: $R_{15} \| R_{16} \| R_{17} = 1/(1/100 + 1/200 + 1/600) = 60\ \Omega$ Calculate R_{DE}: $R_{DE} = 240 \| 60 = (240)(60)/(240 + 60) = 48\ \Omega$. Replace all parallel elements by their equivalent values (Fig. 4).

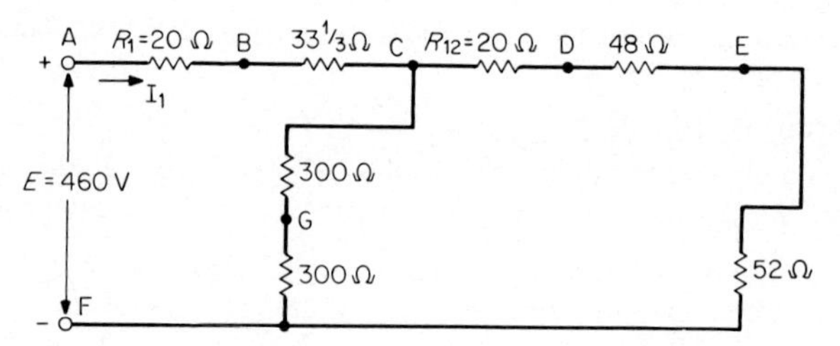

Fig. 4 Parallel elements replaced by their equivalent values.

4. *Combine the Remaining Resistances to Obtain the Total Equivalent Resistance*

Combine the equivalent series resistances of Fig. 4 to obtain the simple series-parallel circuit of Fig. 5: $R_{AB} + R_{BC} = R_{AC} = R_{\text{EQS}} = 20 + 33\tfrac{1}{3} = 53\tfrac{1}{3}\ \Omega$, $R_{CG} + R_{GF} = R_{CF} = R_{\text{EQS}} = 300 + 300 = 600\ \Omega$, $R_{CD} + R_{DE} + R_{EF} = R_{CF} = R_{\text{EQS}} = 20 + 48 + 52 = 120\ \Omega$ Calculate the total equivalent resistance R_{EQT}: $R_{\text{EQT}} = 53\tfrac{1}{3} + (600 \| 120) = 153\tfrac{1}{3}\ \Omega$ The final reduced circuit is illustrated in Fig. 6.

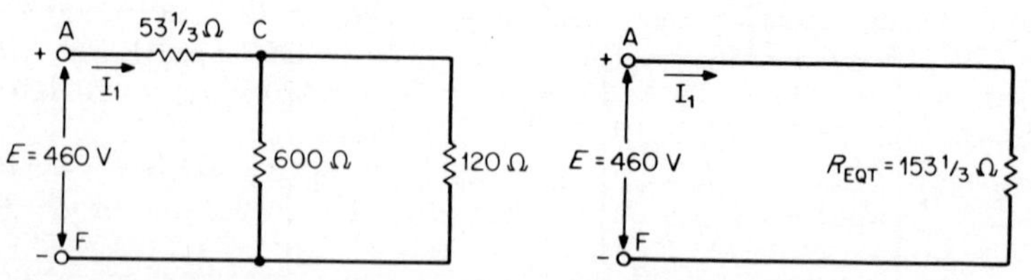

Fig. 5 Circuit of Fig. 4 reduced to a simple series-parallel configuration.

Fig. 6 Final reduced circuit of Fig. 1.

5. *Compute the Total Line Current in Fig. 6 Using Ohm's Law*

$I_1 = E/R_{\text{EQT}}$, where I_1 = total line current, E = line voltage (power-supply voltage), and R_{EQT} = line resistance or total equivalent resistance seen by power supply. Substituting values yields: $I_1 = E/R_{\text{EQT}} = 460/153\frac{1}{3} = 3$ A.

6. *Compute the Current Through, and the Voltage Drop Across, Each Resistor in the Circuit*

Refer to Fig. 2 and Fig. 4; analysis of R_1 yields: $I_1 = 3$ A (calculated in Step 5); $V_1 = V_{AB} = I_1R_1 = (3)(20) = 60$ V; and for R_2, R_3, and R_4 we have: $V_{BC} = V_2 = V_3 = V_4 = I_1R_{BC} = (3)(33\frac{1}{3}) = 100$ V. Current $I_2 = I_3 = I_4 = 100/100 = 1$ A. Hence, V_{CF} can be calculated: $V_{CF} = E - (V_{AB} + V_{BC}) = 460 - (60 + 100) = 300$ V. The current from C to G to F is $300/600 = 0.5$ A.

Kirchhoff's current law (KCL) states: The algebraic sum of the currents entering any node or junction of a circuit is equal to the algebraic sum of the currents leaving that node or junction: ΣI entering $= \Sigma I$ leaving. Applying KCL at node C, we find $I_{12} = 3 - 0.5 = 2.5$ A. Therefore, $V_{12} = V_{CD} = I_{12}R_{12} = (2.5)(20) = 50$ V.

The voltage-divider principle states that the voltage V_N across any resistor R_N in a series circuit is equal to the product of the total applied voltage V_T and R_N divided by the sum of the series resistors, R_{EQS}: $V_N = V_T(R_N/R_{\text{EQS}})$. This equation shows that V_N is directly proportional to R_N and $V_{CG} = V_{GF} = 300 \times (300/600) = 150$ V. Hence, $I_7 = I_8 = 150/600 = 0.25$ A, $V_7 = I_7R_7 = (0.25)(200) = 50$ V, $V_8 = I_8R_8 = (0.25)(400) = 100$ V, $I_{10} = I_{11} = 150/600 = 0.25$ A, $V_{10} = I_{10}R_{10} = (0.25)(400) = 100$ V, $V_{11} = I_{11}R_{11} = (0.25)(200) = 50$ V.

The current-divider principle states that in a circuit containing N parallel branches, the current I_N in a particular branch R_N is equal to the product of the applied current I_T and the equivalent resistance R_{EQP} of the parallel circuit divided by R_N: $I_N = I_T(R_{\text{EQP}}/R_N)$. When there are two resistors R_A and R_B in parallel, the current I_A in R_A is $I_A = I_T[R_B/(R_A + R_B)]$; the current I_B in R_B is $I_B = I_T[R_A/(R_A + R_B)]$. When R_A is equal to R_B, $I_A = I_B = I_T/2$. Refer to Figs. 2, 3, and 4 for the remaining calculations: $(R_5 \| R_6) = R_7 + R_8 = 600\ \Omega$.

From the preceding equations, the value of the current entering the parallel combination of R_5 and R_6 is $I_5 + I_6 = 0.5/2 = 0.25$ A, $I_5 = 0.25 \times (1500/2500) = 0.15$ A, and $I_6 = 0.25 \times (1000/2500) = 0.10$ A. Ohm's law can be used to check the value of V_5 and V_6, which should equal V_{CG} and which was previously calculated to equal 150 V: $V_5 = I_5R_5 = (0.15)(1000) = 150$ V and $V_6 = I_6R_6 = (0.10)(1500) = 150$ V.

The current entering node G equals 0.5 A. Because $R_9 = R_{10} + R_{11}$, $I_9 = I_{10} = I_{11} = 0.5/2 = 0.25$ A. From Ohm's law: $V_9 = I_9R_9 = (0.25)(600) = 150$ V, $V_{10} = I_{10}R_{10} = (0.25)(400) = 100$ V, $V_{11} = I_{11}R_{11} = (0.25)(200) = 50$ V. These values check since $V_{GF} = V_9 = 150\text{ V} = V_{10} + V_{11} = 100 + 50 = 150$ V.

The remaining calculations show that: $V_{DE} = I_{12}R_{DE} = (2.5)(48) = 120$ V, $I_{13} = I_{14} = 120/240 = 0.5$ A, $V_{13} = I_{13}R_{13} = (0.5)(200) = 100$ V, and $V_{14} = I_{14}R_{14} = (0.5)(40) = 20$ V. Since $V_{15} = V_{16} = V_{17} = V_{DE} = 120$ V, $I_{15} = 120/100 = 1.2$ A, $I_{16} = 120/200 = 0.6$ A, and $I_{17} = 120/600 = 0.2$ A.

These current values check, since $I_{15} + I_{16} + I_{17} + I_{13,14} = 1.2 + 0.6 + 0.2 + 0.5 = 2.5$ A, which enters node D and which leaves node E. Because $R_{18} = R_{19}$, $I_{18} = I_{19} = 2.5/2 = 1.25$ A and $V_{EF} = V_{18} = V_{19} = (2.5)(52) = 130$ V.

Kirchhoff's voltage law (KVL) states that the algebraic sum of the potential rises and drops around a closed loop or path is zero. This law can also be expressed as: $\Sigma V_{\text{rises}} =$

ΣV_{drops}. As a final check $E = V_{AB} + V_{BC} + V_{CD} + V_{DE} + V_{EF}$ or 460 V = 60 V + 100 V + 50 V + 120 V + 130 V = 460 V

Related Calculations: Any reducible dc circuit, i.e., any circuit with a single power source that can be reduced to one equivalent resistance, no matter how complex, can be solved in a manner similar to the preceding procedure.

BRANCH-CURRENT ANALYSIS OF A DC NETWORK

Calculate the current through each of the resistors in the dc circuit of Fig. 7 using the branch-current method of solution.

Calculation Procedure:

1. Label the Circuit

Label all the nodes (Fig. 8). There are four nodes in this circuit, indicated by the letters *A*, *B*, *C*, and *D*. A node is a junction where two or more current paths come together. A branch is a portion of a circuit consisting of one or more elements in series. Figure 8 contains three branches, each of which is a current path in the network. Branch *ABC* consists of the power supply E_1 and R_1 in series, branch *ADC* consists of the power supply E_2 and R_2 in series, and branch *CA* consists of R_3 only. Assign a distinct current of arbitrary direction to each branch of the network (I_1, I_2, I_3). Indicate the polarities of each resistor as determined by the assumed direction of current. The polarity of the power-supply terminals is fixed and is therefore not dependent on the assumed direction of current.

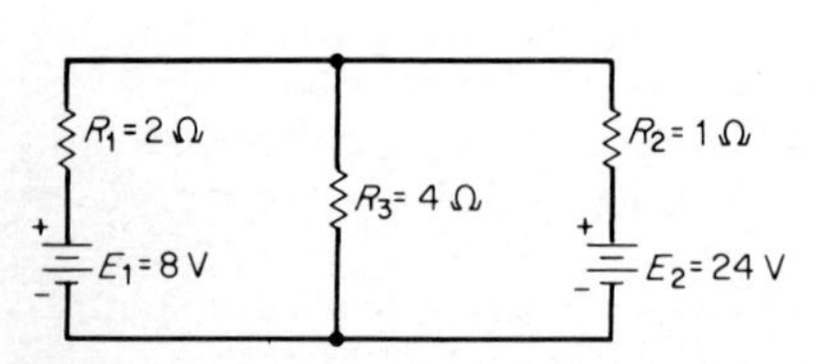

Fig. 7 Circuit to be analyzed by branch currents.

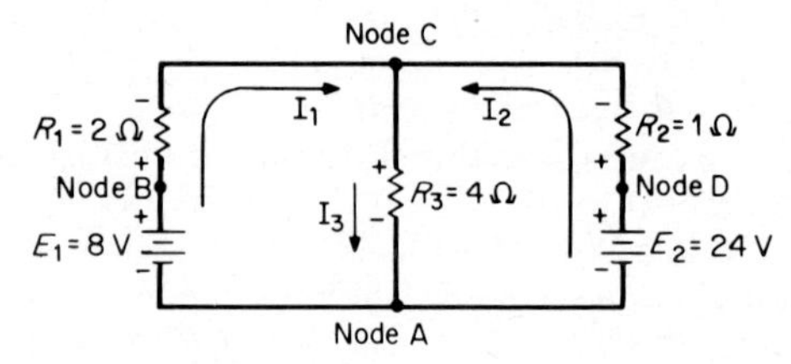

Fig. 8 Labeling the circuit of Fig. 7.

2. Apply KVL and KCL to Network

Apply KVL around each closed loop. A closed loop is any continuous connection of branches that allows us to trace a path which leaves a point in one direction and returns to that same starting point from another direction without leaving the network.

Applying KVL to the minimum number of nodes that will include all the branch currents, one obtains: loop 1 (*ABCA*): $8 - 2I_1 - 4I_3 = 0$; loop 2 (*ADCA*): $24 - I_2 - 4I_3 - 0$. KCL at node *C*: $I_1 + I_2 = I_3$.

3. *Solve the Equations*

The above three simultaneous equations can be solved by the elimination method or by using third-order determinants. The solution yields these results: $I_1 = -4$ A, $I_2 = 8$ A, and $I_3 = 4$ A. The negative sign for I_1 indicates that the actual current flows in the direction opposite to that assumed.

Related Calculations: The above calculation procedure is an application of Kirchhoff's laws to an irreducible circuit. Such a circuit cannot be solved by the method used in the previous calculation procedure because it contains two power supplies. Once the branch currents are determined, all other quantities such as voltage and power can be calculated.

MESH ANALYSIS OF A DC NETWORK

Calculate the current through each of the resistors in the dc circuit of Fig. 9 using mesh analysis.

Calculation Procedure

1. *Assign Mesh or Loop Currents*

The term mesh is used because of the similarity in appearance between the closed loops of the network and a wire mesh fence. One can view the circuit as a "window frame" and the meshes as the "windows." A mesh is a closed pathway with no other closed pathway within it. A loop is also a closed pathway, but a loop may have other closed pathways within it. Therefore all meshes are loops, but all loops are not meshes.

Loop currents I_1 and I_2 are drawn in the clockwise direction in each window (Fig. 10). The loop current or mesh current is a fictitious current that enables us to obtain the actual branch currents more easily. The number of loop currents required is always equal to the number of windows of the network. This assures that the resulting equations are all independent. Loop currents may be drawn in any direction, but assigning a clockwise direction to all of them simplifies the process of writing equations.

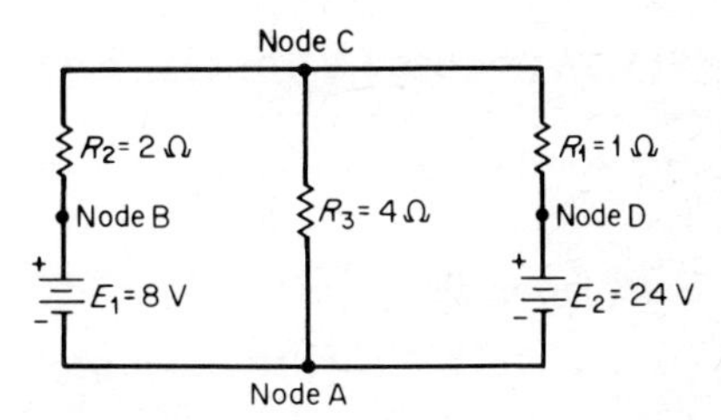

Fig. 9 Circuit to be analyzed using mesh analysis.

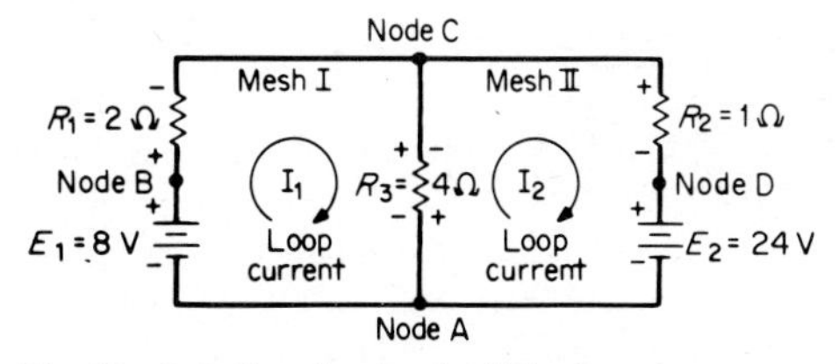

Fig. 10 Labeling the circuit of Fig. 9.

2. *Indicate the Polarities within Each Loop*

Identify polarities to agree with the assumed direction of the loop currents. The polarities across R_3 are the opposite for each loop current. The polarities of E_1 and E_2 are unaffected by the direction of the loop currents passing through them.

3. *Write KVL around Each Mesh*

Write KVL around each mesh in any direction. It is convenient to follow the same direction as the loop current: mesh I: $+8 - 2I_1 - 4(I_1 - I_2) = 0$ mesh II: $-24 - 4(I_2 - I_1) - I_2 = 0$.

4. *Solve the Equations*

Solving the two simultaneous equations gives the following results: $I_1 = -4$ A and $I_2 = -8$ A. The minus signs indicate that the two loop currents flow in a direction opposite to that assumed; i.e., they both flow counterclockwise. Loop current I_1 is therefore 4 A in the direction *CBAC*. Loop current I_2 is 8 A in the direction *ADCA*. The true direction of loop current I_2 through resistor R_3 is from C to A. The true direction of loop current I_1 through resistor R_3 is from A to C. Therefore the current through R_3 equals $(I_2 - I_1)$ or $8 - 4 = 4$ A in the direction *CA*.

Related Calculations: This procedure solved the same network as in Fig. 8. The mesh-analysis solution eliminates the need to substitute KCL into the equations derived by the application of KVL. The initial writing of the equations accomplishes the same result. Mesh analysis is therefore more frequently applied than branch-current analysis.

NODAL ANALYSIS OF A DC NETWORK

Calculate the current through each of the resistors in the dc circuit of Fig. 11 using nodal analysis.

Calculation Procedure:

1. *Label the Circuit*

Label all nodes (Fig. 12). One of the nodes (node A) is chosen as the reference node. It can be thought of as a circuit ground which is at zero voltage or ground potential. Nodes

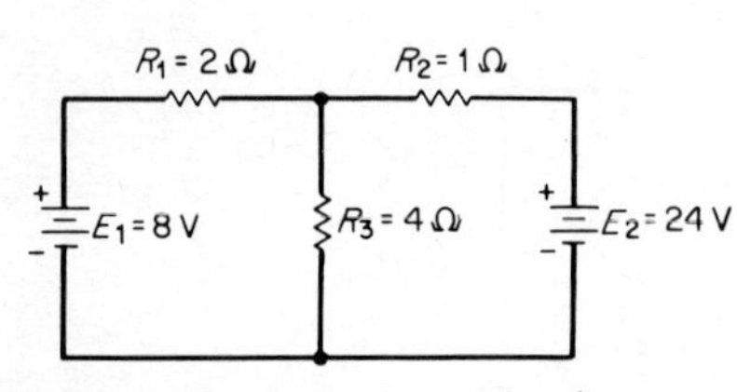

Fig. 11 Circuit to be analyzed by nodal analysis.

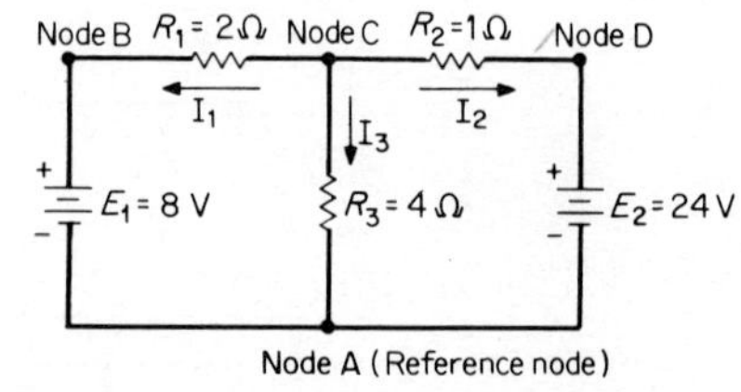

Fig. 12 Labeling the circuit of Fig. 11.

B and D are already known to be at the potential of the source voltages. The voltage at node C (V_C) is unknown.

Assume that $V_C > V_B$ and $V_C > V_D$. Draw all three currents I_1, I_2, and I_3 away from node C, i.e., toward the reference node.

2. ***Write KCL at Node C***
$I_1 + I_2 + I_3 = 0$.

3. ***Express Currents in Terms of Circuit Voltages Using Ohm's Law***
Refer to Fig. 12: $I_1 = V_1/R_1 = (V_C - 8)/2$, $I_2 = V_2/R_2 = (V_C - 24)/1$, and $I_3 = V_3/R_3 = V_C/4$.

4. ***Substitute in KCL Equation of Step 2***
Substituting the current equations obtained in Step 3 into KCL of Step 2, we find $I_1 + I_2 + I_3 = 0$ or $(V_C - 8)/2 + (V_C - 24)/1 + V_C/4 = 0$. Because the only unknown is V_C, this simple equation can be solved to obtain $V_C = 16$ V.

5. ***Solve for All Currents***
$I_1 = (V_C - 8)/2 = (16 - 8)/2 = 4$ A (true direction) and $I_2 = (V_C - 24)/1 = (16 - 24)/1 = -8$ A. The negative sign indicates that I_2 flows toward node C instead of in the assumed direction (away from node C). $I_3 = V_C/4 = 16/4 = 4$ A (true direction).

Related Calculations: Nodal analysis is a very useful technique for solving networks. This procedure solved the same circuits as in Figs. 7 and 9.

DIRECT-CURRENT NETWORK SOLUTION USING SUPERPOSITION THEOREM

Calculate the value of the current through resistor R_3 in the dc network of Fig. 13*a* using the superposition theorem.

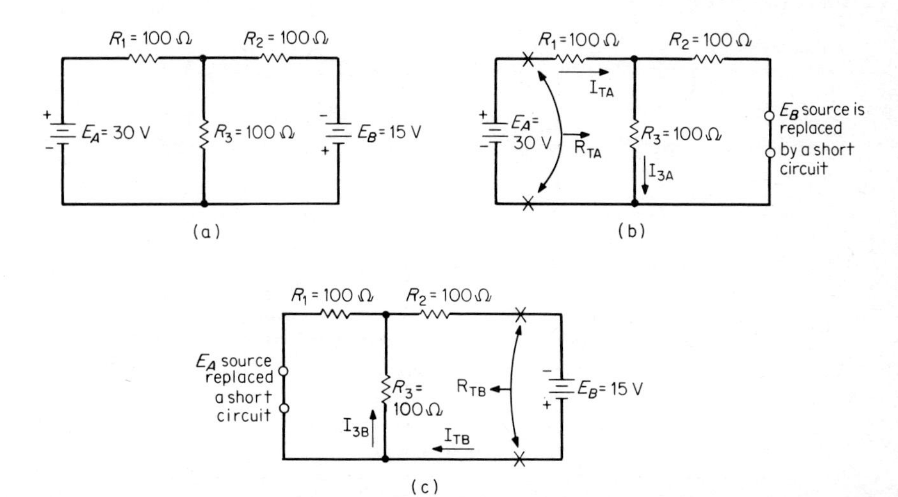

Fig. 13 Application of the superposition theorem. (*a*) Current in R_3 to be determined. (*b*) Effect of E_A alone. (*c*) Effect of E_B alone.

Calculation Procedure:

1. *Consider the Effect of E_A Alone (Fig. 13b)*

The superposition theorem states: In any linear network containing more than one source of electromotive force (emf) or current, the current through any branch is the algebraic sum of the currents produced by each source acting independently.

Because E_B has no internal resistance, the E_B source is replaced by a short circuit. (A current source, if present, is replaced by an open circuit.) Therefore, $R_{TA} = 100 + (100 \| 100) = 150\ \Omega$ and $I_{TA} = E_A/R_{TA} = 30/150 = 200$ mA. From the current-divider rule, $I_{3A} = 200\ \text{mA}/2 = 100$ mA.

2. *Consider the Effect of E_B Alone (Fig. 13c)*

Because E_A has no internal resistance, the E_A source is replaced by a short circuit. Therefore, $R_{TB} = 100 + (100 \| 100) = 150\ \Omega$ and $I_{TB} = E_B/R_{TB} = 15/150 = 100$ mA. From the current-divider rule, $I_{3B} = 100\ \text{mA}/2 = 50$ mA.

3. *Calculate the Value of I_3*

The algebraic sum of the component currents I_{3A} and I_{3B} is used to obtain the true magnitude and direction of I_3: $I_3 = I_{3A} - I_{3B} = 100 - 50 = 50$ mA (in the direction of I_{3A}).

Related Calculations: The superposition theorem simplifies the analysis of a *linear network only* having more than one source of emf. This theorem may also be applied in a network containing both dc and ac sources of emf. This is considered later in the section.

DIRECT-CURRENT NETWORK SOLUTION USING THEVENIN'S THEOREM

Calculate the value of the current I_L through the resistor R_L in the dc network of Fig. 14*a* using Thevenin's theorem.

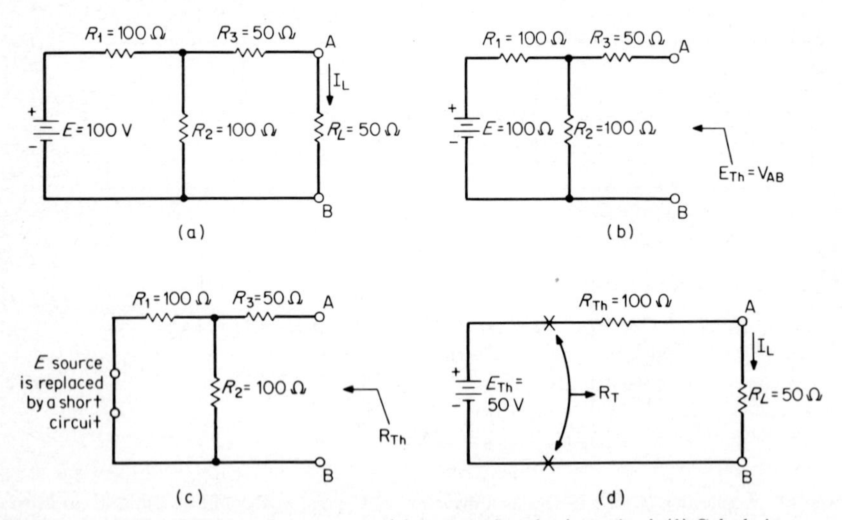

Fig. 14 Application of Thevenin's theorem. (*a*) Current I_L to be determined. (*b*) Calculating E_{Th}. (*c*) Calculating R_{Th}. (*d*) Resultant Thevenin equivalent circuit.

Calculation Procedure:

*1. Calculate the Thevenin Voltage (Fig. 14***b***)*

Thevenin's theorem states: Any two-terminal linear network containing resistances and sources of emf and current may be replaced by a single source of emf in series with a single resistance. The emf of the single source of emf, called E_{Th}, is the open-circuit emf at the network terminal. The single series resistance, called R_{Th}, is the resistance between the network terminals when all of the independent sources are replaced by their internal resistances.

When the Thevenin equivalent circuit is determined for a network, the process is known as "thevenizing" the circuit.

The load resistor is removed as shown in Fig. (14.*b*). The open-circuit terminal voltage of the network is calculated; this value is E_{Th}. Because no current can flow through R_3, the voltage E_{Th} (V_{AB}) is the same as the voltage across resistor R_2. Use the voltage-divider rule to find E_{Th}: $E_{Th} = (100\ \text{V}) \times [100/(100 + 100)] = 50\ \text{V}$.

*2. Calculate the Thevenin Resistance (Fig. 14***c***)*

The network is redrawn with the source of emf replaced by a short circuit. (If a current source were present, it is replaced by an open circuit.) The resistance of the redrawn network as seen by looking back into the network from the load terminals is calculated. This value is R_{Th}, where $R_{Th} = 50\ \Omega + (100\ \Omega)||(100\ \Omega) = 100\ \Omega$

*3. Draw the Thevenin Equivalent Circuit (Fig. 14***d***)*

The Thevenin equivalent circuit consists of the series combination of E_{Th} and R_{Th}. The load resistor R_L is connected across the output terminals of this equivalent circuit. $R_T = R_{Th} + R_L = 100 + 50 = 150\ \Omega$, and $I_L = E_{Th}/R_T = 50/150 = 1/3\ \text{A}$.

Related Calculations: With respect to the terminals only, the Thevenin circuit is equivalent to the original linear network. Changes in R_L do not require any calculations for a new Thevenin circuit. The simple series Thevenin circuit of Fig. 14*d* can be used to solve for load currents each time R_L is changed.

DIRECT-CURRENT NETWORK SOLUTION USING NORTON'S THEOREM

Calculate the value of the current I_L through the resistor R_L in the dc network of Fig. 15*a* using Norton's theorem.

Calculation Procedure:

*1. Calculate the Norton Parallel Resistance R_N (Fig. 15***b***)*

Norton's theorem states: Any two-terminal linear dc network can be replaced by an equivalent circuit consisting of a constant-current source I_N, in parallel with a resistor R_N.

The load resistor is removed (Fig. 15*b*). All sources are set to zero (current sources are replaced by open circuits, and voltage sources are replaced by short circuits). R_N is calculated as the resistance of the redrawn network as seen by looking back into the network from the load terminals A and B: $R_N = 50\ \Omega + (100\ \Omega||100\ \Omega) = 100\ \Omega$. A comparison of Figs. 14*c* and 15*b* shows that $R_N = R_{Th}$.

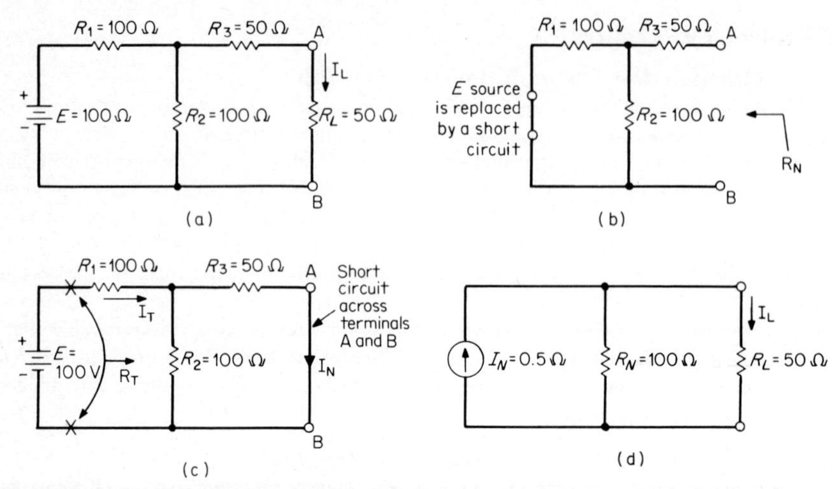

Fig. 15 Application of Norton's theorem. (*a*) Current I_L to be determined. (*b*) Calculating R_N. (*c*) Calculating I_N. (*d*) Resultant Norton equivalent circuit.

2. *Calculate the Norton Constant-Current Source I_N (Fig. 15c).*

I_N is the short-circuit current between terminals A and B. $R_T = 100\ \Omega + 100\ \Omega \| 50\ \Omega) = 133⅓\ \Omega$ and $I_T = E/R_T = (100/133⅓) = ¾$ A. From the current-divider rule: $I_N = (¾\ \text{A})\ (100)/(100 + 50) = 0.5$ A.

3. *Draw the Norton Equivalent Circuit (Fig. 15d)*

The Norton equivalent circuit consists of the parallel combination of I_N and R_N. The load resistor R_L is connected across the output terminals of this equivalent circuit. From the current-divider rule: $I_L = (0.5\ \text{A})[100/(100 + 50)] = ⅓$ A.

Related Calculations: This problem solved the same circuit as in Fig. 14*a*. It is often convenient or necessary to have a voltage source (Thevenin equivalent) rather than a current source (Norton equivalent) or a current source rather than a voltage source. Figure 16 shows the source conversion equations which indicate that a Thevenin equivalent circuit can be replaced by a Norton equivalent circuit, and vice versa, provided that the following equations are used: $R_N = R_{Th}$; $E_{Th} = I_N R_{Th} = I_N R_N$, and $I_N = E_{Th}/R_N = E_{Th}/R_{Th}$.

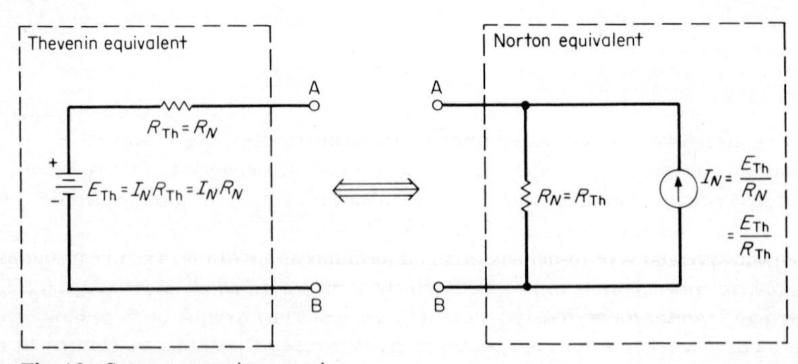

Fig. 16 Source conversion equations.

BALANCED DC BRIDGE NETWORK

Calculate the value of R_x in the balanced dc bridge network of Fig. 17.

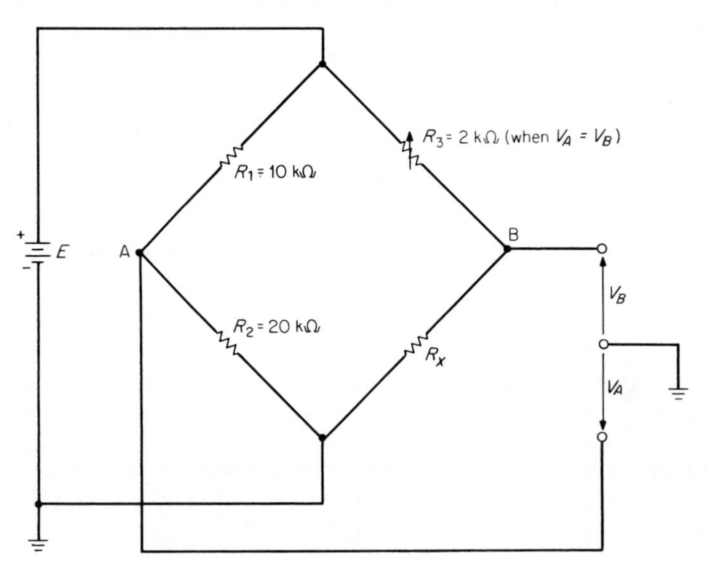

Fig. 17 Analysis of a balanced dc bridge.

Calculation Procedure:

1. Solve for R_x

The bridge network is balanced when R_3 is adjusted so that $V_A = V_B$. Then: $R_1/R_2 = R_3/R_x$. Solving for R_x, we find $R_x = R_2R_3/R_1 = (20)(2)/10 = 4$ kΩ.

Related Calculations: The bridge circuit is used in control systems, dc meters, ac meters, and in electronic circuits for converting an ac input to a unidirectional output. There is a potential drop across terminals A and B when the bridge is not balanced, causing current to flow through any element connected to those terminals. Mesh analysis, nodal analysis, Thevenin's theorem, or Norton's theorem may be used to solve the unbalanced network for voltages and currents.

UNBALANCED DC BRIDGE NETWORK

Calculate the value of R_{EQT} in the unbalanced dc bridge network of Fig. 18.

Calculation Procedure:

1. Convert the Upper Delta to an Equivalent Wye Circuit

Delta-to-wye and wye-to-delta conversion formulas apply to Fig. 19. The formulas for delta-to-wye conversion are: $R_1 = R_AR_C/(R_A + R_B + R_C)$, $R_2 = R_BR_C/(R_A + R_B + R_C)$, and $R_3 = R_BR_A/(R_A + R_B + R_C)$. The formulas for wye-to-delta conversion are: $R_A = (R_1R_2 + R_1R_3 + R_2R_3)/R_2$, $R_B = (R_1R_2 + R_1R_3 + R_2R_3/R_1)$, and $R_C = (R_1R_2 + R_1R_3 + R_2R_3)/R_3$.

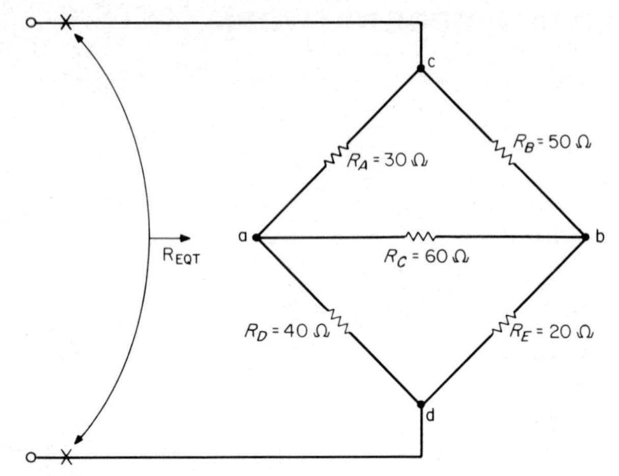

Fig. 18 Analysis of an unbalanced bridge.

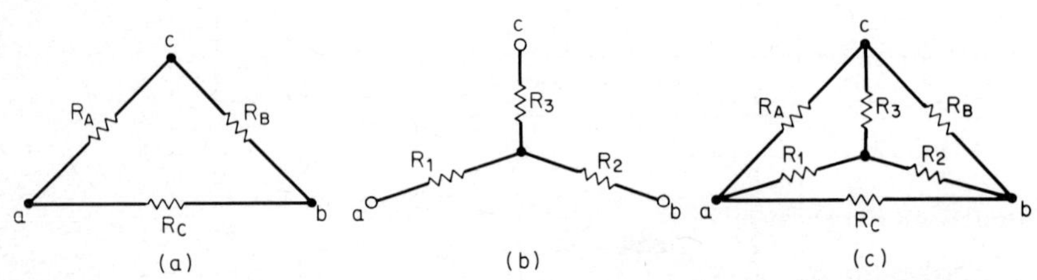

Fig. 19 (*a*) Delta circuit. (*b*) Wye circuit. (*c*) Delta-to-wye and wye-to-delta conversions.

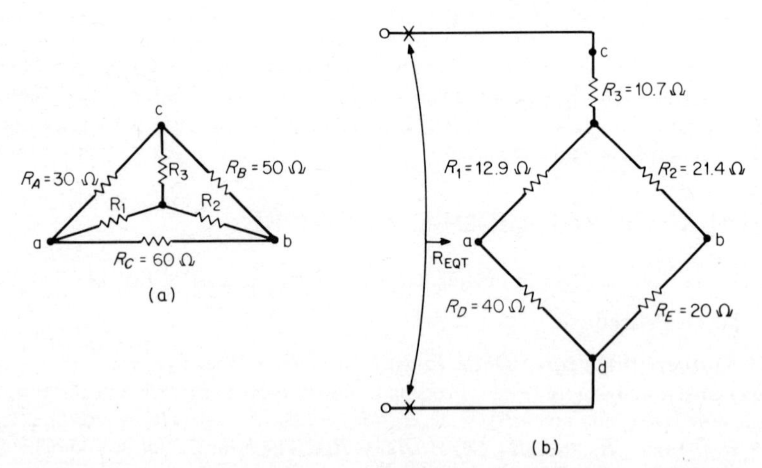

Fig. 20 Converting Fig. 18 to a series-parallel circuit. (*a*) Converting upper delta to a wye circuit. (*b*) Resultant series-parallel circuit.

The upper delta of Fig. 18 is converted to its equivalent wye by the conversion formulas (see Fig. 20): $R_1 = [(30)(60)]/(30 + 50 + 60) = 12.9\ \Omega$, $R_2 = [(50)(60)]/(30 + 50 + 60) = 21.4\ \Omega$, and $R_3 = [(50)(30)]/(30 + 50 + 60) = 10.7\ \Omega$. From the simplified series-parallel circuit of Fig. 20*b*, it can be seen that: $R_{EQT} = 10.7 + [(12.9 + 40)||(21.4 + 20)] = 33.9\ \Omega$.

Related Calculations: Delta-to-wye and wye-to-delta conversion is used to reduce the series-parallel equivalent circuits, thus eliminating the need to apply mesh or nodal analysis. The wye and delta configurations often appear as shown in Fig. 21. They are

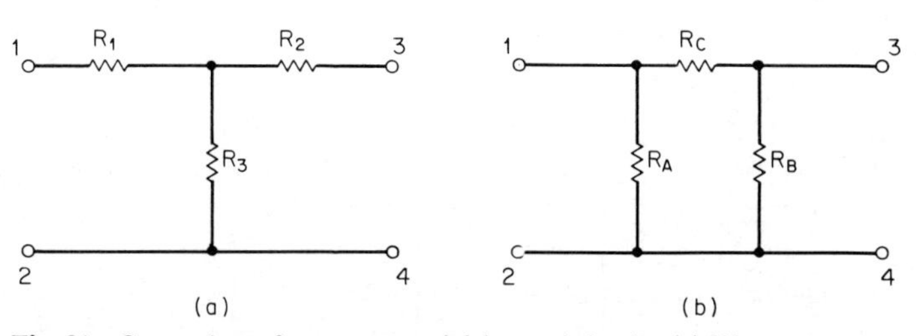

Fig. 21 Comparison of wye to tee and delta to pi circuits. (*a*) Wye or tee configuration. (*b*) Delta or pi configuration.

then referred to as a tee (T) or a pi (π) network. The equations used to convert from a tee to a pi network are exactly the same as those used for the wye and delta transformation.

ANALYSIS OF A SINUSOIDAL WAVE

Given: the voltage $e(t) = 170 \sin 377t$. Calculate the average or dc (E_{dc}), peak (E_m), rms (E), angular frequency (ω), frequency (f), period (T), and peak-to-peak (E_{pp}) values.

Calculation Procedure:

1. Calculate Average Value

$E_{dc} = 0$ because the average value or dc component of a symmetrical wave is zero.

2. Calculate Peak Value

$E_m = 170$ V, which is the maximum value of the sinusoidal wave.

3. Calculate rms Value

$E = 0.707E_m$ where E represents the rms, or effective, value of the sinusoidal wave. Therefore $E = (0.707)(170) = 120$ V.

4. Calculate Angular Frequency

The angular frequency ω equals 377 rad/s.

5. Calculate Frequency

$f = \omega/2\pi = 377/(2 \times 3.1416) = 60$ Hz.

6. Calculate Period

$T = 1/f = 1/60$ s.

7. Calculate Peak-to-Peak Value

$E_{pp} = 2E_m = 2(170) = 340$ V.

Related Calculations: This problem analyzed the sine wave which is standard in the United States, i.e., a voltage wave that has an rms value of 120 V and a frequency of 60 Hz.

ANALYSIS OF A SQUARE WAVE

Find the average and rms values of the square wave of Fig. 22.

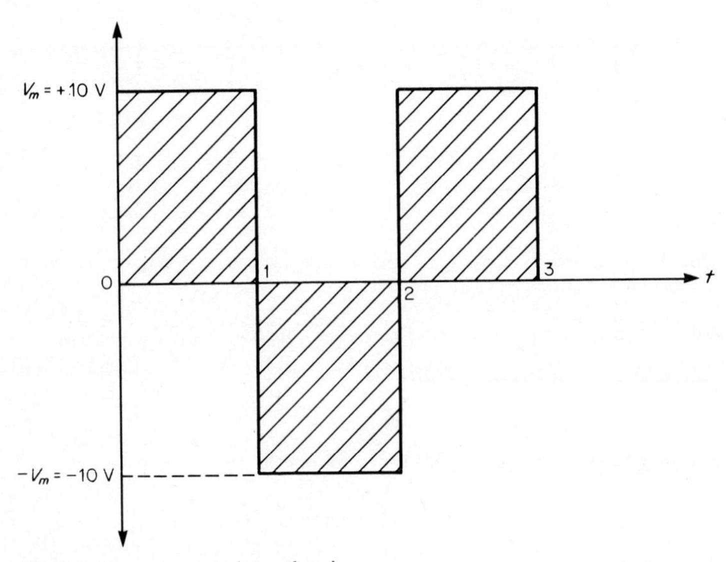

Fig. 22 Square wave to be analyzed.

Calculation Procedure:

1. Calculate the Average Value

The average value, or dc component, of the symmetrical square wave is zero; therefore: $V_{dc} = V_{avg} = 0$.

2. Calculate the rms Value

The rms value is found by squaring the wave over a period of 2 s. This gives a value equal to 100 V^2 which is a constant value over the entire period. The square root of 100 V^2 equals 10 V. Therefore, the rms value is $V = 10$ V.

Related Calculations: The equation for the wave of Fig. 22 is: $v(t) = (4V_m/\pi)(\sin \omega t + \frac{1}{3}\sin 3\omega t + \frac{1}{5}\sin 5\omega t + \cdots + 1/n \sin n\omega t)$. This equation, referred to as a *Fourier series,* shows that a symmetrical square wave beginning at $t = 0$, has no dc component, no even harmonics, and an infinite number of odd harmonics.

ANALYSIS OF AN OFFSET WAVE

Find the average and rms values of the offset wave of Fig. 23.

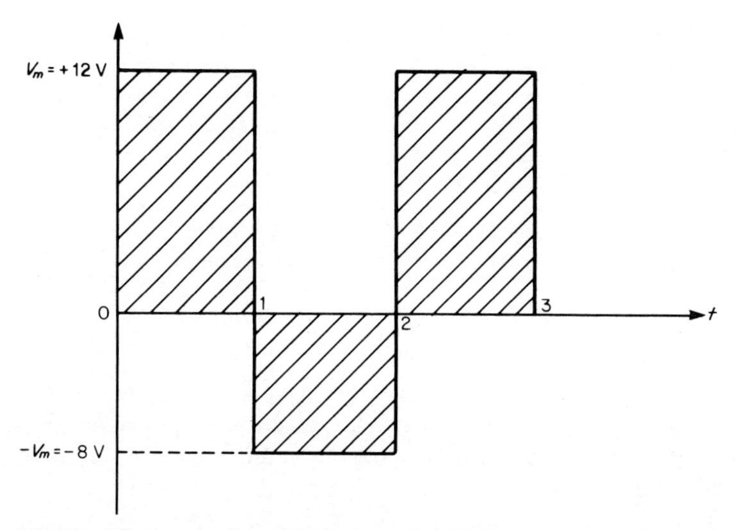

Fig. 23 Offset wave to be analyzed.

Calculation Procedure:

1. Calculate the Average Value

$V_{avg} = V_{dc} =$ net area$/T$ where net area = algebraic sum of areas for one period and T = period of wave. Hence, $V_{avg} = V_{dc} = [(12 \times 1) - (8 \times 1)]/2 = 2$ V.

2. Calculate the rms Value

$$V \text{ (rms value)} = \sqrt{\frac{\text{area}[v(t)^2]}{T}} = \sqrt{\frac{(12 \times 1)^2 + (8 \times 1)^2}{2}}$$
$$= \sqrt{104} = 10.2 \text{ V}.$$

Related Calculations: Figure 23 is the same wave as Fig. 22 except that it has been offset by the addition of a dc component equal to 2 V. The rms, or effective, value of a periodic waveform is equal to the direct current which dissipates the same energy in a given resistor. Since the offset wave has a dc component equal to 2 V, its rms value of 10.2 V is higher than the symmetrical square wave of Fig. 22.

CIRCUIT RESPONSE TO A NONSINUSOIDAL INPUT CONSISTING OF A DC VOLTAGE IN SERIES WITH AN AC VOLTAGE

The input to the circuit of Fig. 24 is $e = 20 + 10 \sin 377t$. (*a*) Find and express i, v_R, and v_C in the time domain. (*b*) Find I, V_R, and V_C. (*c*) Find the power delivered to the circuit. Assume enough time has elapsed that v_C has reached its final (steady-state) value in all three parts of this problem.

Calculation Procedure:

1. Determine the Solution for Part a

This problem can be solved by the application of the superposition theorem, since two separate voltages, one dc and one ac, are present in the circuit. Effect of 20 V dc on circuit:

when v_C has reached its final (steady-state) value $i = 0$, $v_R = iR = 0$ V, and $v_C = 20$ V. Effect of ac voltage (10 sin 377t) on circuit: $X_C = 1/\omega C = 1/(377)(660 \times 10^{-6}) = 4\ \Omega$. Hence, $\mathbf{Z} = 3 - j4 = 5\angle{-53°}\ \Omega$. Then, $\mathbf{I} = \mathbf{E}/\mathbf{Z} = (0.707)(10)\angle{0°}/5\angle{-53°} = 1.414\angle{+53°}$ A.

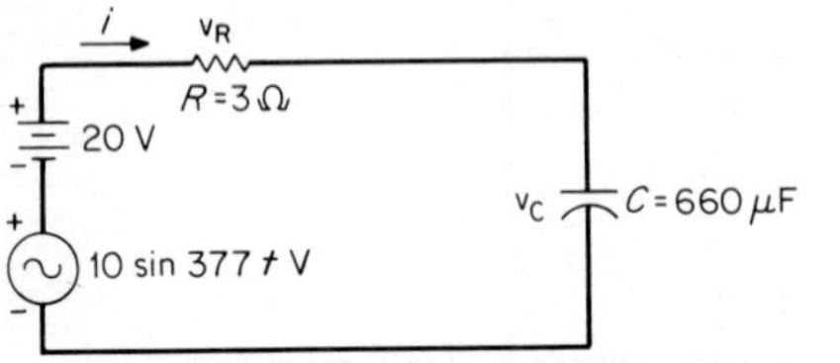

Fig. 24 Analysis of circuit response to a nonsinusoidal input.

Therefore, the maximum value is $I_m = 1.414/0.707 = 2$ A and the current in the time domain is $i = 0 + 2 \sin (377t + 53°)$. $\mathbf{V_R} = \mathbf{I}R = (1.414\angle{+53°})(3\angle{0°}) = 4.242\angle{+53°}$ V. The maximum value for $\mathbf{V_R}$ is $4.242/0.707 = 6$ V, and the voltage v_R in the time domain is $v_R = 0 + 6 \sin (377t + 53°)$.

$\mathbf{V_C} = \mathbf{I}X_C = (1.414\angle{+53°})(4\angle{-90°}) = 5.656\angle{-37°}$. The maximum value for $\mathbf{V_C} = 5.656/0.707 = 8$ V, and the voltage v_C in the time domain is $v_C = 20 + 8 \sin (377t - 37°)$.

2. Determine the Solution for Part b

The effective value of a nonsinusoidal input consisting of dc and ac components can be found from the following equation:

$$V = \sqrt{V_{dc}^2 + \frac{(V_{m1}^2 + V_{m2}^2 + \cdots + V_{mn}^2)}{2}}$$

where V_{dc} = voltage of dc component and V_{m1}, etc. = maximum value of ac components. Therefore $|\mathbf{I}| = \sqrt{0^2 + 2^2/2} = 1.414$ A, $|\mathbf{V_R}| = \sqrt{0^2 + 6^2/2} = 4.24$ A, and $|\mathbf{V_C}| = \sqrt{20^2 + 8^2/2} = 20.8$ V.

3. Determine the Solution for Part c

$P = I^2R = (1.414)^2(3) = 6$ W.

Related Calculations: The concept of a dc component superimposed on a sinusoidal ac component is illustrated in Fig. 25. This figure shows the decay of a dc component because of a short circuit and also shows how the asymmetrical short-circuit current gradually becomes symmetrical when the dc component decays to zero.

ANALYSIS OF A SERIES *RLC* CIRCUIT

Calculate the current in the circuit of Fig. 26*a*.

Calculation Procedure:

1. Calculate **Z**

Angular frequency $\omega = 2\pi f = (2)(3.1416)(60) = 377$ rad/s. But $X_L = \omega L$; therefore $X_L = (377)(0.5) = 188.5\ \Omega$. Also, $X_C = 1/\omega C = 1/[(377)(26.5) \times 10^{-6})] = 100\ \Omega$. Then $\mathbf{Z} = R + j(X_L - X_C) = R + jX_{EQ}$ where $X_{EQ} = X_L - X_C$ = net equivalent reactance.

In polar form the impedance for the series *RLC* circuit is expressed as $\mathbf{Z} = \sqrt{R^2 + X_{EQ}^2}\angle{\tan^{-1}(X_{EQ}/R)} = |\mathbf{Z}|\angle{\theta}$. $\mathbf{Z} = 100 + j(188.5 - 100) = 100 + j88.5$

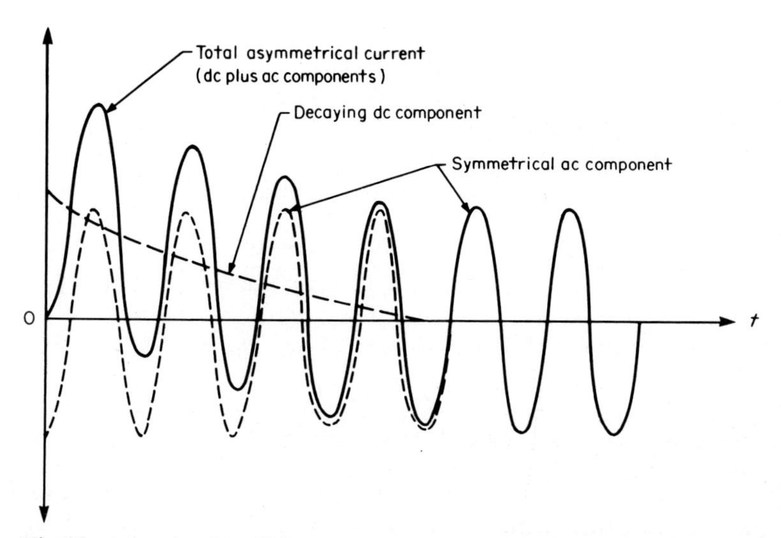

Fig. 25 A decaying sinusoidal wave.

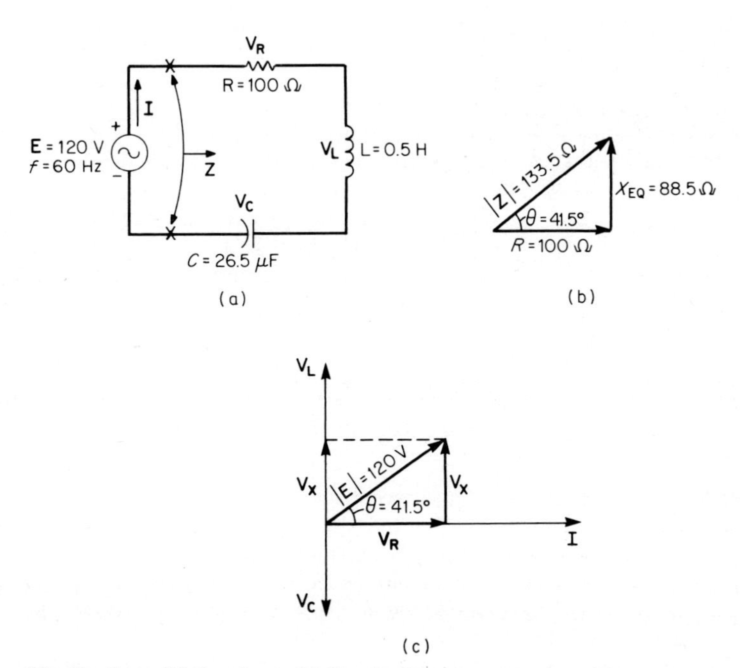

Fig. 26 Series *RLC* ac circuit. (*a*) Circuit with component values. (*b*) Impedance triangle. (*c*) Phasor diagram.

$= \sqrt{(100)^2 + (88.5)^2} \angle \tan^{-1}(88.5/100) = 133.5\angle 41.5°\ \Omega$. The impedance triangle (Fig. 26*b*) illustrates the results of the above solution.

Apply KVL to the circuit: $\mathbf{E} = V_R + jV_L - jV_C = V_R + jV_X$ where $V_X = V_L - V_C$ = net reactive voltage.

2. *Draw the Phasor Diagram*

The phasor diagram of Fig. 26*c* shows the voltage relations with respect to the current as a reference.

3. *Calculate* I

From Ohm's law for ac circuits $|\mathbf{I}| = 120/133.5 = 0.899$ A. Because **I** is a reference it can be expressed in polar form as $\mathbf{I} = 0.899\angle 0°$ A. The angle between the voltage and current in Fig. 26*c* is the same as the angle in the impedance triangle of Fig. 26*b*. Therefore $\mathbf{E} = 120\angle 41.5°$ V.

Related Calculations: In a series *RLC* circuit the net reactive voltage may be zero (when $\mathbf{V_L} = \mathbf{V_C}$), inductive (when $\mathbf{V_L} > \mathbf{V_C}$) or capacitive (when $\mathbf{V_L} < \mathbf{V_C}$). The current in such a circuit may be in phase with, lag, or lead the applied emf. When $\mathbf{V_L} = \mathbf{V_C}$, the condition is referred to as *series resonance.* Voltages $\mathbf{V_L}$ and $\mathbf{V_C}$ may be higher than the applied voltage **E** because the only limiting opposition to current is resistance *R*. A circuit in series resonance has maximum current, minimum impedance, and a power factor of 100 percent.

ANALYSIS OF A PARALLEL *RLC* CIRCUIT

Calculate the impedance of the parallel *RLC* circuit of Fig. 27*a*.

Calculation Procedure:

1. *Calculate the Currents in R, L, and C*

In a parallel circuit, it is convenient to use the voltage as a reference; therefore $\mathbf{E} = 200\angle 0°$ V. Because the *R*, *L*, and *C* parameters of this circuit are the same as in Fig. 26*a* and the frequency (60 Hz) is the same, $X_L = 188.5\ \Omega$ and $X_C = 100\ \Omega$. From Ohm's law: $\mathbf{I_R} = \mathbf{E}/R = 200\angle 0°/100\angle 0° = 2\angle 0°$ A, $\mathbf{I_L} = \mathbf{E}/X_L = 200\angle 0°/188.5\angle 90° = 1.06\angle -90° = -j1.06$ A, and $\mathbf{I_C} = \mathbf{E}/X_C = 200\angle 0°/100\angle -90° = 2\angle 90° = +j2$ A. But $\mathbf{I_T} = I_R - jI_L + jI_C$; therefore $\mathbf{I_T} = 2 - j1.06 + j2 = 2 + j0.94 = 2.21\angle 25.2°$ A.

2. *Calculate* $\mathbf{Z_{EQ}}$

Impedance is $\mathbf{Z_{EQ}} = \mathbf{E}/\mathbf{I_T} = 200\angle 0°/2.21\angle 25.2° = 90.5\angle -25.2°\ \Omega$. $\mathbf{Z_{EQ}}$, changed to rectangular form, is $\mathbf{Z_{EQ}} = 82.6\ \Omega - j39\ \Omega = R_{EQ} - jX_{EQ}$. Figure 27*b* illustrates the voltage-current phasor diagram. The equivalent impedance diagram is given in Fig. 27*c*.

Related Calculations: The impedance diagram of Fig. 27*c* has a negative angle. This indicates that the circuit is an *RC* equivalent circuit. Figure 27*b* verifies this observation because the total circuit current $\mathbf{I_T}$ leads the applied voltage. In a parallel *RLC* circuit the net reactive current may be zero (when $\mathbf{I_L} = \mathbf{I_C}$), inductive (when $\mathbf{I_L} > \mathbf{I_C}$), or capacitive (when $\mathbf{I_L} < \mathbf{I_C}$). The current in such a circuit may be in phase with, lag, or lead the applied emf. When $\mathbf{I_L} = \mathbf{I_C}$, this condition is referred to as *parallel resonance.* Currents $\mathbf{I_L}$ and $\mathbf{I_C}$ may be much higher than the total line current, $\mathbf{I_T}$. A circuit in parallel resonance has a minimum current, maximum impedance, and a power factor of 100 percent. Note in Fig. 1-27*b* that $\mathbf{I_T} = I_R + jI_X$ where $I_X = I_C - I_L$.

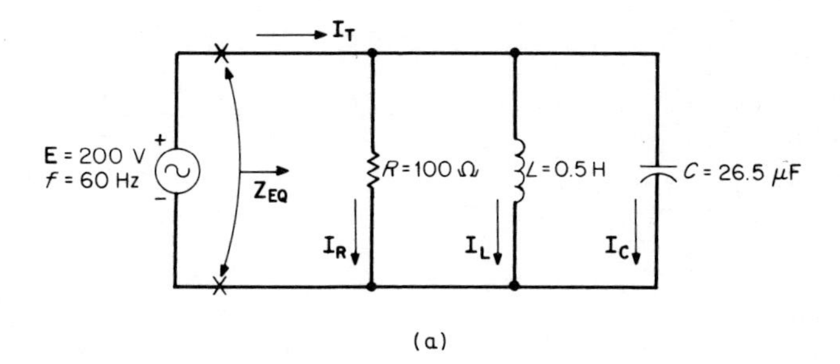

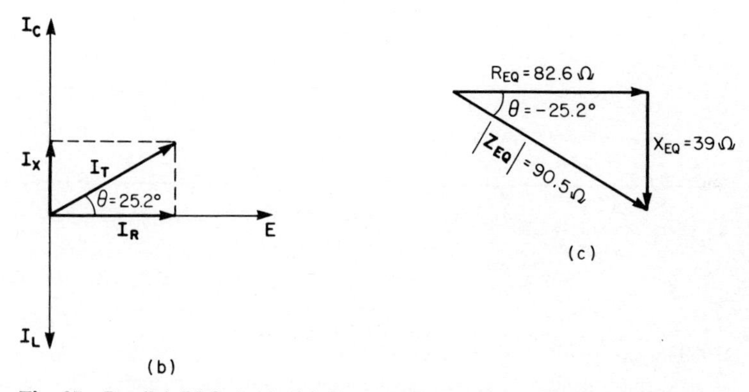

Fig. 27 Parallel *RLC* circuit. (*a*) Circuit with component values. (*b*) Phasor diagram. (*c*) Impedance traingle.

ANALYSIS OF A SERIES-PARALLEL AC NETWORK

A series-parallel ac network is shown in Fig. 28. Calculate $\mathbf{Z_{EQ}}$, $\mathbf{I_1}$, $\mathbf{I_2}$, and $\mathbf{I_3}$.

Calculation Procedure:

1. Combine All Series Impedances

The solution to this problem is similar to that for the first problem in the section, except that vector algebra must be used for the reactances. $\mathbf{Z_1} = 300 + j600 - j200 = 300 + j400 = 500\angle 53.1^\circ\ \Omega$, $\mathbf{Z_2} = 500 + j1200 = 1300\angle 67.4^\circ\ \Omega$, and $\mathbf{Z_3} = 800 - j600 = 1000\angle -36.9^\circ\ \Omega$.

2. Combine All Parallel Impedances

Using the product-over-the-sum rule we find $\mathbf{Z_{BC}} = \mathbf{Z_2 Z_3}/(\mathbf{Z_1}+\mathbf{Z_3}) = (1300\angle 67.4^\circ)(1000\angle -36.9^\circ)/[(500 + j1200) + (800 - j600)] = 908\angle 5.7^\circ = 901 + j90.2\ \Omega$.

3. Combine All Series Impedances to Obtain the Total Impedance $\mathbf{Z_{EQ}}$

$\mathbf{Z_{EQ}} = \mathbf{Z_1} + \mathbf{Z_{BC}} = (300 + j400) + (901 + j90.2) = 1201 + j490 = 1290\angle 22.4^\circ\ \Omega$.

4. Calculate the Currents

$\mathbf{I_1} = \mathbf{E}/\mathbf{Z_{EQ}} = 100\angle 0^\circ/1290\angle 22.4^\circ = 0.0775\angle -22.4^\circ$ A. From the current-divider rule: $\mathbf{I_2} = \mathbf{I_1 Z_3}/(\mathbf{Z_2} + \mathbf{Z_3}) = (0.0775\angle -22.4^\circ)(1000\angle -36.9^\circ)/[(500 +$

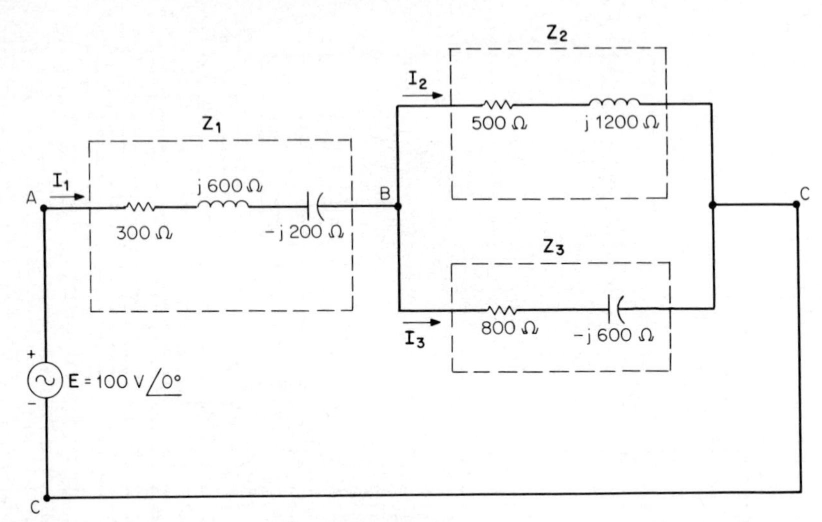

Fig. 28 Series-parallel ac circuit to be analyzed.

$j1200) + (800 - j600)] = 0.0541\angle{-84.1°}$ A. $\mathbf{I}_3 = \mathbf{I}_1\mathbf{Z}_2/(\mathbf{Z}_2 + \mathbf{Z}_3) = (0.0775\angle{-22.4°})\,(1300\angle{67.4°})/[(500 + j1200) + (800 - j600)] = 0.0709\angle{20.2°}$ A.

Related Calculations: Any reducible ac circuit (i.e., any circuit that can be reduced to one equivalent impedance $\mathbf{Z}_{EQ}$ with a single power source), no matter how complex, can be solved in a similar manner to that described above. The dc network theorems used in previous problems can be applied to ac networks except that vector algebra must be used for the ac quantities.

ANALYSIS OF POWER IN AN AC CIRCUIT

Find the total watts, total vars, and total volt-amperes in the ac circuit of Fig. 29*a*.

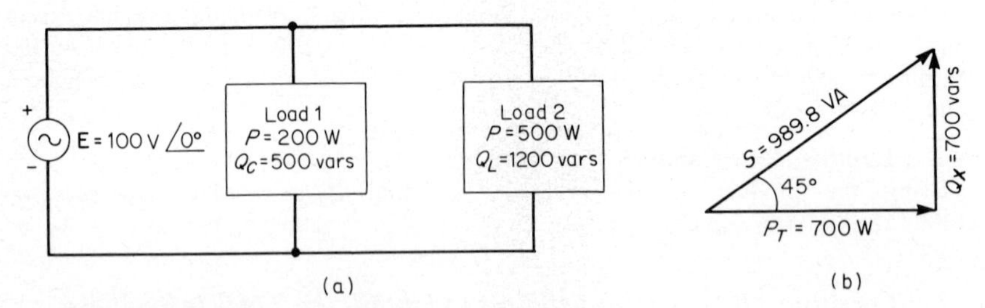

Fig. 29 Calculating ac power. (*a*) Circuit. (*b*) Power triangle.

Calculation Procedure:

1. Study the Power Triangle

Figure 30 shows power triangles for ac circuits. Power triangles are drawn following the standard of drawing inductive reactive power in the $+j$ direction and capacitive reactive power in the $-j$ direction. Two equations are obtained by applying the Pythagorean

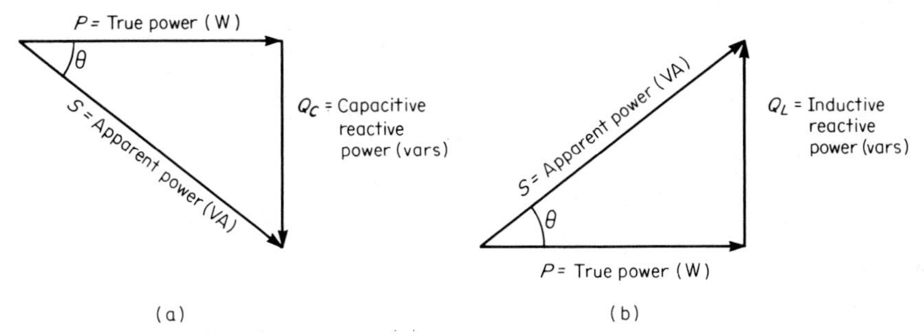

Fig. 30 Power triangles for (*a*) *RC* and (*b*) *RL* equivalent circuits.

theorem to these power triangles: $S^2 = P^2 + Q_L^2$ and $S^2 = P^2 + Q_C^2$. These equations can be applied to series, parallel, or series-parallel circuits.

The net reactive power supplied by the source to an *RLC* circuit is the difference between the positive inductive reactive power and the negative capacitive reactive power: $Q_X = Q_L - Q_C$ where Q_X is the net reactive power, in vars.

2. *Solve for the Total Real Power*

$P_T = P_1 + P_2 = 200 + 500 = 700$ W. Arithmetic addition can be used to find the total real power.

3. *Solve for the Total Reactive Power*

$Q_X = Q_L - Q_C = 1200 - 500 = 700$ vars. Because the total reactive power is positive, the circuit is inductive. (See Fig. 29*b*.)

4. *Solve for the Total Volt-Amperes*

$S = \sqrt{P_T^2 + Q_X^2} = \sqrt{(700)^2 + (700)^2} = 989.8$ VA.

Related Calculations: The principles used in this problem will also be applied to solve the following two problems.

ANALYSIS OF POWER FACTOR AND REACTIVE FACTOR

Calculate the power factor (pf) and the reactive factor (rf) for the circuit shown in Fig. 31.

Calculation Procedure:

1. *Review Power-Factor Analysis*

The power factor of an ac circuit is the numerical ratio between the true power *P* and the apparent power *S*. It can be seen by referring to the power triangles of Fig. 30 that this ratio is equal to the cosine of the power-factor angle θ. The power-factor angle is the same as the phase angle between the voltage across the circuit (or load) and the current through the circuit (or load). $\text{pf} = \cos\theta = P/S$.

2. *Review Reactive-Factor Analysis*

The numerical ratio between the reactive power and the apparent power of a circuit (or load) is called the reactive factor. This ratio is equal to the sine of the power-factor angle (see Fig. 30). $\text{rf} = \sin\theta = Q/S$.

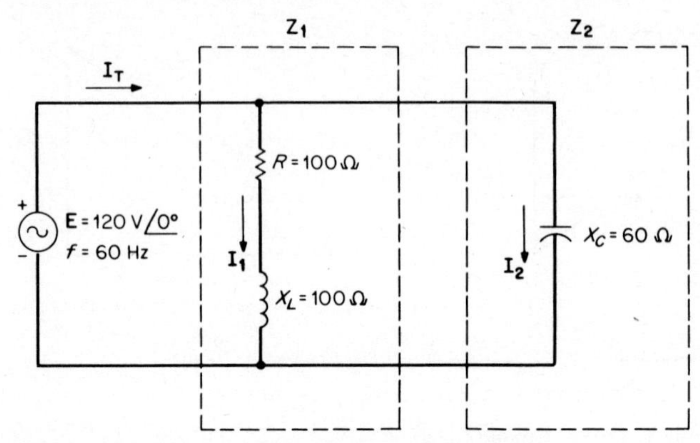

Fig. 31 Calculating power and reactive factors of circuit.

3. *Calculate the Power and Reactive Factors*

$\mathbf{Z}_1 = R + jX_L = 100 + j100 = 141.4\angle 45°$. $\mathbf{I}_1 = \mathbf{E}/\mathbf{Z}_1 = 120\angle 0°/141.4\angle 45° = 0.849\angle -45°$ A. $\mathbf{I}_1 = (0.6 - j0.6)$ A. $\mathbf{I}_2 = \mathbf{E}/X_C = 120\angle 0°/60\angle -90° = 2\angle 90° = (0 + j2)$ A. $\mathbf{I}_T = \mathbf{I}_1 + \mathbf{I}_2 = (0.6 - j0.6) + (0 + j2) = (0.6 + j1.4)$ A $= 1.523\angle 66.8°$ A. $S = |\mathbf{E}||\mathbf{I}_T| = (120)(1.523) = 182.8$ VA. Power factor $= \cos\theta = \cos 66.8° = 0.394$ or 39.4 percent; rf $= \sin\theta = \sin 66.8° = 0.92$ or 92 percent.

Related Calculations: Inductive loads have a lagging power factor; capacitive loads have a leading power factor. The value of the power factor is expressed either as a decimal or as a percentage. This value is always less than 1.0 or less than 100 percent. The majority of industrial loads, such as motors and air conditioners, are inductive (lagging power factor).

POWER-FACTOR CORRECTION

Calculate the value of the capacitor needed to obtain a circuit power factor of 100 percent (Fig. 32).

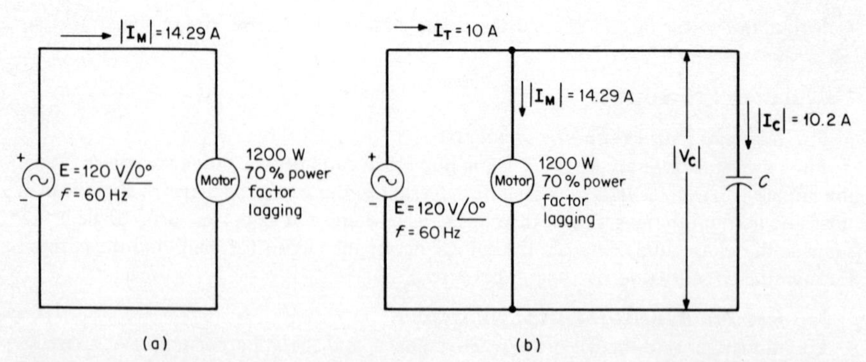

Fig. 32 Power-factor correction. (*a*) Given circuit. (*b*) Adding *C* in parallel to improve power factor.

Calculation Procedure:

1. Calculate the Motor Current

$S = P/\cos\theta = 1200/0.7 = 1714$ VA. Hence, the motor current $|\mathbf{I}|$ is: $|\mathbf{I}| = S/|\mathbf{E}| = (1714\text{ VA})/(120\text{ V}) = 14.29$ A. The active component of this current is the component in phase with the voltage. This component, which results in true power consumption, is: $|\mathbf{I}|\cos\theta = (14.29\text{ A})(0.7) = 10$ A. Because the motor has a 70 percent power factor, the circuit must supply 14.29 A to realize a useful current of 10 A.

2. Calculate the Value of C

In order to obtain a circuit power factor of 100 percent, the inductive apparent power of the motor and the capacitive apparent power of the capacitor must be equal. $Q_L = |\mathbf{E}||\mathbf{I}|\sqrt{1-\cos^2\theta}$ where $\sqrt{1-\cos^2\theta}$ = reactive factor. Hence, $Q_L = (120)(14.29)\sqrt{1-(0.7)^2} = 1714\sqrt{0.51} = 1224$ vars (inductive.) Q_C must equal 1224 vars for 100 percent power factor. $X_C = V_C^2/Q_C = (120)^2/1224 = 11.76\ \Omega$ (capacitive). Therefore, $C = 1/\omega X_C = 1/(377)(11.76) = 225.5\ \mu$F

Related Calculations: The amount of current required by a load determines the sizes of the wire used in the windings of the generator or transformer, and in the conductors connecting the motor to the generator or transformer. Because copper losses depend upon the square of the load current, a power company finds it more economical to supply 10 A at a power factor of 100 percent than to supply 14.29 A at a power factor of 70 percent.

A mathematical analysis of the currents in Fig. 32*b* follows: $|\mathbf{I_C}| = Q_C/|\mathbf{V_C}| = (1220\text{ vars})/(120\text{ V}) = 10.2\text{ A} = (0 + j10.2)$ A. θ (for motor) $= \cos^{-1}0.7 = 45.6°$; therefore, $\mathbf{I_M} = 14.29\angle 45.6° = (10 - j10.2)$ A. Then $\mathbf{I_T} = \mathbf{I_M} + \mathbf{I_C} = (10 - j10.2) + (0 + j10.2) = 10\angle 0°$ A (100 percent power factor).

MAXIMUM POWER TRANSFER IN AN AC CIRCUIT

Calculate the load impedance in Fig. 33 for maximum power to the load.

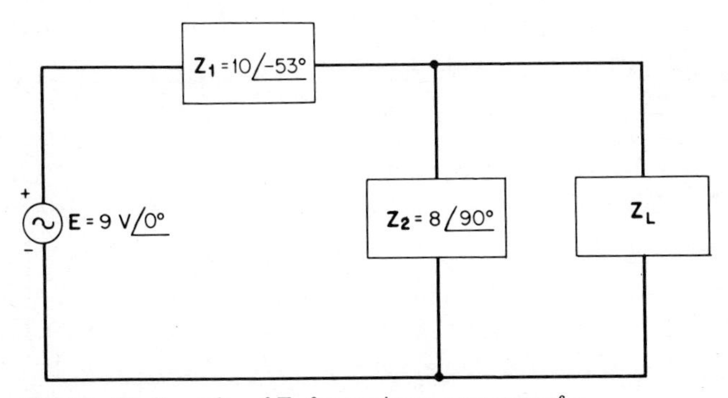

Fig. 33 Finding value of $\mathbf{Z_L}$ for maximum power transfer.

Calculation Procedure:

1. Use the Maximum Power Theorem

The maximum power theorem, when applied to ac circuits, states that maximum power will be delivered to a load when the load impedance is the complex conjugate of the Thevenin impedance across its terminals.

2. Apply Thevenin's Theorem to the Circuit
$\mathbf{Z}_{Th} = \mathbf{Z}_1\mathbf{Z}_2/(\mathbf{Z}_1 + \mathbf{Z}_2) = (10\angle -53°)(8\angle 90)/[(6 - j8) + j8] = 13.3\angle 37°\ \Omega$, or $\mathbf{Z}_{Th} = 10.6 + j8\ \Omega$ where $R = 10.6\ \Omega$ and $X_L = 8\ \Omega$. Then, $\mathbf{Z}_L$ must be $13.3\angle -37° = 10.6 - j8\ \Omega$ where $R_L = 10.6\ \Omega$ and $X_C = -8\ \Omega$.

In order to find the maximum power delivered to the load, $\mathbf{E}_{Th}$ must be found using the voltage-divider rule: $\mathbf{E}_{Th} = \mathbf{E}\mathbf{Z}_2/(\mathbf{Z}_1 + \mathbf{Z}_2) = (9\angle 0°)(8\angle 90°)/[(6 - j8) + j8] = 12\angle 90°$ V. $P_{max} = |\mathbf{E}_{Th}^2|/4R_L$; therefore $P_{max} = (12)^2/(4)(10.6) = 3.4$ W.

Related Calculations: The maximum power transfer theorem, when applied to dc circuits, states that a load will receive maximum power from a dc network when its total resistance is equal to the Thevenin resistance of the network as seen by the load.

ANALYSIS OF A BALANCED WYE-WYE SYSTEM

Calculate the currents in all lines of the balanced three-phase, four-wire, wye-connected system of Fig. 34. The system has the following parameters: $\mathbf{V}_{AN} = 120\angle 0°$ V, $\mathbf{V}_{BN} = 120\angle -120°$ V, $\mathbf{V}_{CN} = 120\angle 120°$ V, and $\mathbf{Z}_A = \mathbf{Z}_B = \mathbf{Z}_C = 12\angle 0°\ \Omega$.

Calculation Procedure:

1. Calculate Currents
$\mathbf{I}_A = \mathbf{V}_{AN}/\mathbf{Z}_A = 120\angle 0°/12\angle 0° = 10\angle 0°$ A. $\mathbf{I}_B = \mathbf{V}_{BN}/\mathbf{Z}_B = 120\angle -120°/12\angle 0° = 10\angle -120°$A. $\mathbf{I}_C = \mathbf{V}_{CN}/\mathbf{Z}_C = 120\angle 120°/12\angle 0° = 10\angle 120°$ A. $\mathbf{I}_N = \mathbf{I}_A + \mathbf{I}_B + \mathbf{I}_C$; hence $\mathbf{I}_N = 10\angle 0° + 10\angle -120° + 10\angle 120° = 0$ A.

Related Calculations: The neutral current in a balanced wye system is always zero. Each load current lags or leads the voltage by the particular power factor of the load. This system, in which one terminal of each phase is connected to a common star point, is often called a star-connected system.

ANALYSIS OF A BALANCED DELTA-DELTA SYSTEM

Calculate the load currents and the line currents of the balanced delta-delta system of Fig. 35. The system has the following load parameters: $\mathbf{V}_{AC} = 200\angle 0°$ V, $\mathbf{V}_{BA} = 200\angle 120°$ V, $\mathbf{V}_{CB} = 200\angle -120°$ V, and $\mathbf{Z}_{AC} = \mathbf{Z}_{BA} = \mathbf{Z}_{CB} = 4\angle 0°\ \Omega$.

Calculation Procedure:

1. Solve for the Load Currents
$\mathbf{I}_{AC} = \mathbf{V}_{AC}/\mathbf{Z}_{AC} = 200\angle 0°/4\angle 0° = 50\angle 0°$ A, $\mathbf{I}_{BA} = \mathbf{V}_{BA}/\mathbf{Z}_{BA} = 200\angle 120°/4\angle 0° = 50\angle 120°$ A, and $\mathbf{I}_{CB} = \mathbf{V}_{CB}/\mathbf{Z}_{CB} = 200\angle 120°/4\angle 0° = 50\angle -120°$ A

2. Solve for the Line Currents
Convert the load currents to rectangular notation: $\mathbf{I}_{AC} = 50\angle 0° = 50 + j0$, $\mathbf{I}_{BA} = 50\angle 120° = -25 + j43.3$, and $\mathbf{I}_{CB} = 50\angle -120° = -25 - j43.3$. Apply KCL at load nodes: $\mathbf{I}_A = \mathbf{I}_{AC} - \mathbf{I}_{BA} = (50 + j0) - (-25 + j43.3) = 86.6\angle -30°$ A $\mathbf{I}_B = \mathbf{I}_{BA} - \mathbf{I}_{CB} = (-25 + j43.3 - (-25 - j43.3) = 86.6\angle 90°$ A $\mathbf{I}_C = \mathbf{I}_{CB} - \mathbf{I}_{AC} = (-25 - j43.3) - (50 + j0) = 86.6\angle -150°$ A

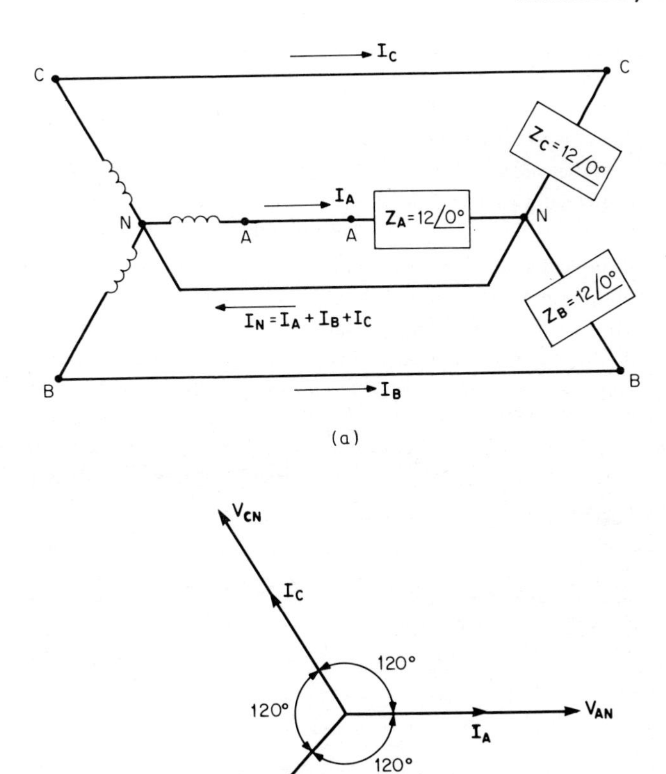

Fig. 34 A balanced three-phase, four-wire, wye-connected system. (*a*) Circuit. (*b*) Load phasor diagram.

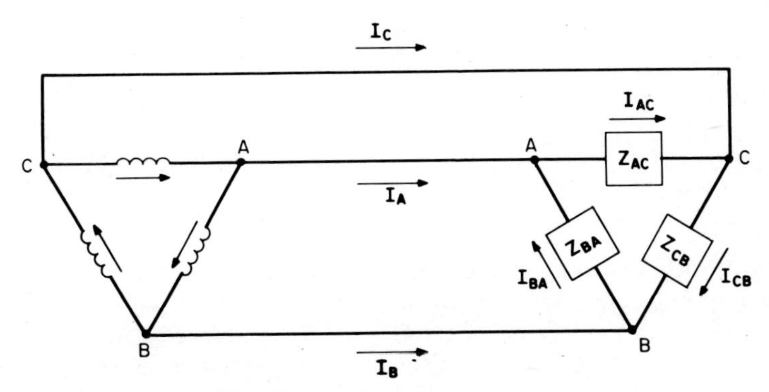

Fig. 35 A balanced delta-delta system.

Related Calculations: In comparing a wye-connected system with a delta-connected system, one can make the following observations:

1. When a load is wye-connected, each arm of the load is connected from a line to the neutral. The impedance **Z** is shown with a single subscript, such as $\mathbf{Z_A}$.
2. When a load is delta-connected, each arm of the load is connected from line to line. The impedance **Z** is shown with a double subscript such as $\mathbf{Z_{AC}}$.
3. In a wye-connected system, the phase current of the source, the line current, and the phase current of the load are all equal.
4. In a delta-connected system, each line must carry components of current for two arms of the load. One current component moves toward the source, and the other current component moves away from the source. The line current to a delta-connected load is the phasor difference between the two load currents at the entering node.
5. The line current in a balanced delta load has a magnitude of $\sqrt{3}$ times the phase current in each arm of the load. The line current is 30° out of phase with the phase current (Fig. 36)

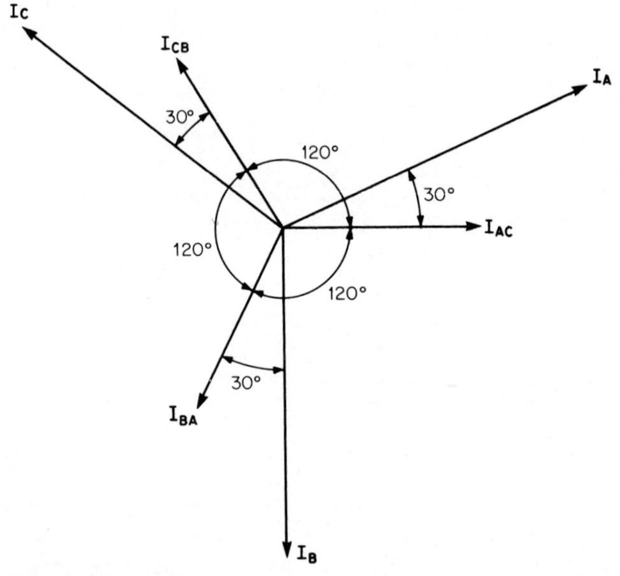

Fig. 36 Relationships between phase and line currents in a balanced delta-connected system.

RESPONSE OF AN INTEGRATOR TO A RECTANGULAR PULSE

A single 10-V pulse with a width of 200 μs is applied to the *RC* integrator of Fig. 37. Calculate the voltage to which the capacitor charges. How long will it take the capacitor to discharge (neglect the resistance of the pulse source)?

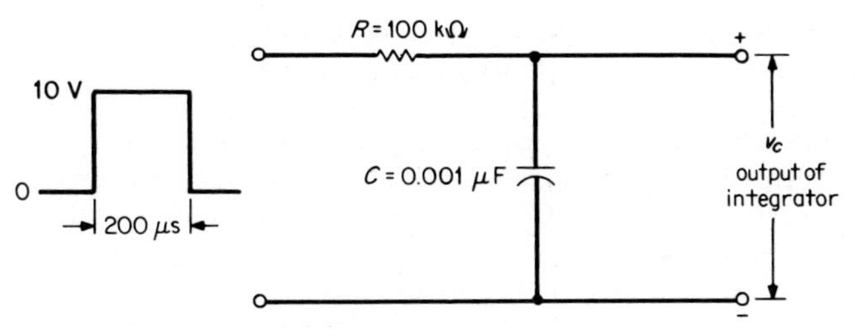

Fig. 37 Pulse input to an *RC* integrator.

Calculation Procedure:

1. Calculate the Voltage to Which the Capacitor Charges

The rate at which a capacitor charges or discharges is determined by the time constant of the circuit. The time constant of a series *RC* circuit is the time interval that equals the

TABLE 1 *RC* Time Constant Charging Characteristics

τ	*% Full charge*
1	63
2	86
3	95
4	98
5	99*

*For practical purposes, five time constants are considered to result in 100 percent charging.

TABLE 2 *RC* Time Constant Discharging Characteristics

τ	*% Full charge*
1	37
2	14
3	5
4	2
5	1*

*For practical purposes, five time constants are considered to result in zero charge or 100 percent discharge.

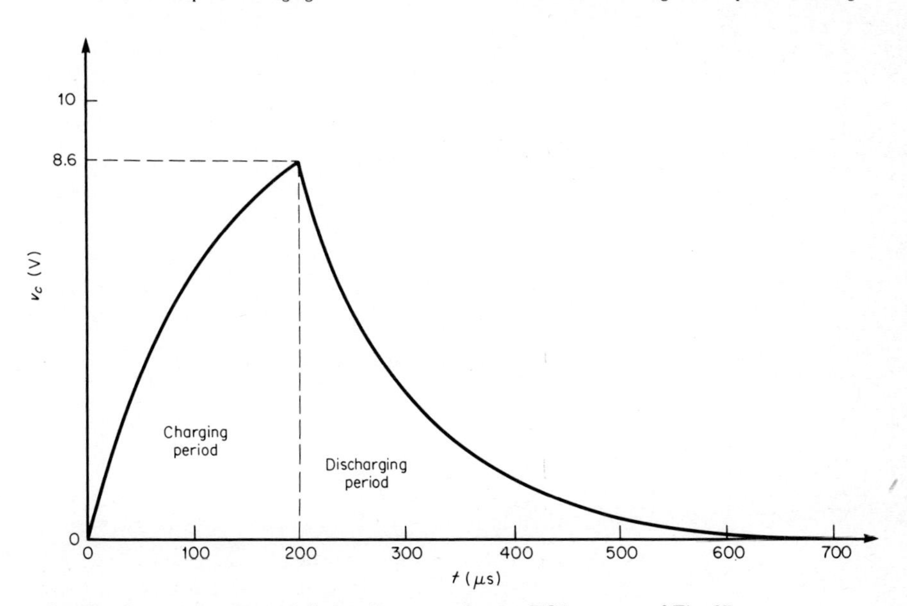

Fig. 38 Output charging and discharging curves for the *RC* integrator of Fig. 37.

product of R and C. The symbol for time constant is τ (Greek letter tau): $\tau = RC$ where R is in ohms, C is in farads, and τ is in seconds.

The time constant of this circuit is: $\tau = RC = (100\ \text{k}\Omega)(0.001\ \mu\text{F}) = 100\ \mu\text{s}$. Because the pulse width equals 200 μs (2 time constants), the capacitor will charge to 86 percent of its full charge, or to a voltage of 8.6 V. The expression for RC charging is: $v_C(t) = V_F(1 - e^{-t/RC})$, where V_F is the final value. In this case the final value, $V_F = 10$ V, would be reached if the pulse had a width of 5 or more time constants. See the RC time constant charging table (Table 1).

2. *Calculate the Discharge Time*

The capacitor discharges back through the source at the end of 200 μs. The total discharge time for practical purposes is 5 time constants or $(5)(100\ \mu\text{s}) = 500\ \mu\text{s}$. The expression for RC discharging is: $v_C(t) = V_i(e^{-t/RC})$, where V_i is the initial value. In this case, the initial value before discharging is 8.6 V. Table 2 shows the RC time constant discharge characteristics.

Related Calculations: Figure 38 illustrates the output charging and discharging curves.

Section 2 INSTRUMENTATION

Edmund G. Trunk, P.E.
Professor of Electrical Technology,
Nassau Community College

REFERENCES Considine—*Process Instruments and Controls Handbook,* McGraw-Hill; Cooper—*Electronic Instrumentation and Measurement Techniques,* Prentice-Hall; Doebelin—*Measurement Systems: Application and Design,* McGraw-Hill; Fink and Carroll—*Standard Handbook for Electrical Engineers,* McGraw-Hill; Kantrowitz, Kousourou, and Zucker—*Electronic Measurements,* Prentice-Hall; Kirk and Rimboi—*Instrumentation,* American Technical Society; Prensky—*Electronic Instrumentation,* Prentice-Hall; Prewitt and Fardo—*Instrumentation: Transducers, Experimentation, and Applications,* Sams; Soisson—*Instrumentation in Industry,* Wiley.

VOLTAGE MEASUREMENT

The line voltage of a three-phase 4160-V power line supplying an industrial plant is to be measured. Choose the appropriate voltmeter and potential transformer for making the measurement.

Calculation Procedure:

1. Select Voltmeter

Self-contained ac voltmeters with scales ranging from 150 to 750 V are available. Where higher voltages are to be measured, a potential transformer is required to produce a voltage suitable for indication on a meter with 150-V full-scale indication. A 150-V meter is therefore selected.

2. Select Potential Transformer

By dividing the line-to-line voltage by the voltmeter full-scale voltage, one obtains an approximate value of transformer ratio: 4160 V/150 V = 27.7. Select the next higher standard value, 40:1. To check the selection, calculate the secondary voltage with the potential transformer: 4160 V/40 = 104 V.

3. Connect Transformer and Voltmeter to the Line

The potential transformer and voltmeter are connected to the three-phase line as shown in Fig. 1.

Related Calculations: Instruments used for measuring electrical quantities in utility or industrial service, such as voltmeters, ammeters, and wattmeters, are referred to as switchboard instruments. Instruments used for measurement of nonelectrical quantities, such as pressure, temperature, and flow rate, involve more complex techniques which include a sensor, transmission line, and receiver or indicator. In these systems, the receiver may also perform a recording function.

Very often, the instrumentation system is part of a process-control system. In such cases, the instrument which receives and indicates also serves as a controller. These industrial measurement systems may involve pneumatic, electrical analog (voltage or current), or electrical digital signal-transmission techniques.

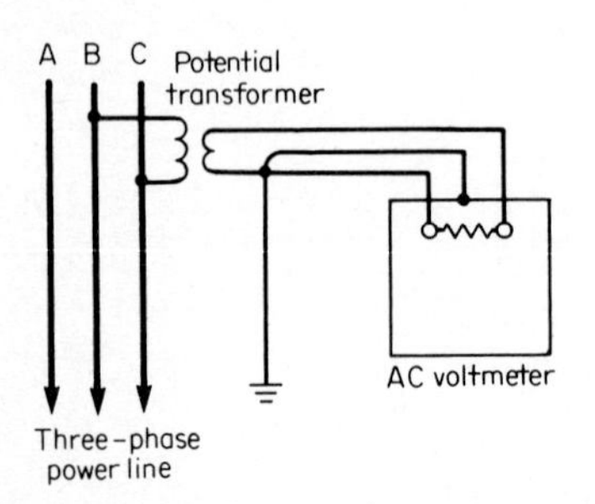

Fig. 1 Voltmeter connections to a three-phase power line.

CURRENT MEASUREMENT

Current is to be measured in a single-phase line which supplies a 240-V, 20-kW load with a 0.8 power factor (pf). Select an appropriate ammeter and current transformer.

Calculation Procedure:

1. Select Ammeter

Direct-reading ammeters are available with full-scale readings ranging from 2 to 20 A. Measurement of larger currents requires the use of a current transformer. Standard practice is to use a 5-A full-scale ammeter with the appropriate current transformer. Ammeters so used are calibrated in accordance with the selected transformer.

2. Calculate Current

$I = P/(V \times \text{pf}) = 20{,}000/(240 \times 0.8) = 104$ A.

3. Select Current Transformer

Because the current is greater than 20 A, a current transformer is required. A transformer is chosen which can accommodate a somewhat higher current; a 150:5 current transformer is therefore selected. The ammeter is a 5-A meter with its scale calibrated from 0 to 150 A.

4. Connect Ammeter and Transformer to Line

The ammeter is connected to the line, through the current transformer, as in Fig. 2.

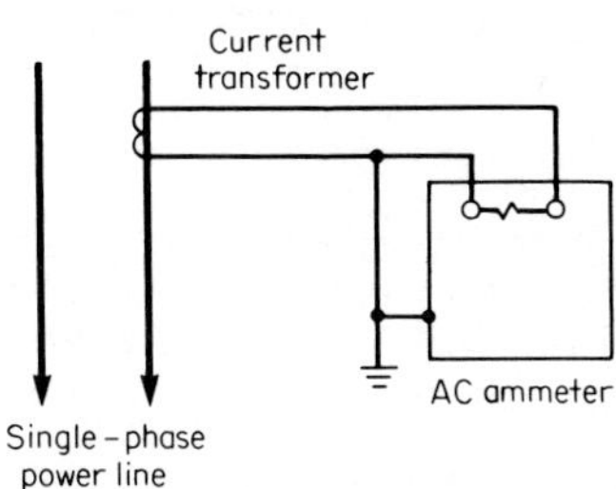

Fig. 2 Ammeter connections to a single-phase power line.

FREQUENCY MEASUREMENT

An independent on-site cogenerating plant supplies an apartment complex and shopping center with power. The alternators are rated at 2400 V, three-phase, 60 Hz. Determine how to measure the frequency of the system.

Calculation Procedure:

1. Select a Scale

An independent power system might be expected to undergo wider frequency deviations than are normal for large utility systems with regional interconnections. Therefore, a range of 55 to 65 Hz is selected for this application.

2. Determine Need for Potential Transformer

If a 120-V line that is derived from the plant system is available, then there is no need for a potential transformer. If such a line is not available, a potential transformer is required. If the primary is connected line-to-line, the voltage ratio is 2400/120 = 20:1.

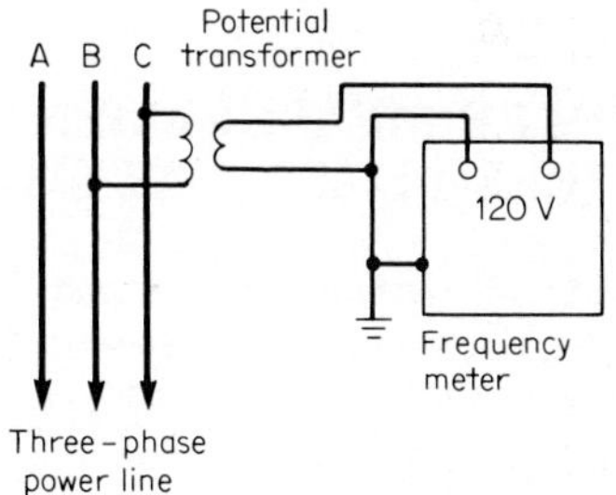

Fig. 3 Frequency meter connections to a power line.

3. Connect the Frequency Meter to Line

The frequency meter is connected to the line through a potential transformer as shown in Fig. 3.

Related Calculations: Frequency meters generally span a narrow range of frequencies centered about the nominal power frequency, such as 50, 60, or 400 Hz. For example, 60-Hz frequency meters have ranges of 59 to 61, 58 to 62, 55 to 65, and 50 to 70 Hz. Frequency meters are designed for 120-V input.

POWER MEASUREMENT USING A SINGLE-PHASE WATTMETER

The power consumption of a load, estimated to be 100 kVA, is to be measured. If the load is supplied by a 2400-V single-phase line, select a suitable wattmeter to make the measurement.

Calculation Procedure:

1. Select Wattmeter

Single-phase, as well as three-phase, wattmeters often require current and/or potential transformers as accessories. Wattmeters are generally designed for 120-V or 480-V operation with a maximum current rating of 5 A. For this application, a 120-V, 5-A wattmeter is selected.

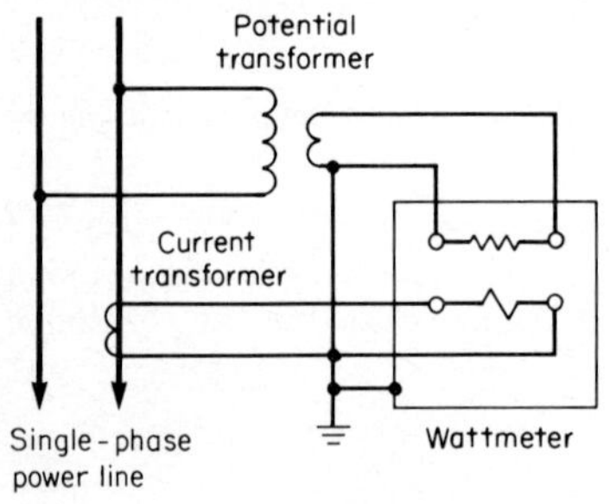

Fig. 4 Single-phase wattmeter circuit.

2. Select Current Transformer

The line current is 100,000/2400 = 41 A. A current transformer is therefore required; a 50:5-A rating is suitable.

3. Select Potential Transformer

The line voltage of 2400 V is required to be stepped down to 120 V. A 20:1 potential transformer is chosen.

4. Connect Wattmeter and Transformers to Line

The wattmeter and transformers are connected to the 2400-V line as indicated in Fig. 4.

POWER MEASUREMENT USING A THREE-PHASE WATTMETER

The power consumption of a load, estimated to be 1500 kVA, is to be measured. The load is supplied by a three-phase, three-wire line, 12,000 V line to line. Specify a suitable wattmeter for the measurement.

Calculation Procedure:

1. Select Wattmeter

Three-phase wattmeters intended for use with three-wire, three-phase lines are available, as well as others intended for use with four-wire, three-phase lines. For this application, a 120-V, 5-A, three-wire wattmeter is a good choice.

2. Select Current Transformer

The line current is $1{,}500{,}000/(1.73 \times 12{,}000) = 72$ A. A current transformer with a 100:5-A ratio is chosen.

3. Select Potential Transformer

The line-to-line voltage of 12,000 V requires that a 12,000/120, or 100:1, potential transformer be used.

4. Connect Wattmeter and Transformers to Line

The wattmeter and transformers are connected to the three-wire line as shown in Fig. 5.

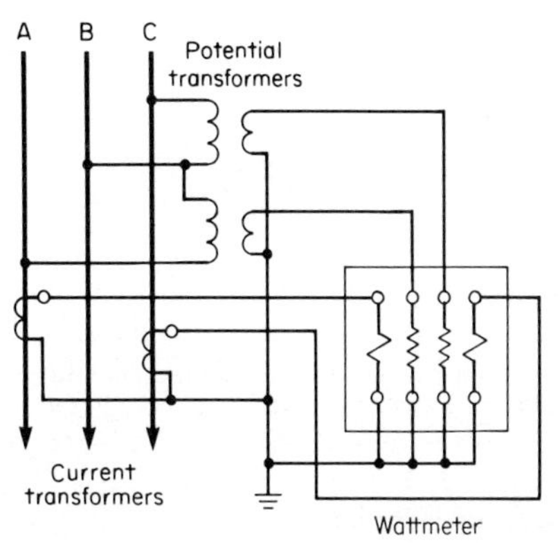

Fig. 5 Three-phase, three-wire wattmeter circuit.

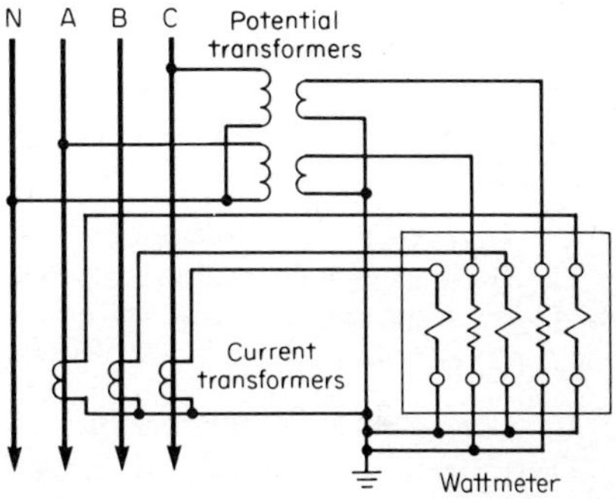

Fig. 6 Three-phase, four-wire wattmeter circuit.

POWER MEASUREMENT ON A FOUR-WIRE LINE

A 500-kVA load is supplied by a three-phase, four-wire, 4160-V line. Select a suitable wattmeter for measuring power consumption.

Calculation Procedure:

1. Select Wattmeter

A four-wire type having a 120-V, 5-A rating is chosen.

2. Select Current Transformer

The line current is $500{,}000/(1.73 \times 4160) = 72$ A. Therefore, a 75:5-A ratio is chosen for each of the current transformers.

3. Select Potential Transformer

The line-to-neutral voltage is $4160/1.73 = 2400$ V. Therefore, a 20:1 ratio is selected for the two potential transformers.

4. Connect the Wattmeter and Transformers to the Lines

The wattmeter and transformers are connected to the 4160-V, three-phase, four-wire line, as in Fig. 6.

REACTIVE-POWER MEASUREMENT

A varmeter is used to measure the reactive power in an industrial plant which is supplied by an 8300-V, three-phase, four-wire line. The plant load is estimated to be 300 kVA at 0.8 power factor. Design a suitable measuring system.

Calculation Procedure:

1. Select Varmeter

The high line voltage and large load dictate the use of a three-phase varmeter (rated at 120 V, 5 A) with current and potential transformers.

2. Select Current Transformer

I_{line} = 300,000/(1.73 × 8300) = 20.8 A   A 25:5-A current transformer is selected.

3. Select Potential Transformer

The potential transformer used with a three-phase, four-wire varmeter is connected line to neutral. The line-to-neutral voltage is 8300/1.73 = 4790 V; 4790/120 = 39.9. A 40:1 potential transformer is selected.

4. Determine a Suitable Scale for Meter

The phase angle θ is equal to $\cos^{-1}$ 0.8 = 36.9°. The reactive power = 300,000 × sin 36.9° = 300,000 × 0.6 = 180,000 vars. A scale providing a maximum reading of 200,000 vars is selected.

5. Connect Varmeter to Line

The varmeter is connected to the power line by means of the current and potential transformers as illustrated in Fig. 7.

Related Calculations: Varmeters are made for single- and three-phase, three- and four-wire systems. Center-zero scales are generally used for a varmeter. The instrument is designed to deflect to the right for a lagging power factor and to the left for a leading power factor. Many varmeters require an external compensator, or phase-shifting transformer. For high-voltage systems with large loads, it is common practice to use current and potential transformers with varmeters designed for 120-V, 5-A operation.

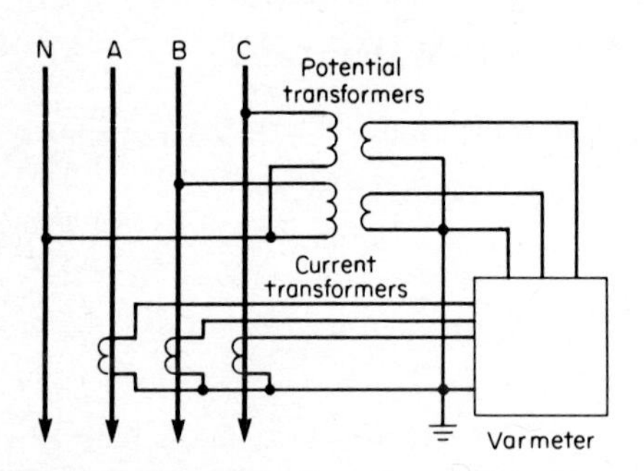

Fig. 7 Varmeter connected to a three-phase, four-wire power line.

POWER-FACTOR MEASUREMENT

The power factor of a group of four 30-hp electric motors in a manufacturing plant is to be measured. The motors are supplied by a 480/277-V, three-phase, three-wire line. The load is estimated to be 160 kVA, 0.85 power factor. Determine how the power factor is to be measured.

Calculation Procedure:

1. Select a Suitable Power-Factor Meter

A three-phase, three-wire power-factor meter is chosen for this application. Although self-contained power-factor meters are available, this application requires the use of current and potential transformers.

2. Select Current Transformer

I_{line} = 160,000/(1.73 × 480) = 192 A   Select a 200:5-A current transformer.

3. *Select Potential Transformer*

For this meter, the potential-transformer primary winding is connected line to line. The potential transformer ratio, therefore, is 480/120 = 4/1. A 4:1 potential transformer is chosen.

4. *Connect Power-Factor Meter to Line*

Figure 8 shows how the current and potential transformers are connected to the meter and line.

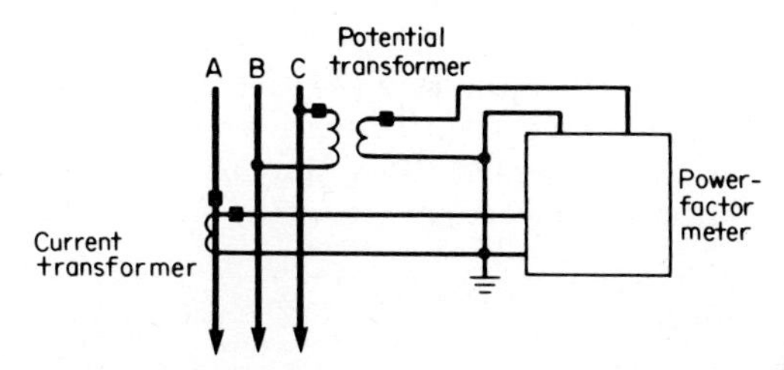

Fig. 8 Power-factor meter connected to a three-phase, three-wire power line.

Related Calculations: Power-factor meters are made for single-phase as well as three-phase systems. Polyphase power-factor meters are designed on the basis of balanced loads.

ELECTRIC ENERGY MEASUREMENT: SINGLE-PHASE WATTHOUR METERING

The electric energy consumption of an office air-conditioning system is to be measured on a monthly basis. The air conditioner is rated at 3500 W, 15.15 A, 230 V, single phase. The supply line is 240-V, three-wire, single-phase, 60-Hz. Choose a suitable meter.

Calculation Procedure:

1. *Select a Suitable Meter*

For this application, a 240-V, three-wire, single-phase wattmeter without a demand register is selected. The meter is capable of carrying 200 A, which is the smallest current rating available. Neither current nor voltage instrument transformers are required for this application.

2. *Connect the Watthour Meter to Line*

A meter socket, or pan, is commonly used as a means of mounting the meter as well as connecting it to the line. The connection to the meter socket, and the internal connections of the meter, are provided in Fig. 9.

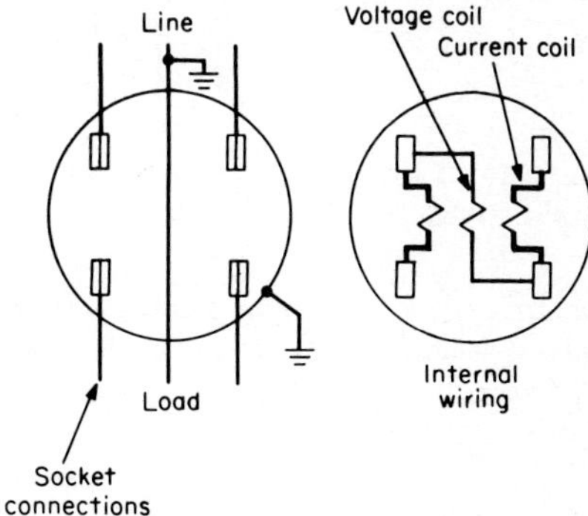

Fig. 9 Single-phase watthour meter connections.

Related Calculations: Watthour meters are available for single- and three-phase loads, with clock-type and cyclometer (digital) readouts, with and without demand registers. The single-phase, three-wire watthour meter contains two current coils and one potential coil. These act as the stator windings of a two-phase induction motor having a solid aluminum disk as its rotor.

ELECTRIC ENERGY MEASUREMENT: THREE-PHASE WATTHOUR METERING

A small building is heated with a pair of 12-kW space heaters containing thermostatic controls. There is a need to monitor the electrical consumption of these heaters on a monthly basis. The heaters are rated for 240 V, three phase, 28.8 A. The line is 240-V, three-phase, three-wire delta, with one line grounded. Determine how the measurement should be made.

Calculation Procedure:

1. Select Meter

Polyphase watthour meters are available for a variety of three-phase systems, both delta and wye, with different grounding arrangements. In this case, the meter chosen is one designed for use with a three-phase, three-wire system. The maximum current rating is 100 A per line, which is the smallest rating available. Neither current nor potential instrument transformers are required for these applications.

2. Connect Watthour Meter to Line

A meter socket is used for connecting the watthour meter to the line, as shown in Fig. 10.

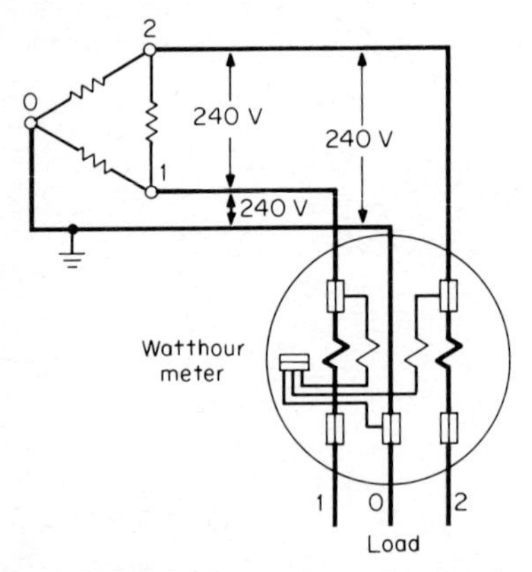

Fig. 10 Three-phase, three-wire watthour meter connections.

ELECTRIC PEAK-POWER DEMAND METERING

It is necessary to measure the peak demand, on a 15-min basis, of an industrial plant for which the peak demand is estimated to be 150 kW. The plant is supplied by a three-phase, four-wire, 7200-V line. The power factor is estimated to be 0.8 at peak demand. Specify how the measurement is to be made.

Calculation Procedure:

1. Select Demand Meter

This application calls for a three-phase, four-wire meter containing a demand register with 15-min time intervals. The application requires use of both current and potential transformers.

2. Select Current Transformer

I_{line} = 150,000/(1.73 × 7200 × 0.8) = 15.1 A Select a 15:5-A current transformer with a 15,000-V insulation rating.

3. Select Potential Transformer

The primary windings of these transformers are connected line to line, and are therefore subject to 7200 V. The secondary winding provides 120 V to the meter. A transformer having a 7200:120, or 60:1, ratio is called for. An insulation rating of 15,000 V is required.

4. Connect Meter to Line

The meter connected to the line is shown in Fig. 11.

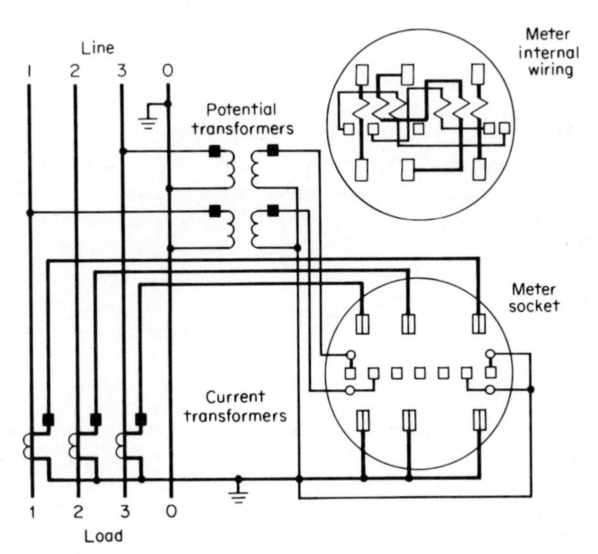

Fig. 11 Connecting a three-phase watthour meter with a demand register to power line.

Related Calculations: Demand meters are incorporated into watthour meters for many commercial, institutional, and industrial plants which are subject to demand charges by the electric utility. Demand meters operate as watthour meters over specified time intervals, usually 15- or 30-min periods. The highest 15- or 30-min consumption of energy is retained as an indication on the demand register, until it is manually reset to zero. This is usually done on a monthly basis. The calibration of the demand register is in kilowatts, and takes into account the period during which energy consumption is accumulated.

The selection of a demand meter is similar to the selection of a watthour meter. It is based on the type of service (single or three phase), the line voltage, anticipated current, and the time interval over which the definition of peak demand is based.

TEMPERATURE MEASUREMENT

The temperature of the cylinder head in a diesel engine is to be remotely indicated at a control-room panel. A temperature range of 0 to 200°C is to be accommodated. Design a system for making the measurement.

Calculation Procedure:

1. Select System Type

Pneumatic and electrical analog systems are used for transmission of nonelectrical quantities. Availability of air supply and environmental and maintenance factors govern the choice of system. In this application, an electrical system is chosen.

2. Select Sensor

A copper-Constantan (type T) thermocouple is selected because of its small size, flexibility, and suitability for the temperature range to be measured.

3. Select Transmitter

Industrial instrumentation systems involve use of electronic or pneumatic instruments, called *transmitters,* to convert the weak signal produced by the sensor to a standard form. In electrical analog systems, this is usually a direct current, ranging from 4 to 20 mA. Transmitters are always used with *receivers,* or *indicators,* from which power is derived. A two-wire transmitter requires only two conductors to be run from transmitter to receiver. In this case a two-wire millivolt-to-current transmitter is selected.

The particular model is designed to accept the output of a type T thermocouple over a temperature range of 0 to 200°C (approximately 0 to 10 mV). The minimum temperature corresponds to 4 mA of output current while the maximum temperature corresponds to 20 mA of output current.

4. Select Indicator

Indicators, or receivers, for electrical analog instrumentation systems are designed to accommodate 4 to 20 mA dc. Calibrations may show 0 to 100 percent or the actual temperature. Recording indicators are used where automatic recording is required. Pointer-type instruments are traditional but digital displays are now available. In this case, a nonrecording pointer-type instrument, with the temperature scale shown on the face of the instrument, is selected.

5. Connect System Components

The connections between thermocouple, transmitter, and indicator are illustrated in Fig. 12.

Related Calculations: Temperature-dependent resistors (TDRs), thermocouples, thermistors, and semiconductor sensors are used for industrial temperature measurement.

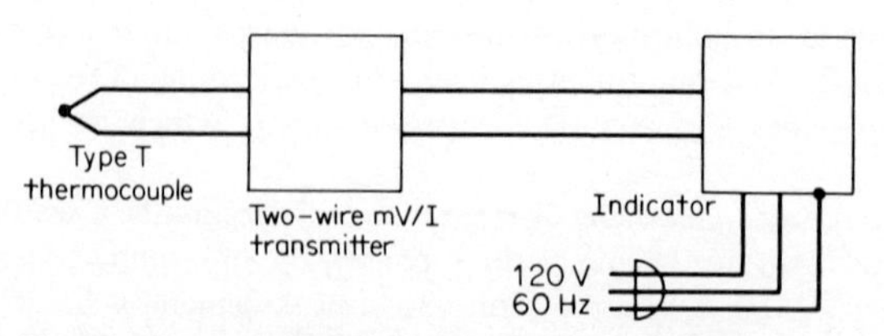

Fig. 12 Temperature-measurement instrumentation.

The temperature range to be accommodated, ease of installation, and environmental factors affect the selection.

PRESSURE MEASUREMENT

The pressure in a retort of a chemical processing plant is to be remotely indicated and recorded. The corrosive and toxic nature of the liquid requires isolation of the pressure-sensing instrument. There is no air supply available for instrumentation purposes. The pressure to be measured is usually between 414 and 552 kPa (60 and 80 psi), but can vary from 0 to 690 kPa (0 to 100 psi). Design a remote-measuring system.

Calculation Procedure:

1. Select Suitable System

The lack of an air supply is a major factor in choosing an electrical analog system. A pressure-to-current transmitter equipped with an isolation diaphragm and capillary tube is required to avoid contact with the hazardous material within the retort. The transmitter is a two-wire transmitter with an output of 4 to 20 mA corresponding to pressures ranging from 0 to 690 kPa (0 to 100 psi). The other major component of the system is the indicator-recorder, which is located in the control room.

2. Connect and Install System

The pressure-to-current transmitter is located at the retort and connected to an isolating diaphragm as shown in Fig. 13. The capillary tube connecting the diaphragm to the transmitter may be prefilled by the instrument manufacturer, or may be done at the site during installation. Care must be exercised to avoid the inclusion of air bubbles, which renders the system unduly temperature sensitive.

A two-wire cable connects the transmitter to the indicator-recorder, located in the control room. This instrument contains the necessary dc power supply and associated electronic circuitry, deriving power from the 120-V, 60-Hz power line.

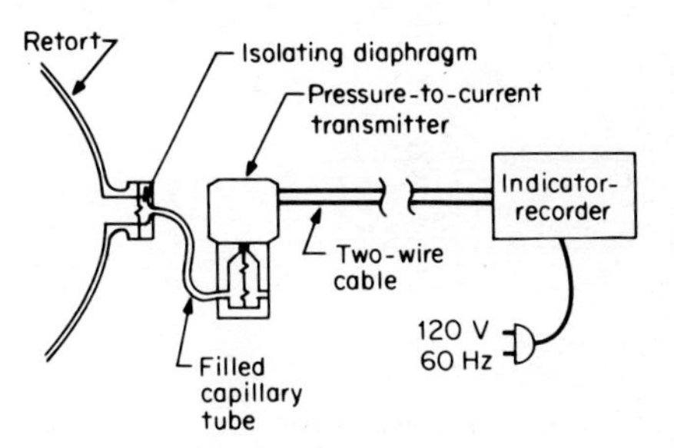

Fig. 13 Pressure instrumentation system.

Related Calculations: Remote indication of pressure is usually accomplished by means of a pressure sensor coupled to a transmitter, which is connected by electrical conductors or pneumatic tubing to a remote indicating instrument. Typically, the systems are either pneumatic, with 20.7 to 103.4 kPa (3 to 15 psi) representing the full span of pressures to be measured, or electrical analog, with 4 to 20 mA dc representing the full span of pressures to be measured. Electrical analog systems use either two- or four-wire transmitters to convert the sensed pressure to an electrical current.

FLOW-RATE MEASUREMENT

A residential apartment and office complex is supplied with heat by a pressurized hot-water system. It is necessary to measure the rate of flow of hot water from the central boiler plant. An air-supply system is available for instrumentation. Design a suitable measurement system.

Calculation Procedure:

1. *Select System*

Cost of installation, complexity, and maintenance factors tend to favor the use of an orifice plate in conjunction with a differential-pressure transmitter and square-root extractor. In this case, an all-pneumatic system is chosen, making use of the available air supply.

The orifice plate is simply a washer-shaped annular disk inserted between two sections of horizontal pipe. Two taps, one upstream and the other downstream of the orifice plate at the vena contracta (contraction of liquid jet), are required to provide for the tubing from these points to connect to the input ports of the differential-pressure transmitter. The differential pressure existing between these points, according to Bernoulli's theorem, is proportional to the square of the rate of flow of the fluid.

2. *Connect and Install System*

The system requires the installation of the orifice plate, the use of high-pressure tubing to make connection from the pipe taps to the input ports of the transmitter, and the connection of a 138-kPa (20-psi) air supply to the transmitter. The transmitter output is then fed to the input port of the square-root extractor, whose output is then connected to the indicator-recorder. The square-root extractor and indicator-recorder also require 138-kPa (20-psi) air for normal operation. The entire system is connected as shown in Fig. 14.

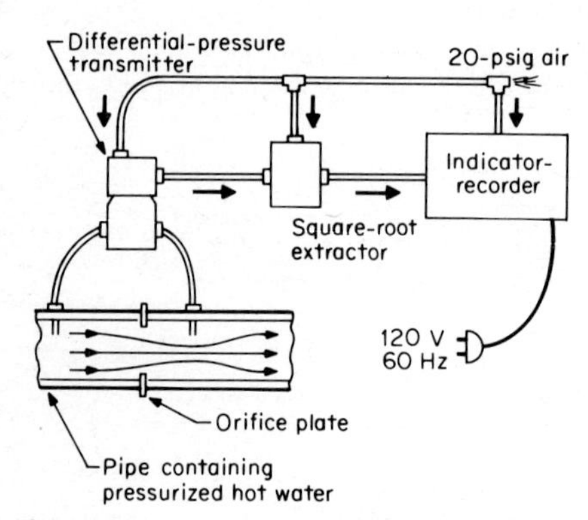

Fig. 14 Pneumatic flow-rate measuring system.

Related Calculations: Turbine, ultrasonic, and magnetic flowmeters are among the wide variety of flowmeters currently available. However, the simplest and most widely used industrial flow-measuring systems involve devices which yield a differential pressure that is indicative of flow. These devices are orifice plates, flow nozzles, venturi sections, and flow elbows. All yield an output that is proportional to the square of the flow rate. It is therefore common practice to employ a *square-root extractor* to provide a final output that is a linear function of flow rate.

Section 3 DC MOTORS AND GENERATORS

Lawrence J. Hollander, P.E.
Associate Dean, Cooper Union

REFERENCES Slemon and Straughen—*Electric Machines,* Addison–Wesley; Matsch—*Electromagnetic and Electromechanical Machines,* Harper and Row; Stein and Hunt—*Electric Power System Components,* Van Nostrand Reinhold; Nasar and Unnewehr—*Electromechanics and Electric Machines,* John Wiley; Fitzgerald and Kingsley—*Electric Machinery,* McGraw-Hill; Kosow—*Electric Machinery and Transformers,* Prentice-Hall; Siskind—*Electrical Machines: Direct and Alternating Currents* McGraw-Hill; McPherson—*An Introduction to Electrical Machines and Transformers,* John Wiley; Fitzgerald and Higginbotham—*Basic Electrical Engineering,* McGraw-Hill; Fitzgerald, Higginbotham, and Grabel—*Basic Electrical Engineering,* McGraw-Hill; Seely—*Electromechanical Energy Conversion,* McGraw-Hill; Smith—*Circuits, Devices, and Systems,* John Wiley; Kloeffler, Kerchner, and Brenneman—*Direct-Current Machinery,* Macmillan; Adkins—*The General Theory of Electrical Machines,* Chapman and Hall.

DIRECT-CURRENT GENERATOR USED AS TACHOMETER, SPEED/VOLTAGE MEASUREMENT

A tachometer consists of a small dc machine having the following features: lap-wound armature, four poles, 780 conductors on the armature (rotor), field (stator) flux per pole $= 0.32 \times 10^{-3}$ Wb. Find the speed calibration for a voltmeter of very high impedance connected to the armature circuit.

Calculation Procedure:

1. Determine the Number of Paths in the Armature Circuit

For a lap winding, the number of paths (in parallel) a is always equal to the number of poles. For a wave winding, the number of paths is always equal to two. Therefore, for a four-pole, lap-wound machine $a = 4$.

2. Calculate the Machine Constant k

Use the equation: machine constant $k = Np/a\pi$, where $N = 780/2 = 390$ turns (i.e., two conductors constitute one turn) on the armature winding, $p = 4$ poles, and $a = 4$ parallel paths. Thus, $(390)(4)/(4\pi) = 124.14$ and $k = 124.14$.

3. Calculate the Induced Voltage as a Function of Mechanical Speed

The average induced armature voltage e_a (proportional to the rate of change of flux linkage in each coil) $= k\phi\omega_m$ where ϕ = flux per pole in webers, ω_m = mechanical speed of the rotor in radians per second, and e_a is in volts. Thus $e_a/\omega_m = k\phi = (124.14)(0.32 \times 10^{-3} \text{ Wb}) = 0.0397 \text{ V}\cdot\text{s/rad}$.

4. Calculate the Speed Calibration of the Voltmeter

Take the reciprocal of the factor e_a/ω_m, finding $1/0.0397 = 25.2 \text{ rad/V}\cdot\text{s}$. Thus, the calibration of the high-impedance voltmeter scale is such that each 1-V division is equivalent to a speed of 25.2 rads/s. With each revolution being equal to 2π rad, the calibration also is equivalent to a speed of $25.2/(2\pi)$ or 4 r/s for each 1-V division, or $(4 \text{ r/s})(60 \text{ s/min}) = 240$ r/min.

Related Calculations: The small dc generator, mechanically coupled to a motor, gives an output voltage that is directly proportional in magnitude to the speed of rotation. Not only may that voltage be read on a voltmeter-scale calibrated in units of speed, but also it may be applied to a control circuit of the larger motor in such a way as to make desired corrections in speed. This type of tachometer affords a simple means for regulating the speed of a motor.

SEPARATELY EXCITED DC GENERATOR'S RATED CONDITIONS FROM NO-LOAD SAURATION CURVE

A separately excited dc generator has a no-load saturation curve as shown in Fig. 1, and an equivalent circuit as shown in Fig. 2. Its nameplate data are as follows: 5 kW, 125 V, 1150 r/min, and armature circuit resistance $R_a = 0.40\ \Omega$. Assume that the generator is driven at 1200 r/min and that the field current is adjusted by field rheostat to equal 2.0 A. If the load is the nameplate rating, find the terminal voltage V_t. Armature reaction and brush losses may be neglected.

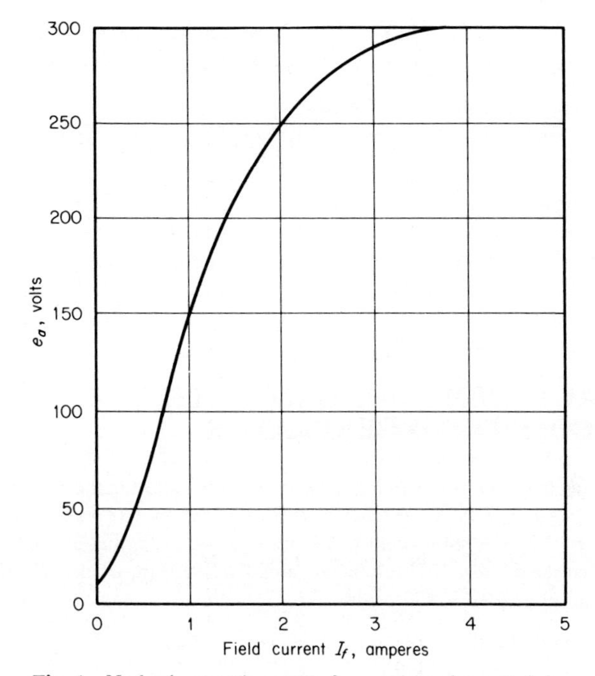

Fig. 1 No-load saturation curve for a separately excited dc generator, at 1150 r/min.

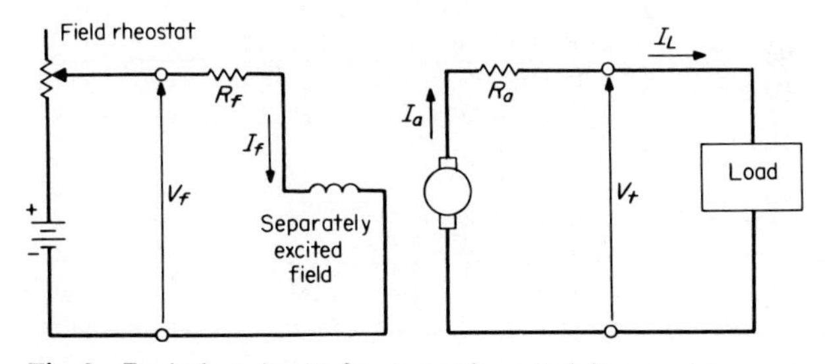

Fig. 2 Equivalent circuit of a separately excited dc generator.

Calculation Procedure:

1. Calculate the Armature Induced Voltage e_a

The no-load saturation curve is obtained at rated speed of 1150 r/min. In this case, however, the driven speed is 1200 r/min. For a given field flux (field current) the induced voltage is proportional to speed. Thus, if I_f is 2.0 A, the induced voltage at 1150 r/min from the no-load saturation curve is equal to 250 V. The induced voltage at the higher speed of 1200 r/min is $e_a = (1200/1150)(250\text{ V}) = 260.9$ V.

2. Calculate the Rated Load Current I_L

The rated load current is obtained from the equation: I_L = nameplate kilowatt rating/nameplate kilovolt rating = 5 kW/0.125 kV = 40 A. Note that the speed does not enter into this calculation; the current, voltage, and power ratings are limited by (or determined by) the thermal limitations (size of copper, insulation, heat dissipation, etc.)

3. Calculate the Terminal Voltage V_t

The terminal voltage is determined from Kirchhoff's voltage law; or simply $V_t = e_a - I_L R_a = 260.9\ \text{V} - (40\ \text{A})(0.4\ \Omega) = 244.9\ \text{V}$.

Related Calculations: This same procedure is used for calculating the terminal voltage for any variation in speed. If there were a series field, or if there were a shunt field connected across the armature, essentially the same procedure is followed, with proper allowance for the additional circuitry and the variation in the total field flux, as in the next three problems.

TERMINAL CONDITIONS CALCULATED FOR DC COMPOUND GENERATOR

A compound dc generator is connected long-shunt. The data for the machine are as follows: 150 kW, 240 V, 625 A, series-field resistance $R_s = 0.0045\ \Omega$, armature-circuit resistance $R_a = 0.023\ \Omega$, shunt-field turns per pole = 1100, and series-field turns per pole = 5. Calculate the terminal voltage at rated load current, when the shunt-field current is 5.0 A and the speed is 950 r/min.

Calculation Procedure:

1. Draw the Equivalent Circuit Diagram

The equivalent circuit diagram is shown in Fig. 3.

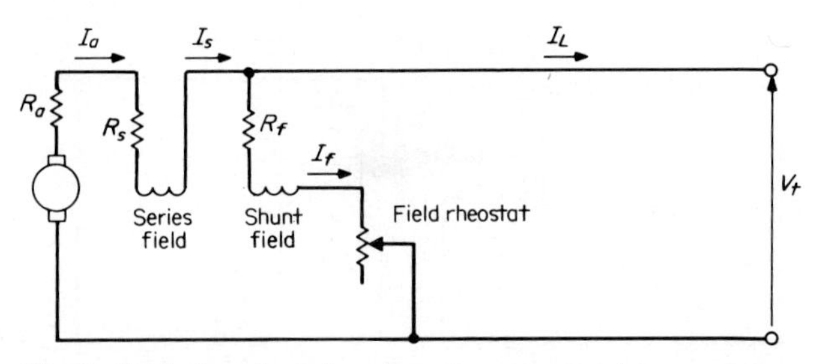

Fig. 3 Equivalent circuit for a long-shunt compound dc generator.

2. Calculate the Current in the Series Field, I_s

With reference to Fig. 3, the current in the series field is $I_s = I_a = I_L + I_f = 625\ \text{A} + 5.0\ \text{A} = 630\ \text{A}$.

3. Convert Series-Field Current to Equivalent Shunt-Field Current

So that the no-load saturation curve (Fig. 4) may be used, the series-field current is converted to equivalent shunt-field current; then, a total field flux is determined from the shunt-field current and the equivalent shunt-field current. The equivalent shunt-field current of the series field is calculated from the relation: actual ampere-turns of series field/actual turns of shunt field = (630 A)(5 turns)/1100 turns = 2.86 A. The total flux in terms of field currents = shunt-field current + equivalent shunt-field current of series field = 5.0 A + 2.86 A = 7.86 A.

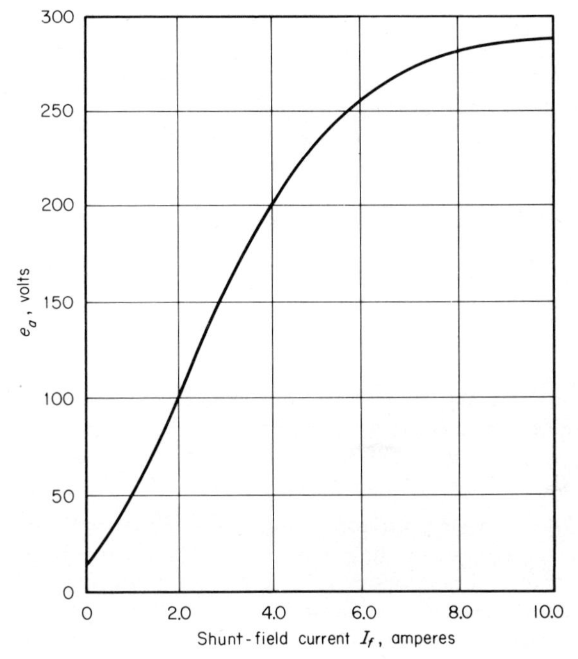

Fig. 4 No-load saturation curve for a 240-V, 1100-r/min dc generator.

4. *Calculate Armature Induced Voltage e_a*

With reference to the no-load saturation curve of Fig. 4, enter the graph at field current of 7.86 A. The no-load induced voltage e_a = 280 V at 1100 r/min. When the machine is driven at 950 r/min, the induced voltage e_a = (280 V)(950/1100) = 242 V.

5. *Calculate the Terminal Voltage V_t*

The terminal voltage is calculated from Kirchhoff's voltage law: $V_t = e_a - I_a(R_a + R_s)$ = 242 V − (630 A)(0.023 Ω + 0.0045 Ω) = 224.7 V.

Related Calculations: In the long-shunt connection the series field is connected on the armature-side of the shunt field; in the short-shunt connection the series field is connected on the terminal-side of the shunt field. In the latter case, the series-field current is equal to I_L rather than I_a; Kirchhoff's voltage equation still is applicable. There is little difference between long-shunt and short-shunt compounding. In both cases it is necessary to determine the net flux per pole; the series-field flux may be additive or subtractive. In this problem it was assumed to be additive.

ADDED SERIES FIELD CALCULATED TO PRODUCE FLAT COMPOUNDING IN DC GENERATOR

A given dc generator is to be flat-compounded. Its data are: 150 kW, 240 V, 625 A, 1100 r/min, armature-circuit resistance R_a = 0.023 Ω, shunt-field turns per pole = 1100, series-field turns per pole = 5, and series-field resistance R_s = 0.0045 Ω. It is driven at a constant speed of 1100 r/min, for which the no-load saturation curve is given in Fig. 4. Determine the shunt-field circuit resistance for rated conditions (assuming long-shunt connections), and calculate the armature current to be diverted from the series field, for flat compounding at 250 V, and the resistance value of the diverter.

Calculation Procedure:

1. Calculate the Value of the Shunt-Field Circuit Resistance

From the no-load saturation curve (Fig. 4), the 250-V point indicates a shunt-field current I_f of 5.6 A. A field-resistance line is drawn on the no-load saturation curve (a straight line from the origin to point 250 V, 5.6 A); this represents a shunt-field circuit resistance of (250 V)/(5.6A) = 44.6 Ω. That is, for the no-load voltage to be 250 V, the shunt-field resistance must be 44.6 Ω.

2. Calculate the Armature Values at Rated Conditions

For an armature current of 625 A, I_aR_a = (625 A)(0.023 Ω) = 14.4 V. The armature current $I_a = I_f + I_L$ = 5.6 A + 625 A = 630.6 A. The induced armature voltage $e_a = V_t + I_a(R_a + R_s)$ = 250 V + (630.6 A)(0.023 Ω + 0.0045 Ω) = 267.3 V.

3. Calculate Needed Contribution of Series Field at Rated Conditions

At rated conditions the current through the series field is 630.6 A (the same as I_a); refer to the long-shunt equivalent circuit shown in Fig. 3. For the induced voltage of 267.3 V (again referring to the no-load saturation curve), the indicated excitation is 6.6 A. Thus, the contribution from the series field = 6.6 A − 5.6 A = 1.0 A (in equivalent shunt-field amperes). The actual series-field current needed to produce the 1.0 A of equivalent shunt-field current = (1.0 A)(1100/5) = 220 A.

4. Calculate Diverter Resistance

Because the actual current through the series field would be 630.6 A, and 220 A is sufficient to produce flat compounding (V_t = 250 V at no load and at full load), 630.6 A − 220 A = 410.6 A must be diverted through a resistor shunted across the series field. The value of that resistor = voltage across the series field/current to be diverted = (220 A × 0.045 Ω)/410.6 A = 0.0024 Ω.

Related Calculations: In this problem, armature reaction was ignored because of lack of data. A better calculation would require saturation curves not only for no-load but also for rated full-load current. That full-load current curve would droop below the no-load curve in the region above the knee of the curve (say above 250 V in this example). With these data taken into account, the needed series-field contribution to the magnetic field would be greater than what was calculated, and the current to be diverted would be less. Effectively, armature reaction causes a weakening of the magnetic field (in addition to distortion of that field). The actual calculations made in this problem would result in slight undercompounding, rather than flat compounding.

CALCULATION OF INTERPOLE WINDINGS FOR A DC GENERATOR

A dc lap-wound generator has the following data: 500 kW, 600 V, 4 poles, 4 interpoles, 464 armature conductors. The magnetomotive force (mmf) of the interpoles is 20 percent greater than that of the armature. Determine the number of turns on the interpoles.

Calculation Procedure:

1. Calculate the Number of Armature Turns per Pole

If there are 464 armature conductors, then the number of turns is one-half of that amount; each two conductors constitute one turn. The machine has 4 poles; thus, armature turns per pole = (armature conductors/2)/number of poles = (464/2) /4 = 58.

2. Calculate the Current per Armature Path

For a lap winding, the number of paths (in parallel) is always equal to the number of poles; in this problem it is 4. The armature current per path = $I_a/4$. This represents the current in each armature conductor, or in each turn.

3. Calculate the Number of Turns on the Interpoles

The interpoles (also known as commutating poles) are for the purpose of providing mmf along the quadrature axis to reverse the flux in that space caused by the cross-magnetizing effect of armature reaction (see Fig. 5). They have a small number of turns

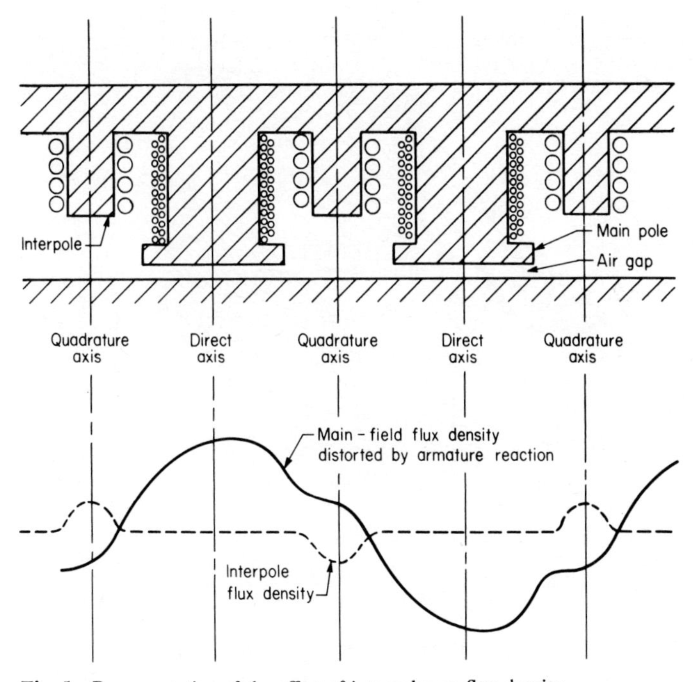

Fig. 5 Representation of the effect of interpoles on flux density.

of large cross section because the interpoles, being in series with the armature, carry armature current. The mmf of the armature per pole is: armature turns per pole × armature current in each conductor = 58 ($I_a/4$) = 14.5I_a ampere-turns. If the interpoles have an mmf 20 percent greater, then each one is (1.20)(14.5)I_a = 17.4I_a ampere-turns. The number of turns on each interpole = 17.4.

Related Calculations: The function of interpoles is to improve commutation, that is, to eliminate sparking of the brushes as they slide over the commutator bars. The mmf of the interpoles usually is between 20 to 40 percent greater than the armature mmf, when the number of interpoles is equal to the number of main poles. If one-half the number of interpoles is used, the mmf design is usually 40 to 60 percent greater than the armature mmf. These calculations are always on a per-pole basis. Because interpoles operate on armature current, their mmf is proportional to the effects of armature reaction.

DESIGN OF COMPENSATING WINDING FOR DC MACHINE

A lap-wound dc machine has the following data: four poles, ratio of pole face to pole span = 0.75, total number of armature slots = 33, number of conductors per slot = 12. Design a compensating winding for each pole face (that is, find the number of conductors to be placed in each pole face and the number of slots). See Fig. 6.

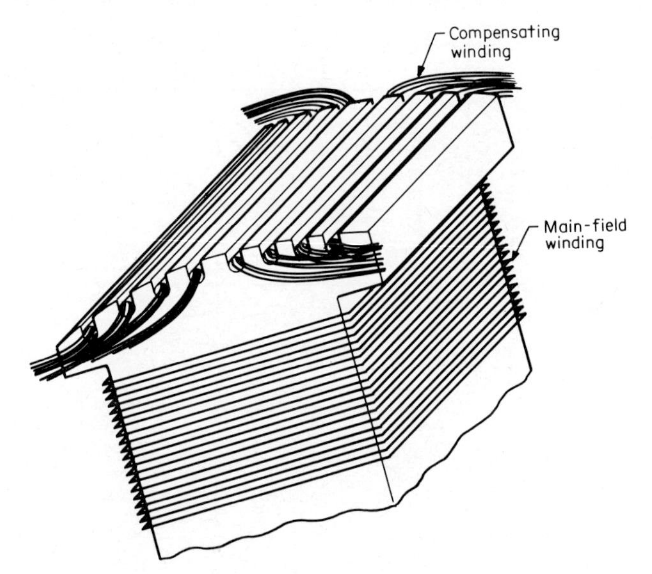

Fig. 6 Pole having compensating winding in face.

Calculation Procedure:

1. Calculate Armature mmf in One Pole Span

The number of armature turns in a pole span is (total number of armature conductors/2)/number of poles = [(12 conductors per slot)(33 slots)/2]/4 poles = 49.5 armature turns. The current in each conductor is the current in each parallel path. For a lap-wound machine the number of parallel paths is the same as the number of poles; in this problem it is 4. The armature mmf in a pole span = (armature turns in a pole span) × (armature current per path) = $(49.5)(I_a/4) = 12.4I_a$ ampere turns.

2. Calculate the Armature Ampere-Turns per Pole Face
The ratio of the pole face to pole span is given as 0.75. Therefore, the armature (mmf) ampere-turns per pole face = (0.75) × ($12.4I_a$) = 9.3 ampere-turns.

3. Calculate the Number of Conductors in the Compensating Winding
The purpose of the compensating winding is to neutralize the armature mmf that distorts the flux density distribution. Under sudden changes in loading there would be a flashover tendency across adjacent commutator bars; the compensating-winding effect overcomes this and is proportional to the armature current because it is connected in the armature circuit. Therefore, to design the compensating winding it is necessary to have it produce essentially the same number of ampere-turns as the armature circuit under each pole face. In this problem each pole face must produce 9.3 ampere-turns; this requires that each pole face have (2)(9.3 turns) = 18.6 conductors (say 18 or 19 conductors).

4. Calculate the Number of Slots in the Pole Face
The slots in the pole face should not have the same spacing as on the armature, otherwise reluctance torques would occur as the opposite slots and teeth went in and out of alignment. The number of slots on the armature opposite a pole face is (33 slots)(0.75 ratio)/4 poles = 6.2 slots. For a different number of slots, 5 could be selected and 4 conductors could be placed in each slot (the total being 20 conductors). Another solution might be 8 slots with 2 conductors in each slot (the total being 16 conductors).

Related Calculations: In many cases, machines will have both interpoles (see previous problem) and compensating windings. The winding connections of the machine are shown in Fig. 7.

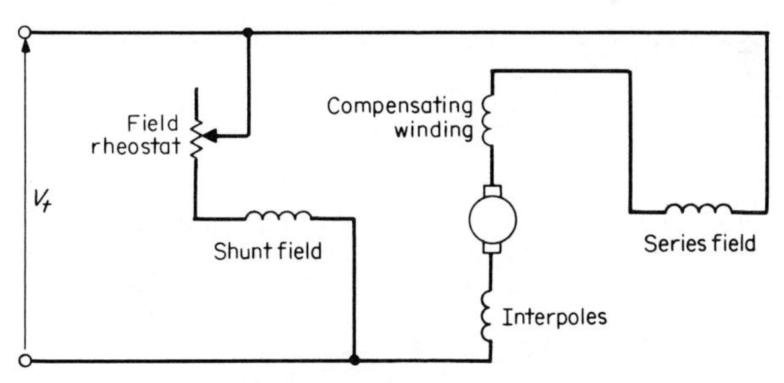

Fig. 7 Connections showing dc compound machine with interpoles and compensating windings.

STATOR AND ARMATURE RESISTANCE CALCULATIONS IN DC SELF-EXCITED GENERATOR

A self-excited dc generator has the following data: 10 kW, 240 V. Under rated conditions, the voltage drop in the armature circuit is determined to be 6.1 percent of the terminal voltage. Also, the shunt-field current is determined to be 4.8 percent of the rated load current. Calculate the armature circuit resistance and the shunt-field circuit resistance.

Calculation Procedure:

1. Calculate the Shunt-Field Current

Use the equation: rated load current I_L = rated load/rated voltage = 10,000 W/240 V = 41.7 A. The shunt-field current $I_f = 0.048I_L = (0.048)(41.7 \text{ A}) = 2.0$ A.

2. Calculate the Resistance of the Shunt-Field Circuit

Use the equation: shunt-field resistance $R_f = V_t/I_f = 240 \text{ V}/2.0 \text{ A} = 120\ \Omega$.

3. Calculate the Resistance of the Armature Circuit

The armature current $I_a = I_L + I_f = 41.7 \text{ A} + 2.0 \text{ A} = 43.7$ A. The voltage drop in the armature circuit, I_aR_a, is given as 6.1 percent of the terminal voltage V_t. Thus, $I_aR_a = (0.061)(240 \text{ V}) = 14.64$ V. But $I_a = 43.7$ A; therefore, $R_a = 14.64 \text{ V}/43.7 \text{ A} = 0.335\ \Omega$.

Related Calculations: Voltage and current relations for dc machines follow from the circuit diagrams as in Figs. 2, 3, and 7. It is a good practice to draw the equivalent circuit to better visualize the combinations of currents and voltages.

EFFICIENCY CALCULATION FOR DC SHUNT GENERATOR

A dc shunt generator has the following ratings: 5.0 kW, 240 V, 1100 r/min, armature-circuit resistance $R_a = 1.10\ \Omega$. The no-load armature current is 1.8 A when the machine is operated as a motor at rated speed and rated voltage. The no-load saturation curve is shown in Fig. 4. Find the efficiency of the generator at rated conditions.

Calculation Procedure:

1. Calculate the Full-Load Current I_L

Use the equation: I_L = rated capacity in watts/rated terminal voltage = 5000 W/240 V = 20.8 A. This is the full-load current when the machine is operated as a generator.

2. Calculate the Armature Induced Voltage e_a at No Load

When the machine is run as a motor, the no-load armature current is 1.8 A. Use the equation: $e_a = V_t - R_aI_a = 240 \text{ V} - (1.10\ \Omega)(1.8 \text{ A}) = 238$ V.

3. Calculate the Rotational Losses $P_{\text{rot(loss)}}$

Use the equation: $P_{\text{rot(loss)}} = e_aI_a = (238 \text{ V})(1.8 \text{ A}) = 428.4$ W.

4. Calculate the Field Current Required at Full Load

Under generator operation at full load, the armature current is the sum of the load current plus the shunt-field current, $I_L + I_f$. Use the equation: $V_t = 240 \text{ V} = e_a - R_a(I_L + I_f)$. Rearranging, find $e_a = 240 \text{ V} + (1.10\ \Omega)(20.8 \text{ A} + I_f) = 262.9 + 1.10I_f$. From observation of the no-load saturation curve (Fig. 4), note that I_f is 7.0 A; the equation is satisfied. Then $e_a = 262.9 \text{ V} + (1.10\ \Omega)(7.0 \text{ A}) = 270.6$ V. It may be necessary to make one or two trial attempts before the equation is satisfied.

5. Calculate the Copper Loss

The armature current I_a is $I_f + I_L = 7.0 \text{ A} + 20.8 \text{ A} = 27.8$ A. The armature-circuit copper loss is $I_a^2R_a = (27.8 \text{ A})^2(1.10\ \Omega) = 850$ W. The shunt-field copper loss is V_tI_f = (240 V)(7.0 A) = 1680 W. The total copper loss is 850 W + 1680 W = 2530 W.

6. Calculate the Efficiency

Use the equation: η = output/(output + losses) = (5000 W)(100 percent)/(5000 W + 2530 W + 428.4 W) = 62.8 percent.

Related Calculations: It should be noted in this problem that the generator was operated as a motor, at no load, in order to determine the armature current; from that current the rotational losses were obtained. The rotational losses include friction, windage, and core losses. The efficiency could be improved by increasing the number of turns on the field winding, reducing I_f.

TORQUE AND EFFICIENCY CALCULATION FOR SEPARATELY EXCITED DC MOTOR

A separately excited dc motor has the following data: 1.5 hp, 240 V, 6.3 A, 1750 r/min, field current I_f = 0.36 A, armature-circuit resstance R_a = 5.1 Ω, field resistance R_f = 667 Ω. Laboratory tests indicate that at rated speed mechanical losses (friction and windage) = 35 W, at rated field current magnetic losses = 32 W, and at rated armature current and rated speed stray-load losses = 22 W. The no-load saturation curve is given

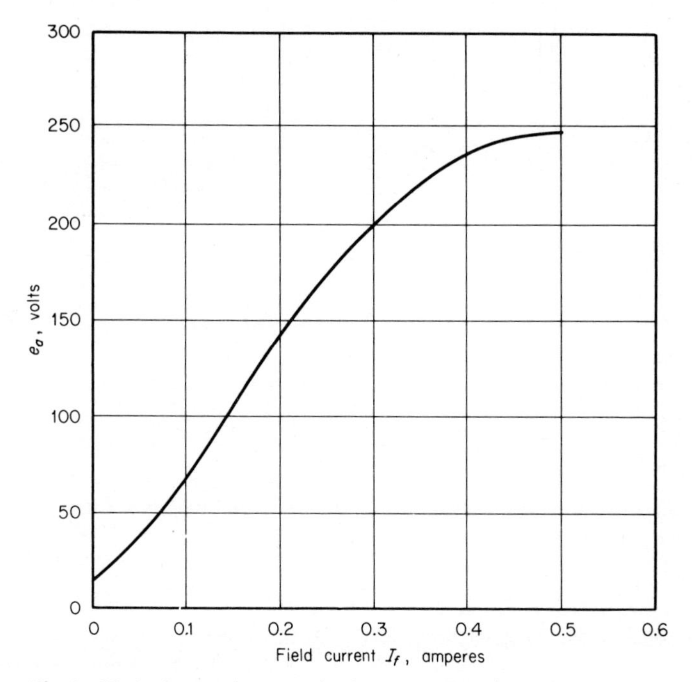

Fig. 8 No-load saturation curve for dc motor, 1750 r/min.

in Fig. 8. Determine rated torque, input armature voltage, input power, and efficiency at rated torque and speed.

Calculation Procedure:

1. Calculate the Rated Torque

Use the equation: P (in horsepower) $= 2\pi nT/33{,}000$ where n = speed in r/min and T = torque in lb·ft. Rearranging, find $T = 33{,}000P/2\pi n = (33{,}000)(1.5 \text{ hp})/(2\pi \times 1750 \text{ r/min}) = 4.50$ lb·ft. Or (4.50 lb·ft)(N·m)/(0.738 lb·ft) = 6.10 N·m.

2. Calculate the Electromagnetic Power

The electromagnetic power P_e is the sum of the (*a*) output power, (*b*) mechanical power including the mechanical losses of friction and windage, (*c*) magnetic power losses, and (*d*) the stray-load losses. Thus $P_e = (1.5 \text{ hp} \times 746 \text{ W/hp}) + 35 \text{ W} + 32 \text{ W} + 22 \text{ W} = 1208 \text{ W}$. See Fig. 9.

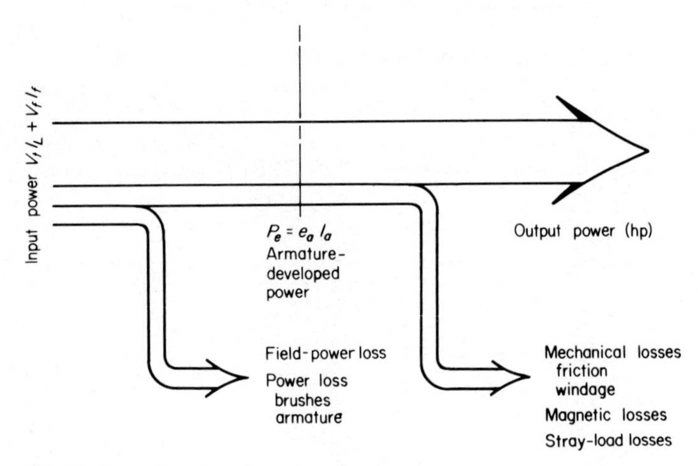

Fig. 9 Power flow in a dc motor.

3. Calculate the Armature Current and Induced Voltage at Rated Load

From the no-load saturation curve, Fig. 8, the induced voltage for a field current of 0.36 A, is 225 V. Use the equation: electromagnetic power $P_e = e_a I_a$. Thus, $I_a = P_e/e_a = 1208 \text{ W}/225 \text{ V} = 5.37$ A. The armature-circuit voltage drop is $I_a R_a = (5.3 \text{ A})(5.1\ \Omega) = 27.4$ V.

4. Calculate the Armature- and Field-Circuit Copper Loss

Use the equation: armature-circuit copper loss $= I_a^2 R_a = (5.37 \text{ A})^2(5.1\ \Omega) = 147$ W. The field-circuit copper loss $= I_f^2 R_f = (0.36 \text{ A})^2(667\ \Omega) = 86$ W.

5. Calculate the Efficiency

Power input = electromagnetic power P_e + armature-circuit copper loss + field-circuit copper loss = 1208 W + 147 W + 86 W = 1441 W. Thus, the efficiency η = (output × 100 percent)/input = (1.5 hp × 746 W/hp)(100 percent)/1441 W = 77.65 percent.

Related Calculations: The mechanical losses of friction and windage include bearing friction (a function of speed and the type of bearing), brush friction (a function of speed, brush pressure, and brush composition), and the windage resulting from air motion in the air spaces of the machine. The magnetic losses are the core losses in the magnetic material. Stray-load loss actually includes many of the previously mentioned losses but is a term accounting for increased losses caused by loading the motor. For example, the distortion of the magnetic field (armature reaction) increases with load and causes increased magnetic loss; it is included among the stray-load losses.

DESIGN OF A MANUAL STARTER FOR A DC SHUNT MOTOR

A dc shunt motor has the following data: 240 V, 18 A, 1100 r/min, and armature-circuit resistance $R_a = 0.33\ \Omega$. The no-load saturation curve for the machine is given in Fig. 4. The motor is to be started by a manual starter, shown in Fig. 10; the field current is set

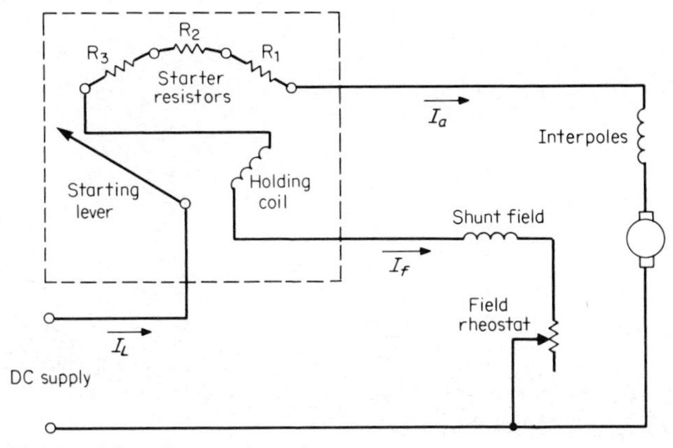

Fig. 10 Manual starter for a shunt motor.

at 5.2 A. If the current during starting may vary between 20 A and 40 A, determine the starting resistors R_1, R_2, and R_3 needed in the manual starter.

Calculation Procedure:

1. Calculate Armature Current at Standstill

At the first contact on the manual starter, 240 V dc is applied to the shunt field; the field rheostat has been set to allow $I_f = 5.2$ A. Thus, the armature current $I_a = I_L - I_f = 40\text{ A} - 5.2\text{ A} = 34.8$ A. The 40 A is the maximum permissible load current.

2. Calculate the Value of the Starter Resistance R_{ST}

Use the equation: $R_{ST} = (V_t/I_a) - R_a = (240\text{ V})/(34.8\text{ A}) - 0.330\ \Omega = 6.57\ \Omega$.

3. Calculate the Induced Armature Voltage e_a when the Machine Accelerates

As the machine accelerates, an induced armature voltage is created, causing the armature current to fall. When the armature current drops to the minimum allowed during starting $I_a = I_L - I_f = 20\text{ A} - 5.2\text{ A} = 14.8\text{ A}$, and the induced voltage is obtained from the equation: $e_a = V_t - I_a(R_a + R_{ST}) = 240\text{ V} - (14.8\text{ A})(0.330\ \Omega + 6.57\ \Omega) = 137.9\text{ V}$.

4. Calculate the Starter Resistance R_{ST1}

Use the equation: $R_{ST1} = (V_t - e_a)/I_a - R_a = (240\text{ V} - 137.9\text{ V})/34.8\text{ A} - 0.330\ \Omega = 2.60\ \Omega$.

5. Calculate the Induced Armature Voltage e_a when Machine Again Accelerates

As in Step 3, $e_a = V_t - I_a(R_a + R_{ST1}) = 240\text{ V} - (14.8\text{ A})(0.330\ \Omega + 2.60\ \Omega) = 196.6\text{ V}$.

6. Calculate the Starter Resistance R_{ST2}

Use the equation: $R_{ST2} = (V_t - e_a)/I_a - R_a = (240\text{ V} - 196.6\text{ V})/34.8\text{ A} - 0.330\ \Omega = 0.917\ \Omega$.

7. Calculate the Induced Armature Voltage e_a when Machine Again Accelerates

As in Steps 3 and 5, $e_a = V_t - I_a(R_a + R_{ST2}) = 240\text{ V} - (14.8\text{ A})(0.330\ \Omega + 0.917\ \Omega) = 221.5\text{ V}$.

8. Calculate the Three Starting Resistances R_1, R_2, and R_3

From the above steps, $R_1 = R_{ST2} = 0.917\ \Omega$, $R_2 = R_{ST1} - R_{ST2} = 2.60\ \Omega - 0.917\ \Omega = 1.683\ \Omega$, and $R_3 = R_{ST} - R_{ST1} = 6.57\ \Omega - 2.60\ \Omega = 3.97\ \Omega$.

Related Calculations: The manual starter is used commonly for starting dc motors. With appropriate resistance values in the starter, the starting current may be limited to any value. The same calculations may be used for automatic starters; electromagnetic relays are used to short-circuit sections of the starting resistance, sensing when the armature current drops to appropriate values.

CONSIDERATION OF DUTY CYCLE USED TO SELECT DC MOTOR

A dc motor is used to operate a small elevator in a factory assembly line. Each cycle of operation runs 5 min and includes four modes of operation as follows: (1) loading of the elevator, motor at standstill for 2 min; (2) ascent of the elevator, motor shaft horsepower required is 75 for 0.75 min; (3) unloading of the elevator, motor at standstill for 1.5 min; and (4) descent of the unloaded elevator, motor shaft horsepower (regenerative braking) required is −25 for 0.75 min. Determine the size of the motor required for this duty cycle.

Calculation Procedure:

1. Draw a Chart to Illustrate the Duty Cycle

See Fig. 11.

2. Calculate the Proportional Heating of the Machine as a Function of Time

The losses (heating) in a machine are proportional to the square of the load current; in turn, the load current is proportional to the horsepower. Thus, the heating of the motor

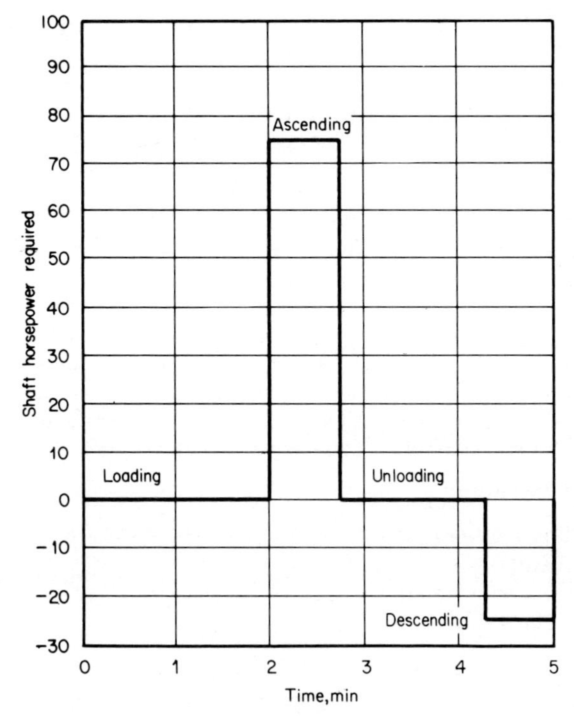

Fig. 11 Duty cycle for an assembly-line elevator motor.

is proportional to the summation of (P^2 × time), where P is the power or horsepower expended (or absorbed during regenerative braking) during the time interval. In this case, the summation is $(75\ \text{hp})^2(0.75\ \text{min}) + (-25\ \text{hp})^2(0.75\ \text{min}) = 4687.5\ \text{hp}^2\cdot\text{min}$.

3. *Consider the Variation in Cooling for Open (NEMA O Classification) or Closed (NEMA TE Classification) Motors*

Allowance is made for the cooling differences between open or closed motors; this is particularly important because at standstill in open motors, without forced-air cooling, poorer cooling results. The running time and the stopped time are separated. At standstill, for open motors the period is divided by 4; for closed motors the period is counted at full value. In this problem, assume a closed motor for which the effective total cycle time is 5 min. If the motor were of the open type, the effective cycle time would be 1.5 min running $+ 3.5/4$ min at standstill $= 2.375$ min.

4. *Calculate Root-Mean-Square Power*

The root-mean-square power $P_{rms} = [(\Sigma P^2 \times \text{time})/(\text{cycle effective time})]^{0.5} = (4687.5\ \text{hp}^2\cdot\text{min}/5\ \text{min})^{0.5} = 30.6\ \text{hp}$ for the case of the closed motor. $P_{rms} = (4687.5\ \text{hp}^2\cdot\text{min}/2.375\ \text{min})^{0.5} = 44.4\ \text{hp}$ for the case of the open motor. Thus, the selection could be a 30-hp motor of the closed type or a 50-hp motor of the open type. It should be noted in this example that the peak power (75 hp used in ascending) is more than twice the 30 hp calculated for the closed motor; it may be desirable to opt for a 40-hp motor; the manufacturer's literature should be consulted for short-time ratings, etc. There is also the possibility of using separate motor-driven fans.

Related Calculations: The concept of the root-mean-square (rms) value used in this problem is similar to that used in the analysis of ac waves wherein the rms value of a current wave produces the same heating in a given resistance as would dc of the same ampere value. The heating effect of a current is proportional to I^2R. The rms value results from squaring the currents (or horsepower in this example) over intervals of time, finding the average value, and taking the square root of the whole.

CALCULATION OF ARMATURE REACTION IN DC MOTOR

A dc shunt motor has the following characteristics: 10.0 kW, 240 V, 1150 r/min, armature-circuit resistance $R_a = 0.72\ \Omega$. The motor is run at no load at 240 V, with an armature current of 1.78 A; the speed is 1225 r/min. If the armature current is allowed to increase to 50 A, because of applied load torque, the speed drops to 1105 r/min. Calculate the effect of armature reaction on the flux per pole.

Calculation Procedure:

1. Draw the Equivalent Circuit and Calculate the Induced Voltage e_a at No Load

The equivalent circuit is shown in Fig. 12. The induced voltage is $e_a = V_t - I_aR_a = 240\text{ V} - (1.78\text{ A})(0.72\ \Omega) = 238.7\text{ V}$.

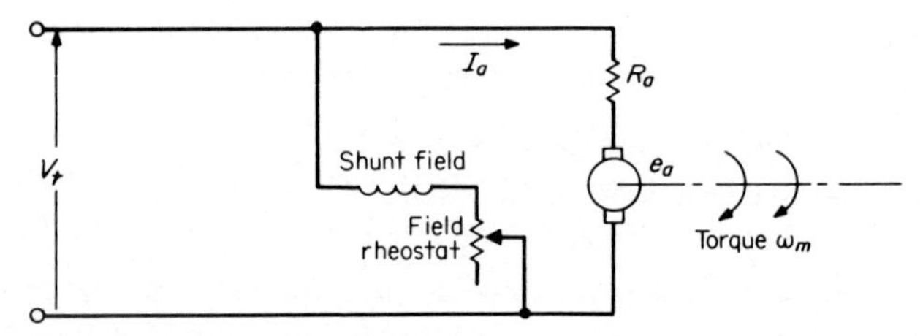

Fig. 12 Equivalent circuit of a dc shunt motor.

2. Determine the No-Load Flux

Knowing the induced voltage at no load, one can derive an expression for the no-load flux from the equation: $e_a = k(\phi_{NL})\omega_m$, where k is the machine constant and ω_m is the mechanical speed in rad/s. The speed $\omega_m = (1225\text{ r/min})(2\pi\text{ rad/r})(1\text{ min}/60\text{ s}) = 128.3\text{ rad/s}$. Thus, $k\phi_{NL} = e_a/\omega_m = 238.7\text{ V}/128.3\text{ rad/s} = 1.86\text{ V}\cdot\text{s}$.

3. Calculate the Load Value of Induced Voltage e_a

The induced voltage under the loaded condition of $V_t = 240$ V and $I_a = 50$ A is $e_a = V_t - I_aR_a = 240\text{ V} - (50\text{ A})(0.72\ \Omega) = 204\text{ V}$. The speed is $\omega_m = (1105\text{ r/min})(2\pi\text{ rad/r})(1\text{ min}/60\text{ s}) = 115.7\text{ rad/s}$. Thus, $k\phi_{\text{load}} = e_a/\omega_m = 204\text{ V}/115.7\text{ rad/s} = 1.76\text{ V}\cdot\text{s}$.

4. Calculate the Effect of Armature Reaction on the Flux per Pole

The term $k\phi$ is a function of the flux per pole; armaure reaction reduces the main magnetic flux per pole. In this case, as the load increased from zero to 50 A (120 percent of rated full-load current of 10,000 W/240 V = 41.7A), the percentage reduction of field flux $= (k\phi_{NL} - k\phi_{load})(100 \text{ percent})/k\phi_{NL} = (1.86 - 1.76)(100 \text{ percent})/1.86 =$ 5.4 percent.

Related Calculations: It is interesting to note the units associated with the equation $e_a = k\phi\omega_m$ where e_a is in V, ϕ in Wb/pole, ω_m in rad/s. Thus, $k = e_a/\phi\ \omega_m = \text{V}/(\text{Wb/pole})(\text{rad/s}) = \text{V}\cdot\text{s/Wb}$. Because $\text{V} = d\phi/dt = \text{Wb/s}$, the machine constant k is unitless. Actually, the machine constant $k = Np/a\pi$, where $N =$ total number of turns on the armature, $p =$ the number of poles, and $a =$ the number of parallel paths on the armature circuit. This combination gives the units of k as turns·poles/paths·radians.

DYNAMIC BRAKING FOR SEPARATELY EXCITED MOTOR

A separately excited dc motor has the following data: 7.5 hp, 240 V, 1750 r/min, no-load armature current $I_a = 1.85$ A, armature-circuit resistance $R_a = 0.19\ \Omega$. When the field and armature circuits are disconnected, the motor coasts to a speed of 400 r/min in 84 s. Calculate a resistance value to be used for dynamic braking, with the initial braking torque at twice the full-speed rated torque, and calculate the time required to brake the unloaded motor from 1750 r/min to 250 r/min.

Calculation Procedure:

1. Calculate the Torque at Rated Conditions

The torque equation is $T = 33{,}000\ P/2\pi n$, where $P =$ rating in hp and $n =$ speed in r/min. $T = (33{,}000)(7.5 \text{ hp})/(2\pi \times 1750 \text{ r/min}) = 22.5 \text{ lb}\cdot\text{ft}$, or $(22.5 \text{ lb}\cdot\text{ft})(1 \text{ N}\cdot\text{m}/0.738 \text{ lb}\cdot\text{ft}) = 30.5 \text{ N}\cdot\text{m}$.

2. Calculate Rotational Losses at Rated Speed

Allow for a 2-V drop across the brushes; the rotational losses are equal to $(V_t - 2 \text{ V})I_a - I_a^2 R_a$, where $I_a =$ no-load current obtained from test. Rotational losses are $(240 \text{ V} - 2\text{V})(1.85 \text{ A}) - (1.85 \text{ A})^2(0.19\ \Omega) = 439.6$ W. This loss power can be equated to a torque: $(439.6 \text{ W})(1 \text{ hp}/746 \text{ W}) = 0.589$ hp, and $(0.589 \text{ hp})(33{,}000)/(2\pi \times 1750 \text{ r/min}) = 1.77 \text{ lb}\cdot\text{ft} = (1.77 \text{ lb}\cdot\text{ft})(1 \text{ N}\cdot\text{m}/0.738 \text{ lb}\cdot\text{ft}) = 2.40 \text{ N}\cdot\text{m}$.

3. Calculate the Initial Braking Torque

The initial braking torque equals twice the full-speed rated torque; the electromagnetic torque, therefore, equals $2 \times$ full-speed rated torque $-$ rotational-losses torque $= 2 \times 30.5 \text{ N}\cdot\text{m} - 2.40 \text{ N}\cdot\text{m} = 58.6 \text{ N}\cdot\text{m}$. This torque, which was determined at no load, is considered proportional to the armature current in the same manner as electromagnetic power equals $e_a I_a$. The electromagnetic *torque* $= k_1 I_a$. Hence, $k_1 =$ no-load torque$/I_a = 2.40 \text{ N}\cdot\text{m}/1.85 \text{ A} = 1.3$.

Under the condition of braking, the electromagnetic torque $= 58.6 \text{ N}\cdot\text{m} = k_1 I_a$, or $I_a = 58.6 \text{ N}\cdot\text{m}/1.3 = 45$ A.

4. Calculate the Value of the Braking Resistance

Use the equation $I_a = (V_t - 2 \text{ V})/(R_a + R_{br}) = (240 \text{ V} - 2 \text{ V})/(0.19\ \Omega + R_{br}) = 45$ A. Solving the equation for the unknown, find $R_{br} = 5.1\ \Omega$.

5. Convert Rotational Speeds into Radians per Second

Use the relation $(\text{r/min})(1 \text{ min}/60 \text{ s})(2\pi \text{ rad/r}) = 1750$ r/min, which multiplied by $\pi/30$ equals 183 rad/s. Likewise, 400 r/min = 42 rad/s, and 250 r/min = 26 rad/s.

6. Calculate Angular Moment of Inertia J

The motor decelerated from 1750 r/min or 183 rad/s to 400 r/min or 42 rad/s in 84 s. Because $T = k_2\omega_0$, $k_2 = T/\omega_0 = (2.40\ \text{N}\cdot\text{m})/(183\ \text{rad/s}) = 0.013$. The acceleration torque is $T_{acc} = J(d\omega/dt)$. Thus, $k_2\omega_0 = J(d\omega/dt)$, or

$$\int_0^{84} dt = (J/k_2)\int_{183}^{42} (1/\omega)\, d\omega$$

Integrating we find

$$[t]\Big|_0^{84} = (J/0.013)(\ln\omega)\Big|_{183}^{42}$$

Thus, $J = (84)(0.013)/(\ln 42 - \ln 183) = -0.742\ \text{kg}\cdot\text{m}^2$.

7. Calculate the Time for Dynamic Braking

From Step 3, the electromagnetic torque $= k_1 I_a$; from Step 4, $I_a = e_a/(R_a + R_{br})$. Thus, because $e_a = k_1\omega$, $I_a = k_1\omega/(R_a + R_{br}) = k_1\omega/5.29$. It follows that $J(d\omega/dt) = k_2\omega + k_1^2\omega/5.29$, or $dt = (5.29J\, d\omega)/(5.29k_2\omega + k_1^2\omega)$. This leads to

$$t = \frac{5.29 \times 0.742}{5.29 \times 0.013 + 1.3^2}\int_{183}^{26} \frac{1}{\omega} d\omega$$

The solution yields $t = 4.35$ s.

Related Calculations: This is an example of braking used on traction motors, whereby the motors act as generators to convert the kinetic energy into electrical energy, which is dissipated through resistance. Voltage generated by the machine is a function of speed and field current; the dissipating resistance is of fixed value. Therefore, the braking effect is related to the armature current and the field current.

A THREE-PHASE SCR DRIVE FOR A DC MOTOR

A three-phase, 60-Hz ac power source (440 V line to line) connected through silicon controlled rectifiers (SCRs) is used to power the armature circuit of a separately excited dc motor of 10-hp rating (see Fig. 13). The data for the motor are: armature-circuit resistance $R_a = 0.42\ \Omega$ and I_f (field current) set so that at 1100 r/min when the machine is run as a generator, the full-load voltage is 208 V, and the machine delivers the equivalent of 10 hp. Determine the average torque that can be achieved for a speed of 1100 r/min if the SCR's firing angle is set at 40° and at 45°.

Calculation Procedure:

1. Calculate the Average Value of SCR Output

The average value of the SCR output is determined from the waveforms shown in Fig. 14; *a*-phase voltage contributes one-third of the time, then *b*-phase voltage, then *c*-phase voltage. Let the firing angle be represented by δ. Use the equation: $V_{avg} = (3\sqrt{3}/2\pi)\, V\cos\delta$ where $V =$ the maximum value of one phase of the ac voltage (the equation

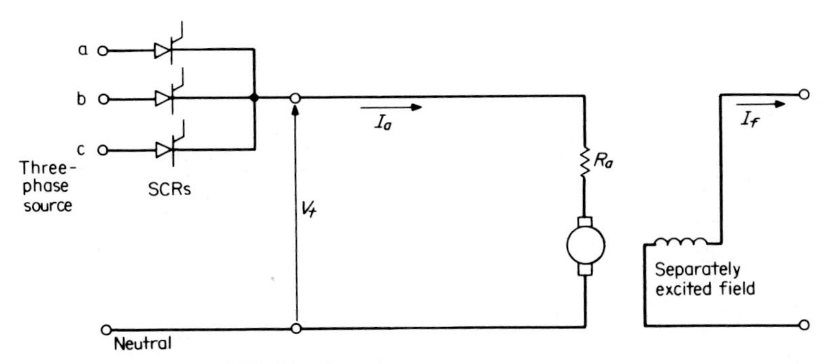

Fig. 13 Three-phase SCR drive for a dc motor.

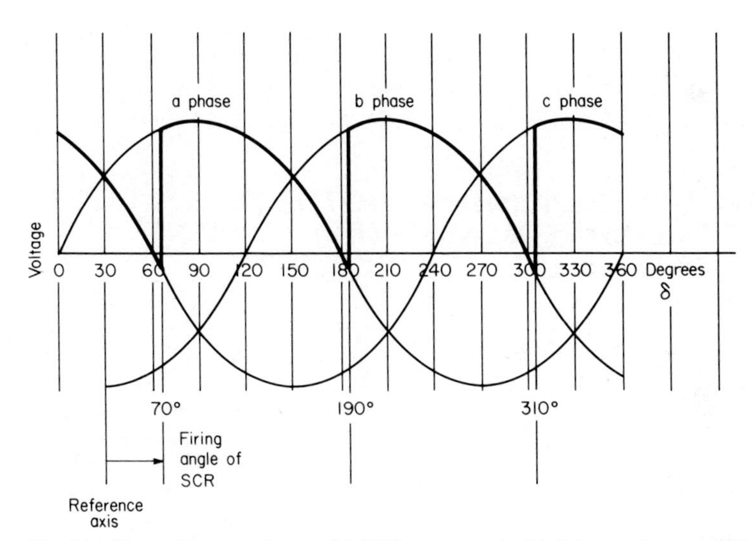

Fig. 14 Three-phase waveforms with SCR output, and with firing angle set at 40°.

assumes that only one SCR is operated at a time). For the case of 440 V line to line: $V = (440/\sqrt{3})\ \sqrt{2} = 359.3$ V.

For $\delta = 40°$, $V_{avg} = 0.827V \cos \delta = (0.827)(359.3 \text{ V})(\cos 40°) = 227.6$ V. For $\delta = 45°$, $V_{avg} = (0.827)(359.3 \text{ V})(\cos 45°) = 210.1$ V.

2. *Convert the Speed to Radians per Second*

The speed is given as 1100 r/min. Use the relation $(1100 \text{ r/min})(2\pi \text{ rad/r})(1 \text{ min}/60 \text{ s}) = 115.2$ rad/s.

3. *Calculate the Torque Constant*

From the equation $T = kI_a$, the torque constant k can be calculated from the results of tests of the machine run as a generator. The armature current is determined from the test results also; thus, $(10 \text{ hp})(746 \text{ W/hp})/208 \text{ V} = 35.9$ A. $T = (33{,}000)(\text{hp})/2\pi n$, where n = speed in r/min. $T = (33{,}000)(10)/(2\pi \times 1100) =$

47.7 lb·ft. Or (47.7 lb·ft)(1 N·m/0.738 lb·ft) = 64.7 N·m. Thus, $k = T/I_a$ = 64.7 N·m/35.9 A) = 1.8 N·m/A.

4. Calculate the Armature Currents for the Two SCR Firing Angles
Use the equation $e_a = k\omega_m$. The units of k, the torque constant or the voltage constant, are N·m/A or V·s/rad. Thus, e_a = (1.8 V·s/rad)(115.2 rad/s) = 208 V. Use the equation $e_a = V_t - I_aR_a$, where the terminal voltage V_t is the average value of the SCR output. Solve for I_a; $I_a = (V_t - e_a)/R_a$. For the case of the 40° angle, I_a = (227.6 V − 208 V)/0.42 = 46.7 A. For the case of the 45° angle, I_a = (210.1 V − 208 V)/0.42 = 5.0 A.

5. Calculate the Torques
Use the equation $T = kI_a$. For the 40° firing angle, T = (1.8 N·m/A)(46.7 A) = 84.1 N·m. For the 45° firing angle, T = (1.8 N·m/A)(5.0 A) = 9 N·m.

Related Calculations: The problem illustrates how a three-phase ac system may be used to power a dc motor. The system would be appropriate for motors larger than 5 hp. By varying the firing angle of the SCRs, the amount of current in the armature circuit may be controlled or limited. In this problem, the SCRs were connected for half-wave rectification. For much larger machines, full-wave rectification circuits may be used.

DIRECT-CURRENT SHUNT-MOTOR SPEED DETERMINED FROM ARMATURE CURRENT AND NO-LOAD SATURATION CURVE

The saturation curve for a dc shunt motor operating at 1150 r/min may be represented by the equation $e_a = (250I_f)/(0.5 + I_f)$. See Fig. 15. The open-circuit voltage is e_a, in volts; the shunt-field current is I_f, in amperes. Data for the machine are as follows: armature-circuit resistance R_a = 0.19 Ω, R_f shunt-field-circuit resistance = 146 Ω, terminal voltage V_t = 220 V. For an armature current of 75 A, calculate the speed. The demagnetizing effect of armature reaction may be estimated as being 8 percent, in terms of shunt-field current at the given load.

Calculation Procedure:

1. Calculate the Effective Shunt-Field Current
In a shunt motor, the terminal voltage is directly across the shunt field. Thus, $I_f = V_t/R_f$ = 220 V/146 Ω = 1.5 A. However, the demagnetizing effect of armature reaction is 8 percent, in terms of shunt-field current. Thus, $I_{f(\text{demag})}$ = (0.08)(1.5 A) = 0.12 A. This amount of demagnetization is subtracted from the shunt-field current to yield $I_{f(\text{effective})} = I_f - I_{f(\text{demag})}$ = 1.5 A − 0.12 A = 1.38 A.

2. Calculate the Induced Voltage at 1150 r/min
The induced voltage e_a is now determined from the given saturation curve, of necessity at 1150 r/min. It may be calculated from the equation, with $I_{f(\text{effective})}$ substituted for I_f: $e_a = 250I_{f(\text{effective})}/(0.5 + I_{f(\text{effective})})$ = (250)(1.38)/(0.5 + 1.38) = 183.5 V. Alternatively, the voltage may be read directly from the graph of the saturation curve (see Fig. 15).

3. Calculate the Induced Voltage at the Unkown Speed
Use the equation: $e'_a = V_t - I_aR_a$ = 220 V − (75 A)(0.19 Ω) = 205.8 V.

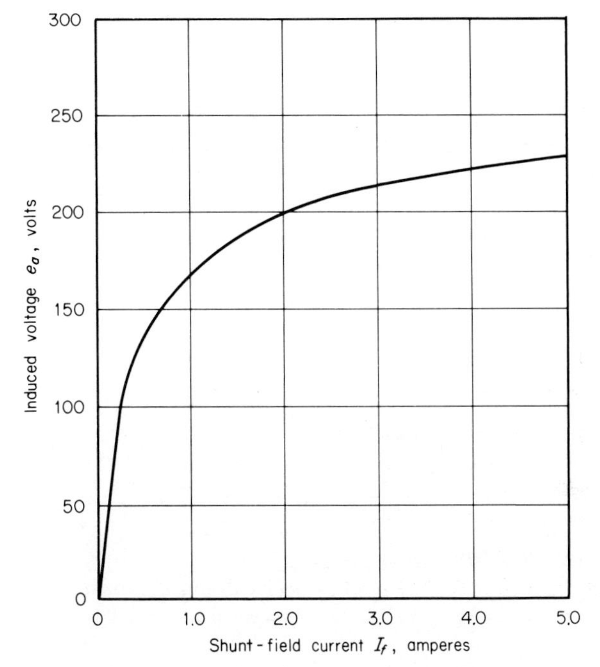

Fig. 15 DC-motor saturation curve, $e_a = 250I_f/(0.5 + I_f)$; $n = 1150$ r/min.

4. Calculate the Unknown Speed

The induced voltage e_a or e_a' is proportional to the speed, provided the effective magnetization is constant. Thus, $e_a/e_a' = (1150 \text{ r/min})/n_x$. The unknown speed is $n_x = (1150 \text{ r/min})(e_a'/e_a) = (1150 \text{ r/min})(205.8 \text{ V})/183.5 \text{ V} = 1290$ r/min.

Related Calculations: DC motor speeds may be determined by referring to the saturation curve and setting proportions of induced voltage to speed, provided that for both sides of the proportion the effective magnetization is constant. The effective magnetization allows for the demagnetization of armature reaction. The procedure may be used with additional fields (e.g., series field) provided the effect of the additional fields can be related to the saturation curve; an iterative process may be necessary.

CHOPPER DRIVE FOR DC MOTOR

A separately excited dc motor is powered through a chopper drive (see Fig. 16), from a 240-V battery supply. A square-wave terminal voltage V_t is developed that is 3 ms ON and 3 ms OFF. The motor is operating at 450 r/min. The machine's torque constant (or voltage constant) $k = 1.8$ N·m/A (or V·s/rad), and the armature-circuit resistance $R_a = 0.42\ \Omega$. Determine the electromagnetic torque that is developed.

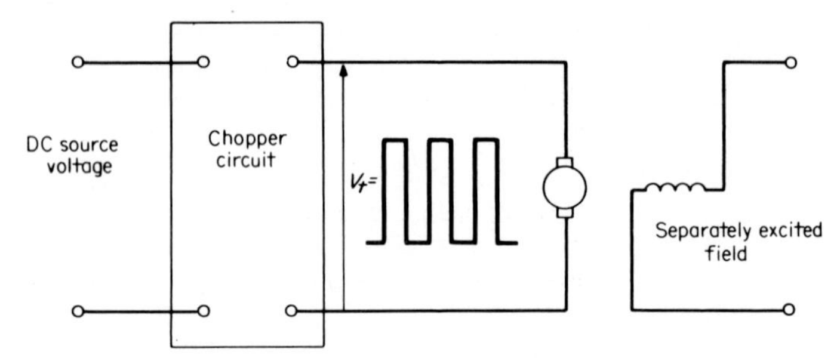

Fig. 16 Chopper drive for a dc motor.

Calculation Procedure:

1. Calculate the Terminal Voltage V_t

The dc value of the pulsating terminal voltage (square wave) is its average value. Use the equation: $V_t = V(\text{time}_{\text{ON}})/(\text{time}_{\text{ON}} + \text{time}_{\text{OFF}}) = (240 \text{ V})(3 \text{ ms})/(3 \text{ ms} + 3 \text{ ms}) = 120 \text{ V}$.

2. Calculate the Induced Voltage e_a

The average induced voltage e_a may be calculated from the equation $e_a = k\omega$, where k = voltage constant in V·s/rad and ω = speed in rad/s. Thus, $e_a = (1.8 \text{ N·m/A})(450 \text{ r/min})(2\pi \text{ rad/r})(1 \text{ min}/60 \text{ s}) = 84.8 \text{ V}$.

3. Calculate the Armature Current I_a

The average armature current is determined from the general equation $e_a = V_t - I_aR_a$. Rearranging, we find $I_a = (V_t - e_a)/R_a = (120 \text{ V} - 84.8 \text{ V})/0.42 \ \Omega = 83.8 \text{ A}$.

4. Calculate the Average Torque

The average torque is determined from the equation $T = kI_a$, where k = the torque constant in N·m/A. Thus, $T = (1.8 \text{ N·m/A})(83.8 \text{ A}) = 150.84 \text{ N·m}$.

Related Calculations: The chopper is a switching circuit, usually electronic, that turns the output of the battery ON and OFF for preset intervals of time, generating a square wave for V_t. The magnitude of the voltage at the terminals of the motor is varied by adjusting the average value of the square wave (either changing the widths of the pulses or the frequency of the pulses). Here we have a method for controlling motors that operate from batteries. Automotive devices operated with battery-driven motors would use chopper circuits for control.

DESIGN OF A COUNTER–EMF AUTOMATIC STARTER FOR A SHUNT MOTOR

A counter-emf automatic starter is to be designed for a dc shunt motor with the following characteristics: 115 V, armature-circuit resistance $R_a = 0.23 \ \Omega$, armature current I_a = 42 A at full load. Determine the values of external resistance for a three-step starter, allowing starting currents of 180 to 70 percent of full load.

Calculation Procedure:

1. Calculate First-Step External Resistance Value R_1

Use the equation R_1 = [line voltage/(1.8 × I_a)] − R_a = [115 V/(1.8 × 42 A)] − 0.23 Ω = 1.29 Ω.

2. Calculate the Counter-emf Generated at First Step

With the controller set for the first step, the motor will attain a speed such that the counter-emf will reach a value to limit the starting current to 70 percent of full-load armature current. Thus, the counter-emf $E_a = V_t - (R_a + R_1)(0.70 \times I_a)$ = 115 V − (0.23 Ω + 1.29 Ω)(0.70 × 42 A) = 70.3 V. The drop across the resistance R_a + R_1 is 44.7 V.

3. Calculate the Second-Step External Resistance Value R_2

If the current is allowed to rise to 1.8 × 42 A through a voltage drop of 44.7 V, the new external resistance value is R_2 = [44.7 V/(1.8 × 42 A)] − 0.23 Ω = 0.36 Ω.

4. Calculate the Counter-emf Generated at the Second Step

As in the previous calculation, with the controller set for the second step, the motor will attain a speed such that the counter-emf $E_a = V_t - (R_a + R_2)(0.70 \times I_a)$ = 115 V − (0.23 Ω + 0.36 Ω)(0.70 × 42 A) = 97.7 V. The drop across the resistance R_a + R_2 = 17.3 V. This last step is sufficient to permit the accelerating motor to reach normal speed without exceeding 1.8 × 42 A.

5. Draw the Circuit for the Three-Step Automatic Starter

See Fig. 17. $R_A = R_1 - R_2$ = 1.29 Ω − 0.36 Ω = 0.93 Ω. $R_B = R_2$ = 0.36 Ω. When the run button is pressed, the closing coil M (main contactor) is ener-

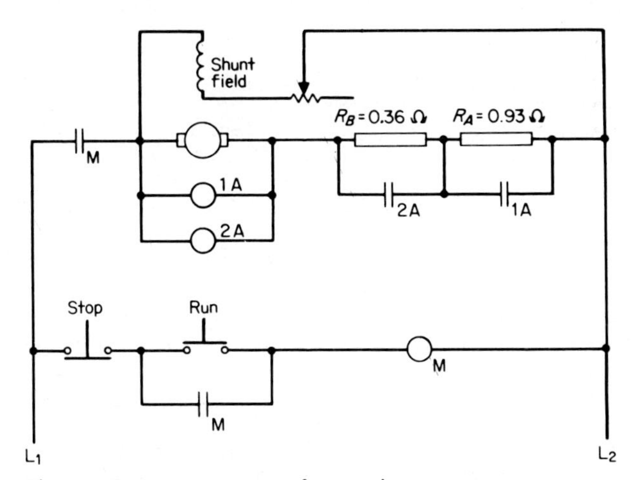

Fig. 17 Three-step counter-emf automatic starter.

gized, the shunt-field circuit is connected across the line, and the armature circuit is connected through the starting resistors to the line. In addition, a contact closes across the run button, serving as an interlock. The coil of contactor 1A is set to operate when the counter-emf = 70 V, thus short-circuiting R_A. The coil of contactor 2A is set to operate when the counter-emf = 98 V, thus short-circuiting R_B. In this manner, the external resistance has three steps: 1.29 Ω, 0.36 Ω, and 0.0 Ω. The starting current is limited to the range of 1.8 × 42 A and 0.7 × 42 A.

Related Calculation: Illustrated here is a common automatic starter employing the changing counter-emf developed across the armature during starting, for operating electromagnetic relays to remove resistances in a predetermined manner. Other starting mechanisms employ current-limiting relays or time-limit relays. Each scheme that is used is for the purpose of short-circuiting resistance steps automatically as the motor accelerates.

Section 4 TRANSFORMERS

Joseph M. DeGuilmo, P.E.
Program Coordinator,
Electronics Technology,
Hudson County Community College

REFERENCES Anderson—*Electric Machines and Transformers,* Reston; DeGuilmo—*Electricity/Electronics: Principles and Applications,* Delmar; Fitzgerald, Kingsley, and Kusko—*Electrical Machinery,* McGraw-Hill; Gingrich—*Electrical Machinery, Transformers, and Controls,* Prentice-Hall; Hicks—*Standard Handbook of Engineering Calculations,* McGraw-Hill; Jackson—*Introduction to Electric Circuits,* Prentice-Hall; Kosow—*Electric Machinery and Transformers,* Prentice-Hall; McPherson—*An Introduction to Electrical Machines and Transformers,* Wiley; Richardson—

Rotating Electric Machines and Transformer Technology, Reston; Wildi—*Electrical Power Technology*, Wiley.

ANALYSIS OF TRANSFORMER TURNS RATIO

Compute the value of the secondary voltage, the voltage per turn of the primary, and the voltage per turn of the secondary of the ideal single-phase transformer of Fig. 1.

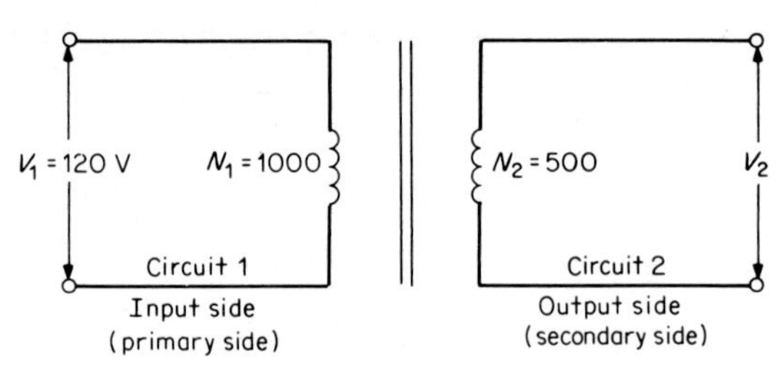

Fig. 1 Ideal single-phase transformer.

Calculation Procedure:

1. Compute the Value of the Secondary Voltage

The turns ratio a of a transformer is the ratio of the primary turns N_1 to the secondary turns, N_2, or the ratio of the primary voltage V_1 to the secondary voltage V_2: $a = N_1/N_2 = V_1/V_2$. Substitution of the values indicated in Fig. 1 for the turns ratio yields $a = 1000/500 = 2$. Secondary voltage $V_2 = V_1/a = 120/2 = 60$ V.

2. Compute the Voltage per Turn of the Primary and Secondary Windings

Voltage per turn of the primary winding $= V_1/N_1 = 120/1000 = 0.12$ V/turn. Voltage per turn of the secondary winding $= V_2/N_2 = 60/500 = 0.12$ V/turn.

Related Calculations: This procedure illustrates a step-down transformer. A characteristic of both step-down and step-up transformers is that the voltage per turn of the primary is equal to the voltage per turn of the secondary.

ANALYSIS OF A STEP-UP TRANSFORMER

Calculate the turns ratio of the transformer in Fig. 1 when it is employed as a step-up transformer.

Calculation Procedure:

1. Compute the Turns Ratio

In the step-up transformer, the low-voltage side is connected to the input, or primary, side. Therefore, $a = N_2/N_1 = V_2/V_1$; hence, $a = 500/1000 = 0.5$.

Related Calculations: For a particular application, the turns ratio a is fixed but is not a transformer constant. In this example $a = 0.5$ when the transformer is used as a step-up transformer. In the previous example, $a = 2$ when the transformer is used as a step-down transformer. The two values of a are reciprocals of each other: $2 = 1/0.5$, and $0.5 = 1/2$.

ANALYSIS OF A TRANSFORMER CONNECTED TO A LOAD

A 25-kVA single-phase transformer is designed to have an induced emf of 2.5 V/turn (Fig. 2). Calculate the number of primary and secondary turns and the full-load current of the primary and secondary windings.

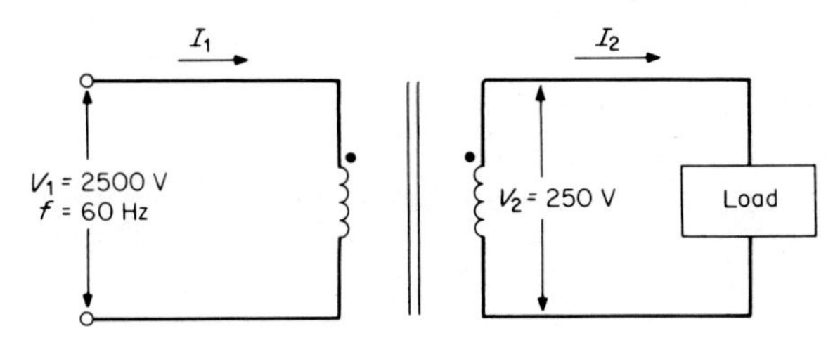

Fig. 2 Single-phase transformer connected to load.

Calculation Procedure:

1. Compute the Number of Primary and Secondary Turns

$N_1 = V_1/(\text{V/turn}) = 2500\text{ V}/(2.5\text{ V/turn}) = 1000$ turns. Similarly, $N_2 = V_2/(\text{V/turn}) = 250/2.5 = 100$ turns. The turns ratio $a = N_1/N_2 = 1000/100 = 10{:}1$.

2. Compute the Full-Load Current of the Primary and Secondary Windings

Primary current $I_1 = (\text{VA})_1/V_1 = 25{,}000\text{ VA}/2500\text{ V} = 10$ A. Secondary current $I_2 = (\text{VA})_2/V_2 = 25{,}000\text{ VA}/250\text{ V} = 100$ A. The current transformation ratio is $1/a = I_1/I_2 = 10\text{ A}/100\text{ A} = 1{:}10$.

Related Calculations: One end of each winding is marked with a dot in Fig. 2. The dots indicate the terminals that have the same relative polarity. As a result of the dot convention, the following rules are established:

1. When current enters a primary which has a dot polarity marking, the current leaves the secondary at its dot polarity terminal.
2. When current leaves a primary terminal which has a dot polarity marking, the current enters the secondary at its dot polarity terminal.

Manufacturers usually mark the leads on the high-voltage sides as H_1, H_2, etc. The leads on the low-voltage side are marked X_1, X_2, etc. Marking H_1 has the same relative polarity as X_1, etc.

SELECTION OF A TRANSFORMER FOR IMPEDANCE MATCHING

Select a transformer with the correct turns ratio to match the 8-Ω resistive load in Fig. 3 to the Thevenin equivalent circuit of the source.

Calculation Procedure:

1. Determine the Turns Ratio

The impedance of the input circuit, Z_i, is 5000 Ω. This value represents the Thevenin impedance of the source. The load impedance Z_L is 8 Ω. To achieve an impedance match, the required turns ratio is

$$a = \sqrt{Z_i/Z_L} = \sqrt{5000\ \Omega/8\ \Omega} = 25$$

Therefore, the impedance-matching transformer must have a turns ratio of 25:1.

Related Calculations: The maximum power transfer theorem (Sec. 1) states that maximum power is delivered by a source to a load when the impedance of the load is equal to the internal impedance of the source. Because the load impedance does not always match the source impedance, transformers are used between source and load to ensure matching. When the load and source impedances are not resistive, maximum power is delivered to the load when the load impedance is the complex conjugate of the source impedance.

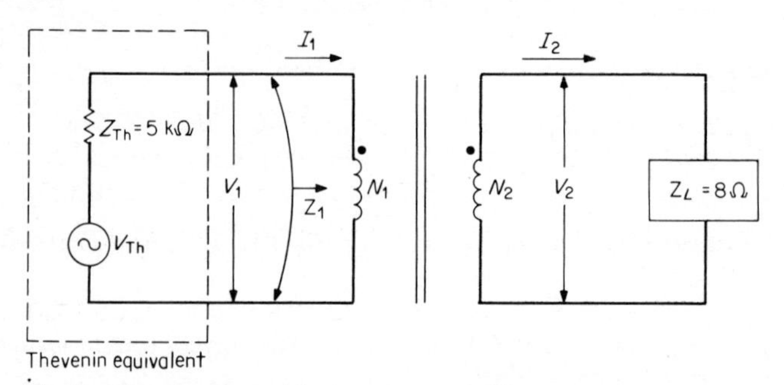

Fig. 3 Transformer used for impedance matching.

PERFORMANCE OF A TRANSFORMER WITH MULTIPLE SECONDARIES

Determine the turns ratio of each secondary circuit, the primary current I_1, and rating in kVA of a transformer with multiple secondaries, illustrated in Fig. 4.

Calculation Procedure:

1. Select the Turns Ratio of Each Secondary Circuit

Let the turns ratio of circuit 1 to circuit 2 be designated as a_2 and the turns ratio of circuit 1 to circuit 3 as a_3. Then, $a_2 = V_1/V_2 = 2000\ \text{V}/1000\ \text{V} = 2{:}1$ and $a_3 = V_1/V_3 = 2000\ \text{V}/500\ \text{V} = 4{:}1$.

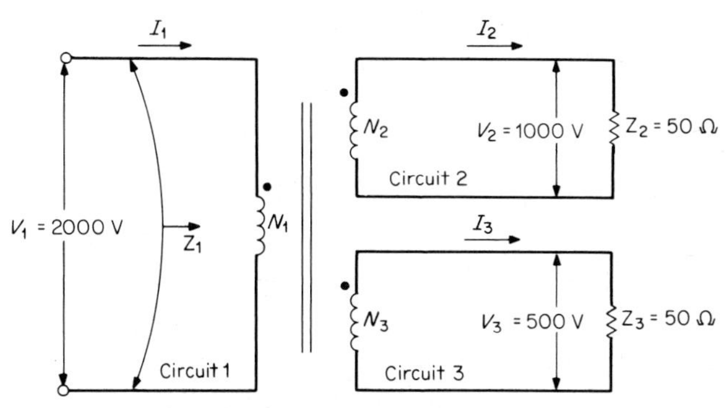

Fig. 4 Transformer with multiple secondaries.

2. Compute the Primary Current I_1

Since $I_2 = V_2/Z_2$ and $I_3 = V_3/Z_3$, then $I_2 = 1000\ \text{V}/500\ \Omega = 20$ A and $I_3 = 500\ \text{V}/50\ \Omega = 10$ A. The ampere-turns of the primary (N_1I_1) of a transformer equals the sum of the ampere turns of all the secondary circuits; therefore, $N_1I_1 = N_2I_2 + N_3I_3$. Solving for I_1, we have $I_1 = (N_2/N_1)I_2 + (N_3/N_1)I_3$. Turns ratio is $N_2/N_1 = 1/a_2$ and $N_3/N_1 = 1/a_3$. Hence, $I_1 = I_2/a_2 + I_3/a_3 = (20\ \text{A})(\frac{1}{2}) + (10\ \text{A})(\frac{1}{4}) = 12.5$ A.

3. Compute the Rating in kVA of the Transformer

The winding ratings are $\text{kVA}_1 = V_1I_1/1000 = (2000\ \text{V})(12.5\ \text{A})/1000 = 25$ kVA, $\text{kVA}_2 = V_2I_2/1000 = (1000\ \text{V})(20\ \text{A})/1000 = 20$ kVA, and $\text{kVA}_3 = V_3I_3/1000 = (500\ \text{V})(10\ \text{A})/1000 = 5$ kVA. The apparent power rating of the primary equals the sum of the apparent power ratings of the secondary. As a check, 25 kVA = (20 + 5) kVA = 25 kVA.

Related Calculations: When the secondary loads have different phase angles, the same equations still apply. The voltages and currents, however, must be expressed as phasor quantities.

IMPEDANCE TRANSFORMATION OF A THREE-WINDING TRANSFORMER

Calculate the impedance Z_1 seen by the primary of the three-winding transformer of Fig. 4 using impedance-transformation concepts.

Calculation Procedure:

1. Calculate Z_1

The equivalent reflected impedance of both secondaries, or the total impedance "seen" by the primary, Z_1, is: $Z_1 = a_2^2Z_2 \| a_3^2Z_3$, where $a_2^2Z_2$ is the reflected impedance of circuit 2 and $a_3^2\, Z_3$ is the reflected impedance of circuit 3. Hence, $Z_1 = (2)^2(50\ \Omega) \| (4)^2(50\ \Omega) = 200 \| 800 = 160\ \Omega$.

2. Check the Value of Z_1 Found in Step 1

$Z_1 = V_1/I_1 = 2000\ \text{V}/12.5\ \text{A} = 160\ \Omega$.

Related Calculations: When the secondary loads are not resistive, the preceding equations still apply. The impedances are expressed by complex numbers and the voltage and current are expressed as phasors.

SELECTION OF A TRANSFORMER WITH TAPPED SECONDARIES

Select the turns ratio of a transformer with tapped secondaries to supply the loads of Fig. 5.

Calculation Procedure:

1. Calculate the Power Requirement of the Primary

When a transformer has multiple, or tapped, secondaries, $P_1 = P_2 + P_3 + \cdots$, where P_1 is the power requirement of the primary and P_2, P_3, $\cdots$ are the power requirements of each secondary circuit. Hence, $P_1 = 5 + 2 + 10 + 3 = 20$ W.

2. Compute the Primary Input Voltage V_1

From $P_1 = V_1^2/Z_1$, one obtains $V_1 = \sqrt{P_1 Z_1} = \sqrt{20 \times 2000} = 200$ V.

3. Compute the Secondary Voltages

$V_2 = \sqrt{5 \times 6} = 5.48$ V, $V_3 = \sqrt{2 \times 8} = 4$ V, $V_4 = \sqrt{10 \times 16} = 12.7$ V, and $V_5 = \sqrt{3 \times 500} = 38.7$ V.

4. Select the Turns Ratio

The ratios are $a_2 = V_1/V_2 = 200/5.48 = 36.5{:}1$, $a_3 = V_1/V_3 = 200/4 = 50{:}1$, $a_4 = V_1/V_4 = 200/12.7 = 15.7{:}1$, and $a_5 = V_1/V_5 = 200/38.7 = 5.17{:}1$.

Related Calculations: A basic transformer rule is that the total power requirements of all the secondaries must equal the power input to the primary of the transformer.

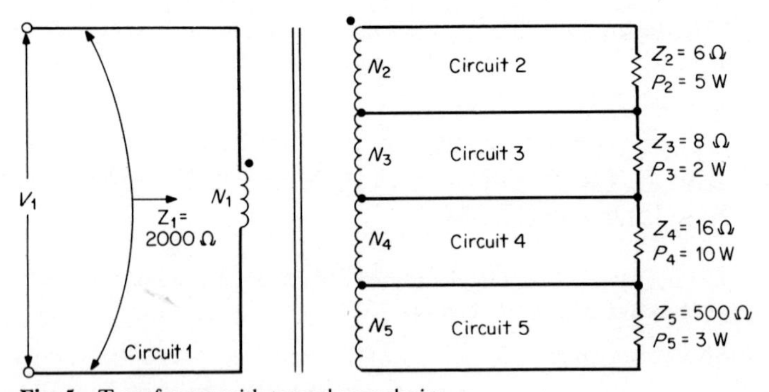

Fig. 5 Transformer with tapped secondaries.

TRANSFORMER CHARACTERISTICS AND PERFORMANCE

Compute the number of turns in each secondary winding, the rated primary current at unity-power-factor loads, and the rated current in each secondary winding of the three-winding transformer of Fig. 6.

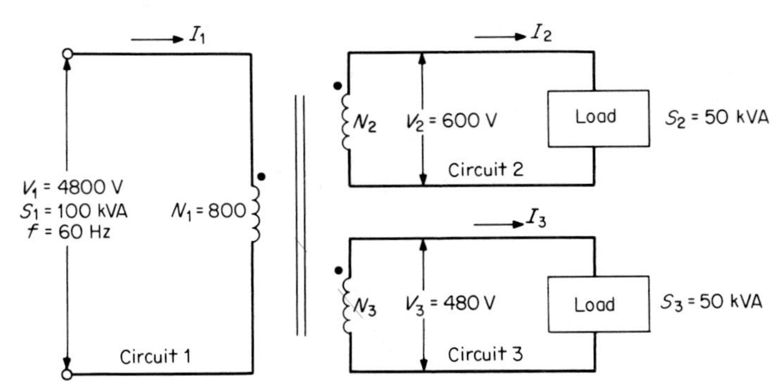

Fig. 6 Three-winding transformer.

Calculation Procedure:

1. Compute the Number of Turns in Each Secondary Winding

The turns ratio in winding 2 is $a_2 = V_1/V_2 = N_1/N_2 = 4800/600 = 8{:}1$; hence, $N_2 = N_1/a_2 = 800/8 = 100$ turns. Similarly, $a_3 = V_1/V_3 = N_1/N_3 = 4800/480 = 10{:}1$ from which $N_3 = N_1/a_3 = 800/100 = 80$ turns.

2. Compute the Rated Primary Current

$I_1 = (\text{VA})_1/V_1 = 100{,}000/4800 = 20.83$ A.

3. Compute the Rated Secondary Currents

$I_2 = (\text{VA})_2/V_2 = 50{,}000/600 = 83.8$ A, and $I_3 = (\text{VA})_3/V_3 = 50{,}000/480 = 104.2$ A.

Related Calculations: This method may be used to analyze transformers with one or more secondary windings for power, distribution, residential, or commercial service. When the loads are not at unity power factor, complex algebra and phasors are used where applicable.

PERFORMANCE AND ANALYSIS OF A TRANSFORMER WITH A LAGGING POWER-FACTOR LOAD

Calculate the primary voltage required to produce rated voltage at the secondary terminals of a 100 kVA, 2400/240-V, single-phase transformer operating at full load. The power factor (pf) of the load is 80 percent lagging.

Calculation Procedure:

1. Analyze the Circuit Model of Fig. 7

The model of Fig. 7 includes winding resistances, inductive reactances, and core and copper losses. The symbols are defined as follows:

V_1 = supply voltage applied to the primary circuit
R_1 = resistance of the primary circuit
X_1 = inductive reactance of the primary circuit
I_1 = current drawn by the primary from the power source
I_{EXC} = exciting current

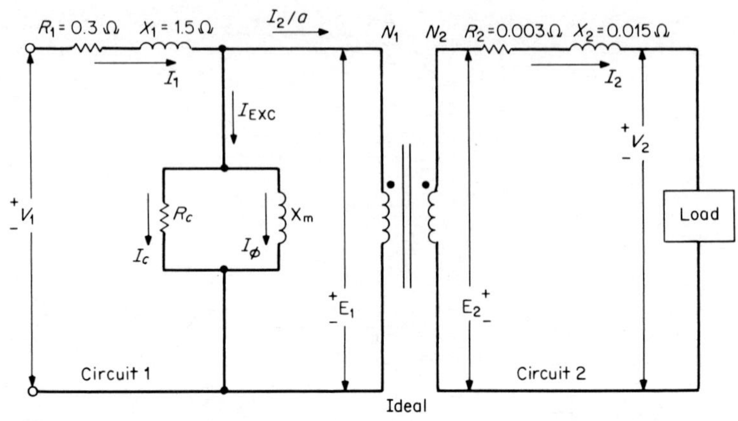

Fig. 7 Circuit model of a practical transformer.

I_c = core loss component of the exciting current; this component accounts for the hysteresis and eddy-current losses

I_ϕ = magnetizing component of exciting current

R_c = equivalent resistance representing core loss

X_m = primary self-inductance which accounts for magnetizing current

I_2/a = load component of primary current

E_1 = voltage induced in the primary coil by all the flux linking the coil

E_2 = voltage induced in the secondary coil by all the flux linking the coil

R_2 = resistance of the secondary circuit, excluding the load

X_2 = inductive reactance of the secondary circuit

I_2 = current delivered by the secondary circuit to the load

V_2 = voltage which appears at the terminals of the secondary winding, across the load

2. *Draw Phasor Diagram*

The phasor diagram for the model of Fig. 7 is provided in Fig. 8. The magnetizing current I_ϕ is about 5 percent of full-load primary current. The core loss component of the

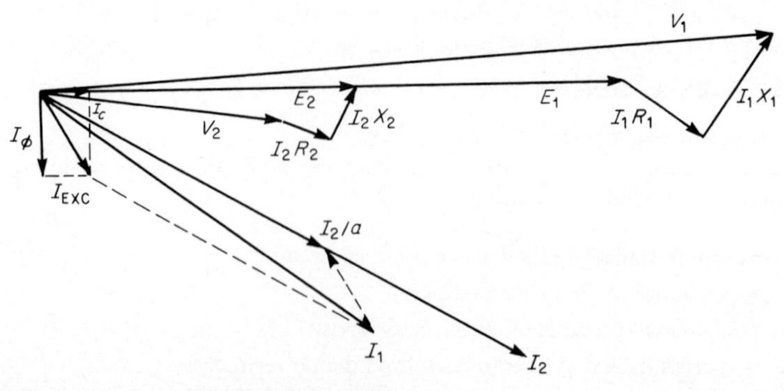

Fig. 8 Phasor diagram for circuit of Fig. 7.

exciting current I_c is about 1 percent of full-load primary current. Current I_c is in phase with E_1, and I_ϕ lags E_1 by 90°. The power factor of the exciting current I_{EXC} is quite low, and lags E_1 by approximately 80°.

3. Simplify the Circuit of Fig. 7

The exciting current is neglected in the approximate transformer model of Fig. 9. The primary parameters R_1 and X_1 are referred to the secondary side as R_1/a^2 and X_1/a^2, respectively. Hence $R_{EQ2} = R_1/a^2 + R_2$ and $|X_{EQ2}| = X_1/a^2 + X_2$. The equivalent impedance of the transformer referred to the secondary side is: $\mathbf{Z_{EQ2}} = R_{EQ2} + jX_{EQ2}$.

4. Solve for $\mathbf{V_1}$

Transformer manufacturers agree that the ratio of rated primary and secondary voltages of a transformer is equal to the turns ratio: $a = V_1/V_2 = 2400/240 = 10{:}1$. $\mathbf{Z_{EQ2}} = (R_1/a^2 + R_2) + j(X_1/a^2 + X_2) = (0.3/100 + 0.003) + j(1.5/100 + 0.015) = 0.03059\underline{/78.69°}$. But $|\mathbf{I_2}| = (\text{VA})_2/V_2 = 100{,}000/240 = 416.67$ A. Using $\mathbf{V_2}$ as a reference, $\mathbf{I_2} = 416.67\underline{/-36.87°}$ A. But $\mathbf{V_1}/a = \mathbf{V_2} + \mathbf{I_2Z_{EQ2}} = 240\underline{/0°} + (416.67\underline{/-36.87°})(0.03059\underline{/78.69°}) = 249.65\underline{/1.952°}$. However, $|\mathbf{V_1}| = a|\mathbf{V_1}/a| = 10 \times 249.65 = 2496.5$ V.

Related Calculations: For lagging power-factor loads, $\mathbf{V_1}$ must be greater than rated value (in this case 2400 V) in order to produce rated voltage across the secondary (240 V).

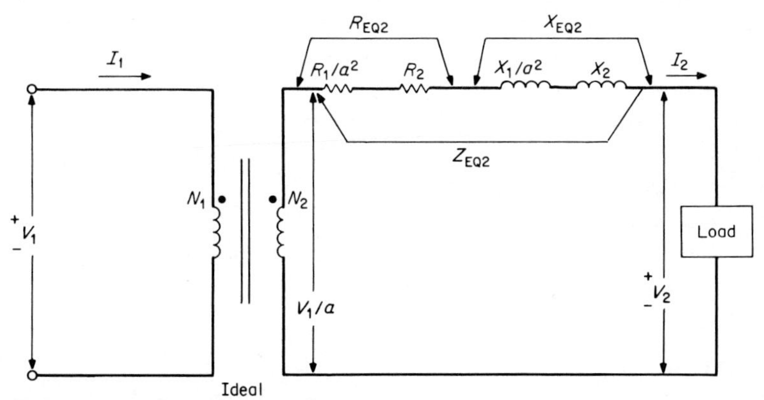

Fig. 9 Simplified version of Fig. 7.

PERFORMANCE AND ANALYSIS OF A TRANSFORMER WITH A LEADING POWER-FACTOR LOAD

Calculate the primary voltage required to produce rated voltage at the secondary terminals of the transformer of Fig. 7. The transformer is operating at full load with an 80 percent leading power factor.

Calculation Procedure:

1. Solve for $\mathbf{V_1}$

With $\mathbf{V_2}$ as a reference, $\mathbf{I_2} = 416.67\underline{/36.87°}$ A. Substituting, we have $\mathbf{V_1}/a = \mathbf{V_2} + \mathbf{I_2Z_{EQ2}} = 240\underline{/0°} + (416.67\underline{/36.87°})(0.03059\underline{/78.69°}) = 234.78\underline{/2.81°}$. The magnitude $|\mathbf{V_1}| = a|\mathbf{V_1}/a| = (10)(234.78) = 2347.8$ V.

Related Calculations: When the load is sufficiently leading, as in this case, $|\mathbf{V}_1| = 2347.8$ V. This value is less than the rated primary voltage of 2400 V to produce the rated secondary voltage.

CALCULATION OF TRANSFORMER-VOLTAGE REGULATION

Calculate the full-load voltage regulation of the transformer shown in Fig. 7 for 80 percent leading and lagging power factors.

Calculation Procedure:

1. Solve for the Full-Load Voltage Regulation at 80 Percent Lagging Power Factor

Transformer voltage regulation is defined as the difference between the full-load and the no-load secondary voltages (with the same impressed primary voltage for each case). Expressed as a percentage of the full-load secondary voltage, voltage regulation is: $VR = [(|\mathbf{V}_1/a| - |\mathbf{V}_2|)/|\mathbf{V}_2|] \times 100$ percent. From the results of the 80 percent lagging power factor example, $VR = (249.65 - 240)/240 \times 100$ percent $= 4.02$ percent.

The phasor diagram for a lagging power-factor condition, which is illustrated in Fig. 10, indicates positive voltage regulation.

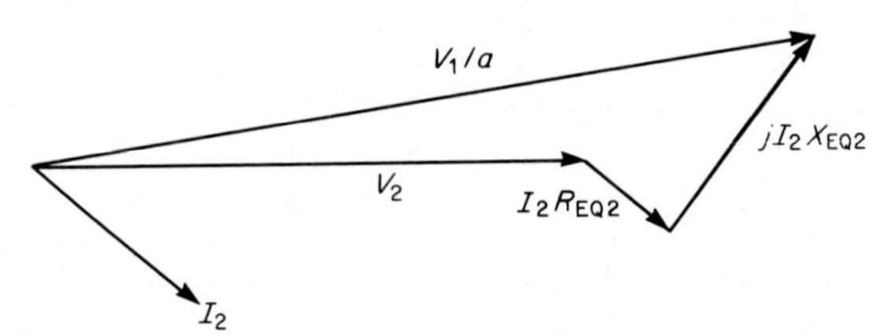

Fig. 10 Phasor diagram of transformer for lagging power factor.

2. Solve for the Full-Load Voltage Regulation at 80 Percent Leading Power Factor

From the 80 percent leading power-factor example, $VR = (234.78 - 240)/240 \times 100$ percent $= -2.18$ percent. The phasor diagram for a leading power-factor condition, shown in Fig. 11, illustrates negative voltage regulation.

Related Calculations: Negative regulation denotes that the secondary voltage increases when the transformer is loaded. This stems from a partial resonance condition between the capacitance of the load and the leakage inductance of the transformer. Zero

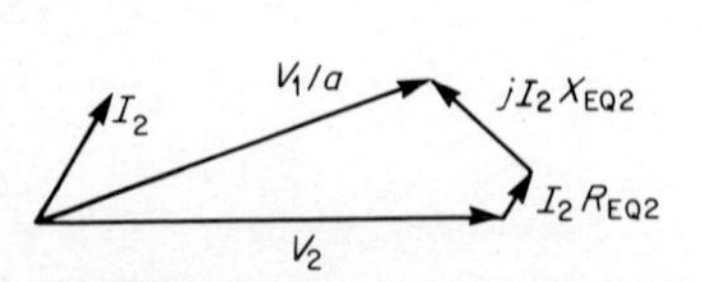

Fig. 11 Phasor diagram of transformer for leading power factor.

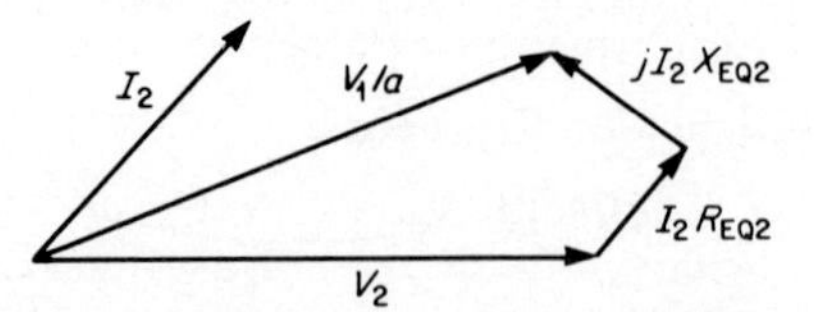

Fig. 12 Phasor diagram for zero transformer-voltage regulation.

transformer voltage regulation occurs when $V_1/a = V_2$. This condition, which occurs with a slightly leading power-factor load, is illustrated in Fig. 12.

CALCULATION OF EFFICIENCY

A 10-kVA transformer has 40-W iron loss at rated voltage and 160-W copper loss at full load. Calculate the efficiency for a 5-kVA, 80 percent power-factor load.

Calculation Procedure:

1. Analyze the Losses

The sum of the hysteresis and eddy-current losses is called the core, or iron, loss of the transformer; this will be designated P_i. The core loss is a constant loss of the transformer.

The sum of the primary and secondary I^2R losses is called the copper loss P_{cu}: $P_{cu} = I_1^2R_1 + I_2^2R_2$. This shows that copper losses vary with the square of the current.

2. Solve for Efficiency

Efficiency η can be found from: $\eta = P_{out}/P_{in} = P_{out}/(P_{out} + \text{losses}) = P_{out}/(P_{out} + P_i + P_{cu}) = (\text{VA}_{load})(\text{pf})/[(\text{VA}_{load})(\text{pf}) + P_i + P_{cu}(\text{VA}_{load}/\text{VA}_{rating})^2]$. From this equation, $\eta = (5000 \times 0.8)/[5000 \times 0.8 + 40 + 160(5000/10{,}000)^2] = 0.98$ or 98 percent.

Related Calculations: The iron losses of a transformer are determined quite accurately by measuring the power input at rated voltage and frequency under no-load conditions. Although it makes no difference which winding is energized, it is usually more convenient to energize the low-voltage side. (It is essential to use rated voltage for this test.)

The copper loss is measured by short-circuiting the transformer and measuring the power input at rated frequency and full-load current. It is usually convenient to perform the short-circuit test by shorting out the low-voltage side and energizing the high-voltage side; however, it does not matter if the procedure is reversed.

Because changing the power factor of the load does not change the losses, raising the load power factor will improve the efficiency of the transformer. The losses then become a smaller proportion of the total power input. The no-load efficiency of the transformer is zero. High loads increase the copper losses, which vary with the square of the current, thereby decreasing the efficiency. Maximum efficiency operation occurs at some intermediate value of load.

ANALYSIS OF TRANSFORMER OPERATION AT MAXIMUM EFFICIENCY

Calculate the load level at which maximum efficiency occurs for the transformer of the previous example. Find the value of maximum efficiency with a 100 percent power-factor load and a 50 percent power-factor load.

Calculation Procedure:

1. Calculate the Load Level for Maximum Efficiency

Maximum efficiency occurs when the copper losses equal the iron losses. Then, $P_i = P_{cu}(\text{kVA}_{load}/\text{kVA}_{rating})^2$ or $40 = 160(\text{kVA}_{load}/10\ \text{kVA})^2$. Solving, find $\text{kVA}_{load} = 5\ \text{kVA}$.

2. *Calculate Maximum Efficiency with 100 Percent Power-Factor Load*

With the copper and core losses equal, $\eta = P_{out}(P_{out} + P_i + P_{cu}) = 5000/(5000 + 40 + 40) = 0.9842$ or 98.42 percent.

3. *Calculate Maximum Efficiency with 50 Percent Power-Factor Load*

Maximum efficiency is $\eta = (5000 \times 0.5)/(5000 \times 0.5 + 40 + 40) = 0.969$ or 96.9 percent.

Related Calculations: Maximum efficiency occurs at approximately half load for most transformers. In this procedure, it occurs at exactly half load. Transformers maintain their high efficiency over a wide range of load values above and below the half-load value. Maximum efficiency for this transformer decreases with lower power factors from a value of 98.42 percent at 100 percent power factor to a value of 96.9 percent at 50 percent power factor. Efficiencies of transformers are higher than those of rotating machinery, for the same capacity, because rotating electrical machinery has additional losses, such as rotational and stray load losses.

CALCULATION OF ALL-DAY EFFICIENCY

A 50-kVA transformer has 180-W iron loss at rated voltage and 620-W copper loss at full load. Calculate the all-day efficiency of the transformer when it operates with the following unity-power-factor loads: full load, 8 h; half load, 5 h; one-quarter load, 7 h; no load, 4 h.

Calculation Procedure:

1. *Find the Total Energy Iron Losses*

Since iron losses exist for the entire 24 h the transformer is energized, the total iron loss is $W_{i(total)} = P_i t = (180 \times 24)/1000 = 4.32$ kWh.

2. *Determine the Total Energy Copper Losses*

Energy copper losses $W_{cu} = P_{cu}t$. Because copper losses vary with the square of the load, the total-energy copper losses are found as follows: $W_{cu(total)} = (1^2 \times 620 \times 8 + 0.5^2 \times 620 \times 5 + 0.25^2 \times 620 \times 7)/1000 = 6.006$ kWh over the 20 h that the transformer supplies a load.

3. *Calculate the Total Energy Loss*

The total energy loss over the 24-h period is: $W_{loss(total)} = W_{i(total)} + W_{cu(total)} = 4.32 + 6.006 = 10.326$ kWh.

4. *Solve for the Total Energy Output*

$W_{out(total)} = 50 \times 8 + 50 \times ½ \times 5 + 50 \times ¼ \times 7 = 612.5$ kWh.

5. *Compute the All-Day Efficiency*

All-day efficiency $= W_{out(total)}/[W_{out(total)} + W_{loss(total)}] \times 100$ percent $= 612.5/(612.5 + 10.236) \times 100$ percent $= 98.3$ percent.

Related Calculations: All-day efficiency is important when the transformer is connected to the supply for the entire 24 h, as is typical for ac distribution systems. It is usual to calculate this efficiency at unity power factor. At any other power factor, the all-day efficiency would be lower because the power output would be less for the same losses.

The overall energy efficiency of a distribution transformer over a 24-h period is high in spite of varying load and power-factor conditions. A low all-day efficiency exists only when there is a complete lack of use of the transformer, or during operation at extremely low power factors.

SELECTION OF TRANSFORMER TO SUPPLY A CYCLIC LOAD

Select a minimum-size transformer to supply a cyclic load that draws 100 kVA for 2 min, 50 kVA for 3 min, 25 kVA for 2 min, and no load for the balance of its 10-min cycle.

Calculation Procedure:

1. Solve for the Apparent-Power Rating in kVA of the Transformer

When the load cycle is sufficiently short so that the temperature of the transformer does not change appreciably during the cycle, the minimum transformer size is the rms value of the load. Hence $S = \sqrt{(S_1^2 t_1 + S_2^2 t_2 + S_3^2 t_3)[t(\text{cycle})]} = \sqrt{(100^2 \times 2 + 50^2 \times 3 + 25^2 \times 2)/10} = 53.62$ kVA.

Related Calculations: When selecting a transformer to supply a cyclic load, it is essential to verify that the voltage regulation is not excessive at peak load. The method used to select a transformer in this example is satisfactory provided the load cycle is short. If the load cycle is long (several hours), this method cannot be used. In that case the thermal time constant of the transformer must be considered.

ANALYSIS OF TRANSFORMER UNDER SHORT-CIRCUIT CONDITIONS

A transformer is designed to carry 30 times its rated current for 1 s. Determine the length of time that a current of 20 times the rating can be allowed to flow. Find the maximum amount of current that the transformer can carry for 2 s.

Calculation Procedure:

1. Calculate the Time for 20 Times Rating

Transformers have a definite I^2t limitation because heat equals I^2R_{EQ} and R_{EQ} is constant for a particular transformer. (R_{EQ} represents the total resistance of the primary and secondary circuits.) Hence, $30^2 \times 1 = 20^2 t$; solving, $t = 2.25$ s.

2. Solve for the Maximum Permissible Current for 2 s

Since $30^2 \times 1 = I^2 \times 2$, $I = 21.21$ times full-load current.

Related Calculations: The thermal problem is basically a matter of how much heat can be stored in the transformer windings before an objectionable temperature is reached. The method followed in this example is valid for values of t below 10 s.

PERFORMANCE OF A STEP-UP AUTOTRANSFORMER

Calculate the full-load currents of the 2400-V and 120-V windings of the 2400/120-V isolation transformer of Fig. 13*a*. When this transformer is connected as a step-up trans-

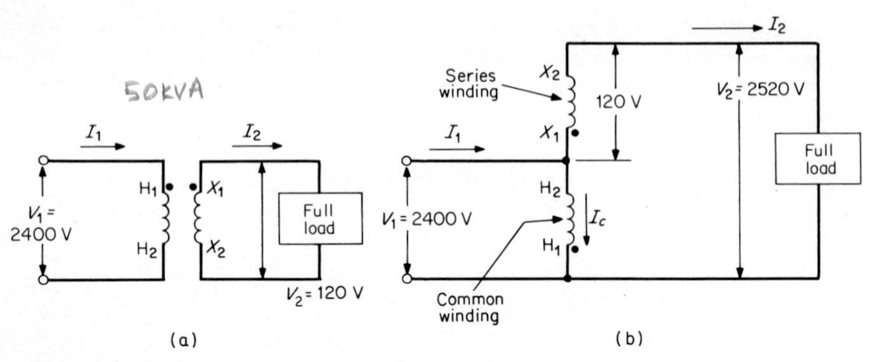

Fig. 13 Application of an (*a*) isolation transformer and (*b*) autotransformer.

former in Fig. 13*b*, calculate its apparent power rating in kVA, the percent increase in apparent power capacity in kVA, and the full-load currents.

Calculation Procedure:

1. Find the Full-Load Currents of the Isolation Transformer

$I_1 = (\text{VA})_1/V_1 = 50{,}000/2400 = 20.83$ A, and $I_2 = (\text{VA})_2/V_2 = 50{,}000/120 = 416.7$ A.

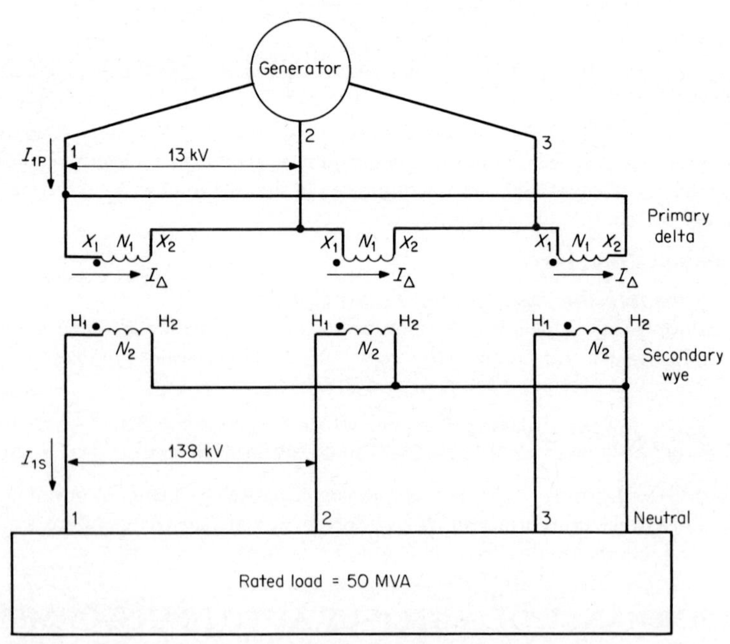

Fig. 14 Delta-wye three-phase transformer bank used as a generator step-up transformer.

2. ***Determine the Power Rating in kVA of the Autotransformer***
Because the 120-V winding is capable of carrying 416.7 A, the power rating in kVA of the autotransformer is $VA_2 = (2520 \times 416.7)/1000 = 1050$ kVA.

3. ***Calculate the Percent Increase in Apparent Power Using the Isolation Transformer as an Autotransformer***
$kVA_{auto}/kVA_{isolation} = 1050/50 \times 100$ percent $= 2100$ percent.

4. ***Solve for the Full-Load Currents of the Autotransformer***
Because the series winding (X_1 to X_2) has a full-load rating of 416.7 A, $I_2 = 416.7$ A. Current $I_1 = (VA)_1/V_1 = (1050 \times 1000)/2400 = 437.5$ A. The current in the common winding is $I_c = I_1 - I_2 = 437.5 - 416.7 = 20.8$ A.

Related Calculations: With the circuit as an autotransformer, the power in kVA has increased to 2100 percent of its original value with the low-voltage winding operating at its rated capacity. The effect on the high-voltage winding is negligible because $I_c = 20.8$ A while I_1, with the circuit as an isolation transformer, is 20.83 A.

The increase in apparent-power capacity produced by connecting an isolation transformer as an autotransformer accounts for the smaller size in autotransformers of the same capacity in kVA compared with ordinary isolation transformers. However, this marked increase in capacity only occurs as the ratio of primary to secondary voltages in the autotransformer approaches unity.

ANALYSIS OF A DELTA-WYE THREE-PHASE TRANSFORMER BANK USED AS A GENERATOR-STEP-UP TRANSFORMER

Calculate the line current in the primary, the phase current in the primary, the phase-to-neutral voltage of the secondary, and the turns ratio of the 50-MVA, three-phase transformer bank of Fig. 14 when used as a generator-step-up transformer and operating at rated load.

Calculation Procedure:

1. ***Find the Line Current in the Primary***
$I_{1P} = S/(\sqrt{3} \times V_{LP}) = 50{,}000{,}000/(\sqrt{3} \times 13{,}000) = 2221$ A.

2. ***Determine the value of the Phase Current in the Primary***
$I_\Delta = I_1P/\sqrt{3} = 2221/\sqrt{3} = 1282$ A.

3. ***Calculate the Phase-to-Neutral Voltage of the Secondary***
$V_{1N} = V_{LS}/\sqrt{3} = 138{,}000/\sqrt{3} = 79{,}677$ V.

4. ***Solve for the Line Current in the Secondary***
$I_{1S} = S/(\sqrt{3} \times V_{LS}) = 50{,}000{,}000/(\sqrt{3} \times 138{,}000) = 209$ A.

5. ***Compute the Transformer Turns Ratio***
The turns ratio is $a = N_1/N_2 = 13{,}000/79{,}677 = 0.163 = 1{:}6.13$.

Related Calculations: The line current in the secondary is related to the line current in the primary as follows: $I_{1S} = aI_{1P}/\sqrt{3} = (0.163 \times 2221)/\sqrt{3} = 209$ A which

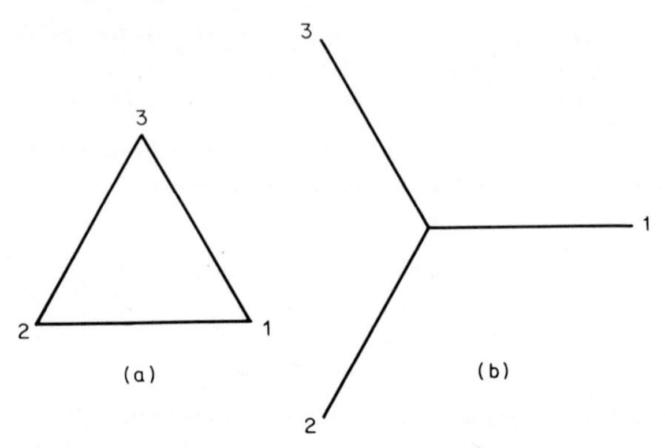

Fig. 15 Voltage phasor relations. (*a*) Delta primary at 13 kV. (*b*) Wye secondary at 138 kV.

checks the value of I_{1S} calculated in Step 4. The voltage phasor relations (Fig. 15) show that the secondary voltages of the wye side lead the primary voltages of the delta side by 30°

PERFORMANCE OF AN OPEN DELTA OR VEE-VEE SYSTEM

Each of the delta-delta transformers of Fig. 16 is rated at 40 kVA, 2400/240 V. The bank supplies an 80-kVA load at unity power factor. If transformer C is removed for repair, calculate, for the resulting vee-vee connection: load in kVA carried by each transformer, percent of rated load carried by each transformer, total apparent power rating in kVA of the transformer bank in vee-vee, ratio of vee-vee bank to delta-delta bank transformer ratings, and the percent increase in load on each transformer.

Calculation Procedure:

1. Find the Load in kVA Carried by Each Transformer

Load per transformer = total kVA/$\sqrt{3}$ = 80 kVA/$\sqrt{3}$ = 46.2 kVA.

2. Determine the Percent of Rated Load Carried by Each Transformer

Percent transformer load = (load in kVA/transformer)/(rating in kVA per transformer) × 100 percent = 46.2 kVA/40 kVA × 100 percent = 115.5 percent.

3. Calculate the Total Rating in kVA of the Transformer Bank in Vee-Vee

Rating in kVA of vee-vee bank = $\sqrt{3}$(rating in kVA per transformer) = $\sqrt{3} \times 40$ = 69.3 kVA.

4. Solve for the Ratio of the Vee-Vee Bank to the Delta-Delta Bank Transformer Ratings

Ratio of ratings = vee-vee bank/delta-delta bank = 69.3 kVA/120 kVA × 100 percent = 57.7 percent.

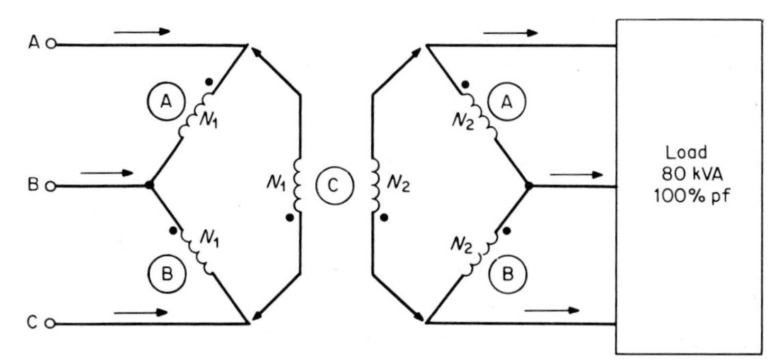

Fig. 16 Removal of transformer C from a delta-delta system results in a vee-vee system.

5. Compute the Percent Increase in Load on Each Transformer

Original load in delta-delta per transformer is 80 kVA/$\sqrt{3}$ = 26.67 kVA per transformer. Percent increase in load = (kVA per transformer in vee-vee)/(kVA per transformer in delta-delta) = 46.2 kVA/26.67 kVA × 100 percent = 173.2 percent.

Related Calculations: This example demonstrates that while the transformer load increases by 173.2 percent in a vee-vee system, each transformer is only slightly overloaded (115.5 percent). Because each transformer in a vee-vee system delivers line current and not phase current, each transformer in open delta supplies 57.7 percent of the total volt-amperes.

ANALYSIS OF A SCOTT-CONNECTED SYSTEM

A two-phase, 10-hp, 240-V, 60-Hz motor has an efficiency of 85 percent and a power factor of 80 percent. It is fed from the 600-V, three-phase system of Fig. 17 by a Scott-connected transformer bank. Calculate the apparent power drawn by the motor at full load, the current in each two-phase line, and the current in each three-phase line.

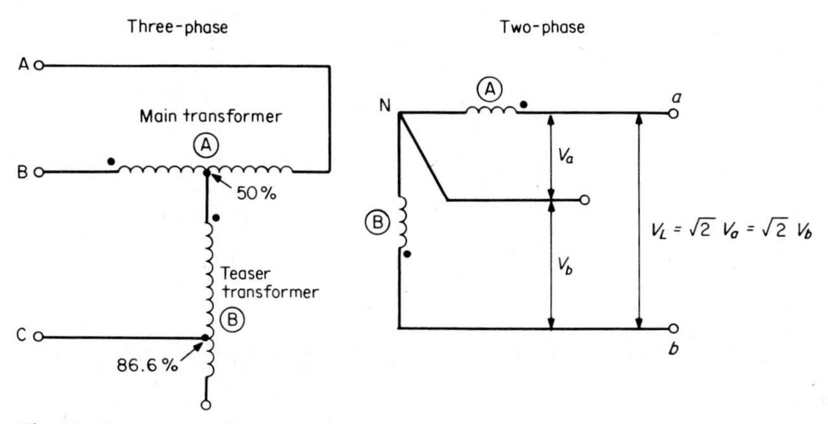

Fig. 17 Scott-connected system, three-phase to two-phase or two-phase to three-phase.

Calculation Procedure:

1. Find the Apparent Power Drawn by the Motor

Rated output power of the motor is P_o = 746 W/hp × 10 hp = 7460 W. Active power P drawn by the motor at full load is P_o/η = 7460 W/0.85 = 8776 W where η = efficiency. Apparent power drawn by the motor at full load is $S = P/\text{pf}$ = (8776 W)/0.8 = 10,970 VA.

2. Determine the Current in Each Two-Phase Line

Apparent power per phase = 10,970/2 = 5485 VA; hence $I = S/V$ = 5485/240 = 22.85 A.

3. Calculate the Current in Each Three-Phase Line

$I = S/(\sqrt{3} \times V)$ = 10,970/($\sqrt{3}$ × 600) = 10.56 A.

Related Calculations: The Scott connection isolates the three-phase and two-phase systems and provides the desired voltage ratio. It can be used to change three-phase to two-phase or two-phase to three-phase. Because transformers are less costly than rotating machines, this connection is very useful when industrial concerns wish to retain their two-phase motors even though their line service is three-phase.

Section 5 THREE-PHASE INDUCTION MOTORS

Sanjit K. Bardhan
Engineer, PSE&G

REFERENCES Del Toro—*Electromechanical Devices for Energy Conversion and Control Systems,* Prentice-Hall; Fink and Beaty—*Standard Handbook for Electrical Engineers,* McGraw-Hill; Fitzgerald et al.—*Electrical Machinery,* McGraw-Hill; Hicks—*Standard Handbook of Engineering Calculations,* McGraw-Hill; IEEE—*Test Procedure for Polyphase Induction Motors and Generators,* Std. 112A, Institute of Electrical and Electronics Engineers; Libby—*Motor Selection and Application,* McGraw-Hill; Matsch—*Electromagnetic and Electromechanical Machines,* IEP: Smeaton—*Motor Application and Maintenance Handbook,* McGraw-Hill; Smeaton—*Switchgear and Control Handbook,* McGraw-Hill.

LINE CURRENT AND POWER FACTOR

A 100-hp, 460-V, wye-connected, three-phase, 60-Hz, four-pole, 1754-r/min squirrel-cage induction motor has the following specifications (referred to the stator): $R_1 = 0.083$

Ω, R_2 = 0.058 Ω (running) and 0.087 Ω (starting), X_1 = 0.181 Ω, X_2 = 0.271 Ω (running) and 0.212 Ω (starting), X_m = 7.6 Ω. The core loss P_c = 1200 W, friction and windage loss P_{fw} = 1600 W, and stray loss P_{stray} = 785 W. Calculate the line current and power factor at full load and at starting.

Calculation Procedure:

1. Draw the per-Phase Equivalent Circuit

The equivalent circuit is drawn in Fig. 1*a*.

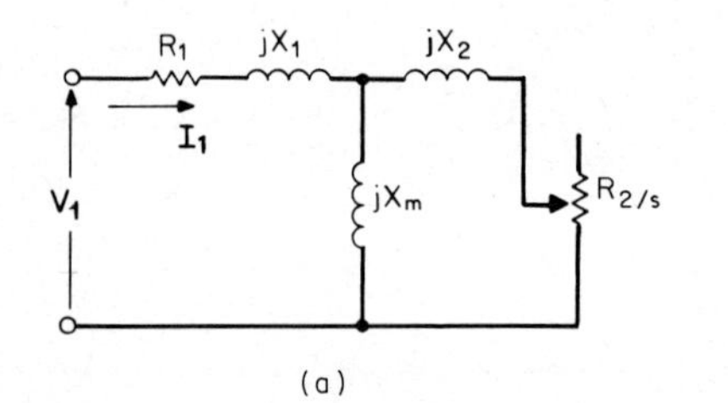

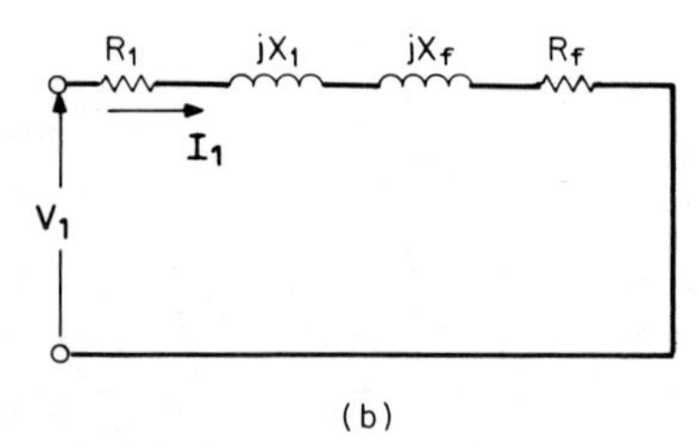

Fig. 1 Three-phase induction motor. (*a*) Per-phase approximate equivalent circuit. (*b*) Simplified equivalent circuit.

2. Replace the Parallel Circuit with Equivalent Series Impedance

$\mathbf{Z_f} = R_f + jX_f = jX_m || (R_2/s + jX_2)$; see Fig. 1*b*.

3. Compute Synchronous Speed n_s

The speed $n_s = 120f/p = (120 \times 60)/4 = 1800$ r/min.

4. Compute Slip s at Full Load

The slip $s = (n_s - n)/n_s = (1800 - 1754)/1800 = 0.0255 = 2.55$ percent.

5. Calculate $\mathbf{Z_f}$

$\mathbf{Z_f} = R_f + jX_f = j7.6 || (0.058/0.0255 + j0.271) = 1.957 + j0.827\ \Omega$ at full load. At start, $\mathbf{Z_f} = j7.6 || (0.087 + j0.212) = 0.082 + j0.207\ \Omega$.

6. Calculate $\mathbf{Z}$

$\mathbf{Z} = \mathbf{Z_1} + \mathbf{Z_f} = (0.083 + j0.181) + (1.957 + j0.827) = 2.276\angle 26.3°\ \Omega$ at full load. At start, $\mathbf{Z} = (0.083 + j0.181) + (0.082 + j0.207) = 0.422\angle 66.9°\ \Omega$.

7. Calculate the per-Phase Voltage $\mathbf{V_1}$

$|\mathbf{V_1}| = 460/\sqrt{3}$; $\mathbf{V_1} = 266\angle 0°$ V (reference).

8. Calculate Input Current $\mathbf{I_1}$

$\mathbf{I_1} = \mathbf{V_1}/\mathbf{Z} = 266\angle 0°/2.276\angle 26.3° = 116.9\angle 26.3°$ A at full load. At start, $\mathbf{I_1} = 266\angle 0°/0.422\angle 66.9° = 630.5\angle -66.9°$ A.

9. Calculate the power factor pf

The power factor is pf = cos θ = cos (−26.3°) = 0.896 lagging at full load, and at start, pf = cos (−66.9°) = 0.392 lagging.

TORQUE AND POWER

Determine the developed electromagnetic (internal) torque T, developed electromagnetic power P, output (shaft) power P_o, and output torque T_o at full load and starting for the motor in the previous example.

Calculation Procedure:

1. Compute Air-Gap Power

Air-gap power is $P_{g1} = q_1|\mathbf{I}_1|^2R_f = 3 \times 116.9^2 \times 1.957 = 80{,}231$ W at full load; at starting, $P_{g1} = 3 \times 630.5^2 \times 0.082 = 98{,}150$ W.

2. Compute Electromagnetic Torque

$T = P_{g1}/\omega_s = P_{g1}/(2\pi n_s/60) = 80{,}231/188.5 = 425.6$ N·m at full load. Similarly, at starting, $T = 520.7$ N·m (or 128 percent of the full-load torque).

3. Compute Electromagnetic Power at Full Load

$P = (1 - s)P_{g1} = (1 - 0.0255) \times 80{,}231 = 78{,}185$ W

4. Compute Output Power at Full Load

$P_o = P - (P_c + P_{fw} + P_{stray}) = 78{,}185 - (1200 + 1600 + 785) = 746$ kW = 100 hp.

5. Compute Output Torque at Full Load

$T_o = P_o/\omega_m = P_o/(1 - s)\omega_s = 746/[(1 - 0.0255) \times 188.5] = 406.1$ N·m.

Related Calculations: An alternate method to find output torque is to use $T_o = T - (\Sigma \text{ rot. losses})/\omega_m = 425.6 - 3585/(0.97 \times 188.5) = 406$ N·m.

LOSSES AND EFFICIENCY

Determine the losses, rotor efficiency, and motor efficiency at full load for the motor of the preceding example.

Calculation Procedure:

1. Compute Stator Copper Loss P_{cus}

$P_{cus} = q_1|\mathbf{I}_1|^2R_1 = 3 \times 116.9^2 \times 0.083 = 3403$ W.

2. Compute Rotor Copper Loss P_{cur}

$P_{cur} = sP_{g1} = sq_1|\mathbf{I}_1|^2R_f = = 0.0255 \times 80{,}231 = 2046$ W.

3. Compute Rotor Efficiency η_r

Efficiency is $\eta_r = (1 - s)P_{g1}/P_{g1} = (1 - s) = 1 - 0.0255 = 0.9745 = 97.45$ percent.

4. Compute Total Losses

Total losses $= P_{cus} + P_{cur} + P_{rot} = 3403 + 2046 + 3585 = 9034$ W.

5. Compute Input Power P_{in}

$P_{in} = P_o + \Sigma \text{ losses} = 74{,}600 + 9034 = 83{,}634$ W.

6. Compute Motor Efficiency

$\eta = P_o/P_{in} = 1 - (\Sigma \text{ losses})/P_{in} = 1 - 9034/83{,}634 = 0.892 = 89.2$ percent.

Related Calculations: The rotor efficiency is the ideal limit of an induction motor efficiency. Therefore, higher rated slip implies lower rated efficiency and higher rotor copper losses. A power-flow diagram based on the equivalent circuits of Fig. 1 is illustrated in Fig. 2.

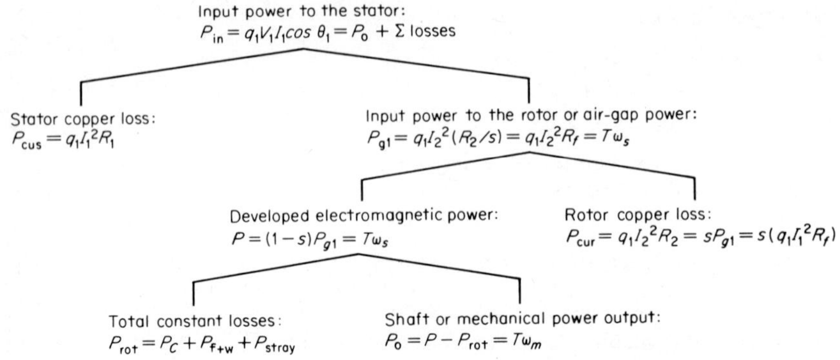

Fig. 2 Power-flow diagram for an induction motor based on the equivalent circuits of Fig. 1.

MAXIMUM (PULLOUT, BREAKDOWN) TORQUE

Calculate the maximum torque T_{max} of a 1000-hp, three-phase, 2300-V, wye-connected, 60-Hz, 227-A, wound-rotor induction motor. The following data apply: $R_1 = 0.0721$ Ω, $X_1 = 0.605$ Ω, $R_2 = 0.0234$ Ω, $X_2 = 0.151$ Ω, $X_m = 17.8$ Ω, and turns ratio $b = 2$.

Calculation Procedure:

1. Draw the per-Phase Equivalent Thevenin Circuit

The equivalent circuit is drawn in Fig. 3.

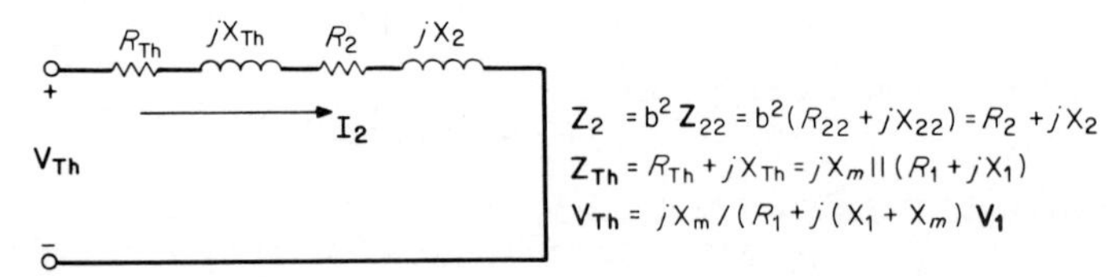

Fig. 3 Per-phase Thevenin equivalent circuit of a three-phase induction motor to calculate maximum torque.

2. Compute the Stator Phase Voltage V_1

$V_1 = 2300/\sqrt{3} = 1328$ V.

3. Determine the Synchronous Speed ω_s

The speed is $\omega_s = 2\pi \times 2 \times 60/16 = 47.12$ rad/s.

4. Find the Thevenin Voltage

$V_{Th} = (17.8 \times 1328)/\sqrt{0.0721^2 + (0.605 + 17.8)^2} = 1284$ V.

5. Find the Thevenin Impedance

$\mathbf{Z_{Th}} = [j17.8(0.0721 + j0.605)]/(0.0721 + j18.405) = 0.589\angle 83.4° = 0.0677 + j0.5851$ Ω.

6. Determine $\mathbf{Z_2}$

$\mathbf{Z_2} = R_2 + jX_2 = (0.0234 + j0.151) \times 2^2 = 0.0946 + j0.605$ Ω.

7. *Find the Slip $s_{\max T}$ for Maximum Torque*

The slip is $s_{\max T} = R_2/\sqrt{R_{\rm Th}^2 + (X_{\rm Th} + X_2)^2} = 0.0946/\sqrt{0.0677^2 + (0.605 + 0.585)^2} = 0.07945 = 7.945$ percent.

8. *Compute the Maximum Torque*

$T_{\max} = \frac{3}{2} \times 1/\omega_s \times V_{\rm Th}/[R_1 + \sqrt{R_{\rm Th}^2 + (X_{\rm Th} + X_2)^2}] = \frac{3}{2} \times 1/47.12 \times 1284^2/(0.721 + 1.192) = 41{,}518\ \text{N}\cdot\text{m}$.

PLUGGING AND BRAKING TORQUE

Calculate the plugging torque T_p and the braking torque T_b for the first motor (whose equivalent circuit is supplied in Fig. 1*a*) running at full load.

Calculation Procedure:

1. *Review Motor Data*

Developed electromagnetic torque at full load is $T_{\rm fl} = 425.6\ \text{N}\cdot\text{m}$; full-load slip $s_1 = 0.0255$; $\mathbf{V}_1 = 266\angle 0°$; $\omega_s = 188.5$ rad/s; $\mathbf{Z}_1 = 0.083 + j0.181\ \Omega$; and $\mathbf{Z}_2 = 0.058 + j0.271\ \Omega$.

2. *Compute $\mathbf{Z}_f$ at $2 - s_1$*

At a very high slip of $2 - s_1$, $\mathbf{Z}_f \cong \mathbf{Z}_2 = R_2/(2 - s_1) + jX_2 = 0.058/(2 - 0.0255) + j0.271 = 0.0294 + j0.271\ \Omega$.

3. *Compute $\mathbf{Z}$*

$\mathbf{Z} = \mathbf{Z}_1 + \mathbf{Z}_2 = (0.083 + j0.181) + (0.0294 + j0.271) = 0.1124 + j0.452 = 0.4658\angle 76°\ \Omega$.

4. *Compute $\mathbf{I}_1$*

$\mathbf{I}_1 = \mathbf{V}_1/\mathbf{Z} = 266\angle 0°/0.4658\angle 76° = 571.1\angle -76°$ A.

5. *Determine Air-Gap Power*

$P_{g1} = 3|\mathbf{I}_1|^2 R_f \cong 3|\mathbf{I}_1| R_2/(2 - s_1) = 3 \times 571.1^2 \times 0.0294 = 28{,}762.9$ W.

6. *Compute Plugging Torque*

$T_p = P_{g1}/\omega_s = 28{,}762.9/188.5 = 152.6\ \text{N}\cdot\text{m}$.

7. *Compute Braking Torque*

$T_b = T_p + T_{\rm fl} = 152.6 + 425.6 = 578.2\ \text{N}\cdot\text{m}$.

Related Calculations: If two of the three supply leads to a motor running at any speed are interchanged, a countertorque is generated. To stop the motor, it is necessary to disconnect it from the line at zero speed before it starts rotating in the reverse direction. The magnitude of the braking torque can be varied by any means which changes the winding current. Plugging produces very high transient braking torque for which additional winding bracings may be required. Losses incurred may be as high as 3 times that in a single start and 4 times if the motor is allowed to accelerate to full speed in the reverse direction.

BRAKE TORQUE RATING

Calculate the torque rating T_b of a brake mounted on a 200-hp, 450-r/min motor and the stopping time t_d (after the brake shoes have been applied) for 50 percent more than the motor rated torque. The motor inertia $J_M = 785\ \text{lb}\cdot\text{ft}^2$.

Calculation Procedure:

1. *Calculate the Starting Torque*

$T_b = 1.5 \times 746 \text{ hp}/\omega_m = 1.5 \times 746 \times 200/(2\pi \times 450/60) = 4749 \text{ N}\cdot\text{m}$.

2. *Express Motor Inertia in kg·m²*

J_M in $\text{kg}\cdot\text{m}^2 = \text{lb}\cdot\text{ft}^2/23.72 = 785/23.72 = 33 \text{ kg}\cdot\text{m}^2$.

3. *Calculate the Stopping Time*

For constant braking torque,

$t_d = J_M\omega_m/T_b = 33 \times 2\pi \times (450/600/4749) = 0.33$ s.

Related Calculations: To stop an induction motor, one can always disconnect it from the lines. The stopping time will vary depending upon the inertia, friction, and windage. Often the motor will be required to stop quickly and deceleration will then have to be controlled as required, for example, in a crane or hoist.

The function of braking is to provide means to dissipate the stored kinetic energy of the combined motor-system inertia during the period of deceleration from a higher to a lower, or zero, speed. The dissipation of kinetic energy may be external to the windings (external braking) or within them (internal braking). Sometimes a combination of internal and external braking may be employed to optimize braking performance. A summary of braking features is provided in Table 1.

TABLE 1 Summary of Braking Features

	Feature		*Braking*	
No.	*Description*	*Desirable*	*External*	*Internal*
1	Holding action	Yes	Yes	No
2	Motor-winding heating	No	No	Yes
3	Electric power for braking	No	No	Yes
4	Heat dissipation rate	Higher	Not applicable	Decreases with lower speed
5	Countertorque during deceleration	Yes	No	Yes
6	Overspeed control	Yes	No	Yes
7	Positive retardation	Yes	No	Yes
8	Increased inertia	No	Yes	No
9	Extra space	No	Yes	No

External braking uses a form of mechanical brake which is coupled to the motor. It can be used for all types of motors and requires special mechanical coupling. Examples of external braking include friction (a shoe- or disk-type brake that can be activated mechanically, electrically, hydraulically, or pneumatically), eddy-current (a direct-connected magnetically coupled unit), hydraulic (coupled hydraulic pump), and magnetic-particle (a direct-connected magnetic-particle coupling) braking.

Methods for producing internal braking can be divided into two groups: countertorque (plugging) and generating (dynamic and regenerative braking). In the countertorque group, torque is developed to rotate the rotor in the opposite direction to that which existed before braking. In the generating group, torque is developed from the rotor speed. Dynamic torque is developed by motor speeds that are less than the synchronous speed and energy is dissipated within the motor or in a connected load. Developed torque is zero at zero rotor speed.

Regenerative braking torque is developed by motor speeds that are higher than the synchronous speed. The motor is always connected to the power system and the generated power is returned to the power system.

For large motors, forced ventilation may be required during braking.

DYNAMIC BRAKING TORQUE

Calculate the dynamic braking torque T_{db} when a motor is running at rated load and a dc voltage is used to produce a rated equivalent ac in the stator (Fig. 4). Also, determine

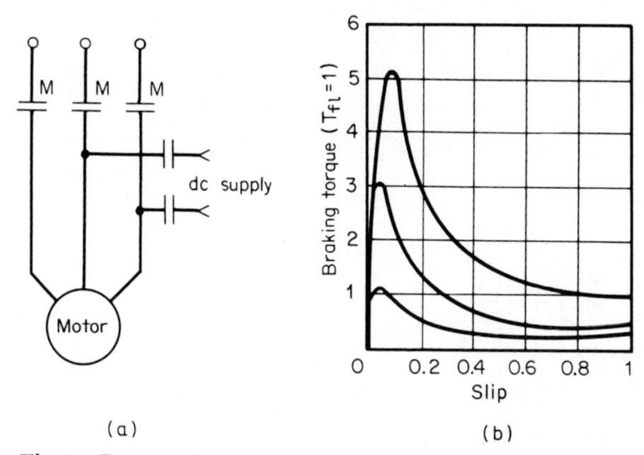

Fig. 4 Dynamic braking using dc. (*a*) Circuit. (*b*) Torque-speed characteristics.

the maximum braking torque $T_{db(max)}$ and the corresponding slip s_{max}. For the motor, $|\mathbf{I}_1| = 116.9$ A, $\omega_s = 188.5$ rad/s, $s = 0.0255$, $T_{fl} = 425.6$ N·m, $\mathbf{Z}_2$ (starting) $= R_2 + jX_2 = 0.087 + j0.212\ \Omega$, and $\mathbf{Z}_2$ (running) $= 0.058 + j0.271\ \Omega$. The saturated magnetizing reactance $X_m = 2.6\ \Omega$.

Calculation Procedure:

1. *Compute* T_{db}

$T_{db} = -(3/\omega_s)|\mathbf{I}_1|^2 X_m{}^2/\{[R_2/(1 - s)]^2 + (X_m + X_2)^2\}R_2/(1 - s) = -(3/188.5)(116.9)^2(2.6)^2/\{[0.087/(1 - 0.0255)]^2 + (2.6 + 0.212)^2\} \times 0.087/(1 - 0.0255) = -16.58$ N·m or $-(16.58/425.6) \times 100$ percent $= -3.9$ percent of full-load torque T_{fl}.

2. *Compute* $T_{db(max)}$

$T_{db(max)} = -(3/\omega_s)|\mathbf{I}_1|^2 X_m^2/2(X_m + X_2) = -(3/188.5)(116.9)^2 \times 2.6^2/(2.6 + 0.271) = -255.2$ N·m or 60 percent of T_{fl}.

3. Compute s_{max}

The slip is $s_{max} = 1 - R_2/(X_m + X_2) = 1 - 0.058/(2.6 + 0.271) = 0.98.$

Related Calculations: In this example, the braking torque was obtained by connecting a dc source to one phase of the motor after it is disconnected from the lines (Fig. 4*a*). With solid-state electronics, the problem of a separate dc source has been solved economically. The braking torque is low at the higher initial speed and increases to a high peak value as the motor decelerates; however, it drops rapidly to zero speed (Fig. 4*b*). The losses are approximately the same as that for a single start. Higher braking torque can be obtained by using higher values of dc voltage and by inserting external resistance in the rotor circuit for a wound-rotor motor. Approximately 150 percent rated current (dc) is required to produce an average braking torque of 100 percent starting torque.

A form of dc braking using a capacitor-resistor-rectifier circuit where a variable dc voltage is applied as the capacitor discharges through the motor windings instead of a fixed dc voltage, as in dc braking, is illustrated in Fig. 5. The dc voltage and motor speed decrease together to provide a more nearly constant braking torque. The energy stored in the capacitor is all that is required for the braking power. This method, however, requires costly, large capacitors.

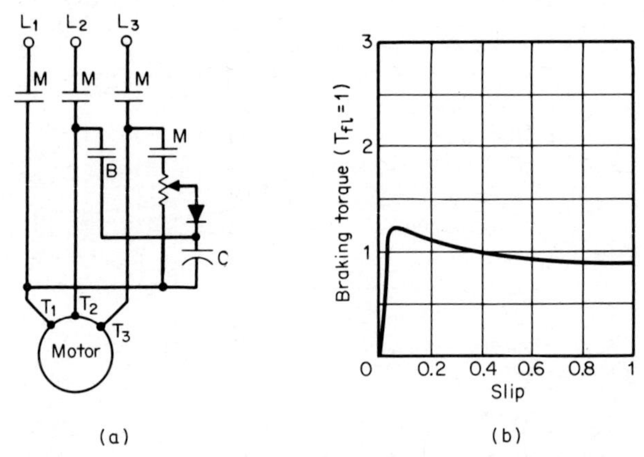

Fig. 5 Dynamic braking using a capacitor-rectifier-resistor circuit. (*a*) Circuit. (*b*) Torque-speed characteristics.

An induction motor once disconnected from the lines can produce braking torque by generator action if a bank of capacitors is connected to the motor terminals (Fig. 6). The braking torque may be increased by including loading resistors. If capacitors sized for power-factor correction are used, the braking torque will be small. To produce an initial peak torque of twice the rated torque, capacitance equal to about 3 times the no-load magnetizing apparent power in kVA is required. (This may produce a high transient voltage that could damage the winding insulation.) At a fairly high speed, the braking torque reduces to zero.

In ac dynamic braking (Fig. 7), "single-phasing" a three-phase induction motor will not stop it, but will generate a low braking torque with zero torque at zero speed. It is a very simple and comparatively inexpensive method. Losses within the motor winding may

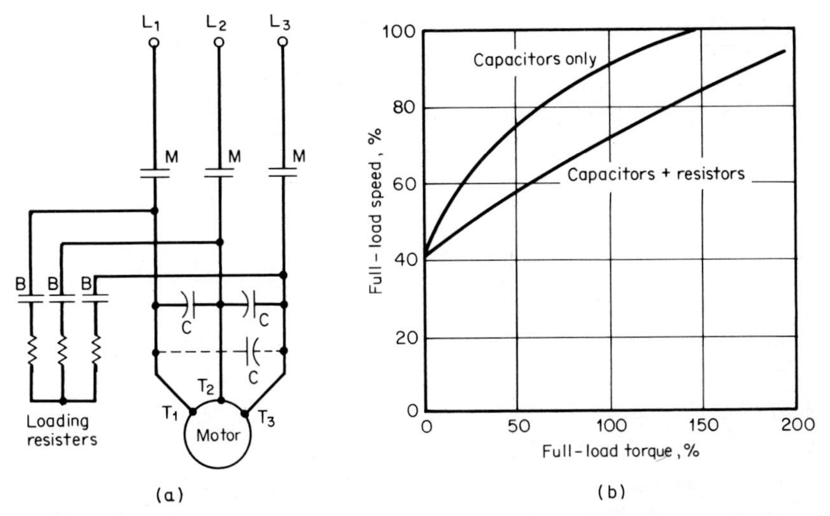

Fig. 6 Dynamic braking using capacitors. (*a*) Circuit. (*b*) Torque-speed characteristics.

require a larger motor size to dissipate the heat. Braking torque in the case of a wound rotor may be varied by the insertion of external resistance in the rotor circuit. Sometimes, a separate braking winding is provided.

In regenerative braking, torque is produced by running the motor as an induction generator and the braking power is returned to the lines. For example, regenerative braking can be applied to a two or four-pole squirrel-cage motor initially running at, say, 1760 r/min by changing the number of poles from two to four. The synchronous speed now is 900 r/min and the machine runs as a generator. Regenerative braking is mainly used for squirrel-cage motors. The method becomes too involved when poles are changed for a wound-rotor motor.

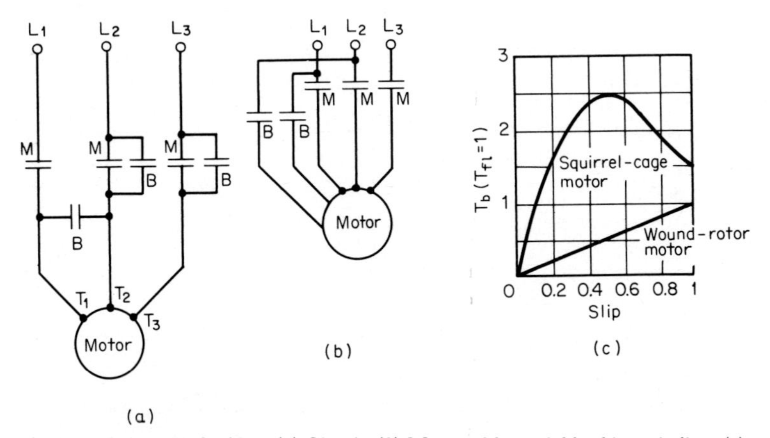

Fig. 7 AC dynamic braking. (*a*) Circuit. (*b*) Motor with special braking winding. (*c*) Torque-speed characteristics.

EQUIVALENT CIRCUIT PARAMETERS

Determine the equivalent circuit parameters of a 10-hp, 230-V, three-phase, wye-connected squirrel-cage induction motor with a double squirrel cage (NEMA design C) from the following test data:

1. *No-load test at stator frequency* f_1 = *60 Hz:* V_{NL} = 229.9 V, I_{NL} = 6.36 A, P_{NL} = 512 W.
2. *Locked-rotor test at stator frequency* f_1 = *15 Hz:* V'_L = 24 V, I'_L = 24.06 A, P'_L = 721 W. Average dc stator resistance measured between stator terminals immediately following test is 0.42 Ω.
3. *Locked-rotor test at stator frequency* f_1 = *60 Hz:* V_L = 230 V, I_L = 110 A, P_L = 27,225 W, T_L = 188.5 N·m.

Calculation Procedure:

1. Determine Values for $|\mathbf{Z}_{NL}|$, R_{NL}, X_{NL}, and R_1

From the no-load test in item 1: $|\mathbf{Z}_{NL}| = V_{NL}/(\sqrt{3}I_{NL}) = 229.9/(\sqrt{3} \times 6.36) = 20.87\ \Omega$, $R_{NL} = P_{NL}/(3I_{NL}{}^2) = 512/(3 \times 6.36^2) = 4.22\ \Omega$; $X_{NL} = \sqrt{|\mathbf{Z}_{NL}|^2 - R_{NL}^2} = \sqrt{20.87^2 - 4.22^2} = 20.44\ \Omega$. From the dc stator resistance measurement in item 2, $R_1 = 0.42/2 = 0.21\ \Omega$.

2. Compute Rotational Losses P_{rot}

$P_{rot} = P_c + P_{fw} = P_{NL} - 3I_{NL}^2R_1 = 512 - 3 \times 6.36^2 \times 0.22 = 487$ W.

3. Compute Z'_L, R'_L, and X'_L

From the locked-rotor test in item 2: $Z'_L = V'_L/(\sqrt{3}I'_L) = 24/(\sqrt{3} \times 24) = 0.577\ \Omega$; $R'_L = P'_L/(3I'^2_L) = 771/(3 \times 24^2) = 0.466\ \Omega$; $X'_L = \sqrt{Z'^2_L - R'^2_L} = \sqrt{0.577^2 - 0.466^2} = 0.366\ \Omega$.

4. Determine the Leakage Reactances at 60 Hz

$X_L = (60/15)X'_L = 4 \times 0.366 = 1.464\ \Omega$. For a NEMA design C motor, from Table 2: $X_1 = 0.3X_L = 0.3 \times 1.464 = 0.44\ \Omega$; $X_2 = 0.7X_L = 0.7 \times 1.464 = 1.024\ \Omega$.

TABLE 2 IEEE Test Code for Empirical Ratios of Leakage Reactances

	Squirrel-cage: design class				*Wound rotor*
	A	*B*	*C*	*D*	
X_1/X_L	0.5	0.4	0.3	0.5	0.5
X_2/X_L	0.5	0.6	0.7	0.5	0.5

5. Find the Magnetizing Reactance

$X_m = X_{NL} - X_1 = 20.44 - 0.44 = 20\ \Omega$.

6. Compute the Rotor Resistance Referred to the Stator Side

$R_f = R'_L - R_1 = 0.466 - 0.21 = 0.236\ \Omega$; $R_2 = R_f[(X_2 + X_m)/X_m]^2 = 0.236 \times [(1.025 + 20)/20]^2 = 0.26\ \Omega$.

7. Determine the Locked Rotor Torque

From the locked rotor test in item 3: $P_{g1} = P_L - 3I_L^2R_1 = 27{,}225 - 3 \times 110^2 \times 0.21 = 19{,}608$ W. $\omega_s = 2\pi(2f/p) = 2\pi(2 \times 60/4) = 188.5$ rad/s. $T_L = 19{,}608/188.5 = 104$ N·m. (*Note:* the measured value of the torque is somewhat less than the calculated value. This difference stems from neglecting the core and stray-load losses in calculating P_{g1}.)

8. Compute the Rotor Leakage Reactance

From the locked rotor test in item 3: $Z_L = V_L/(\sqrt{3}I_L = 230/(\sqrt{3} \times 110) = 1.21$ Ω; $R_L = P_L/3I_L^2) = 27{,}225/(3 \times 110^2) = 0.75$ Ω; $X_L = \sqrt{Z_L^2 - R_L^2} = \sqrt{1.21^2 - 0.75^2} = 0.95$ Ω; $X_{2L} = X_L - X_1 = 0.95 - 0.44 = 0.51$ Ω.

9. Determine the Rotor Resistance at Start

$R_{fl} = R_L - R_1 = 0.75 - 0.21 = 0.54$ Ω; $R_{2L} = R_{fl}[(X_{2L} + X_m)/X_m]^2 = 0.54 \times [(0.51 + 20)/20]^2 = 0.568$ Ω.

Related Calculations: Both rotor resistance and leakage reactance at start have different values from those at running. Higher starting resistance and lower starting reactance are characteristics of a double-cage NEMA class C motor.

AUTOTRANSFORMER STARTING

Referring to Fig. 8, determine the line current I_{LL}, motor current I_{LA}, and starting torque T_{LA} for autotransformer starting at taps $\alpha = 0.5$, 0.65, and 0.8. At full voltage, the motor has I_L, the locked-rotor current in percent full-load current, equal to 600 percent and T_L, the locked-rotor torque in percent full-load torque, equal to 130 percent.

Select a transformer tap for the motor if the line current is not to exceed 300 percent of full-load current I_{fl} and starting torque is not to be less than 35 percent of full-load torque T_{fl} at start.

Calculation Procedure:

1. Calculate Starting Currents and Torque

$I_{LL} = \alpha^2 I_L$ for three transformers. For $\alpha = 0.5$, $I_{LL} = 0.5^2 \times 600 = 150$ percent. If two transformers are used, $I_{LL} = \alpha^2 I_L + 15$ percent $= 165$ percent, $I_{LA} = \alpha I_L = 0.5 \times 600 = 300$ percent, and $T_{LA} = \alpha^2 T_L = 0.5^2 \times 130 = 32.5$ percent.

2. Use the Following Table

Values for the other taps are obtained as in Step 1.

No.	Tap, α	Line current, I_{LL}: Two transformers		Line current, I_{LL}: Three transformers		Motor current, I_{LA}		Starting torque, T_{LA}	
1	0.5	27.5	165	25	150	50	300	25	32.5
2	0.65	44.8	269	42.3	254	65	390	42.3	55
3	0.8	66.5	399	64	384	80	480	64	83.2
		$\%I_L$	$\%I_{fl}$	$\%I_L$	$\%I_{fl}$	$\%I_L$	$\%I_{fl}$	$\%T_L$	$\%T_{fl}$

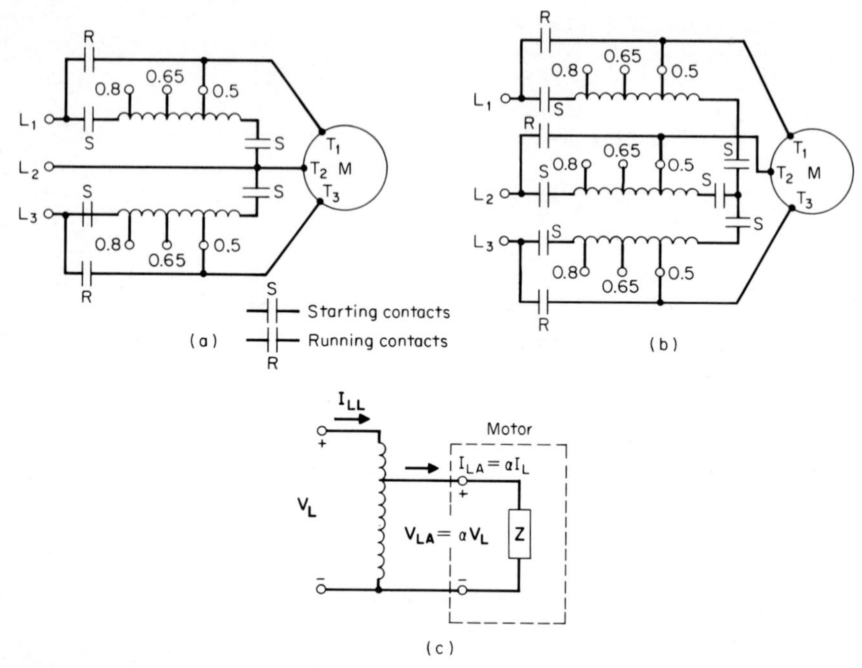

Fig. 8 Autotransformer reduced-voltage starting. (*a*) Open-circuit transition. (*b*) Closed-circuit transition. (*c*) Equivalent circuit.

$$\begin{aligned} \alpha &= \text{voltage tap} = V_{\text{LA}}/V_g \\ I_{\text{LA}} &= \alpha I_L = \text{ motor current} \\ I_{\text{LL}} &= \alpha I_{\text{LA}} = \alpha^2 I_L = \text{line current for } (b) \\ &= \alpha^2 I_l + 0.15\ I_{\text{fl}} \text{ for } (a) \\ T_A &= \alpha^2 T = \text{ motor torque with autotransformer.} \end{aligned}$$

3. *Select Tap*

To meet the current constraint that $I_{\text{LA}} \leq 300$ percent of I_{fl}, the tap may be either 0.5 or 0.65. To satisfy the torque constraint, $T_{\text{LA}} \geq 35$ percent of T_{fl}, the tap may be 0.65 or 0.8. Select tap 0.65.

Related Calculations: Autotransformer (or compensator) starting is the most commonly used reduced-voltage starter. A voltage equal to the tap voltage is maintained at the stator terminals during acceleration from the start to the transition. Usually, three taps, 0.5, 0.65, and 0.8, are provided.

Because of lower cost, two autotransformers in open-delta (vee connection) are employed. The result is a higher current in the third phase as well as a momentary open circuit during the transition. A very high transient current of short duration occurs during this momentary open circuit. The use of three autotransformers allows a closed-circuit transition, thereby eliminating the open-circuit transient and the unbalanced line current; this, however, adds to the cost of an additional transformer.

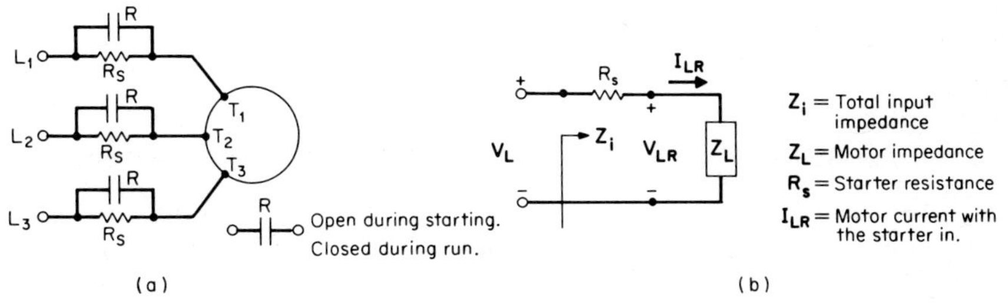

Fig. 9 Resistance starting. (*a*) Connection diagram. (*b*) Equivalent circuit.

RESISTANCE STARTING

Referring to Fig. 9*a*, size the resistor R_S of a two-step resistance starter having taps T_1, T_2, and T_3 at 0.5, 0.65, and 0.8, respectively. Also, calculate the corresponding line current, pf, and torque. For the motor, $\mathbf{Z_L} = 0.165 + j0.283\ \Omega$, $I_L = 539$ percent of I_{fl}, and $T_L = 128$ percent of T_{fl}.

Calculation Procedure:

1. *Use the Following Table*

Symbols V_{LR}, I_{LR}, and T_{LR} represent the motor locked-rotor quantities with the starter in.

No.	Equation		Tap α value 0.50	0.65	0.80
1	$I_{LR} = \frac{V_{LR}}{Z_L} = \frac{\alpha V_L}{Z_L} = \alpha I_L$	in $\%I_{fl}$	269.5	350.4	431.2
		$\%I_L$	50	65	80
2	$\lvert \mathbf{Z}_{iL} \rvert = \lvert R_L + jX_L \rvert = \frac{V_L}{I_{LR}} = \frac{V_L}{\alpha I_L} = \left(\frac{Z}{\alpha}\right)\Omega$		0.844	0.649	0.528
3*	$R_{iL} = R_S + R_L = \sqrt{Z_{iL}^2 - X_L^2}\ \Omega$		0.810	0.604	0.471
4	$R_S = (R_{iL} - R_L)\ \Omega$		0.645	0.439	0.306
5	$\text{pf} = \cos(-\theta_{iL}) = R_{iL}/\lvert \mathbf{Z}_{iL} \rvert$		0.96	0.931	0.892
6	$\text{pf angle } \theta_{iL} = -\cos^{-1}(R_{iL}/\lvert \mathbf{Z}_{iL} \rvert)$		−16.3°	−21.4°	−26.9°
7	$T_{LR} = \alpha^2 T_L$	in $\%T_{fl}$	32	54.1	81.9
		$\%T_L$	25	42.3	64

*R_S = starter resistance (Fig. 9*b*).

2. *Select Tap*

If I_{LR} is less than 400 percent, the tap may be 0.5 or 0.65 (see above table). If T_{LR} is greater than 35 percent, the tap may be 0.65 or 0.8. Hence, the selected tap is 0.65.

3. *Determine R_S (Fig. 9b)*

For $\alpha = 0.65$, from the preceding table $R_S = 0.439\ \Omega$.

TABLE 3 Comparison of Resistance and Reactance Starting

No.	*Feature*	*Resistance type*	*Reactor type*
1	Starting pf	Higher	Lower
2	Developed torque around 75 to 85% n_s	Lower (15 to 20% T_{fl})	Higher (15 to 20% T_{fl})
3	Heat loss (I^2R)	Very high	Very small
4	Size	Larger	Smaller
5	Cost	Lower	Higher

Related Calculations: During starting, the effective voltage at the motor terminals can be reduced through the use of a series resistance or reactance. With an increase in motor speed, the motor impedance increases and the starter impedance remains constant. This causes a greater voltage to appear at the motor terminals as the motor speed increases, resulting in higher developed torque. It provides "closed-circuit transition" and balanced line currents. These starters are designed for a single fixed-tap value of either 0.5, 0.65, or 0.8 of the line voltage.

Because of its flexibility and lower cost, the most widely used form of impedance starting is resistance starting. It could be designed to form a stepless starter through the aid of an automatically controlled variable resistance. For larger motors (greater than 200 hp) or for high voltages (greater than 2300 V), reactance starting is preferred. The features of both starting methods are summarized in Table 3.

REACTANCE STARTING

Repeat the previous example using the reactance starter of Fig. 10.

Calculation Procedure:

1. Use the Following Table

Symbols V_{LX}, I_{LX}, and T_{LX} represent the motor locked-rotor quantities with the starter in.

No.	Equation		Tap α value		
			0.50	**0.65**	**0.80**
1	$I_{LX} = \dfrac{V_{LX}}{Z_L} = \dfrac{\alpha V_L}{Z_L} = \alpha I_L$	in $\%I_{fl}$	269.5	350.4	431.2
		$\%I_L$	50	65	80
2	$\lvert \mathbf{Z}_{iL} \rvert = \lvert R_L + jX_{iL} \rvert = V_L/I_{LX} = V_L/\alpha I_L = (Z/\alpha)\ \Omega$		0.844	0.649	0.528
3	$X_{iL} = X_r + X_L = \sqrt{\lvert \mathbf{Z}_{iL} \rvert - \mathbf{R}_L^2}\ \Omega$		0.828	0.628	0.502
4	$X_r = (X_{iL} - X_L)\ \Omega$		0.590	0.390	0.264
5	$\text{pf} = \cos(-\theta_{iL}) = R_L/\lvert \mathbf{Z}_{iL} \rvert$		0.195	0.254	0.313
6	pf angle: $\theta_{iL} = -\cos^{-1}(R_L/Z_{iL})$		−78.8°	−75.3°	−71.8°
7	$T_{LX} = \alpha^2 T_L$	in $\%T_{fl}$	32	54.1	81.9
		$\%T_L$	25	42.3	64

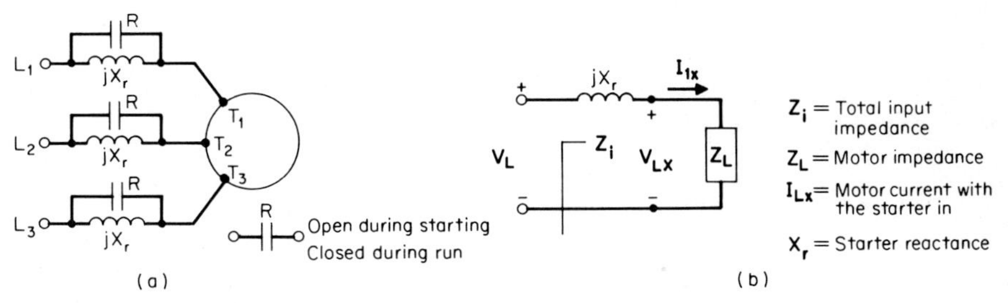

Fig. 10 Reactance starting. (*a*) Connection diagram. (*b*) Equivalent circuit.

2. *Select Tap*

Since the starting current and torque are the same in both reactance and resistance starting, select the same tap of 0.65.

3. *Compute Starter Inductance*

From the above table, for $\alpha = 0.65$, $X_r = 0.390\ \Omega$. Hence, $L_S = X_r/(2\pi f_1) = 0.39/(2\pi \times 60) = 103$ mH.

SERIES-PARALLEL STARTING

In this method of starting (Fig. 11), dual-voltage windings are connected for higher voltage on a lower supply voltage during starting. Then, the connection is switched over to the normal lower-voltage connection.

Calculate the starting line current I_{LS} and starting torque T_{LS} of a 230/460-V motor to be connected for 460-V service with 230 V applied.

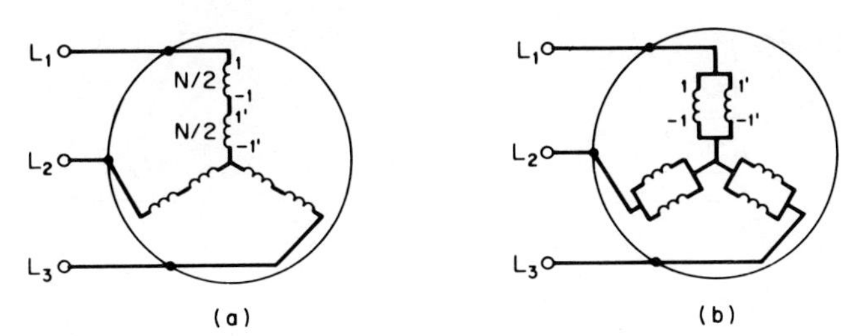

Fig. 11 Series-parallel starting. (*a*) Series connection for start. (*b*) Parallel connection for run.

Calculation Procedure:

1. *Compute Line Current*

$I_{LS} = (230/460)I_L = 0.5I_L$.

2. *Compute Starting Torque*

$T_{LS} = (230/460)^2 T_L = 0.25T_L$.

Related Calculations: Although it is occasionally employed for two-step starting, series-parallel starting is primarily used as the first step in a multistep increment starter because of its very low starting torque. Any 230/460-V motor can be connected for 230-V series-parallel starting.

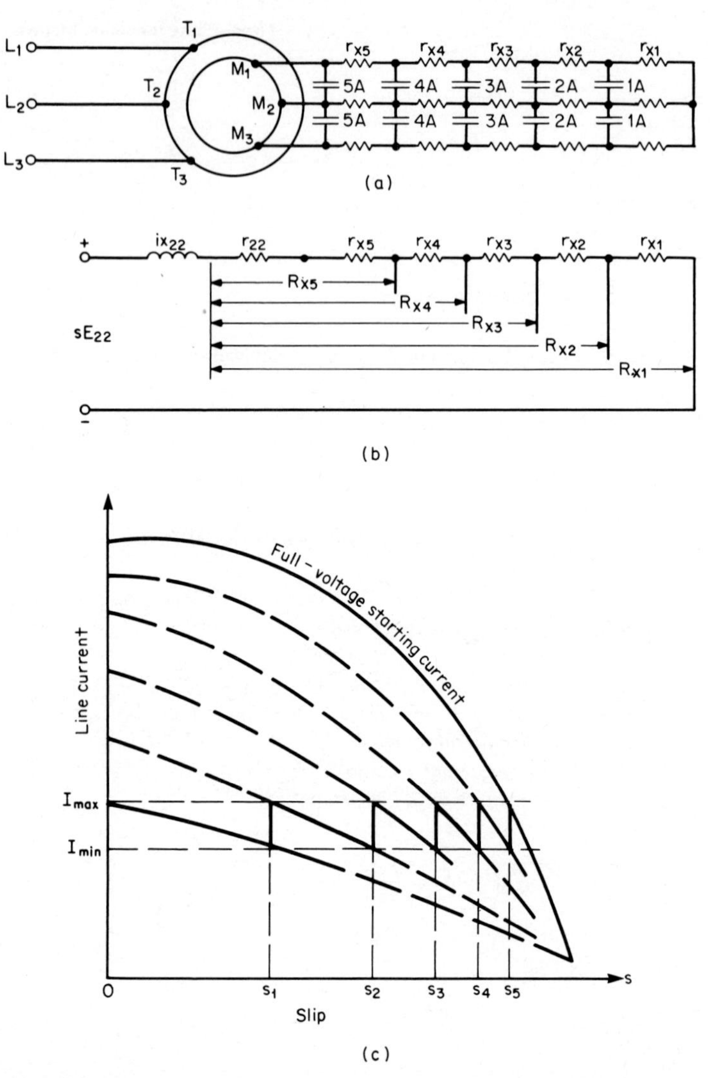

Fig. 12 Five-step wound-rotor resistance starter. (*a*) Connection diagram. (*b*) Rotor equivalent circuit. (*c*) Acceleration curve.

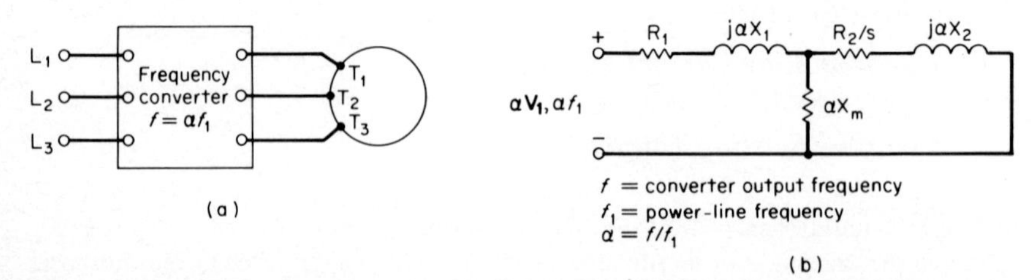

Fig. 13 Employing a frequency converter for speed control of an induction motor. (*a*) Circuit diagram. (*b*) Equivalent circuit.

FIVE-STEP STARTER

Calculate the resistance of a five-step starter for a wound-rotor motor (Fig. 12*a*) with maximum starting torque; the inrushes shall be equal. Also, find the minimum motor current I_{min}. Motor data: $R_2/s_{maxT} = 1.192\ \Omega$, $s_{maxT} = 0.0795$, and $I_{maxT} = 768$ A.

Calculation Procedure:

1. *Compute Resistance Values*

From Fig. 12*b*, $R_{x1} = R_2/s_{maxT}$ (for $I_{max} = I_{maxT}$) $= 1.192\ \Omega$ as given. Slip $s_1 = \sqrt[n]{s_{maxT}} = \sqrt[5]{0.0795} = 0.6026$. $R_{x2} = s_1R_{x1} = 0.6026 \times 1.192 = 0.7183\ \Omega$. Similarly, $R_{x3} = s_1R_{x2} = 0.4328\ \Omega$, $R_{x4} = s_1R_{x3} = 0.2608\ \Omega$, $R_{x5} = s_1R_{x4} = 0.1572\ \Omega$, $r_{x1} = R_{x1} - R_{x2} = 0.4737\ \Omega$, $r_{x2} = R_{x2} - R_{x3} = 0.2855\ \Omega$, $r_{x3} = R_{x3} - R_{x4} = 0.172\ \Omega$, $r_{x4} = R_{x4} - R_{x5} = 0.1036\ \Omega$, and $r_{x5} = R_{x5} - R_2 = 0.0625\ \Omega$. The acceleration curve is provided in Fig. 12*c*.

2. *Compute* I_{min}

$I_{min} = I_{maxT}s_1 = 760 \times 0.6026 = 458$ A.

Related Calculations: Wound-rotor motors are universally started on full voltage. Reduction in the starting current is achieved through the addition of several steps of balanced resistors in the rotor circuit. As the motor accelerates, the rotor resistances can be lowered steadily in steps until they are completely shorted out at the rated speed. Thus, the developed torque may be shaped to meet the requirements of the load within the limits on the current set by the power company. Reactors can be employed instead of resistors to reduce the starting current, though they are seldom used.

VARIABLE-FREQUENCY OPERATION

Calculate the maximum torque T_{max} and the corresponding slip s_{maxT}, of a 1000-hp, wye-connected, 2300-V, 16-pole, 60-Hz wound-rotor motor fed by a frequency converter (Fig. 13) at frequencies $f = 40$, 50, and 60 Hz. Motor data: $R_1 = 0.0721\ \Omega$, $X_1 = 0.605\ \Omega$, $R_2 = 0.0947\ \Omega$, and $X_2 = 0.605\ \Omega$. X_m (17.8 Ω) may be neglected.

Calculation Procedure:

1. *Determine Synchronous Speed*

The synchronous speed corresponding to the line frequency $f_l = 60$ Hz is $\omega_s = 2 \times 2 \times 60/16 = 47.12$ rad/s.

2. *Compute Slip*

The slip corresponding to the maximum torque is given by

$$s_{maxT} \cong R_2/\sqrt{R_1^2 + \alpha^2(X_1 + X_2}$$

for $\alpha = f/f_l$. For $f_l = 60$ Hz, we obtain

$$s_{maxT} = 0.0947/\sqrt{0.0721^2 + 1^2(0.605 + 0.605)^2} = 0.078.$$

Similarly, for $f_l = 50$ Hz, $s_{maxT} = 0.0937$ and for $f_l = 40$ Hz, $s_{maxT} = 0.117$.

3. *Compute Maximum Torque*

The maximum torque is given by

$$T_{max} \cong \tfrac{3}{2}(V_1^2/\omega_s)\,\{\alpha/[R_1 + \sqrt{R_1^2 + \alpha^2(X_1 + X_2)^2}]\}.$$

Substitution of values yields T_{max} = 43,649 N·m for f_l = 60 Hz, 43,207 N·m for f_l = 50 Hz, and 42,458 N·m for 40 Hz. The maximum torque changed very little for a 33 percent change in speed.

Related Calculations: The synchronous speed of an induction motor is directly related to the frequency of the power to the motor. Hence, continuous speed control over a wide range for both squirrel-cage and wound-rotor motors may be realized by controlling the frequency. In addition to speed control, lower-frequency starting of an induction motor has the advantages of lower starting current and higher starting power factor. To maintain the air-gap flux density fairly constant, the motor input voltage per unit frequency is kept constant. This results in the maximum torque being very nearly constant over the high end of the speed range.

To overcome the resistive drops, the input voltage is not lowered below a value corresponding to a specified low frequency. A variable-frequency supply may be obtained from (1) a variable-frequency alternator, (2) a wound-rotor induction motor acting as a frequency changer, or (3) a solid-state static frequency converter. Method (3) is becoming more popular because of its lower cost and size.

LINE-VOLTAGE CONTROL

Calculate the speed and load torque as a percentage of the rated torque for a 10-hp, three-phase, four-pole squirrel-cage motor driving a fan when the supply voltage is reduced to one-half the normal value. The fan torque varies as the square of its speed. The slip at rated load is 4 percent and the slip corresponding to maximum torque is 14 percent.

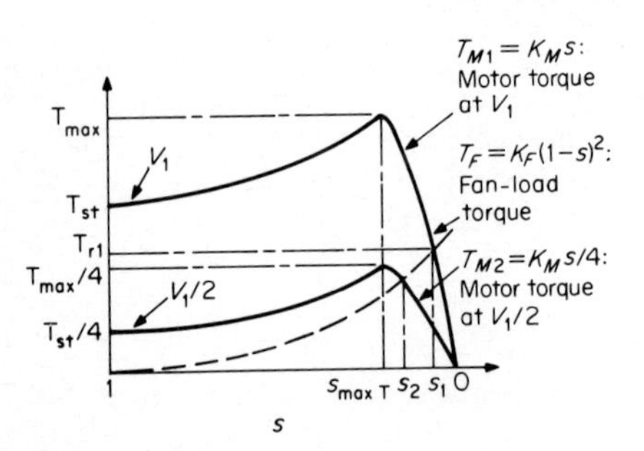

Fig. 14 Using line-voltage control to vary speed of an induction motor.

Calculation Procedure:

1. *Determine the Operating Slip*

At two different operating speeds n_1 and n_2 corresponding to two different supply voltages V_1 and $0.5V_1$, $T_{M1}/T_{M2} = s_1/(s_2/4)$ and $T_{F1}/T_{F2} = n_1^2/n_2^2 = (1 - s_1)^2/(1 - s_2)^2$. Equating the motor and load torques, we find $s_1/(s_2/4) = (1 - s_1)^2/(1 - s_2)^2$. Therefore, $(1 - s_2)^2/s_2 = (1 - s_1)^2/4s_1 = (1 - 0.04)^2/(4 \times 0.04) = 5.76$. Then, $s_2 = 0.13$. Because s_2 is less than $s_{maxT} = 0.14$, the operation is in the stable region of the torque-speed characteristic (Fig. 14).

2. *Compute the Operating Speed*

Operating speed is $n_2 = n_s(1 - s_2) = 120 \times 60 \times (1 - 0.13)/4 = 1566$ r/min.

3. Calculate Percentage Operating Torque

The load torque is $T_{F2} = T_{M2} = T_{r1}(s_2/4s_1) = T_{r1}(0.13/0.16) = 0.813\ T_{r1}$ (= 81.3 percent of rated torque).

Related Calculations: The electromagnetic torque developed by an induction motor is proportional to the square of the impressed stator voltage. Line-voltage control is used mainly for small cage-rotor motors driving fan-type loads which are a function of speed. Disadvantages of this method include:

1. Torque is reduced with a reduction in stator voltage.
2. Range of speed control is limited.
3. Operation at higher than rated voltage is restricted by magnetic saturation.

TABLE 4 Typical Alternating-Current Motor Starters*

		Typical range	
Motor type	*Starter type*	*Voltage*	*Horsepower*
Squirrel cage	Magnetic, full-voltage	110–550	1.5–600
	With fusible or nonfusible disconnect or circuit breaker	208–550	2–200
	Reversible	110–550	1.5–200
	Manual, full-voltage	110–550	1.5–7.5
	Manual, reduced-voltage, autotransformer	220–2500	5–150
	Magnetic, reduced-voltage, autotransformer	220–5000	5–1750
	Magnetic, reduced-voltage, resistor	220–550	5–600
Wound rotor	Magnetic, primary and secondary control	220–4500	5–1000
	Drums and resistors for secondary control	1000 max.	5–750
Synchronous	Reduced-voltage, magnetic	220–5000	25–3000
	Reduced-voltage, semimagnetic	220–2500	20–175
	Full-voltage, magnetic	220–5000	25–3000
High-capacity induction	Magnetic, full-voltage	2300–4600	Up to 2250
	Magnetic, reduced-voltage	2300–4600	Up to 2250
High-capacity synchronous	Magnetic, full-voltage	2300–4600	Up to 2500
	Magnetic, reduced-voltage	2300–4600	Up to 2500
High-capacity wound rotor	Magnetic primary and secondary	2300–4600	Up to 2250

*Based on Allis-Chalmers, General Electric, and Westinghouse units. This table is taken from *Standard Handbook of Engineering Calculations* by Hicks.

TABLE 5 Adjustable-Speed Drives*

Drive features	Drive types						
	Constant-voltage dc	*Adjustable-voltage dc motor-generator set*	*Adjustable-voltage rectifier*	*Eddy-current clutch*	*Wound-rotor ac, standard*	*Wound-rotor thyratron*	*Wound-rotor dc-motor set*
Power units required	Rectifier, dc motor	AC motor, dc generator, dc motor	Rectifier, reactor,[a] dc motor	AC motor, eddy-current clutch	AC motor,	AC motor thyratrons	AC motor, dc motor, rectifier
Normal speed range	4–1	8–1 c-t+[b] 4–1 c-hp[c]	8–1 c-t+ 4–1 c-hp[c]	34–1, 2 pole; 17–1, 4 pole	3–1	10–1[c]	3–1
Low speed for jogging	No[d]	Yes	Yes	Yes	Yes	Yes	Yes
Torque available	c-hp	c-t	c-t	c-t	c-t	c-t	c-t, c-hp
Speed regulation	10–15%	5% with regulator	5% with regulator	2% with regulator	Poor	±3%	5–7½%
Speed control	Field rheostat	Rheostats or pots	Rheostats or pots	Rheostats or pots	Steps, power contactors	Rheostats or pots	Rheostats or pots
Enclosures available	All	All	All	Open[e]	All	All	All
Braking: Regenerative	No	Yes	No	No	Yes	Yes	No
Dynamic	Yes	Yes	Yes	No[f]	Yes	Yes	Yes

Multiple operation	Yes	Yes	Yes	Yes	Yes	Yes	No
Parallel operation	Yes	Yes	Yes	Yes	No	Yes	Yes
Controlled acceleration, deceleration	Yes	Yes	Yes	Yes	No	Yes	No
Efficiency	80–85%	63–73%	70–80%	80–85%	80–85%	80–85%	80–85%
Top speed at maximum torque	83–87%	60–67%	60–70%	29%	29%	85–90%	73–78%
Rotor inertia[g]	100%[h]	100%	100%	75%	90%	90%	175%
Starting torque	200–300%	200–300%	200–300%	200–300%	200%	200–300%	200–300%
Number of comm., rings	1 comm.	2 comm.	1 comm.	None	1 set rings	1 set rings	1 comm., 1 set rings

[a]Used only in saturable-reactor designs.

[b]c-t—constant-torque; c-hp—constant horsepower.

[c]Units of 200 to 1 speed range are available.

[d]Low speed can be obtained using armature resistance.

[e]Totally enclosed units must be water- or oil-cooled.

[f]Eddy-current brake may be integral with unit.

[g]Based on standard dc motor.

[h]Normally is a larger dc motor since it has slower base speed.

*This table is taken from *Standard Handbook of Engineering Calculations* by Hicks.

TABLE 6 Direct-Current Motor Starters*

Type of starter	*Typical uses*
Across-the-line	Limited to motors of less than 2 hp
Reduced-voltage, manual-control (face-plate type)	Used for motors up to 50 hp where starting is infrequent
Reduced-voltage, multiple-switch	Motors of more than 50 hp
Reduced-voltage, drum-switch	Large motors; frequent starting and stopping
Reduced-voltage, magnetic-switch	Frequent starting and stopping; large motors

*This table is taken from *Standard Handbook of Engineering Calculations* by Hicks.

SELECTING ELECTRIC-MOTOR STARTING AND SPEED CONTROLS*

Choose a suitable starter and speed control for a 500-hp wound-rotor ac motor that must have a speed range of 2 to 1 with a capability for low-speed jogging. The motor is to operate at about 1800 r/min with current supplied at 4160 V, 60 Hz. An enclosed starter and controller is desirable from the standpoint of protection. What is the actual motor speed if the motor has four poles and a slip of 3 percent?

Calculation Procedure:

1. Select the Type of Starter to Use

Table 4 shows that a magnetic starter is suitable for wound-rotor motors in the 220 to 4500-V and 5 to 1000-hp range. Since the motor is in this range of voltage and horsepower, a magnetic starter will probably be suitable. Also, the magnetic starter is available in an enclosed cabinet, making it suitable for this installation.

Table 5 (page 5-19) shows that a motor starting torque of approximately 200 percent of the full-load motor torque and current are obtained on the first point of acceleration.

2. Compute the Full-Load Speed of the Motor

Use the relation $n = [(100 - s)/100]120f/p$, where n = motor full-load speed, r/min; s = slip, percent; f = frequency of supply current, Hz; p = number of poles in the motor. For this motor, $n = [(100 - 3)/100]120(60)/4 = 1750$ r/min.

3. Choose the Type of Speed Control to Use

Table 5 summarizes the various types of adjustable-speed drives available today. This listing shows that power-operated contactors used with wound-rotor motors will give a 3 to 1 speed range with low-speed jogging. Since a 2 to 1 speed range is required, the proposed controller is suitable because it gives a wider speed range than needed.

Note from Table 5 that if a wider speed range were required, a thyratron control could produce a range up to 10 to 1 on a wound-rotor motor. Also, a wound-rotor

*Adapted from Hicks, *Standard Handbook of Engineering Calculations,* McGraw-Hill, with permission of McGraw-Hill, Inc.

direct-current motor set might be used too. In such an arrangement, an ac and dc motor are combined on the same shaft. The rotor current is converted to dc by external silicon rectifiers and fed back to the dc armature through the commutator.

Related Calculations: Use Tables 6 and 7 as a guide to the selection of starters and controls for alternating-current motors serving industrial, commercial, marine, portable, and residential applications.

To choose a direct-current motor starter, use Table 6 as a guide.

Speed controls for dc motors can be chosen using Table 7 as a guide. DC motors are finding increasing use in industry. They are also popular in marine service.

TABLE 7 Direct-Current Motor-Speed Controls*

Type of motor	*Speed characteristic*	*Type of control*
Series-wound	Varying; wide speed regulation	Armature shunt and series resistors
Shunt-wound	Constant at selected speed	Armature shunt and series resistors; field weakening; variable armature voltage
Compound-wound	Regulation is about 25%	Armature shunt and series resistors; field weakening; variable armature voltage

*This table is taken from *Standard Handbook of Engineering Calculations* by Hicks.

MOTOR SELECTION FOR A CONSTANT LOAD

Select a motor driving a load requiring a torque of 55 N·m at 1764 r/min.

Calculation Procedure:

1. Calculate Horsepower

Use $hp = 2\pi(n/60)T/746$. With values substituted, $hp = (2\pi \times 1764/60 \times 55)/746 = 13.6$ hp.

2. Select Motor

A NEMA design B 15-hp motor for continuous duty is selected.

Related Calculations: The horsepower rating and NEMA design class of an induction motor depend on the load-torque speed characteristic, duty cycle, inertia, temperature rise and heat dissipation, environmental conditions, and auxiliary drives. In the vast majority of applications, the selection process is relatively simple. But in a few applications, the selection process may involve criteria such as high starting torque, low starting current, intermittent duty, torque pulsations, high load inertia, or a duty cycle with frequent starting and stopping.

Selection of the proper horsepower rating will minimize initial and maintenance costs and improve operating efficiency and life expectancy. An undersized motor may result in overloads and a consequent reduction in life expectancy. Too large a motor horsepower rating will cause lower efficiency and power factor, greater space requirements, and higher initial cost.

TABLE 8 Summary of Motor Characteristics and Applications*

Speed regulation	Speed control	Starting torque	Breakdown torque	Applications
		Polyphase motors		
General-purpose squirrel-cage (design B):				
Drops about 3% for large to 5% for small sizes	None, except multispeed types, designed for two to four fixed speeds	100% for large, 275% for 1-hp, four-pole unit	200% of full load	Constant-speed service where starting torque is not excessive. Fans, blowers, rotary compressors, and centrifugal pumps
High-torque squirrel-cage (design C):				
Drops about 3% for large to 6% for small sizes	None, except multispeed types, designed for two to four fixed speeds	250% of full load for high-speed to 200% for low-speed designs	200% of full load	Constant-speed service where fairly high starting torque is required infrequently with starting current about 550% of full load. Reciprocating pumps and compressors, crushers, etc.
High-slip squirrel-cage (design D):				
Drops about 10 to 15% from no load to full load	None, except multispeed types, designed for two to four fixed speeds	225 to 300% of full load, depending on speed with rotor resistance	200%. Will usually not stall until loaded to maximum torque, which occurs at standstill	Constant speed starting torque, if starting is not too frequent, and for high-peak loads with or without flywheels. Punch presses, shears, elevators, etc.
Low-torque squirrel-cage (design F):				
Drops about 3% for large to 5% for small sizes	None, except multispeed types, designed for two to four fixed speeds	50% of full load for high-speed to 90% for low-speed designs	135 to 170% of full load	Constant-speed service where starting duty is light. Fans, blowers, centrifugal pumps and similar loads

Wound-rotor:				
With rotor rings short-circuited drops about 3% for large to 5% for small sizes	Speed can be reduced to 50% by rotor resistance. Speed varies inversely as load	Up to 300% depending on external resistance in rotor circuit and how distributed	300% when rotor slip rings are short-circuited	Where high starting torque with low starting current or where limited speed control is required. Fans, centrifugal and plunger pumps, compressors, conveyors, hoists, cranes, etc.
Synchronous:				
Constant	None, except special motors designed for two fixed speeds	40% for slow- to 160% for medium-speed 80% pf. Specials develop higher	Unity-pf motors, 170%; 80% pf motors, 225%. Specials up to 300%	For constant-speed service, direct connection to slow-speed machines and where power-factor correction is required
Series:				
Varies inversely as load. Races on light loads and full voltage	Zero to maximum depending on control and load	High. Varies as square of voltage. Limited by commutation, heating, capacity	High. Limited by commutation, heating and line capacity	Where high starting torque is required and speed can be regulated. Traction, bridges, hoists, gates, car dumpers, car retarders
Shunt:				
Drops 3 to 5% from no load to full load	Any desired range depending on design type and type of system	Good. With constant field, varies directly as voltage applied to armature	High. Limited by commutation, heating, and line capacity	Where constant or adjustable speed is required and starting conditions are not severe. Fans, blowers, centrifugal pumps, conveyors, wood- and metal-working, machines, and elevators
Compound:				
Drops 7 to 20% from no load to full load depending on amount of compounding	Any desired range, depending on design and type of control.	Higher than for shunt, depending on amount of compounding	High. Limited by commutation, heating, and line capacity	Where high starting torque and fairly constant speed is required. Plunger pumps, punch presses, shears, bending rolls, geared elevators, conveyors, hoists

TABLE 8 Summary of Motor Characteristics and Applications* (continued)

390

DC and single-phase motors				
Speed regulation	***Speed control***	***Starting torque***	***Breakdown torque***	***Applications***
Split-phase:				
Drops about 10% from no load to full load	None	75% for large to 175% for small sizes	150% for large to 200% for small sizes	Constant-speed service where starting is easy. Small fans, centrifugal pumps and light-running machines, where polyphase is not available
Capacitor:				
Drops about 5% for large to 10% for small sizes	None	150 to 350% of full load depending on design and size	150% for large to 200% for small sizes	Constant-speed service for any starting duty and quiet operation, where polyphase current cannot be used
Commutator:				
Drops about 5% for large to 10% for small sizes	Repulsion induction, none. Brush-shifting types, 4:1 at full load.	250% for large to 350% for small sizes	150% for large to 250% for small sizes	Constant-speed service for any starting duty where speed control is required and polyphase current cannot be used

*This table is from *Standard Handbook of Engineering Calculations* by Hicks.

MOTOR SELECTION FOR A VARIABLE LOAD

Select a motor for a load cycle lasting 3 min and having the following duty cycle: 1 hp for 35 s, no load for 50 s, and 4.4 hp for 95 s.

Calculation Procedure:

1. Compute Horsepower

Use:

$$hp_{rms} = \sqrt{(\Sigma\, hp^2 \times \text{time})/(\text{running time} + \text{standstill time}/K)}.$$

where $K = 4$ for an enclosed motor and $K = 3$ for an open motor. Assume $K = 3$. Substituting the given values, we find:

$$hp_{rms} = \sqrt{(1^2 \times 35 + 0 + 4.4^2 \times 95)/(35 + 95 + 50/3)} = 3.57 \text{ hp}.$$

2. Select Motor

A 3-hp design B open-type motor is selected.

Related Calculations: A summary of various motor characteristics and applications is provided in Table 8.

Section 6 SINGLE-PHASE MOTORS

Lawrence J. Hollander, P.E.
Associate Dean, Cooper Union

REFERENCES: Slemon and Straughen—*Electric Machines,* Addison-Wesley; Matsch—*Electromagnetic and Electromechanical Machines,* Harper and Row; Stein and Hunt—*Electric Power System Components,* Van Nostrand Reinhold; Nasar and Unnewehr—*Electromechanics and Electric Machines,* Wiley; Fitzgerald and Kingsley—*Electric Machinery,* McGraw-Hill; Kosow—*Electric Machinery and Transformers,* Prentice-Hall; Siskind—*Electrical Machines: Direct and Alternating Currents,* McGraw-Hill; McPherson—*An Introduction to Electrical Machines and Transformers,* Wiley; Fitzgerald and Higginbotham—*Basic Electrical Engineering,* McGraw-Hill; Fitzgerald, Higginbotham, and Grabel—*Basic Electrical Engineering,* McGraw-Hill; Seely—*Electromechanical Energy Conversion,* McGraw-Hill; Smith—*Circuits, Devices, and Systems,* Wiley; Adkins—*The General Theory of Electrical Machines,* Chapman and Hall; Puchstein, Lloyd and Conrad—*Alternating-Current Machines,* Wiley; Langsdorf—*Theory of Alternating Current Machines,* McGraw-Hill; White and Woodson—*Electromechanical Energy Conversion,* Wiley.

EQUIVALENT CIRCUIT OF A SINGLE-PHASE INDUCTION MOTOR DETERMINED FROM NO-LOAD AND LOCKED-ROTOR TESTS

A single-phase induction motor has the following data: 1 hp, two pole, 240 V, 60 Hz, stator-winding resistance $R_s = 1.6\ \Omega$. The no-load test results are $V_{NL} = 240$ V, $I_{NL} = 3.8$ A, $P_{NL} = 190$ W; the locked-rotor test results are $V_{LR} = 88$ V, $I_{LR} = 9.5$ A, $P_{LR} = 418$ W. Establish the equivalent circuit of the motor.

Calculation Procedure:

1. *Calculate the Magnetizing Reactance X_ϕ*

The magnetizing reactance essentially is equal to the no-load reactance. From the no-load test, $X_\phi = V_{NL}/I_{NL} = 240\ \text{V}/3.8\ \text{A} = 63.2\ \Omega$. One-half of this (31.6 Ω) is assigned to the equivalent circuit portion representing the forward-rotating mmf waves and the other half to the portion representing the backward-rotating mmf waves. See Fig. 1. The no-load power, 190 W, represents the rotational losses.

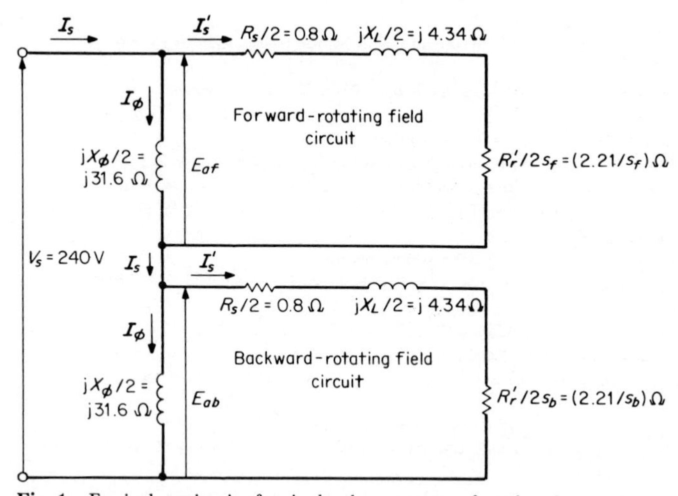

Fig. 1 Equivalent circuit of a single-phase motor, referred to the stator.

2. *Calculate the Impedance Values from Locked-Rotor Test*

For the locked-rotor test in induction machines, the magnetizing branch of the equivalent circuit is considered to be an open circuit, because the ratio of impedances X_L/X_ϕ is very small. Considering V_{LR} as reference (phase angle = 0), use the equation $P_{LR} = V_{LR}I_{LR}\cos\theta_{LR}$. Thus, $\theta_{LR} = \cos^{-1}(P_{LR}/V_{LR}I_{LR}) = \cos^{-1}[418\ \text{V}/(88\ \text{V})(9.5\ \text{A})] = \cos^{-1} 0.5 = 60°$. The effective stator current is $\mathbf{I}'_s = \mathbf{I}_{LR} - \mathbf{I}_\phi = \mathbf{I}_{LR} - \mathbf{V}_{LR}/jX_\phi = 9.5\angle{-60°} - 88\ \text{V}/j63.2\Omega = 4.75 - j8.23 + j1.39 = 4.75 - j6.84 = 8.33\angle{-55.2°}$ A. The rotor impedance referred to the stator is $\mathbf{V}_{LR}/\mathbf{I}'_s = 88\angle{0°}\ \text{V}/8.33\angle{-55.2°}\ \text{A} = 10.56\angle{55.2°}\ \Omega$.

Also, from the locked-rotor test, $P_{LR} = I_{LR}^2(R_s + R'_r)$; $R_s + R'_r = P_{LR}/I_s'^2 = 418\ \text{W}/(8.33\ \text{A})^2 = 6.02\ \Omega$. $R'_r = (R_s + R'_r) - R_s = 6.02 - 1.6 = 4.42\ \Omega$. Finally, leakage reactance is $\mathbf{X_L} = \mathbf{Z'_r} - (R_s + \mathbf{R'_r}) = 10.56\underline{/+55.2^\circ} - 6.02 = 6.02 + j8.67 - 6.02 = j8.67\ \Omega$.

3. Draw the Equivalent Circuit

See Fig. 1, wherein $R_s/2 = 1.6/2 = 0.8\ \Omega$, $X_L/2 = 8.67/2 = 4.34\ \Omega$, and $R'_r/2 = 4.42/2 = 2.21\ \Omega$. Also:

R_s = resistance of stator winding, Ω
R'_r = resistance of rotor winding referred to the stator, Ω
X_ϕ = magnetizing reactance, Ω
X_L = leakage reactance, Ω
I_ϕ = magnetizing current, A
I_s = actual stator current, A
I'_s = effective stator current $(I_s - I_\phi)$, A
I_{LR} = locked-rotor current, A
V_{LR} = locked-rotor voltage, V
s_f = forward slip
s_b = backward slip $= 2 - s_f$
E_{af} = counter-emf developed by forward rotating fields of stator and rotor mmf's, V
E_{ab} = counter-emf developed by backward-rotating fields of stator and rotor mmf's, V
V_s = terminal voltage applied to the stator, V
Z'_r = impedance of rotor circuit referred to the stator, Ω

Related Calculations: The open-circuit (no-load) and the short-circuit (locked-rotor) tests are similar to those for a transformer, or for a polyphase induction motor. The data are used in a similar fashion and the equivalent circuits have many similarities. In a single-phase machine the single waveform is divided into two half-amplitude rotating fields: a forward-rotating wave and a backward-rotating wave. This device leads to the division of the equivalent circuit into a forward-designated section and a backward-designated section. The use of this equivalent circuit is demonstrated in subsequent problems.

TORQUE AND EFFICIENCY CALCULATIONS FOR SINGLE-PHASE INDUCTION MOTOR

For the 1-hp single-phase induction motor in the previous problem, calculate the shaft torque and efficiency when the motor is operating at a speed of 3470 r/min.

Calculation Procedure:

1. Calculate the Forward and Backward Slips

The synchronous speed in r/min for a two-pole, 60-Hz machine is calculated from the equation $n_{sync} = 120f/p$ where f = frequency in Hz and p = number of poles; $n_{sync} = (120)(60\text{ Hz})/2\text{ poles} = 3600$ r/min. To calculate the forward slip use the equation $s_f = (n_{sync} - n_{actual})/n_{sync} = (3600\text{ r/min} - 3470\text{ r/min})/(3600\text{ r/min}) = 0.036$. The backward slip is $s_b = 2 - 0.036 = 1.964$.

2. *Calculate the Total Impedance of the Forward Equivalent Circuit, $\mathbf{Z_f}$*

The total impedance of the forward equivalent circuit is $\mathbf{Z_f} = R_s/2 + R'_r/2s_f + jX_L/2$ in parallel with $jX_\phi/2$. Thus, $R_s/2 + R'_r/2s_f + jX_L/2 = 0.8 + 2.21/0.036 + j4.34 = 62.2 + j4.34 = 62.4\angle 3.99°\ \Omega$. $\mathbf{Z_f} = (62.4\angle 3.99°)(31.6\angle 90°)/(62.2 + j4.34 + j31.6) = 27.45\angle 63.97° = 12.05 + j24.67\ \Omega$.

3. *Calculate the Total Impedance of the Backward Equivalent Circuit, $\mathbf{Z_b}$*

The total impedance of the backward equivalent circuit is $\mathbf{Z_b} = R_s/2 + R'_r/2s_b + jX_L/2$ in parallel with $jX_\phi/2$. Thus, $R_s/2 + R'_r/2s_b + jX_L/2 = 0.8 + 2.21/1.1964 + j4.34 = 1.93 + j4.34 = 4.75\angle 66.03°\ \Omega$. $\mathbf{Z_b} = (4.75\angle 66.03°)(31.6\angle 90°)/(1.93 + j4.34 + j31.6) = 4.17\angle 69.1° = 1.49 + j3.89\ \Omega$.

4. *Calculate the Total Circuit Impedance $\mathbf{Z}$*

The total circuit impedance is $\mathbf{Z} = \mathbf{Z_f} + \mathbf{Z_b} = (12.05 + j24.67) + (1.49 + j3.89) = 13.54 + j28.56 = 31.61\angle 64.63°\ \Omega$.

5. *Calculate the Power Factor pf and the Source Current $\mathbf{I_s}$*

The power factor pf $= \cos 64.63° = 0.43$. The source current (stator current) is $\mathbf{V_s}/\mathbf{Z} = 240\angle 0°\ \text{V}/31.61\angle 64.63°\ \Omega = 7.59\angle -64.63°\ \text{A}$.

6. *Calculate the Forward and Backward Counter-emf's*

The forward and backward components of the counter-emf's are each pro rated on the source voltage, according to the ratio of the forward and backward equivalent impedances $\mathbf{Z_f}$ and $\mathbf{Z_b}$ to the combined impedance $\mathbf{Z}$. Thus, $\mathbf{E_{af}} = \mathbf{V_s}(\mathbf{Z_f}/\mathbf{Z}) = (240\angle 0°\ \text{V})(27.45\angle 63.97°/31.61\angle 64.63°) = 208.4\angle -0.66°\ \text{V}$. Similarly, $\mathbf{E_{ab}} = \mathbf{V_s}(\mathbf{Z_b}/\mathbf{Z}) = (240\angle 0°\ \text{V})(4.17\angle 69.1°/31.61\angle 64.63°) = 31.67\angle 4.47°\ \text{V}$.

7. *Calculate the Forward- and Backward-Component Currents $\mathbf{I'_s}$*

The forward-component current is $\mathbf{I'_{sf}} = \mathbf{E_{af}}/\mathbf{Z'_{rf}} = 208.4\angle -0.66°\ \text{V}/62.4\angle 3.99°\ \Omega = 3.34\angle -63.06°\ \text{A}$. The backward-component current is $\mathbf{I'_{sb}} = \mathbf{E_{ab}}/\mathbf{Z'_{rb}} = 31.67\angle 4.47°\ \text{V}/4.75\angle 66.03°\ \Omega = 6.67\angle -61.56°\ \text{A}$.

8. *Calculate the Internally Developed Torque T_{int}*

The internally developed torque is T_{int} = forward torque − backward torque = $(30/\pi)(1/n_{sync})[(I'_{sf})^2R'_r/2s_f) - (I'_{sb})^2(R'_r/2s_b)] = (30/\pi)(1/3600\ \text{r/min})[3.34^2(2.21/0.036) - (6.67)^2(2.21/1.964)] = 1.684\ \text{N}\cdot\text{m}$.

9. *Calculate the Lost Torque in Rotational Losses, T_{rot}*

The lost torque in rotational losses is $T_{rot} = P_{NL}/\omega_m$, where ω_m = rotor speed in rad/s $= (3470\ \text{r/min})(\pi/30) = 363.4$ rad/s. $T_{rot} = 190\ \text{W}/(363.4\ \text{rad/s}) = 0.522\ \text{N}\cdot\text{m}$.

10. *Calculate the Shaft Torque T_{shaft}*

The shaft torque is $T_{shaft} = T_{int} - T_{rot} = 1.684\ \text{N}\cdot\text{m} - 0.522\ \text{N}\cdot\text{m} = 1.162\ \text{N}\cdot\text{m}$.

11. *Calculate Power Input and Power Output*

To calculate the power input, use the equation $P_{in} = V_sI_s\cos\theta = (240\ \text{V})(7.59\ \text{A})(0.43) = 783.3$ W. The power output is calculated from the equation $P_{out} = T_{shaft}(n_{actual}\ \text{r/min})(\pi/30) = (1.162\ \text{N}\cdot\text{m})(3470\ \text{r/min})(\pi/30) = 422.2$ W.

12. Calculate the Efficiency η

Use the equation for efficiency: η = (output × 100 percent)/input = (422.2 W)(100 percent)/783.3 W = 53.9 percent.

Related Calculations: It should be observed from Step 8, where T_{int} = forward torque − backward torque, that if the machine is at standstill both s_f and s_b = 1. By study of the equivalent circuit it is seen that $R_r'/2s_f = R_r'/2s_b$, which leads to the conclusion that the net torque is 0. In subsequent problems, consideration will be given to methods used for starting, that is, the creation of a net starting torque.

DETERMINATION OF INPUT CONDITIONS AND INTERNALLY DEVELOPED POWER FROM THE EQUIVALENT CIRCUIT FOR SINGLE-PHASE INDUCTION MOTORS

A four-pole, 220-V, 60-Hz, ¼-hp, single-phase induction motor is operating at 13 percent slip. The equivalent circuit is shown in Fig. 2 (different from the previous two problems in that the stator resistance and leakage reactance are separated from the forward and backward circuits). Find the input current, power factor, and the internally developed power.

Calculation Procedure:

1. Calculate the Forward- and Backward-Circuit Impedances $\mathbf{Z_f}$ *and* $\mathbf{Z_b}$

Use the equation $\mathbf{Z_f} = (R_r'/2s_f + jX_r'/2)(jX_\phi/2)/(R_r'/2s_f + jX_r'/2 + jX_\phi/2) = (185 + j5.0)(j140)/(185 + j5.0 + j140) = 110.2\angle 53.6^\circ = 65.5 + j88.6\ \Omega$. Similarly, $\mathbf{Z_b} = (13 + j5.0)(j140)/(13 + j5.0 + j140) = 13.4\angle 26.1^\circ = 12.0 + j5.9\ \Omega$

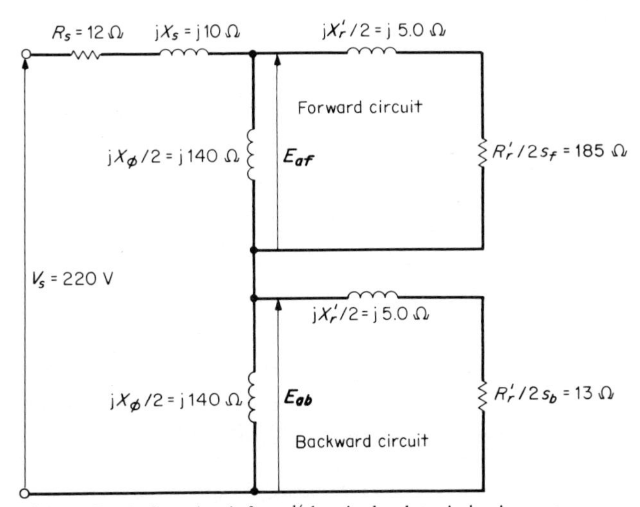

Fig. 2 Equivalent circuit for a ¼-hp single-phase induction motor.

2. *Calculate the Total Impedance of the Equivalent Circuit* Z

The total impedance of the equivalent circuit is $\mathbf{Z} = \mathbf{Z}_s + \mathbf{Z}_f + \mathbf{Z}_b = 12 + j10 + 65.5 + j88.6 + 12.0 + j5.9 = 89.5 + j104.5 = 137.6\angle 49.4^\circ\ \Omega$.

3. *Calculate the Power Factor* pf *and Input Current* $\mathbf{I}_s$

The power factor is $\text{pf} = \cos 49.4^\circ = 0.65$. The input current is $\mathbf{I}_s = \mathbf{V}_s/\mathbf{Z} = 220\angle 0^\circ\ \text{V}/137.6\angle 49.4^\circ\ \Omega = 1.6\angle -49.4^\circ$ A.

4. *Calculate the Forward and Backward Counter-emf's*

As in the previous problem, $\mathbf{E}_{af} = V_s(Z_f/Z) = (220\angle 0^\circ\ \text{V})(110.2\angle 53.6^\circ/137.6\angle 49.4^\circ) = 176.2\angle 4.2^\circ$ V. $\mathbf{E}_{ab} = \mathbf{V}_s(\mathbf{Z}_b/\mathbf{Z}) = (220\angle 0^\circ\ \text{V})(13.4\angle 26.1^\circ/137.6\angle 49.4^\circ) = 21.4\angle -23.3^\circ$ V.

5. *Calculate the Forward- and Backward-Component Currents* $\mathbf{I}'_s$

The forward-component current is $\mathbf{I}'_{sf} = \mathbf{E}_{af}/\mathbf{Z}'_{rf} = 176.2\angle 4.2^\circ/185\angle 1.55^\circ = 0.95\angle 2.65^\circ$ A. The backward-component current is $\mathbf{I}'_{sb} = \mathbf{E}_{ab}/\mathbf{Z}'_{rb} = 21.4\angle -23.2^\circ/13.9\angle 21^\circ = 1.54\angle -44.3^\circ$ A.

6. *Calculate the Internally Developed Power* P_{int}

Use the equation $P_{int} = [(R'_r/2s_f)I'^2_{sf} - R'_r/2s_b)I'^2_{sb}](1 - s) = [(185)(0.95)^2 - (13)(1.54)^2](1 - 0.013) = 118.4$ W; rotational losses must be subtracted from this in order to determine the power output P_{out}. The power input is $P_{in} = V_s I_s \cos\theta = (220\ \text{V})(1.6\ \text{A})(0.65) = 228.8$ W.

Related Calculations: This problem, compared with the previous problem, illustrates a slightly different version of the equivalent circuit for a single-phase induction motor. It should be noted that in the backward portion of the circuit the magnetizing reactance $X_\phi/2$ is so much larger than the values of X and R in the circuit that it may be neglected. Likewise, the forward portion of the circuit $X'_r/2$ becomes insignificant. An approximate equivalent circuit is used in the next problem. These variations of equivalent circuits give similar results depending, of course, on what precise information is being sought.

DETERMINATION OF INPUT CONDITIONS AND INTERNALLY DEVELOPED POWER FROM THE APPROXIMATE EQUIVALENT CIRCUIT FOR SINGLE-PHASE INDUCTION MOTORS

An approximate equivalent circuit is given in Fig. 3, for the ¼-hp single-phase induction motor having the following data: 60 Hz, $V_s = 220$ V, $s_f = 13$ percent. Find the input current, power factor, and the internally developed power.

Calculation Procedure:

1. *Calculate the Forward- and Backward-Circuit Impedances* $\mathbf{Z}_f$ *and* $\mathbf{Z}_b$

Notice that in the approximate equivalent circuit, the resistance element in the backward portion is $R'_r/(2)(2)$ rather than $R'_r/(2)(2 - s_f)$, where $(2 - s_f) = s_b$. $R'_r = 2s_f(185\ \Omega) = (2)(0.13)(185) = 48.1\ \Omega$. Thus, $R'_r/(2)(2) = 48.1\ \Omega/4 = 12.0\ \Omega$.

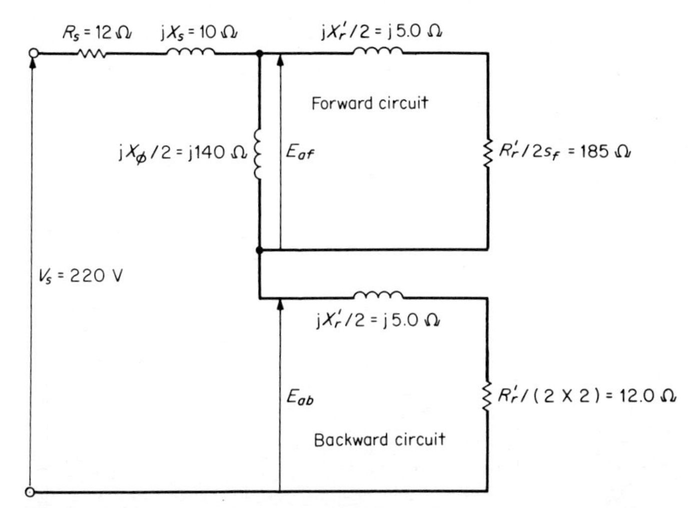

Fig. 3 Approximate equivalent circuit for a ¼-hp single-phase induction motor.

This differs from the 13.0 Ω in the previous problem. Because of the magnitude of the difference between $jX_\phi/2$ and $jX_r'/2$, the latter will be neglected in calculating $\mathbf{Z_f}$. Thus, $\mathbf{Z_f} = (jX_\phi/2)(R_r'/2s_f)/(jX_\phi/2 + R_r'/2s_f) = (140\angle 90°)(185\angle 0°)/(j140 + 185\angle 0°) = 111.6\angle 52.9° = 67.3 + j89.0$ Ω. The calculation for $\mathbf{Z_b}$ simply is taken directly from the approximate equivalent circuit: $\mathbf{Z_b} = R_r'/4 + j(X_r'/2) = 12.0 + j5.0 = 13.0\angle 22.6°$ Ω.

2. Calculate the Total Impedance of the Equivalent Circuit **Z**

The total impedance of the equivalent circuit is $\mathbf{Z} = \mathbf{Z_s} + \mathbf{Z_f} + \mathbf{Z_b} = 12 + j10 + 67.3 + j89.0 + 12 + j5.0 = 91.3 + j104 = 138.4\angle 48.7°$ Ω.

3. Calculate the Power Factor pf *and Input Current* $\mathbf{I_s}$

The power factor is pf = cos 48.7° = 0.66. The input current is $\mathbf{I_s} = \mathbf{V_s}/\mathbf{Z} = 220\angle 0°$ V$/138.4\angle 48.7°$ Ω $= 1.59\angle -48.7°$ A.

4. Calculate the Forward and Backward Counter-emf's

As in the previous two problems, $\mathbf{E_{af}} = \mathbf{V_s}(\mathbf{Z_f}/\mathbf{Z}) = (220\angle 0°$ V$)(111.6\angle 52.9°/138.4\angle 48.7°) = 177.4\angle 4.2°$ V. $\mathbf{E_{ab}} = \mathbf{V_s}(\mathbf{Z_b}/\mathbf{Z}) = (220\angle 0°$ V$)(13.0\angle 22.6°/138.4\angle 48.7°) = 20.7\angle -26.1°$ V.

5. Calculate the Forward- and Backward-Component Currents $\mathbf{I_s'}$

The forward-component current is $\mathbf{I'_{sf}} = \mathbf{E_{af}}/\mathbf{Z'_{rf}} = 177.4\angle 4.2°$ V$/185\angle 0°$ Ω $= 0.96\angle 4.2°$ A. (*Note:* $jX_r'/2$ was neglected.) The backward-component current is $\mathbf{I'_{sb}} = \mathbf{E_{ab}}/\mathbf{Z'_{rb}} = 207\angle -26.1°$ V$/13.0\angle 22.6°$ Ω $= 1.59\angle -48.7°$ A.

6. Calculate the Internally Developed Power P_{int}

Use the equation $P_{int} = [(R_r'/2s_f)I_{sf}'^2 - (R_r'/4)I_{sb}'^2](1 - s) = [(185\ \Omega)(0.96\ \text{A})^2 - (13\ \Omega)(1.59\ \text{A})^2](1 - 0.13) = 119.7$ W. As previously, rotational losses must be subtracted from this value in order to determine the power output P_{out}. The power input is $P_{in} = V_s I_s \cos\theta = (220\ \text{V})(1.59\ \text{A}) \cos 48.7° = 230.9$ W.

7. *Compare the Results of the General and Approximate Equivalent Circuits*

The comparison follows. (The general equivalent circuit was used in the previous problem.)

Component	General equivalent circuit	Approximate equivalent circuit
$\mathbf{Z}_f$, Ω	$110.2\angle 53.6°$	$111.6\angle 52.9$
$\mathbf{I}_s$, A	$1.6\angle -49.4°$	$1.59\angle -48.7°$
$\mathbf{Z}_b$, Ω	$13.4\angle 26.1°$	$13.0\angle 22.6°$
$\mathbf{Z}$, Ω	$137.6\angle 49.4°$	$138.4\angle 48.7°$
$\mathbf{E}_{af}$, V	$176.2\angle 4.2°$	$177.4\angle 4.2°$
$\mathbf{E}_{ab}$, V	$21.4\angle -23.3°$	$20.7\angle -26.1°$
$\mathbf{I}'_{sf}$, A	$0.95\angle 2.65°$	$0.96\angle 4.2°$
$\mathbf{I}'_{sb}$, A	$1.54\angle -44.3°$	$1.59\angle -48.7°$
P_{int}, W	118.4	119.7
P_{in}, W	228.8	230.9

Related Calculations: A reasonable amount of calculations are simplified by using the approximate equivalent circuit for single-phase machine analysis, even in this example for a slip of 13 percent. Lower slips will result in smaller differences between the two methods of analysis.

LOSS AND EFFICIENCY CALCULATIONS FROM THE EQUIVALENT CIRCUIT OF THE SINGLE-PHASE INDUCTION MOTOR

For the ¼-hp, single-phase induction motor of the previous two problems, calculate losses and efficiency from the general equivalent circuit. Assume that the core loss = 30 W and the friction and windage loss = 15 W.

Calculation Procedure:

1. *Calculate the Stator-Copper Loss*

The stator-copper loss is calculated from the general equation for heat loss $P = I^2R$. In this case, stator-copper loss is $P_{s(loss)} = I_s^2 R_s = (1.6\ A)^2(12\ \Omega) = 30.7$ W.

2. *Calculate Rotor-Copper Loss*

The rotor-copper loss has two components: loss resulting from the forward-circuit current $P_{f(loss)}$ and loss resulting from the backward-circuit current $P_{b(loss)}$. Thus, $P_{f(loss)} = s_f I'^2_{sf}(R'_r/2s_f) = (0.13)(0.95\ A)^2(185\ \Omega) = 21.7$ W, and $P_{b(loss)} = s_b I'^2_{sb}(R'_r/2s_b)$, where $s_b = 2 - s_f$. $P_{b(loss)} = (2 - 0.13)(1.54\ A)^2(13\ \Omega) = 57.7$ W.

3. *Calculate the Total Losses*

Stator-copper loss	30.7 W
Rotor-copper loss (forward)	21.7 W

Rotor-copper loss (backward)	57.7 W
Core loss (given)	30.0 W
Friction and windage loss (given)	15.0 W
Total losses	155.1 W

4. Calculate the Efficiency

Use the equation η = [(input − losses)(100 percent)]/input = [(228.8 W − 155.1 W)(100 percent)]/228.8 W = 32 percent. This is a low efficiency indicative of the machine running at a slip of 13 percent, rather than a typical 3 to 5 percent. Alternatively, the output power may be calculated from: P_{int} − friction and windage losses − core loss = 118.4 W − 15.0 W − 30.0 W = 73.4 W. Here, efficiency = (output power)(100 percent)/input power = (73.4 W)(100 percent)/228.8 W = 32 percent.

Related Calculations: The establishment of an equivalent circuit makes it possible to make many different calculations. Also, once the equivalent circuit is created, these calculations may be made for any presumed value of input voltage and operating slip.

STARTING-TORQUE CALCULATION FOR A CAPACITOR MOTOR

A four-pole capacitor motor (induction motor) has the following data associated with it: stator main-winding resistance R_{sm} = 2.1 Ω, stator auxiliary-winding resistance R_{sa} = 7.2 Ω, stator main-winding leakage reactance X_{sm} = 2.6 Ω, stator auxiliary-winding leakage reactance X_{sa} = 3.0 Ω, reactance of capacitance inserted in series with the stator auxiliary winding X_{sc} = 65 Ω, rotor-circuit resistance referred to main winding R'_r = 3.9 Ω, rotor-circuit leakage reactance referred to stator main winding X'_r = 2.1 Ω, magnetizing reactance referred to stator main winding X_ϕ = 75.0 Ω. Calculate the starting torque; assume that the motor is rated ¼-hp at 115 V, 60 Hz, and that the effective turns ratio of the auxiliary winding to the main winding is 1.4. See Fig. 4.

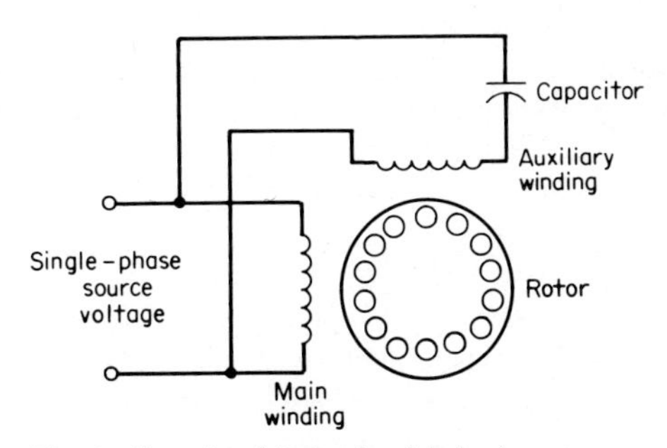

Fig. 4 Capacitor (single-phase) induction motor.

Calculation Procedure:

1. Calculate the Forward- and Backward-Circuit Impedances $\mathbf{Z_f}$ *and* $\mathbf{Z_b}$

A capacitor motor is similar to a two-phase induction motor, except that the two phases are unbalanced or unsymmetrical; their current relationship is usually not at 90°, but is determined by the capacitance in series with the auxiliary winding. At start (or standstill), $s = 1$; forward slip $s_f = 1$; and backward slip $s_b = (2 - s_f) = 1$.

At standstill, the forward- and backward-circuit impedances are equal: $\mathbf{Z_f} = \mathbf{Z_b} = (jX_\phi)(R'_r + jX'_r)/(jX_\phi + R'_r + jX'_r) = (j75)(3.9 + j2.1)/(j75 + 3.9 + j2.1) = 4.28\angle 31.2° = 3.66 + j2.22\ \Omega$. The analysis departs from that for a single-phase machine, and follows that for a two-phase machine; the rotor values are not divided by 2 as was done in the previous problems. The rotating fields of constant amplitude have one-half amplitude for single phase, unity amplitude for two phase, and three-halves amplitude for three phase.

2. *Calculate the Total Impedance* $\mathbf{Z_m}$ *Referred to the Stator Main Winding*

Use the equation $\mathbf{Z_m} = \mathbf{Z_{sm}} + \mathbf{Z_f} = 2.1 + j2.6 + 3.66 + j2.22 = 5.76 + j4.82 = 7.51\angle 39.9°\ \Omega$.

3. *Calculate the Stator Main-Winding Current* $\mathbf{I_{sm}}$

If the terminal voltage is the reference phasor, $\mathbf{I_{sm}} = \mathbf{V_t}/\mathbf{Z_m} = 115\angle 0°\ \text{V}/7.51\angle 39.9°\ \Omega = 15.3\angle -39.9°\ \text{A} = 11.74 - j9.81\ \text{A}$.

4. *Calculate the Total Impedance* $\mathbf{Z_a}$ *Referred to the Stator Auxiliary Winding*

The impedances used to calculate $\mathbf{Z_f}$ were referred to the main winding. In order to make the calculations relating to the auxiliary winding, the impedances now must be referred to *that* winding. This is done by multiplying $\mathbf{Z_f}$ by a^2, where a = effective turns ratio of the auxiliary winding to the main winding = 1.4. Thus, $\mathbf{Z_{fa}} = a^2\mathbf{Z_f} = (1.4)^2(3.66 + j2.22) = 7.17 + j4.35\ \Omega$.

The total impedance of the auxiliary stator winding is $\mathbf{Z_a} = \mathbf{Z_{sa}} + \mathbf{Z_{sc}} + \mathbf{Z_f} = 7.2 + j3.0 - j65 + 7.17 + j4.35 = 14.37 - j57.65 = 59.4\angle -76.0°\ \Omega$.

5. *Calculate the Stator Auxiliary-Winding Current* $\mathbf{I_{sa}}$

As in Step 3, $\mathbf{I_{sa}} = \mathbf{V_t}/\mathbf{Z_a} = 115\angle 0°\ \text{V}/59.4\angle -76.0°\ \Omega = 1.94\angle 76.0°\ \text{A} = 0.47 + j1.88\ \text{A}$.

6. *Calculate the Starting Torque* T_{start}

The starting torque is $T_{start} = (2/\omega_s)I_{sm}\,aI_{sa}R \sin\phi$ where $\omega_s = 4\pi f/p$ rad/s, f = frequency in Hz, p = number of poles, R = the resistance component of Z_f, and ϕ = angle separating I_m and I_a. Thus, $\omega_s = (4\pi)(60\ \text{Hz})/4 = 188.5$ rad/s. And $T_{start} = (2/188.5)(15.3\ \text{A})(1.4)(1.94\ \text{A})(3.66) \sin(39.9° + 76.03°) = 1.45\ \text{N}\cdot\text{m}$.

Related Calculations: It would appear that an unusually large capacitor has been used, resulting in an angle difference between I_{sm} and I_{sa} greater than 90°. However, in this case the capacitor remains in the circuit during running; when s_f and s_b are both different from unity, the angle would not be greater than 90°. This problem illustrates one method of handling currents from the two windings. Another method is with the use of symmetrical components of two unbalanced currents in two-phase windings.

STARTING TORQUE FOR A RESISTANCE-START SPLIT-PHASE MOTOR

A resistance-start split-phase motor has the following data associated with it: ½ hp, 120 V, starting current in the auxiliary stator winding I_a = 8.4 A at a lagging angle of 14.5°, starting current in the main stator winding I_m = 12.65 A at a lagging angle of 40°. Determine the total starting current and the starting torque, assuming the machine constant is 0.185 V·s/A.

Calculation Procedure:

1. Divide the Starting Currents into in-Phase and Quadrature Components

The line voltage ($\mathbf{V_L} = 120\angle 0°$ V) is used as the reference phasor. Thus, at starting (as at locked-rotor) the current in the auxiliary stator winding is $\mathbf{I_a} = 8.4\angle -14.5°$ A $= 8.13 - j2.10$ A. Similarly, the current in the main stator winding is $\mathbf{I_m} = 12.65\angle -40.0°$ A $= 9.69 - j8.13$ A.

2. Calculate the Total in-Phase and Quadrature Starting Currents

The total in-phase starting current is $\mathbf{I}_{in-phase} = 8.13 + 9.69 = 17.82$ A. The total quadrature starting current is $\mathbf{I}_{quad} = -j2.10 - j8.13 = -j10.23$ A. Thus, the starting current is $\mathbf{I}_{start} = 17.82 - j10.23$ A $= 20.55\angle 29.86°$ A. The starting-current power factor is $pf_{start} = \cos 29.86° = 0.867$ lagging. See Fig. 5.

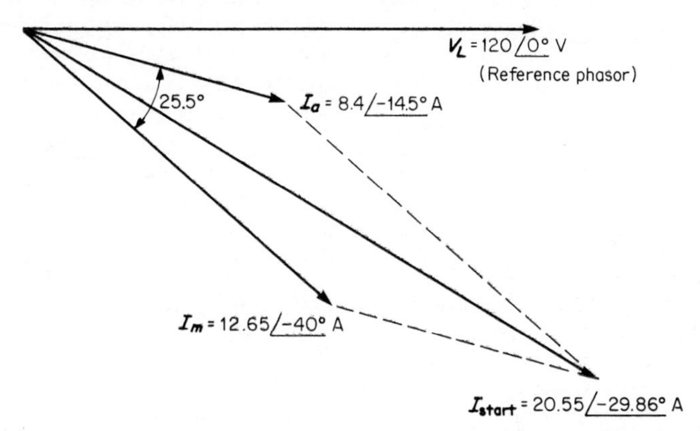

Fig. 5 Phase relationships of currents in main and auxiliary windings of a resistance-start split-phase motor.

3. Calculate the Starting Torque

Use the equation $T_{start} = KI_mI_a \sin \theta_{ma}$, where K is the machine constant in V·s/A and θ_{ma} is the phase angle in degrees between $\mathbf{I_m}$ (main-winding current) and $\mathbf{I_a}$ (auxiliary-winding current). T_{start} = (0.185 V·s/A)(8.4 A)(12.65 A) sin 25.5° = 8.46 V·A·s or 8.46 N·m.

Related Calculations: Split-phase motors achieve the higher resistance in the auxiliary windings by using fewer turns than in the main winding, and using a smaller diameter wire. Therefore the phase difference (usually less than 40°) does not allow for as high a starting torque as would be the case with a capacitor-start auxiliary winding. This type of starting mechanism is less expensive than the capacitor type for starting. Once started, the auxiliary winding is opened, the motor runs on the main winding only, and there is no difference in operation from the split-phase (resistance-start) or capacitor motor (capacitor-start).

SHADED-POLE MOTOR LOSSES AND EFFICIENCY

A four-pole shaded-pole motor (Fig. 6) has the following data associated with it: 120 V, full-load delivered power of 2.5 mhp (millihorsepower), 60 Hz, 350-mA full-load current, 12-W full-load power input, 1525-r/min full-load speed, 1760-r/min no-load speed, 6.6-W no-load power input, 235-mA no-load current, and stator resistance measured with dc of 30 Ω. Calculate the losses and efficiency at full load.

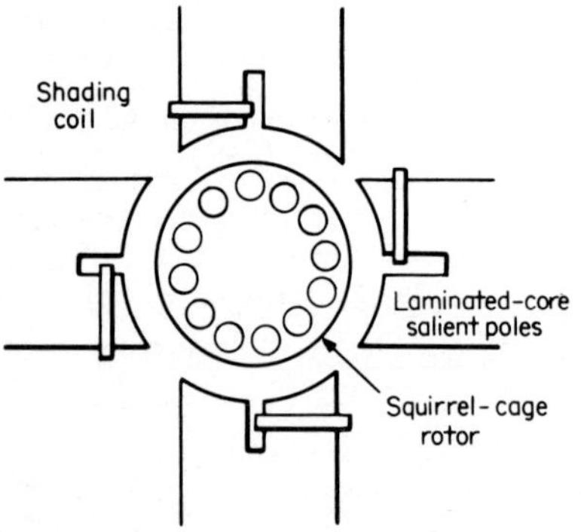

Fig. 6 Four-pole shaded-pole motor with a squirrel-cage rotor.

Calculation Procedure:

1. Calculate the Rotational Losses

From the no-load conditions, consider that the rotational losses of friction and windage are equal to the power input less the stator-copper loss. The stator resistance, measured with dc, was found to be 30 Ω; this is not the same as the effective ac value of resistance, which is influenced by nonuniform distribution of current over the cross section of the conductors (skin effect). The increase of resistance to ac as compared with dc may vary from 10 to 30 percent, the lower values being for small or stranded conductors and the higher values for large solid conductors. Assume a value of 15 percent for this problem.

The rotational losses equal $P_{fw} = P_{NL} - I^2_{NL}(R_{dc})$(dc-to-ac resistance-correction factor) $= 6.6\ W - (235 \times 10^{-3}\ A)^2(30\ \Omega)(1.15) = 6.6\ W - 1.905\ W = 4.69\ W$.

2. Calculate Stator-Copper Loss at Full Load

At full load the stator-copper loss is $P_{scu} = I^2_{FL}(R_{dc})$(dc-to-ac resistance-correction factor) $= (350 \times 10^{-3}\ A)^2(30\ \Omega)(1.15) = 4.23\ W$.

3. Calculate the Slip

The synchronous speed for a four-pole machine is obtained from the relation $n = 120 f/p$, where f = frequency in Hz and p = the number of poles. Thus, $n = (120)(60)/4 = 1800$ r/min. Since the actual speed is 1525 r/min at full load, the slip speed is 1800 − 1525 = 275 r/min, and the slip is (275 r/min)/(1800 r/min) = 0.153 or 15.3 percent.

4. Calculate Rotor-Copper Loss at Full Load

In induction machines, the rotor-copper loss is equal to the power transferred across the air gap multiplied by the slip. The power transferred across the air gap equals the input power minus the stator-copper loss. Thus, at full load, $P_{rcu} = (12\ W - 4.23\ W)(0.153) = 1.2\ W$.

5. Summarize the Full-Load Losses

Stator-copper loss	4.23 W
Rotor-copper loss	1.2 W
Friction and windage loss	4.69 W
Total losses	10.12 W

6. *Calculate the Efficiency*

The motor delivers 2.5 mhp; the input is 12 W or (12 W)(1 hp/746 W) = 16.1 mhp. Therefore, the efficiency is η = (output)(100 percent)/input = (2.5 mhp)(100 percent)/16.1 mhp = 15.5 percent.

Alternatively, the developed mechanical power is equal to the power transferred across the air gap multiplied by $1 - s$, or (12 W − 4.23 W)(1 − 0.153) = 6.58 W. From this must be subtracted the friction and windage losses: 6.58 W − 4.69 W = 1.89 W, or (1.89 W)(1 hp/746 W) = 2.53 mhp. The efficiency is η = (input − losses)(100 percent)/input = (12 W − 10.12 W)(100 percent)/12 W = 15.7 percent.

Related Calculations The calculations in this problem are the same as for any induction motor. The rotor in this problem is of the squirrel-cage type, and part of each of the four salient stator poles is enclosed by heavy, short-circuited, single-turn copper coils.

SYNCHRONOUS SPEED AND DEVELOPED TORQUE FOR A RELUCTANCE MOTOR

A singly fed reluctance motor having an eight-pole rotor (Fig. 7) has a sinusoidal reluctance variation as shown in Fig. 8. The power source is 120 V at 60 Hz, and the 2000-turn coil has negligible resistance. The maximum reluctance is $\mathcal{R}_q = 3 \times 10^7$ A/Wb and the minimum reluctance is $\mathcal{R}_d = 1 \times 10^7$ A/Wb. Calculate the speed of the rotor and the developed torque.

Calculation Procedure:

1. *Calculate the Speed*

Use the equation for synchronous speed: $n = 120f/p$ r/min, where n = speed in r/min, f = frequency in Hz, and p = number of poles on the rotor. A reluctance motor operates at synchronous speed. Thus, n = (120)(60 Hz)/8 poles = 900 r/min.

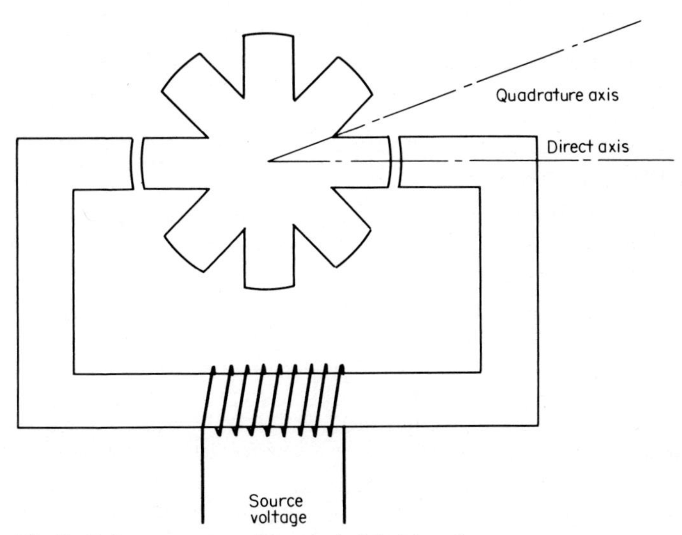

Fig. 7 Reluctance motor with a singly fed eight-pole rotor.

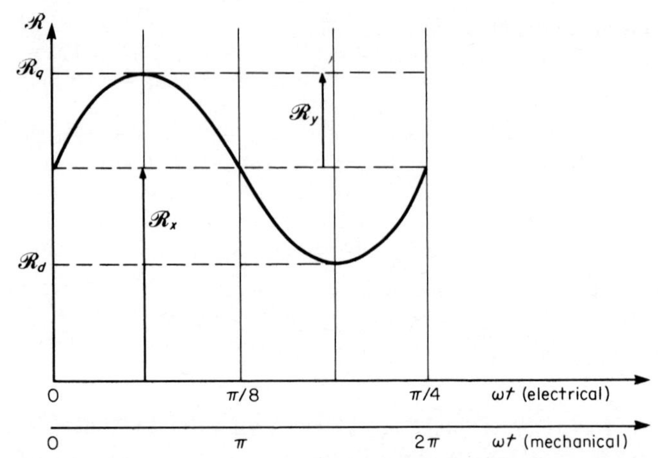

Fig. 8 Sinusoidal variation of reluctance for an eight-pole reluctance motor.

2. *Write the Equation for Variation of Reluctance*

With reference to Fig. 8, $\mathcal{R}_x = (\mathcal{R}_q + \mathcal{R}_d)/2 = (3 \times 10^7$ A/Wb $+ 1 \times 10^7$ A/Wb)/2 $= 2.0 \times 10^7$ A/Wb. $\mathcal{R}_y = (\mathcal{R}_q - \mathcal{R}_d)/2 = (3 \times 10^7$ A/Wb $- 1 \times 10^7$ A/Wb)/2 $= 1.0 \times 10^7$ A/Wb. Therefore, $\mathcal{R} = \mathcal{R}_x + \mathcal{R}_y \sin(p\omega t)$ where p = number of poles on the rotor; $\mathcal{R} = 2 \times 10^7 + 1 \times 10^7 \sin(8\omega t)$ A/Wb.

3. *Calculate the Average Maximum Torque*

Use the general equation for maximum time-average torque $T_{avg} = \frac{1}{8} p\phi_{max}^2 \mathcal{R}_y$ N·m where p = number of poles on the rotor. The maximum flux is found from the equation $\phi_{max} = \sqrt{2}V_{source}/2\pi fn$ Wb, on the assumption that the n-turn excitation coil has negligible resistance, where f = frequency of the source voltage in Hz. Further, for synchronous speed to occur, and that there be an average torque, the electrical radians of the supply voltage, ω_e, must be related to the mechanical radians of the rotor, ω_m, such that $\omega_m = (2/p)\omega_e$.

The maximum flux is $\phi_{max} = (\sqrt{2})(120\text{ V})/(2\pi 60)(2000\text{ turns}) = 0.000225$ Wb $= 2.25 \times 10^{-4}$ Wb. $T_{avg} = \frac{1}{8}(8\text{ poles})(2.25 \times 10^{-4}\text{ Wb})^2(1.0 \times 10^7\text{ A/Wb}) = 5.066 \times 10^{-1}$ Wb·A or 0.507 N·m.

4. *Calculate the Average Mechanical Power Developed*

Use the equation $P_{mech} = T_{avg}\omega_m = (0.507\text{ N·m})(900\text{ r/min})(2\pi\text{ rad/r})(1\text{ min}/60\text{ s}) = 47.8$ N·m/s or 47.8 W.

Related Calculations: A necessary condition for reluctance motors, wherein the number of poles on the rotor differs from the number of stator poles, is that the electrical radians of the supply voltage relate to the mechanical radians of the rotor, as indicated in this problem. Also, it is fundamental that this solution is based on the variation of the reluctance being sinusoidal; this requirement is met approximately by the geometric shape of the iron. Also, it is assumed here that the electric-power source voltage is sinusoidal.

These calculations and those of the next problem make reference to average torque and power because in reluctance machines the instantaneous torque and power are not constant but vary. Only over a complete time cycle of the electrical frequency is there average or net torque.

MAXIMUM VALUE OF AVERAGE MECHANICAL POWER FOR A RELUCTANCE MOTOR

The singly fed reluctance motor shown in Fig. 9 has an iron stator and rotor with cross-sectional area 1 in by 1 in; the length of the magnetic path in the rotor is 2 in. The length of each air gap is 0.2 in. The coil of 2800 turns is connected to a source of 120 V at 60 Hz; the resistance is neglible. The reluctance varies sinusoidally, and quadrature-axis reluctance $\mathcal{R}_q$ is equal to 3.8 times direct-axis reluctance $\mathcal{R}_d$. Determine the maximum value of average mechanical power.

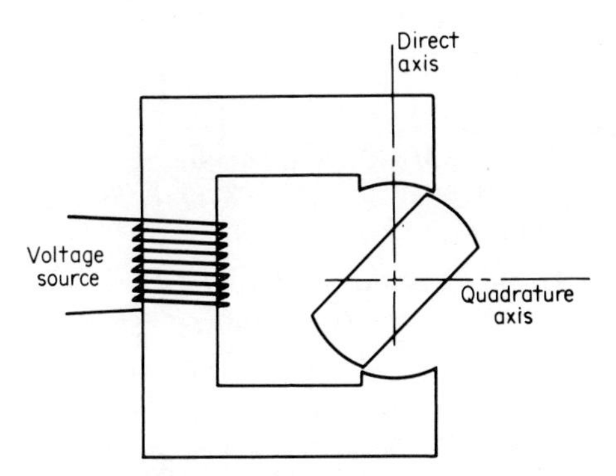

Fig. 9 Singly fed reluctance motor.

Calculation Procedure:

1. Calculate Direct-Axis Reluctance

Use the equation: $\mathcal{R}_d = 2g/\mu_0 A$ A/Wb, where g = air gap length in meters, $\mu_0 = 4\pi \times 10^{-7}$ N·m/A^2, and A = cross section of iron in m^2. Converting the given dimensions of inches to meters, 1 in/(39.36 in/m) = 0.0254 m, 2 in/(39.36 in/m) = 0.0508 m, and 0.2 in/(39.36 in/m) = 0.00508 m. Thus, $\mathcal{R}_d = (2)(0.00508 \text{ m})/[(4\pi \times 10^{-7})(0.0254 \text{ m})^2] = 1.253 \times 10^7$ A/Wb. And $\mathcal{R}_q$, being 3.8 times $\mathcal{R}_d$, is equal to $(3.8)(1.253 \times 10^7$ A/Wb) or 4.76×10^7 A/Wb.

2. Write the Equation for the Variation of Reluctance

Refer to Fig. 10. $\mathcal{R}_x = (\mathcal{R}_d + \mathcal{R}_q)/2 = (1.253 \times 10^7 \text{ A/Wb} + 4.76 \times 10^7 \text{ A/Wb})/2 = 3.0 \times 10^7$ A/Wb. $\mathcal{R}_y = (\mathcal{R}_q - \mathcal{R}_d)/2 = (4.76 \times 10^7 \text{ A/Wb} - 1.253 \times 10^7 \text{ A/Wb})/2 = 1.75 \times 10^7$ A/Wb. Therefore, $\mathcal{R} = \mathcal{R}_x + \mathcal{R}_y \sin(2\omega t) = 3.0 \times 10^7 + 1.75 \times 10^7 \sin(2\omega t)$ A/Wb.

3. Calculate the Maximum Value of the Flux

The induced voltage in the coil is sinusoidal (resistance is negligible) and is equal to $N d\phi/dt = -N\omega\phi_{max} \sin \omega t = V_{source}$. Thus, $\phi_{max} = V_{source}\sqrt{2}/N\omega$, where N = number of turns, $\omega = 2\pi f$, and f = frequency in Hz; $\phi_{max} = (120 \text{ V})\sqrt{2}/(2800 \text{ turns})(377 \text{ rad/s}) = 16 \times 10^{-5}$ Wb.

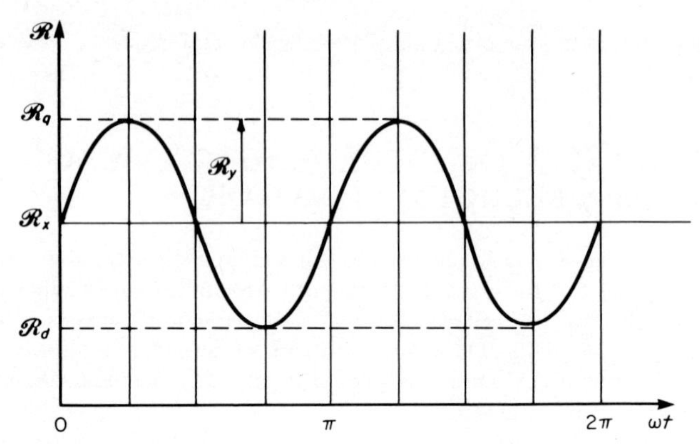

Fig. 10 Sinusoidal variation of reluctance for one full revolution of rotor of a reluctance motor.

4. *Calculate the Average Maximum Torque*

Use the equation $T_{avg} = \mathcal{R}_y \phi_{max}^2/4 = (1.75 \times 10^7 \text{ A/Wb})(16 \times 10^{-5} \text{ Wb})^2/4 = 0.112 \text{ Wb}\cdot\text{A}$ or $0.112 \text{ N}\cdot\text{m}$.

5. *Calculate the Mechanical Power Developed*

Use the equation $P_{mech} = T_{avg}\omega = (0.112 \text{ N}\cdot\text{m})(2\pi \times 60 \text{ rad/s}) = 42.2 \text{ N}\cdot\text{m/s}$ or 42.2 W.

Related Calculations: The reluctance motor is a synchronous motor; average torque exists only at synchronous speed. This motor finds application in electric clocks and record players. It has the characteristic of providing accurate and constant speed. The starting conditions may be similar to induction starting; any method employing the principle of induction starting may be used.

BREAKDOWN TORQUE-SPEED RELATIONSHIP FOR A FRACTIONAL-HORSEPOWER MOTOR

A given shaded-pole, 60-Hz motor with a rating of 12.5 mhp, has four poles. The breakdown torque is 10.5 oz·in. The shaded-pole speed-torque curve for this machine is given in Fig. 11. Determine (1) the speed and horsepower at breakdown, (2) the speed at rated horsepower, and (3) the rated horsepower converted to watts.

Calculation Procedure:

1. *Determine the Speed at Breakdown Torque*

The synchronous speed is calculated from the equation: $n_{sync} = 120f/p$ r/min, where f = frequency in Hz and p = number of poles. Thus, $n_{sync} = (120)(60)/4 = 1800$ r/min. Refer to Fig. 11. At the point where the torque in percent of breakdown torque is 100, the speed in percent of synchronous speed is 80. Thus, the rotor speed is $n_{rot} = 0.80 n_{sync} = (0.80)(1800) = 1440$ r/min.

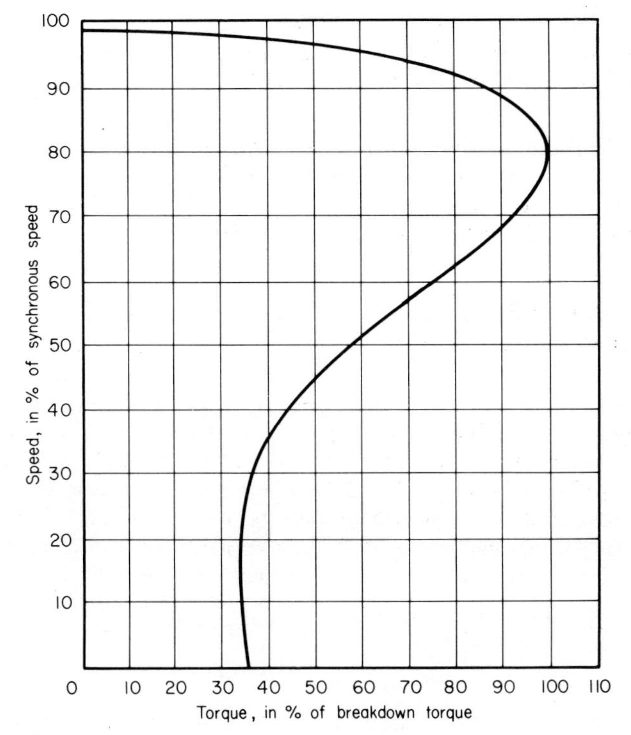

Fig. 11 Shaded-pole speed-torque curve for a fractional-horsepower motor.

2. *Calculate the Horsepower at Breakdown*

The general equation is horsepower = force × distance × speed. Thus, hp = (lb)(ft)(r/min)(2π rad/r)(hp·min)/33,000 ft·lb = (torque in lb·ft)(n_{rot} in r/min)/5252.1. Further, since 16 oz = 1 lb and 12 in = 1 ft, hp = (torque in oz·in)(n_{rot} in r/min)/1,008,403.2, or, approximately, hp = (torque in oz·in)(n_{rot} in r/min) × 10^{-6}. At breakdown, hp = (10.5 oz·in)(1440 r/min) × 10^{-6} = 15.12 × 10^{-3}hp = 15.12 mhp.

3. *Calculate the Speed at Rated Horsepower*

Convert the rated horsepower to torque, assuming a speed; referring to Fig. 11, use 95 percent of synchronous speed as a first approximation. Then torque in oz·in = hp × 10^6/(n_{rot} in r/min) = (12.5 × 10^{-3} × 10^6)/(0.95)(1800 r/min) = 7.31 oz·in or 0.051 N·m. From this, (rated torque)(100 percent)/breakdown torque = (7.31)(100 percent)/10.5 = 69.6 percent. Refer to Fig. 11; read the percentage speed as 94 percent. Notice that in the range of torque from zero to 80 percent of breakdown torque, the speed variation is small, from about 98 to 92 percent of synchronous speed. Using the 94 percent value, the actual rotor speed at rated horsepower is (0.94)(1800 r/min) = 1692 r/min. Note that at this speed the rated horsepower converts to a torque of 7.39 oz·in or 0.052 N·m.

4. *Convert the Rated Horsepower to Watts*

Use the relation: 746 W = 1 hp. The general equation is watts = (torque in oz·in)(n_{rot} in r/min)/(1.352×10^3) = (7.39 oz·in)(1692 r/min)/(1.352×10^3) = 9.26 W.

Related Calculations: Shaded-pole machines low torque and horsepower values; the ounce-inch is a common unit of torque, and the millihorsepower (i.e., hp $\times$ 10^{-3}) is a common unit of power. This problem illustrates the conversion of these units. These equations are applicable to fractional-horsepower machines.

FIELD- AND ARMATURE-WINDING DESIGN OF A REPULSION MOTOR

A $\frac{1}{3}$-hp, 60-Hz, two-pole repulsion motor (Fig. 12) of the enclosed fan-cooled type operating at 3600 r/min has an efficiency of 65 percent. Other data for the machine are as follows: flux per pole = 3.5 mWb, terminal voltage to stator (field winding) = 240 V, stator-winding distribution factor K_s = 0.91, number of stator slots = 20 (with conductors distributed in eight slots per pole), number of rotor slots = 24 (with two coils per slot). Calculate (1) the number of conductors per slot in the stator (field winding), (2) the number of conductors and coils for the rotor (armature winding), and (3) the operating voltage of the armature winding.

Calculation Procedure:

1. *Calculate the Number of Conductors per Field Pole (Stator Winding)*

Use the general equation $V = 4.44K_sZ_sf\phi_{pole}$ where V = terminal voltage, K_s = winding distribution factor (stator), Z_s = number of stator conductors, f = frequency in Hz, and ϕ_{pole} = flux per pole in Wb. The constant 4.44 represents $\sqrt{2}\pi$, where $\sqrt{2}$ is used to cause the voltage to be the rms value of ac, rather than the instantaneous value.

Rearranging the equation, find $Z_s = V/4.44K_sf\phi_{pole}$ = 240 V/(4.44)(0.91)(60 Hz)(3.5×10^{-3} Wb) = 283 conductors per pole.

2. *Determine the Number of Stator Conductors per Slot*

Although the stator has 20 slots, the conductors are distributed in eight slots per pole. If the 283 conductors per pole is divided by 8, the quotient is 35.375. Assume an even whole number such as 36

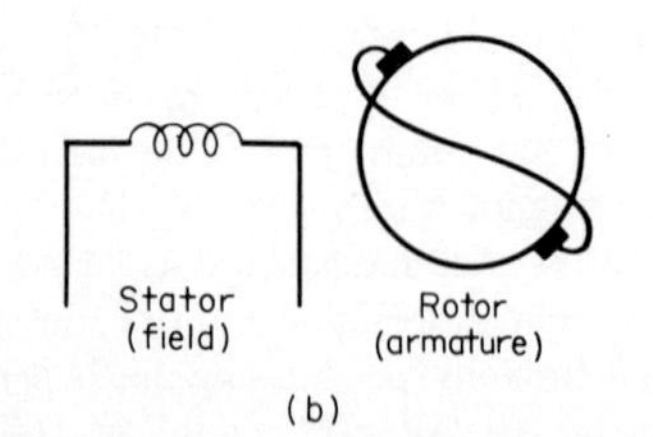

Fig. 12 Repulsion motor showing stator and rotor arrangement. (*a*) Physical arrangement. (*b*) Simplified diagram.

conductors per slot, yielding 288 conductors per pole (there being eight slots per pole wherein the conductors are to be distributed).

3. ***Calculate the Number of Conductors per Armature Pole (Rotor Winding)***

Use the general equation $E = \phi_{pole} Z_r np/\sqrt{2}fa$ where E = rotational emf in volts, Z_r = rotor (armature) conductors per pole, n = speed in r/min, p = number of poles, and a = number of parallel paths in the armature winding (for a lap winding, a = the number of poles; for a wave winding, $a = 2$).

Rearranging the equation, find $Z_r = \sqrt{2}Efa/\phi_{pole}np$. In this case a voltage E is to be assumed as a first try; this voltage should be somewhat low, say 80 V, since the brushes are short-circuited in a repulsion motor. Thus, $Z_r = (\sqrt{2})(80 \text{ V})(60 \text{ Hz})(2)/(3.5 \times 10^{-3} \text{ Wb})(3600 \text{ r/min})(2) = 538$ conductors per pole.

4. ***Calculate the Coils-per-Slot Arrangement***

Because there are 24 slots on the armature, try an arrangement of two coils per slot = (24)(2) = 48 coils. The number of conductors must be a multiple of the number of coils. Try 12 conductors per coil (i.e., six turns per coil); this yields Z_r(12 conductors per coil)(48 coils) = 576 conductors per pole.

Using this combination, calculate the rotational emf to compare with the first assumption. $E = (3.5 \times 10^{-3} \text{ Wb})(576 \text{ conductors})(3600 \text{ r/min})(2 \text{ poles})/(\sqrt{2})(60 \text{ Hz})(2 \text{ paths}) = 85.5 \text{ V}$. This is a reasonable comparison.

Related Calculations: The distribution factor for the stator winding in this problem accounts for the pitch factor and the breadth factor. A winding is of fractional pitch when the span from center to center of the coil sides that form a phase belt is less than the pole pitch. The breadth factor is less than unity when the coils forming a phase are distributed in two or more slots per pole.

The repulsion motor is similar in structure to the series motor; it has a nonsalient field (uniform air gap), and the commutated armature (rotor) has a short circuit across the brushes.

AC/DC TORQUE COMPARISON AND MECHANICAL POWER FOR A UNIVERSAL MOTOR

A universal motor operating on dc has the following conditions at starting: line current $I_L = 3.6$ A, starting torque $T_{start} = 2.3$ N·m. The motor is now connected to an ac source at 120 V, 60 Hz. It may be assumed that the total resistance of the motor circuit is 2.7 Ω, the inductance is 36 mH, rotational losses are negligible, and the magnetic field strength varies linearly with the line current. Calculate for the ac condition: (1) the starting torque, (2) the mechanical power produced at 3.6 A, and (3) the operating power factor.

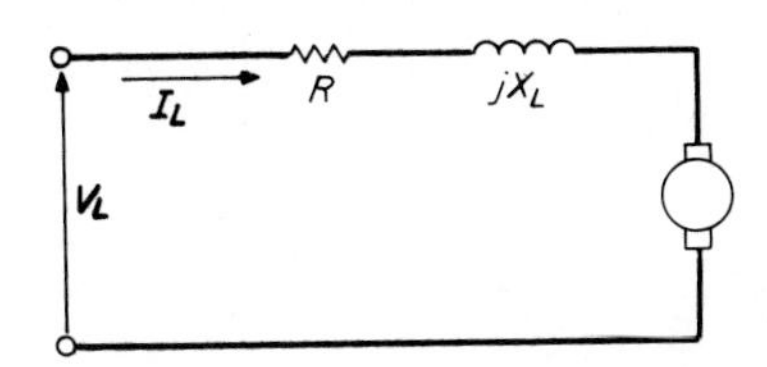

Fig. 13 Equivalent circuit of a universal motor (ac operation).

Calculation Procedure:

1. ***Calculate the AC Impedance of the Motor Circuit***

Refer to Fig. 13. The inductive reactance is calculated from the relation X_L

$= \omega L = 2\pi fL$ where f = frequency in Hz and L = inductance in henrys. Thus $X_L = (2\pi)(60\text{ Hz})(36 \times 10^{-3}\text{ H}) = 13.6\ \Omega$. The impedance of the motor circuit is $R + jX_L = 2.7 + j13.6\ \Omega = 13.9\angle 78.8°\ \Omega$.

2. *Calculate the Flux*

The flux in the magnetic circuit is proportional to the current in the circuit; the field circuit is in series with the armature. Use the general relation $T = k\phi I_L$ where T = torque in N·m and $k\phi$ = flux in Wb. For the dc condition at starting $k\phi = T_{\text{start}}/I_L = 2.3\text{ N·m}/3.6\text{ A} = 0.639$ Wb.

For the ac condition at starting $I_L = V_L/Z = 120\text{ V}/13.9\ \Omega = 8.63$ A, where Z = circuit impedance. Thus, $k\phi$, being proportional to current, becomes for the ac case $(0.639\text{ Wb})(8.63\text{ A})/3.6\text{ A} = 1.53$ Wb.

3. *Calculate the AC Starting Torque*

Use the equation again: $T = k\phi I_L$. $T_{\text{start(ac)}} = k\phi I_L = (1.53\text{ Wb})(8.63\text{ A}) = 13.2$ Wb·A, or 13.2 N·m.

4. *Calculate the Counter-emf, $\mathbf{e_a}$*

Use the general equation: $\mathbf{e_a} = \mathbf{V_L} - \mathbf{I_L Z}$. Because the phasor relation of $\mathbf{I_L}$ to $\mathbf{V_L}$ is not known, except that it should be a small angle, and e_a should be approximately close to the value of $\mathbf{V_L}$ (say about 85 percent), a rough first calculation will assume $\mathbf{V_L}$ and $\mathbf{I_L}$ are in phase. Thus, $e_a = 120\angle 0°\text{ V} - (3.6\text{ A})(13.9\angle 78.8°\ \Omega) = 120 - 9.7 - j49.1 = 110.3 - j49.1 = 120.7\angle -24.0°$ V. Now, repeat the calculation assuming I_L lags V_L by 24°. Thus, $e_a = 120\angle 0° - (3.6/-24°)(13.9\angle 78.8°) = 120 - 50.04\angle 54.8° = 120 - 28.8 - j40.9 = 91.2 - j40.9 = 99.95\angle -24.2°$ V. Notice that the angle of $-24°$ remains almost exactly the same. See Fig. 14.

5. *Calculate the Mechanical Power P_{mech}*

Use the equation $P_{\text{mech}} = e_a I_L = (99.95\text{ V})(3.6\text{ A}) = 359.8$ W.

6. *Calculate the Power Factor* pf *of the Input*

The power factor of the input is pf $= \cos\theta$ where θ is the angle by which I_L lags V_L; pf $= \cos 24.2° = 0.912$. From this information $P_{\text{mech}} = V_L I_L \cos\theta$ − copper loss $= V_L I_L \cos\theta - I_L^2 R = (120\text{ V})(3.6\text{ A})(0.912) - (3.6\text{ A})^2(2.7\ \Omega) = 393.98 - 34.99 - 359.0$ W, which compares with 359.8 W calculated in the previous step (allowing for rounding off of numbers).

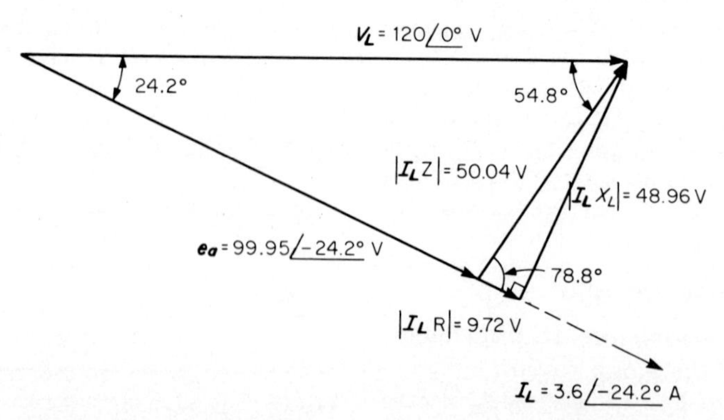

Fig. 14 Phasor diagram of a universal motor circuit (ac operation).

Related Calculations: The universal motor is the same as a series-wound dc motor; it is capable of being operated on ac or dc. The equations used are similar in form, except for the consideration of inductive reactance and phase relations of voltage and current.

SINGLE-PHASE SERIES MOTOR (UNIVERSAL) EQUIVALENT CIRCUIT AND PHASOR DIAGRAM

A 400-hp ac series motor operating at full load has the following data associated with it: 240 V, 25 Hz, 1580 A, 350 kW, 1890 lb·ft (torque), 1111 r/min. The resistances and reactance of the elements are as follows: main series field, $R_{SF} = 0.0018\ \Omega$; armature and brushes, $R_{ab} = 0.0042\ \Omega$; interpole circuit, $R_{int} = 0.0036\ \Omega$; compensating field, $R_{comp} = 0.0081\ \Omega$; total series-circuit reactance (armature, main series field, interpole circuit, compensating field), $X_L = 0.046\ \Omega$. See Fig. 15. Calculate (1) the horsepower, (2) the efficiency, (3) the power factor, and (4) the counter-emf.

Calculation Procedure:

1. *Calculate the Horsepower Output*

Use the equation: hp = (torque in lb·ft)(speed n in r/min)(2π rad/r)(hp·min/33,000ft·lb) = (torque in lb·ft)(n in r/min)/5252 = (1890 lb·ft)(1111 r/min)/5252 = 399.8 hp.

2. *Calculate the Efficiency*

Use the equation for efficiency: η = (power output)(100 percent)/power input = (399.8 hp)(746 W/hp)(100 percent)/350 × 10^3 W) = 85.2 percent.

Alternatively, the efficiency may be approximated from the known losses; the copper losses = I^2R. The total circuit resistance is $R_{SF} + R_{ab} + R_{int} + R_{comp} = 0.0018 + 0.0042 + 0.0036 + 0.0081 = 0.0177\ \Omega$. The copper losses = $(1580\ \text{A})^2(0.0177\ \Omega)$ = 44,186 W. The efficiency is η = (output × 100 percent)/(output + losses) = (399.8 hp × 746 W/hp)(100 percent)/(399.8 hp × 746 W/hp + 44,186 W) = 87.1 percent. Of course, the alternative method does not account for the friction and windage losses, nor for the magnetic losses. These may be calculated from the difference in efficiencies of the two methods. Thus, 350 × 10^3 W − (399.8 hp)(746 W/hp) − 44,186 W = 7563 W, which may be taken as the friction, windage, and magnetic losses.

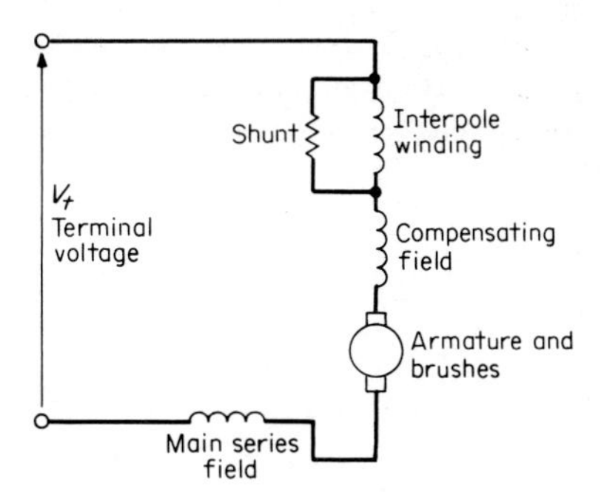

Fig. 15 Equivalent circuit of a series motor.

The calculation of efficiency becomes η = (399.8 hp)(746 W/hp)(100 percent)/(399.8 hp)(746 W/hp + 44,186 W + 7563 W) = 85.2 percent.

3. *Calculate the Power Factor* pf

Use the equation for power: $W = VA\cos\theta$ where the power factor = $\cos\theta$. Thus, power factor is (hp)(746 W/hp)/VA = (399.8 hp)(746 W/hp)/(240 V)(1580 A) = 0.786. The power factor angle = $\cos^{-1} 0.786 = 38.1°$

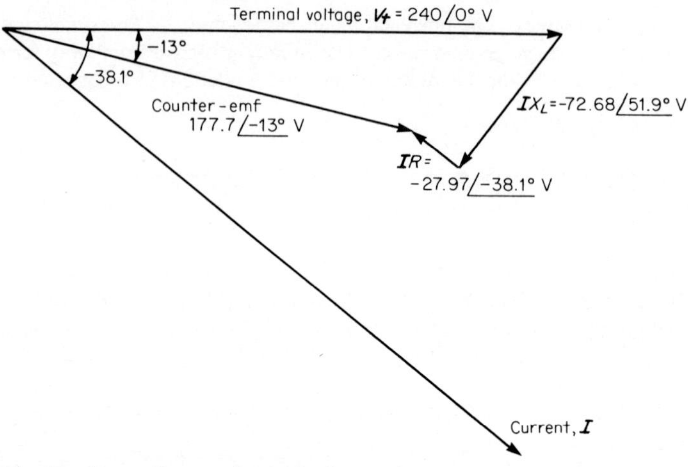

Fig. 16 Phasor diagram of a single-phase series motor.

4. Calculate the Counter-emf

The counter-emf equals (terminal voltage V_t) − (resistive voltage drop) − (reactive voltage drop) = $240\angle 0°$ V − ($1580\angle -38.1°$ A)(0.0177 Ω) − j($1580\angle -38.1°$ A)(0.046 Ω) = 240 − 22.0 + j17.3 − 44.9 − j57.2 = 173.1 − j39.9 = $177.7\angle -13°$ V. See Fig. 16.

Related Calculations: Just as in the case of dc motors, series motors used for ac or dc usually have interpoles and compensating windings. The former are for the purpose of improving commutation, and the latter for neutralizing the field-distorting effects of armature reaction. Each of these elements contributes inductive reactance which is important to consider in the case of ac operation.

Section 7 SYNCHRONOUS MACHINES

Omar S. Mazzoni, P.E.
Supervising Engineer, Gibbs & Hill, Inc.

Marco W. Migliaro
Associate Consulting Engineer,
Ebasco Services Incorporated

REFERENCES ANSI Standard C50.10—*General Requirements for Synchronous Machines;* ANSI Standard C50.13—*Requirements for Cylindrical-Rotor Synchronous Generators;* ANSI/NEMA Publication MG 1—*Motors and Generators;* Fink and Beaty—*Standard Handbook for Electrical Engineers,* McGraw-Hill; Fitzgerald and Kingsley—*Electric Machinery,* McGraw-Hill; Gross—*Power System Analysis,* Wiley; Libby—*Motor Selection and Application,* McGraw-Hill; Smeaton—*Motor Application and Maintenance Handbook,* McGraw-Hill; Stevenson—*Elements of Power System Analysis,* McGraw-Hill; Weedy—*Electric Power Systems,* Wiley; Westinghouse—*Electrical Transmission and Distribution Reference Book,* Westinghouse Electric.

The material contained in this section represents the authors' approach and does not necessarily represent Gibbs & Hill practice. The authors express their appreciation to the Gibbs & Hill editorial and automated typing service staffs for their help in the preparation of this section.

PER-UNIT BASE QUANTITIES

Calculate the per-unit (p.u.) base quantities for a 150-MVA, 13.8-kV, 60-Hz, three-phase, two-pole synchronous machine that has the following constants: *d*-axis mutual inductance between rotor and stator, $L_{ad} = 0.0056$ H; *d*-axis mutual inductance between stator winding *a* and rotor, $L_{afd} = 0.0138$ H; mutual inductance between stator winding *a* and *d*-axis amortisseur, $L_{akd} = 0.0054$ H; *q*-axis mutual inductance between rotor and stator, $L_{aq} = 0.0058$ H; and mutual inductance between stator winding *a* and *q*-axis amortisseur, $L_{akq} = 0.0063$ H. The per-unit system used should be the reciprocal mutual per-unit system. This denotes a per-unit system where the per-unit mutual inductances between the rotor and stator circuits are reciprocal. This also implies that $L_{ad} = L_{afd} = L_{akd}$ and $L_{aq} = L_{akq}$.

Calculation Procedure:

1. Select Base Values

Select $VA_{base} = 150$ MVA, $V_{base} = 13.8$ kV, and $f_{(base)} = 60$ Hz. From these values, other base quantities may be derived.

2. Calculate RMS Stator Phase Current Base $I_{s(base)}$

$I_{s(base)} = (MVA_{base} \times 1000)/(\sqrt{3} \times kV_{base}) = (150)(1000)/(\sqrt{3})(13.8) = 6276$ A.

3. Calculate Peak Stator Phase Current Base $i_{s(base)}$

The current is $i_{s(base)} = \sqrt{2}I_{s(base)} = (\sqrt{2})(6276) = 8876$ A.

4. Calculate Stator Base Impedance $Z_{s(base)}$

$Z_{s(base)} = kV^2_{(base)}/MVA_{(base)} = 13.8^2/150 = 1.270\ \Omega$.

5. Calculate Stator Base Inductance $L_{s(base)}$

$L_{s(base)} = Z_{s(base)}/\omega_{(base)} = 1.270/377 = 3.37 \times 10^{-3}$ H.

6. Calculate Field Base Current $i_{fd(base)}$

The current is $i_{fd(base)} = (L_{ad}/L_{afd})i_{s(base)} = (0.0056/0.0138)\ 8876 = 3602$ A.

7. Calculate Field Base Impedance $Z_{fd(base)}$

$Z_{fd(base)} = (MVA_{(base)} \times 10^6)/i^2_{fd(base)} = (150 \times 10^6)/3602^2 = 11.56\ \Omega$.

8. Calculate Field Base Inductance $L_{fd(base)}$

$L_{fd(base)} = Z_{fd(base)}/\omega_{(base)} = 11.56/377 = 30.66 \times 10^{-3}$ H.

9. Calculate Field Base Voltage $e_{fd(base)}$

The voltage is
$e_{fd(base)} = (MVA_{(base)} \times 10^6)/i_{fd(base)} = (150 \times 10^6)/3602 = 41{,}644$ V.

10. Calculate Direct-Axis Amortisseur Base Current $i_{kd(base)}$

The current is $i_{kd(base)} = (L_{ad}/L_{akd})i_{s(base)} = (0.0056/0.0054)(8876) = 9204$ A.

11. Calculate Direct-Axis Amortisseur Base Impedance $Z_{kd(base)}$

$Z_{kd(base)} = (MVA_{(base)} \times 10^6)/i^2_{kd(base)} = (150 \times 10^6)/9204^2 = 1.77\ \Omega$.

12. Calculate Direct-Axis Amortisseur Base Inductance $L_{kd(base)}$

$L_{kd(base)} = Z_{kd(base)}/\omega_{(base)} = 1.77/377 = 4.70 \times 10^{-3}$ H.

13. Calculate Quadrature-Axis Amortisseur Base Current $i_{kq(base)}$

The current is $i_{kq(base)} = (L_{aq}/L_{akq})i_{s(base)} = (0.0058/0.0063)(8876) = 8172$ A.

14. Calculate Quadrature-Axis Amortisseur Base Impedance $Z_{\text{kq(base)}}$
$Z_{\text{kq(base)}} = (\text{MVA}_{\text{(base)}} \times 10^6)/i^2_{\text{kq(base)}} = (150 \times 10^6)/8172^2 = 2.246\ \Omega$.

15. Calculate Quadrature-Axis Amortisseur Base Inductance $L_{\text{kq(base)}}$
$L_{\text{kq(base)}} = Z_{\text{kq(base)}}/\omega_{\text{(base)}} = 2.246/377 = 5.96 \times 10^{-3}$ H.

16. Calculate Base Mutual Inductance Between Amortisseur and Field, $L_{\text{fkd(base)}}$
$L_{\text{fkd(base)}} = (i_{\text{fd(base)}}/i_{\text{kd(base)}})L_{\text{fd(base)}} = (3602/9204)(30.66 \times 10^{-3}) = 12 \times 10^{-3}$ H.

17. Calculate Base Flux Linkage $\phi_{\text{s(base)}}$
$\phi_{\text{s(base)}} = L_{\text{s(base)}} i_{\text{s(base)}} = (3.37 \times 10^{-3})\ 8876 = 29.9$ Wb·turns.

18. Calculate Base Rotation Speed in r/min
The base speed is $120f_{\text{(base)}}/P = (120)(60/2) = 3600$ r/min, where P is the number of poles.

19. Calculate Base Torque $T_{\text{(base)}}$
$T_{\text{(base)}} = (7.04\text{MVA}_{\text{(base)}} \times 10^6)/\text{r/min}_{\text{(base)}} = (7.04)(150)(10^6)/3600 = 293$ klb·ft (397.2 kN·m).

PER-UNIT DIRECT-AXIS REACTANCES

Calculate the synchronous, transient, and subtransient per-unit reactances for the direct axis of the machine in the previous example for which the field resistance $r_{\text{fd}} = 0.0072$ Ω, stator resistance $r_s = 0.0016$ Ω, stator leakage inductance $L_l = 0.4 \times 10^{-3}$ H, field self-inductance $L_{\text{ffd}} = 0.0535$ H, d-axis amortisseur self-inductance $L_{\text{kkd}} = 0.0087$ H, d-axis amortisseur resistance $r_{\text{kd}} = 0.028$ Ω, and the q-axis amortisseur resistance $r_{\text{kq}} = 0.031$ Ω. Assume the leakage inductances in the d and q axes are equal, which is generally a reasonable assumption for wound-rotor machines.

Calculation Procedure:

1. Calculate Per-Unit Values for L_{ad} *and* L_l
From the previous example $L_{\text{s(base)}} = 0.00337$ H. Therefore $\overline{L}_{\text{ad}} = L_{\text{ad}}/L_{\text{s(base)}} = 0.0056/0.00337 = 1.66$ p.u. and $\overline{L}_l = L_l/L_{\text{s(base)}} = 0.0004/0.00337 = 0.12$ p.u. (*Note:* Bar over symbol designates per-unit value.)

2. Calculate Per-Unit Value of d-Axis Synchronous Inductance $\overline{L}_d$
$\overline{L}_d = \overline{L}_{\text{ad}} + \overline{L}_l = 1.66 + 0.12 = 1.78$ p.u.

3. Calculate Per-Unit Value of d-axis Synchronous Reactance $\overline{X}_{\text{d}}$
$\overline{X}_d = \omega\overline{L}_d$. By choosing $f_{\text{(base)}} = 60$ Hz, the rated frequency of the machine is $\bar{f} = 60/f_{\text{(base)}} = 60/60 = 1$. It also follows that $\overline{\omega} = 1$ and $\overline{X}_d = \overline{L}_d = 1.78$ p.u.

4. Calculate $\overline{L}_{\text{ffd}}$, $\overline{L}_{\text{kkd}}$, $\overline{L}_{\text{afd}}$, *and* $\overline{L}_{\text{akd}}$
$\overline{L}_{\text{ffd}} = L_{\text{ffd}}/L_{\text{fd(base)}} = 0.0535/(30.66 \times 10^{-3}) = 1.74$ p.u. $\overline{L}_{\text{kkd}} = L_{\text{kkd}}/L_{\text{kd(base)}} = 0.0087/(4.7 \times 10^{-3}) = 1.85$ p.u. $\overline{L}_{\text{afd}} = L_{\text{afd}}/L_{\text{s(base)}}(i_{\text{s(base)}}/i_{\text{fd(base)}}) = 0.0138/(0.00337)(8876/3602) = 1.66$ p.u. $\overline{L}_{\text{akd}} = L_{\text{akd}}/[\tfrac{2}{3}L_{\text{kd(base)}}(i_{\text{kd(base)}}/i_{\text{s(base)}})] = 0.0054/(\tfrac{2}{3})(4.7 \times 10^{-3})(9204/8876) = 1.66$ p.u.

5. Calculate $\overline{L}_{fd}$

$\overline{L}_{fd} = \overline{L}_{ffd} - \overline{L}_{afd} = 1.74 - 1.66 = 0.08$ p.u.

6. Calculate $\overline{L}_{kd}$

$\overline{L}_{kd} = \overline{L}_{kkd} - \overline{L}_{akd} = 1.85 - 1.66 = 0.19$ p.u.

7. Calculate Per-Unit Values of d-axis Transient Inductance $\overline{L}'_d$ and $\overline{X}'_d$

$\overline{L}'_d = \overline{X}'_d = \overline{L}_{ad}\overline{L}_{fd}/\overline{L}_{ffd} + L_l = (1.66)(0.08)/1.74 + 0.12 = 0.196$ p.u.

8. Calculate Per-Unit Values of d-Axis Subtransient Inductance $\overline{L}''_d$ and $\overline{X}''_d$

$$\overline{L}''_d = \overline{X}''_d = (1/\overline{L}_{kd} + 1/\overline{L}_{ad} + 1/\overline{L}_{fd}) + L_l$$
$$= (1/0.19) + (1/1.66) + (1/0.08) + 0.12 = 0.174 \text{ p.u.}$$

PER-UNIT QUADRATURE-AXIS REACTANCES

Calculate the synchronous and subtransient per-unit reactances for the quadrature axis of the machine in the first example. Additional data for the machine are: q-axis amortisseur self-inductance $L_{kkq} = 0.0107$ H and the mutual inductance between stator winding a and q-axis, $L_{akq} = 6.3 \times 10^{-3}$ H.

Calculation Procedure:

1. Calculate $\overline{L}_{aq}$

From values obtained in the two previous examples, $\overline{L}_{aq} = L_{aq}/L_{s(base)} = 0.0058/(3.37 \times 10^{-3}) = 1.72$ p.u.

2. Calculate $\overline{L}_q$

$\overline{L}_q = \overline{L}_{aq} + \overline{L}_l = 1.72 + 0.12 = 1.84$ p.u.

3. Calculate $\overline{L}_{kkq}$ and $\overline{L}_{akq}$

$\overline{L}_{kkq} = L_{kkq}/L_{kq(base)} = 0.0107/(5.96 \times 10^{-3}) = 1.80$ p.u. $\overline{L}_{akq} = L_{akq}/(L_{s(base)})(i_{s(base)}/i_{kq(base)}) = (6.3 \times 10^{-3})/(3.37 \times 10^{-3})(8876/8172) = 1.72$ p.u.

4. Calculate Per-Unit Value of q-axis Leakage Inductance $\overline{L}_{kq}$

$\overline{L}_{kq} = \overline{L}_{kkq} - \overline{L}_{akq} = 1.80 - 1.72 = 0.08$ p.u.

5. Calculate Per-Unit Transient Inductance $\overline{L}'_q$

$\overline{L}'_q = \overline{L}_{aq}\overline{L}_{kq}/(\overline{L}_{aq} + \overline{L}_{kq}) + \overline{L}_l = (1.72)(0.08)/(1.72 + 0.08) + 0.12 = 0.196$ p.u.

Related Calculations: Transient inductance L'_q is sometimes referred to as the q-axis subtransient inductance.

PER-UNIT OPEN-CIRCUIT TIME CONSTANTS

Calculate the per-unit field and subtransient open-circuit time constants for the direct axis of the machine in the first example. Use results obtained in previous examples.

Calculation Procedure:

1. Calculate $\bar{r}_{fd}$ and $\bar{r}_{kd}$

The quantities are $\bar{r}_{fd} = r_{fd}/Z_{fd(base)} = 0.0072/11.56 = 6.23 \times 10^{-4}$ p.u. and $\bar{r}_{kd} = r_{kd}/Z_{kd(base)} = 0.028/1.77 = 0.0158$ p.u.

2. Calculate Field Open-Circuit Time Constant $\bar{T}'_{do}$

$\bar{T}'_{do} = \bar{L}_{ffd}/\bar{r}_{fd} = 1.74/(6.23 \times 10^{-4}) = 2793$ p.u.

3. Calculate Subtransient Open-Circuit Time Constant $\bar{T}''_{do}$

$\bar{T}''_{do} = (1/\bar{r}_{kd})(\bar{L}_{kd} + \bar{L}_{fd}\bar{L}_{ad}/\bar{L}_{ffd}) = (1/0.0158)[0.19 + (0.08)(1.66/1.74)] = 16.9$ p.u.

PER-UNIT SHORT-CIRCUIT TIME CONSTANTS

Calculate the per-unit transient and subtransient short-circuit time constants for the direct axis of the machine in the first example. Also, calculate the direct-axis amortisseur leakage time constant.

Calculation Procedure:

1. Calculate Per-Unit Transient Short-Circuit Time Constant $\bar{T}'_d$

$\bar{T}'_d = (1/\bar{r}_{fd})\,[\bar{L}_{fd} + \bar{L}_l\,\bar{L}_{ad}/(\bar{L}_l + \bar{L}_{ad})] = [1/(6.23 \times 10^{-4})][0.08 + (0.12)(1.66)/(0.12 + 1.66)] = 308$ p.u.

2. Calculate Per-Unit Subtransient Short-Circuit Time Constant $\bar{T}''_d$

$$\frac{1}{\bar{r}_{kd}}\left[\bar{L}_{fd} + \frac{1}{(1/\bar{L}_{ad}) + (1/\bar{L}_{fd}) + (1/\bar{L}_l)}\right]$$

$$= \frac{1}{0.0158}\left[0.19 + \frac{1}{(1/1.66) + (1/0.08) + (1/0.12)}\right] = 15 \text{ p.u.}$$

3. Calculate Per-Unit Amortisseur Leakage Time Constant $\bar{T}_{kd}$

$\bar{T}_{kd} = \bar{L}_{kd}/\bar{r}_{kd} = 0.19/0.158 = 12$ p.u.

Related Calculations: Per-unit open-circuit and short-circuit time constants for the quadrature axis of the machine may be calculated with procedures similar to those used in the two previous examples.

To calculate a time constant in seconds, multiply the per-unit quantity by its time base, $1/\omega_{(base)} = 1/377$ s; $T = \bar{T}/377$. For example, $T'_d = 308/377 = 0.817$ s.

STEADY-STATE PHASOR DIAGRAM

Calculate the per-unit values and plot the steady-state phasor diagram for a synchronous generator rated at 100 MVA, 0.8 pf lagging, 13.8 kV, 3600 r/min, 60 Hz, operating at rated load and power factor. Important machine constants are $\bar{X}_d = 1.84$ p.u., $\bar{X}_q =$

1.84 p.u., and $\overline{X}'_d$ = 0.24 p.u. The effects of saturation and machine resistance may be neglected.

Calculation Procedure:

1. *Determine Reference*

If the VA and V base values are equal to the machine ratings, $I_{s(\text{base})} = (\text{MVA} \times 1000)/\sqrt{3}\text{kV} = (100)(1000)/(\sqrt{3})(13.8) = 4184$ A. Because the base voltage has been taken as 13.8 kV, the per-unit rms terminal voltage is $\overline{E}_t = 1.0$ p.u. From this, the per-unit peak voltage, $\bar{\mathbf{e}}_t = \overline{\mathbf{E}}_t = 1.0$ p.u. Quantity $\bar{\mathbf{e}}_t$ will be chosen as the reference phasor: $\bar{\mathbf{e}}_t = 1.0\underline{/0°}$ p.u.

2. *Locate q Axis*

Calculate a fictitious voltage $\overline{\mathbf{E}}_q = |\overline{\mathbf{E}}_q|\underline{/\delta}$ where δ is the machine internal power angle. $\overline{\mathbf{E}}_q$ may be calculated by $\overline{\mathbf{E}}_q = \bar{\mathbf{e}}_t + \bar{\mathbf{i}}_t(\bar{r} + j\overline{X}_q)$. But $\bar{\mathbf{i}}_t = 1.0\underline{/\theta°}$ p.u., where $\theta = \cos^{-1} 0.8 = -36.9°$. Therefore, $\overline{\mathbf{E}}_q = 1\underline{/0°} + 1.0\underline{/-36.9°} \times j1.84 = 2.38\underline{/35°}$ p.u. Power angle $\delta = 35°$.

3. *Calculate the d- and q-axis Components*

The d- and q-axis components may now be found by resolving $\bar{\mathbf{e}}_t$ and $\bar{\mathbf{i}}_t$ into components along the d and q axes, respectively: $\bar{\mathbf{e}}_q = |\bar{\mathbf{e}}_t| \cos\delta = (1.0)(0.819) = 0.819$ p.u. $\bar{\mathbf{e}}_d = |\bar{\mathbf{e}}_t| \sin\delta = (1.0)(0.574) = 0.574$ p.u. and $\bar{\mathbf{i}}_q = |\bar{\mathbf{i}}_t| \cos(\delta - \theta) = 1.0 \cos(35° + 36.9°) = 0.311$ p.u. $\bar{\mathbf{i}}_d = |\bar{\mathbf{i}}_t| \sin(\delta - \theta) = 0.951$ p.u.

4. *Calculate* $\overline{\mathbf{E}}_I$

Voltage $\overline{\mathbf{E}}_I$ lies on the q axis and represents the d-axis quantity, field current. $\overline{\mathbf{E}}_I = \overline{X}_{ad}\bar{\mathbf{i}}_{fd} = \bar{\mathbf{e}}_q + \overline{X}_d\bar{\mathbf{i}}_d + \bar{r}\bar{\mathbf{i}}_d = 0.819 + (1.84)(0.951) = 2.57$ p.u.

5. *Draw Phasor Diagram*

The phasor diagram is drawn in Fig. 1.

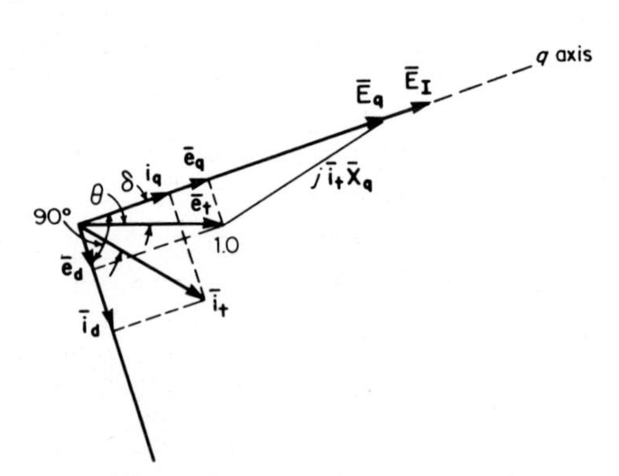

Fig. 1 Phasor diagram for a synchronous generator.

GENERATOR-CAPABILITY CURVE

The generator-capability curve, supplied by the manufacturer, is used to determine the ability of a generator to deliver real (MW) and reactive (Mvar) power to a network. Determine the capability curve, in per-unit values, of a generator with the following characteristics: 980 kVA, pf = 0.85, synchronous reactance X_d = 1.78 p.u., maximum value of generator internal voltage $\overline{E}_{\max}$ = 1.85 p.u., terminal voltage $\overline{V}$ = 1.0 p.u., δ = load angle, and the system reactance, external to generator is $\overline{X}_e$ = 0.4 p.u. Consider steady-state stability as the limit of operation for a leading power factor.

Calculation Procedure:

1. *Calculate Stator-Limited Portion*

The stator-limited portion is directly proportional to the full power output, which is an arc of a circle with radius $\overline{R}_s$ = 1.0 p.u. (curve *ABC* of Fig. 2).

2. *Calculate Field-Limited Portion*

The field-limited portion is obtained from the following expression: $\overline{P} = (3\overline{V}\overline{E}_{\max}/\overline{X}_d) \sin \delta + j[(\sqrt{3}\overline{V}\overline{E}_{\max}/\overline{X}_d \cos \delta - \sqrt{3}\overline{V}^2/\overline{X}_d]$ which is a circle with the center at 0, $-\sqrt{3}\overline{V}^2/\overline{X}_d$ and the radius is $\overline{R}_F = \sqrt{3}\overline{V}\overline{E}_{\max}/\overline{X}_d = \sqrt{3}(1.0)(1.85)/1.78 = 1.8$ p.u. Because $\overline{V}$ = 1.0 p.u., the center is located at 0, $(-1.0 \times \sqrt{3})/1.78$ = 0, −0.97 (curve *DEF,* Fig. 2).

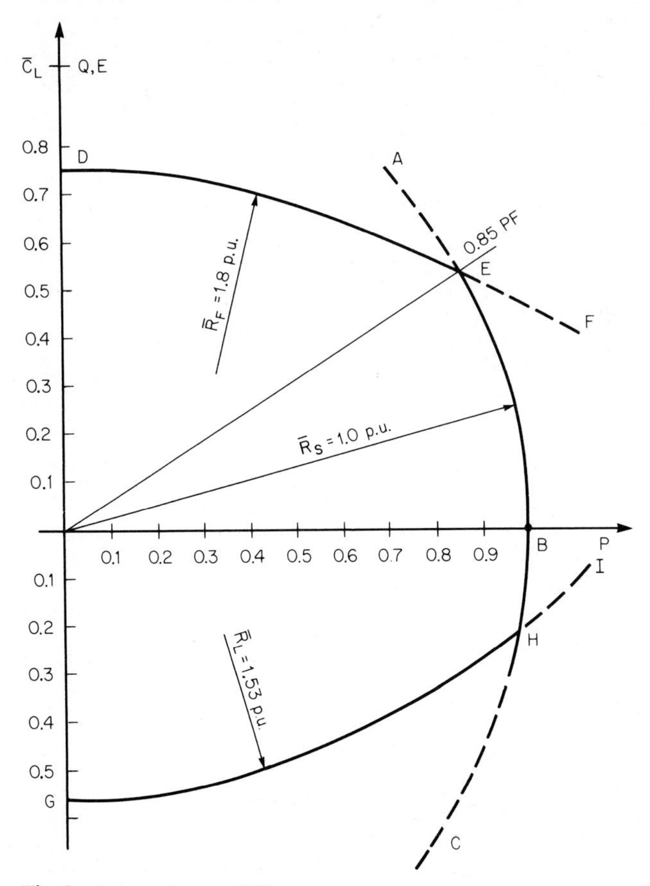

Fig. 2 A generator-capability curve.

3. *Calculate Steady-State Stability Curve*

The steady-state stability curve is given by an arc of the circle defined by: $\overline{C}_L$ = center = $j\overline{V}^2/2(1/\overline{X}_e - 1/\overline{X}_d) = j/2(1/0.4 - 1/1.78) = j0.97$. $\overline{R}_L$ = radius = $\overline{V}^2/2(1/\overline{X}_e + 1/\overline{X}_d) = 1/2(1/0.4 + 1.78) = 1.53$ p.u. (curve *GHI*, Fig. 2).

Related Calculations: Synchronous machines are capable of producing and consuming megawatts and megavars. When a generator is overexcited, it generates kilovars and delivers them to the system. When a generator is underexcited, negative megavars flow from the system into the machine. When the machine operates at unity power factor, it is just self-sufficient in its excitation.

Because operation of a generator is possible at any point within the area bounded by the capability curve, operators make use of the capability curve to control machine output within safe limits. The stator-limited portion relates to the current-carrying capacity of the stator-winding conductors. The field-limited portion relates to the area of operation under overexcited conditions, where field current will be higher than normal. The steady state stability portion relates to the ability of the machine to remain stable.

The curves in Fig. 3 are typical capability characteristics for a generator. A family of

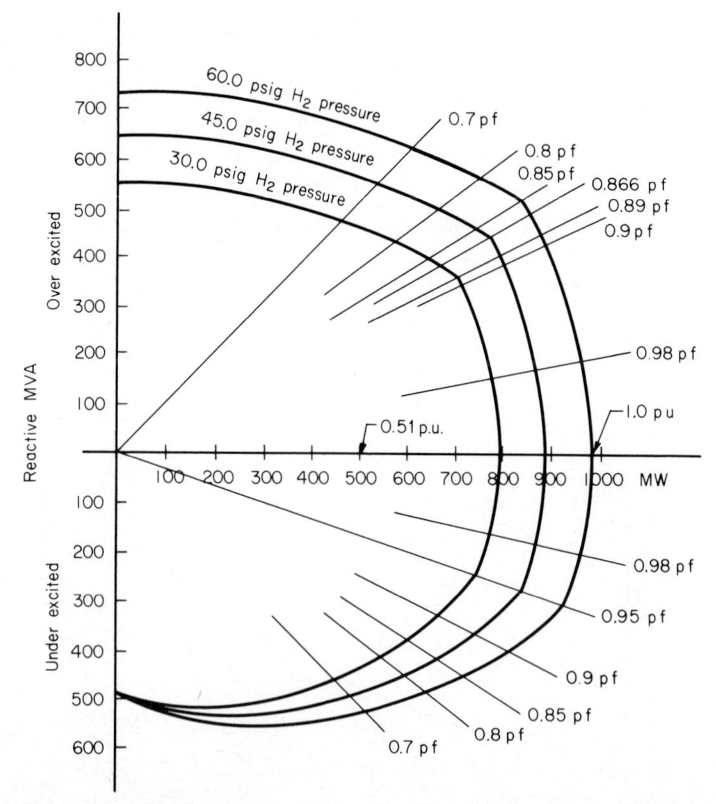

Fig. 3 Capability curves for a hydrogen inner-cooled generator rated at 983 MVA, 0.85 pf, 22 kV.

curves corresponding to different hydrogen cooling pressures is shown. The rated power factor is 0.8; rated apparent power is taken as 1.0 p.u. on its own rated MVA base. This means that the machine will deliver rated apparent power down to 0.85 lagging power factor. For a lower power factor, the apparent power capability is lower than rated.

GENERATOR REGULATION

Determine the regulation of a generator with the following characteristics: armature resistance $\overline{R}_a$ = 0.00219 p.u.; power factor = 0.975; and open-circuit, zero power-factor, and short-circuit saturation as in Fig. 4.

Calculation Procedure:

1. Calculate Potier Reactance $\overline{X}_p$

Use the relation $E_0 = E + \sqrt{3} I_a X_p$ from zero power-factor conditions where E_0 = voltage at no load, I_a = armature current at full load, and E is the terminal voltage.

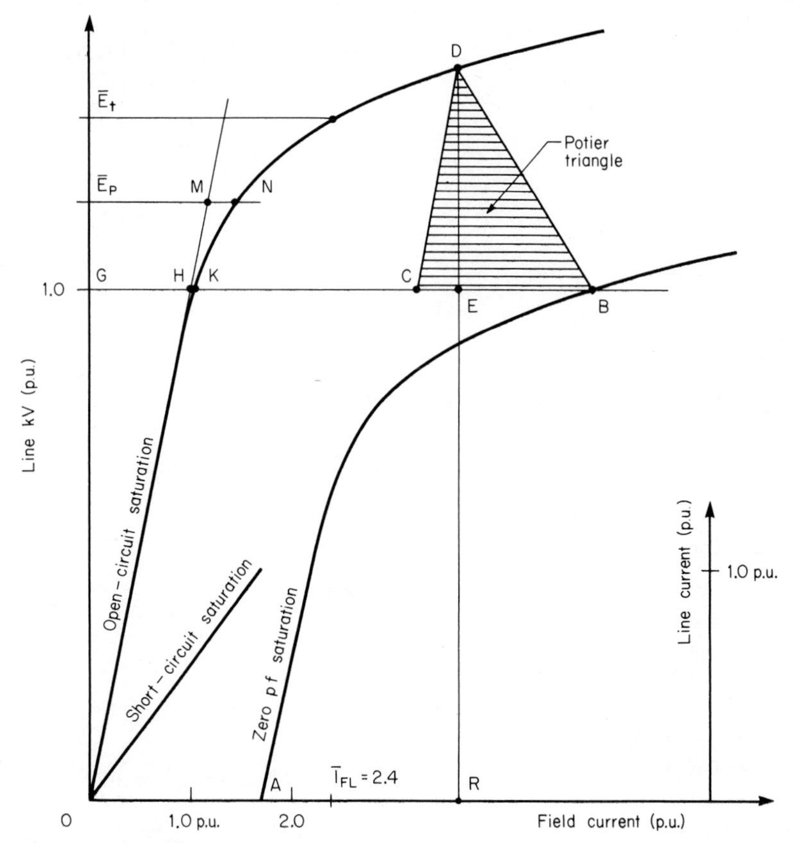

Fig. 4 Determination of generator regulation.

In Fig. 4, RD = RE + DE; therefore, DE = $\sqrt{3}I_aX_p$. By definition, $\overline{X}_p = I_aX_p/(V_{LL}/\sqrt{3})$ = DE/$\overline{V}_{LL}$ = DE/RE = 0.43 p.u.

2. Calculate Voltage behind Potier Reactance $\overline{E}_p$

From Fig. 4 $\overline{E}_p$ = 1.175 p.u.

3. Determine Excitation Required to Overcome Saturation $\overline{I}_{FS}$

From Fig. 4 $\overline{I}_{FS} = MN$ = 0.294 p.u.

4. Determine Excitation for Air-Gap Line $\overline{I}_{FG}$

From Fig. 4 $\overline{I}_{FG} = GH$ = 1.0 p.u.

5. Determine Excitation for Full-Load Current on Short-Circuit Saturation Curve $\overline{I}_{FS1}$

From Fig. 4 $\overline{I}_{FS1} = 0A$ = 1.75 p.u.

6. Calculate Excitation for Full-Load Current $\overline{I}_{FL}$

From Fig. 5 $\overline{I}_{FL} + \overline{I}_{FS} + [(\overline{I}_{FG} + \overline{I}_{FS1}\sin\phi)^2 + (\overline{I}_{FS1}\cos\phi)^2]^{1/2} = 0.294 + [(1 + 1.75 \times 0.2223)^2 + (1.75 \times 0.975)^2]^{1/2} = 2.4$ p.u.

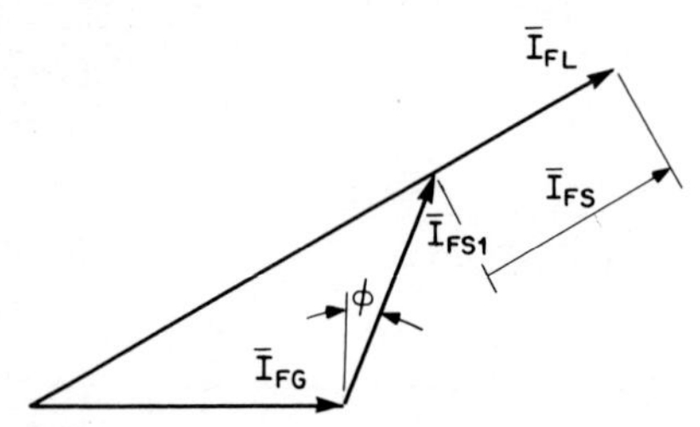

Fig. 5 Generator excitation at full-load current.

7. Determine $\overline{E}_t$ on Open-Circuit Saturation Curve Corresponding to Full-Load Field Excitation I_{FL}

From Fig. 4 $\overline{E}_t$ = 1.33 p.u.

8. Calculate Percent Regulation %R

$\%R = (\overline{E}_t - \overline{E})/\overline{E} = (1.33 - 1.0)/1.0 = 33\%$.

GENERATOR SHORT-CIRCUIT RATIO

Calculate the short-circuit ratio SCR for a generator having characteristic curves shown in Fig. 4.

Calculation Procedure:

1. Determine Excitation Value for Full-Load Current $\overline{I}_{FS1}$ on the Short-Circuit Curve

$\overline{I}_{FS1} = 0A$ = 1.75 p.u.

2. Determine Excitation $\overline{I}_{FV}$ Required to Produce Full Voltage on Open-Circuit Curve

$\overline{I}_{FV} = GK$ = 1.06 pu.

3. Calculate Short-Circuit Ratio SCR

SCR = $GK/0A$ = 1.06/1.75 = 0.6.

POWER OUTPUT AND POWER FACTOR

Calculate the maximum output power for an excitation increase of 20 percent for a 13.8-kV wye-connected generator having a synchronous impedance of 3.8 Ω/phase. It is connected to an infinite bus and delivers 3900 A at unity power factor.

Calculation Procedure:

1. Draw Phasor Diagrams

See Fig. 6. Subscript *o* indicates initial conditions. Voltage **V** is the line-to-neutral terminal voltage and **E** is the line-to-neutral voltage behind the synchronous reactance. Angle δ is the machine internal power angle and ϕ is the angle between the phase voltage and phase current.

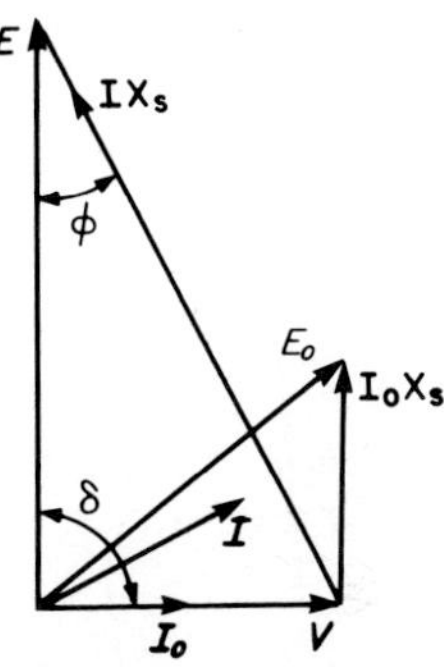

Fig. 6 Generator phasor diagrams: power factor and power output.

2. Calculate Voltage behind Synchronous Reactance

$E = [(IX_s)^2 - V^2]^{1/2} = [(3900 \times 3.8/1000)^2 - (13.8/\sqrt{3})^2]^{1/2} = 14.54$ kV

3. Calculate Maximum Power P_{max}

$P_{max} = 3EV/X_s \sin\delta$ where $\sin\delta = 1$ for maximum power. For a 20 percent higher excitation, $P_{max} = [(3)(1.2)(14.54)(13.8)/\sqrt{3}]/3.8 = 110$ MW.

4. Calculate Power Factor

The power factor is $\cos\phi = E/IX_s = 14.54/(3.9)(3.8) = 0.98$ lagging.

GENERATOR EFFICIENCY

Determine the efficiency of a generator having the same basic characteristics of the generator in the "Generator Regulation" example. Additional data include armature full-load current $I_a = 28{,}000$ A; core and short-circuit losses as in Fig. 7; friction and windage loss, 500 kW (from drive motor input); armature resistance $R_a = 0.0011$ Ω/phase; excitation voltage at rated load, 470 V, excitation current for air-gap line, 3200 A; and output voltage 25 kV.

Calculation Procedure:

1. Compute Core Loss

From Fig. 7 and the Potier voltage E_p (Fig. 4), for $\overline{E}_p = 1.175$ p.u., core loss = 2100 kW.

2. Determine Short-Circuit Loss

From Fig. 7, for 1.0 p.u. line current, short-circuit loss = 4700 kW.

3. Calculate Armature-Copper Loss

Armature-copper loss $= I_a^2R_a = (28{,}000)^2(0.0011) = 862{,}000$ W = 862 kW.

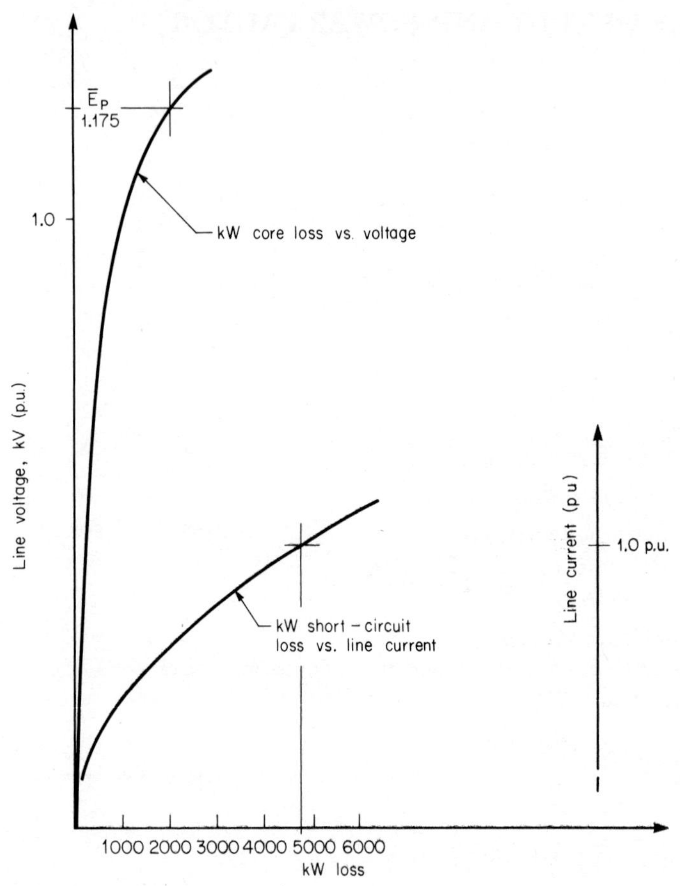

Fig. 7 Generator loss curves.

4. *Calculate Stray-Current Loss*

Stray-current loss = short-circuit loss − armature-copper loss = 4700 − 862 = 3838 kW.

5. *Calculate Power for Excitation*

Power for excitation = field voltage × I_{FL} = (470)(2.4)(3200)/1000 = 3610 kW, where I_{FL} is from the "Generator Regulation" example.

6. *Determine the Total Losses*

Friction and windage	500 kW
Core loss	2,100
Armature copper loss	862
Stray current loss	3,838
Power for excitation	3,610
Total losses	10,910 kW

7. Calculate Generator Output P_o
$P_o = \sqrt{3} \times \text{kV} \times \text{A} \times \text{pf} = (\sqrt{3})(25)(28{,}000)(0.975) = 1{,}182{,}125$ kW

8. Calculate Generator Efficiency
Efficiency = (power output)/(power output + total losses) = 1,182,125/(1,182,125 + 10,910) = 0.99 or 99 percent.

SYNCHRONIZING POWER COEFFICIENT

Calculate the synchronizing power coefficient at rated load for the following generator: 75,000 kW, terminal voltage $\overline{V} = 1.0$ p.u., armature current $\overline{I}_a = 1$ p.u., quadrature axis reactance $\overline{X}_q = 1.8$ p.u., and pf = 0.80 lagging. Neglect the resistive component of armature.

Calculation Procedure:

1. Calculate Rated Load Angle δ
The angle is $\delta = \tan^{-1} [\overline{X}_q \cos \phi \, \overline{I}_a/(\overline{I}_a\overline{X}_q \sin \phi + \overline{V})] = \tan^{-1} \{(1.8)(0.80)(1.0)/[(1.0)(1.80)(0.6) + 1]\} = 35°$.

2. Calculate Synchronizing Power Coefficient P_r
P_r = (rated kW)/(rated load angle $\times$ $2\pi/360$) = $(75)(1000)/(35)(2\pi/360)$ = 122,780 kW/rad.

GENERATOR GROUNDING TRANSFORMER AND RESISTOR

Determine the size of a transformer and resistor required to adequately provide a high-resistance ground system for a wye-connected generator rated 1000 MVA, 26 kV, 60 Hz. In addition, generator capacitance = 1.27 μF, main transformer capacitance = 0.12 μF, generator lead capacitance = 0.01 μF, and auxiliary transformer capacitance = 0.024 μF.

Calculation Procedure:

1. Calculate Generator Line-to-Neutral Voltage $V_{\text{L-N}}$
$V_{\text{L-N}} = (26 \text{ kV})/\sqrt{3} = 15$ kV.

2. Calculate Total Capacitance C_T
$C_T = (1.27 + 0.12 + 0.01 + 0.024)\ \mu\text{F} = 1.424\ \mu\text{F}$.

3. Calculate Total Capacitive Reactance X_{CT}
$X_{\text{CT}} = 1/2\pi f C_T = 1/(6.28)(60)(1.424 \times 10^{-6}) = 1864\ \Omega$.

4. Select $R = X_{CT}$ to Limit Transient Overvoltage during a Line-to-Ground Fault
Assume a 19.92/0.480-kV transformer. The resistance reflected to the primary is $R' = N^2R$, where R is the required resistor. Solve $R = R'/N^2$ where $N = 19.92/0.480 = 41.5$, and find $R = 1864/41.5^2 = 1.08\ \Omega$.

5. Calculate Transformer Secondary Voltage V_s during a Line-to-Ground Fault
$V_s = V/N = 15{,}000/41.5 = 361$ V.

6. Calculate Current I_s through Grounding Resistor

$I_s = V_s/R = 361/1.08 = 334.3$ A

7. Calculate Required Continuous Rating in kVA of Grounding Transformer

The rating is kVA $= I_sV_s = (334.3)(361) = 120.7$ kVA.

8. Select Short-Time Rated Transformer

From ANSI standards, a 50-kVA transformer may be used if a 9-min rating is adequate.

9. Calculate Generator Line-to-Ground Fault Current I_f

$I_f = V/X_{CT} = 15{,}000/1864 = 8.05$ A

POWER-FACTOR IMPROVEMENT

An industrial plant has a 5000-hp induction motor load at 4000 V with an average power factor of 0.8, lagging, and an average motor efficiency of 90 percent. A new synchronous motor rated at 3000 hp is installed to replace an equivalent load of induction motors. The synchronous motor efficiency is 90 percent. Determine the synchronous motor current and power factor for a system current of 80 percent of the original system and unity power factor.

Calculation Procedure:

1. Calculate Initial System Rating kVA_o

The rating is $\text{kVA}_o = (\text{hp}/\eta)(0.746/\text{pf}) = (5000/0.9)\ (0.746/0.8) = 5181$ kVA, where η is efficiency.

2. Calculate Initial System Current I_o

$I_o = \text{kVA}/\sqrt{3}V = 5181/(\sqrt{3})(4) = 148$ A

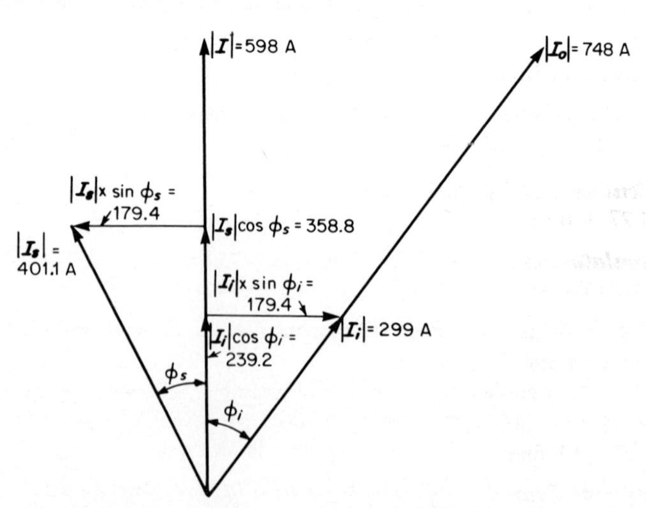

Fig. 8 Synchronous-motor power-factor improvement.

3. Calculate New System Current I

$I = 0.8I_o = (0.8)(746) = 598$ A (see Fig. 8).

4. Calculate New Induction-Motor Current I_i

$I_i = 0.746\text{hp}/\sqrt{3}V\eta \text{pf}_i = (0.746)(2000)/(\sqrt{3})(4)(0.9)(0.8) = 299$ A

5. Calculate Synchronous-Motor Current I_s

$I_s = [(I_i \sin \phi_i)^2 + (I - I_i \cos \phi_i)^2]^{1/2} = \{[(299)(0.6)^2 + [598 - (299)(0.8)]^2\}^{1/2} = (149.4^2 + 358.8^2)^{1/2} = 401.1$ A

6. Calculate Synchronous-Motor Power Factor pf_s

The power factor is $\text{pf}_s = 358.8/401.1 = 0.895$.

Related Calculations: For verification, the synchronous motor horsepower hp_s should equal 3000 hp, where $\text{hp}_s = 3VI_s\eta\text{pf}_s/0.746 = (\sqrt{3})(4)(401.1)(0.9)(0.895)/0.746 = 3000$ hp.

Section 8 GENERATION OF ELECTRIC POWER

Glen E. Jablonka, P.E.
Senior Generation Planning Engineer, Wisconsin Power & Light Company

REFERENCES: Babcock & Wilcox Co.—*Steam—Its Generation and Use,* Babcock & Wilcox; Baldwin and Hoffman—"System Planning by Simulation with Mathematical Models," *Proc. of the American Power Conf.,* March 1960; Bartlett—*Steam Turbine Performance and Economics,* McGraw-Hill; Baumeister et al.—*Marks' Standard Handbook for Mechanical Engineers,* McGraw-Hill; Bernhardt et al.—*Power Station Engineering and Economy,* McGraw-Hill; Billington—*Power System Reliability Evaluation,* Gordon and Breach; Considine—*Energy Technology Handbook,* McGraw-Hill; Day—"Forecasting Minimum Production Costs with Linear Programming," *IEEE Trans. on Power Apparatus and Systems,* March/April 1971; DeGarmo—*Engineering Economy,* Macmillan; Fink and Beatty—*Standard Handbook for Electrical Engineers,* McGraw-Hill; Jeynes—*Profitability and Economic Choice,* Iowa State University Press; Kirchmayer—*Economic Operation of Power Systems,* Wiley; Lapedes—*Encyclopedia of Energy,* McGraw-Hill; Loftness—*Energy Handbook,* Van Nostrand Reinhold; Powell—*Principles of Electric Utility Engineering,* MIT Press; Shepard—*Introduction to Energy Technology,* Ann Arbor Science; Stevenson—*Elements of Power System Analysis,* McGraw-Hill.

MAJOR PARAMETER DECISIONS

The major parameter decisions which must be made for any new electric power-generating plant or unit include the choices of energy source (fuel), type of generation system, unit and plant rating, and plant site. These decisions must be based upon a number of technical, economic, and environmental factors which are to a large extent interrelated (see Table 1). Evaluate the parameters for a new power-generating plant or unit.

TABLE 1 Major Parameter Decisions for New Plant

Parameter	*Some alternatives*
Energy source or fuel	Common fossil fuels (coal, oil, natural gas)
	Nuclear fuels (uranium, thorium)
	Elevated water (hydroelectric)
	Geothermal steam
	Other renewable, advanced technology, or nonconventional sources
Generation system type	Steam-cycle (e.g., steam-turbine) systems (with or without cogeneration steam for district heating and industrial steam loads)
	Hydroelectric systems
	Combustion-turbine (e.g., gas-turbine) systems
	Combined-cycle (i.e., combined steam and gas turbine) systems
	Internal-combustion engine (e.g., diesel) systems
	Advanced technology or nonconventional sources
Unit and plant rating	Capable of serving the current expected maximum electrical load and providing some spinning reserve for reliability and future load growth considerations
	Capable of serving only the expected maximum electrical load (e.g., peaking unit)
	Capable of serving most of the expected maximum load (e.g., using conservation or load management to eliminate the load which exceeds generation capacity)
Plant site	Near electrical load
	Near fuel source
	Near water source (water availability)
	Near existing electrical transmission system
	Near existing transportation system
	Near or on existing electrical generation plant site

Calculation Procedure:

1. Consider the Energy Source and Generating System

As indicated in Table 2, a single energy source or fuel (e.g., oil) is often capable of being used in a number of different types of generating systems. These include steam cycles, combined steam- and gas-turbine cycles (systems where the hot exhaust gases are delivered to a heat-recovery steam generator to produce steam which is used to drive a steam turbine), and a number of advanced technology processes such as fuel cells (i.e., systems having cathode and anode electrodes separated by a conducting electrolyte which convert liquid or gaseous fuels to electric energy without the efficiency limits of the Carnot cycle).

Similarly, at least in the planning stage, a single generic type of electric-power generating system (e.g., a steam cycle) can be designed to operate on any one of a number of fuels. Conversion from one fuel to another after plant construction does, however, generally entail significant capital costs and operational difficulties.

As Table 3 indicates, each combination of energy source and power-generating-system type has technical, economic, and environmental advantages and disadvantages which are unique. Often, however, in a particular situation there are other unique considerations which make the rankings of the various systems quite different from the typical values listed in Table 3. In order to make a determination of the best system it is necessary to quantify and evaluate all factors in Table 3. Generally, this involves a complicated trade-off process and a considerable amount of experience and subjective judgment. Usually, there is no one system which is best on the basis all of the appropriate criteria.

For example, in a comparison between coal and nuclear energy, nuclear energy generally has much lower fuel costs but higher capital costs. This makes an economic choice dependent to a large extent on the expected capacity factor (or equivalent full-load hours of operation expected per year) for the unit. Coal and nuclear-energy systems have, however, significant but vastly different environmental impacts. It may well turn out that one system is chosen over another largely on the basis of a subjective perception of the risks or of the environmental impacts of the two systems.

Similarly, a seemingly desirable and economically justified hydroelectric project (which has the additional attractive features of using a renewable energy source, and in general having high system availability and reliability) may not be undertaken because of the adverse environmental impacts associated with the construction of a dam required for the project. The adverse environmental impacts might include the effects that the dam would have on the aquatic life in the river, or the need to permanently flood land above the dam which is currently being farmed, or is inhabited by people who do not wish to be displaced.

2. Select the Plant, Unit Rating, and Site

The choice of plant, unit rating, and site is a similarly complex, interrelated process. As indicated in Table 3 (and 9), the range of unit ratings which are commercially available is quite different for each of the various systems. If, for example, a plant is needed with a capacity rating much above 100 MW, combustion turbine, diesel, and geothermal units could not be used unless multiple units were considered for the installation.

Similarly, the available plant sites can have an important impact upon the choice of fuel, power-generating system, and rating of the plant. Fossil-fuel or nuclear-energy steam-cycle units require tremendous quantities of cooling water [50.5 to 63.1 m^3/s (800,000 to 1,000,000 gal/min)] for a typical 1000-MW unit whereas gas-turbine units

TABLE 2 Generic Types of Electric-Generating Systems

Energy source or fuel	*Approximate percentage of total electric generation*	*Steam cycle 85%*	*Hydro-electric 13%*	*Combined steam and gas turbine cycle 1%*	*Combustion turbines 1%*	*Internal-combustion engine (diesel) 1%*	*Photo-voltaic*	*Wind turbine*	*Fuel cell*	*Magneto-hydro-dynamic*	*Thermo-electric*	*Therm-ionic*	*Open steam or closed ammonia cycle*
Coal	44	X							X	X	X	X	
Oil	16	X		X	X	X			X	X	X	X	
Natural gas	14	X		X	X				X	X	X	X	
Elevated water supply	13		X										
Nuclear fission (uranium or thorium)	13	X											
Geothermal	0.15	X											
Refuse-derived fuels		X											
Shale oil		X		X	X				X	X	X	X	
Tar sands		X		X	X	X			X	X	X	X	
Coal-derived liquids and gases		X		X	X	X			X	X	X	X	
Wood		X											
Vegetation (biomass)		X											
Hydrogen		X				X			X	X	X	X	
Solar		X					X						
Wind								X					
Tides			X										
Waves			X										
Ocean thermal gradients													X
Nuclear fusion		X											

require essentially no cooling water. Coal-fired units rated at 1000 MW would typically require over 2.7 million tonnes (3 million tons) of coal annually whereas nuclear units rated at 1000 MW would typically require only 32.9 tonnes (36.2 tons) of enriched uranium dioxide (UO_2) fuel annually.

Coal-fired units require disposal of large quantities of ash and scrubber sludge, whereas natural-gas–fired units require no solid-waste disposal whatsoever. From each of these comparisons it is easy to see how the choice of energy source and power-generating system can have an impact on the appropriate criteria to be used in choosing a plant site. The location and physical characteristics of the available plant sites (such as proximity to and availability of water, proximity to fuel or fuel transportation, and soil characteristics) can have an impact on the choice of fuel and power-generating system.

3. Examine the Alternatives

Each of the more conventional electric-power generating systems indicated in Table 3 is available in a variety of ratings. In general, installed capital costs (on a dollars per kilowatt basis) and system efficiencies (heat rates) are quite different for the different ratings. Similarly, each of the more conventional systems is available in many variations of equipment types, equipment configurations, system parameters, and operating conditions.

For example, there are both pulverized-coal and cyclone boilers which are of either the drum or once-through type. Steam turbines used in steam cycles can be of either the tandem-compound or cross-compound type, with any number of feedwater heaters, and be either of the condensing, back-pressure, or extraction (cogeneration) type. Similarly, there are a number of standard inlet and reheat steam conditions (i.e., temperatures and pressures). Units may be designed for base-load, intermediate-load, cycling, or peaking operation. Each particular combination of equipment type, equipment configuration, system parameters, and operating conditions has associated cost and operational advantages and disadvantages, which for a specific application must be evaluated and determined in somewhat the same manner that the fuel and electrical generation system choice is made.

4. Consider the Electrical Load

The electrical load, or demand, on an electric-power system of any size generally fluctuates considerably on a daily basis, as shown by the shapes of typical daily load curves for the months April, August, and December in Fig. 1. In addition, on an annual basis, the system electrical load varies between a minimum load level, below which the electrical demand never falls, and a maximum, or peak, load level, which occurs for only a few hours per year. The annual load duration curve of Fig. 1*a* graphically shows the number of hours per year that the load on a particular power system exceeds a certain level.

For example, if the peak-power system load in the year (100 percent load) is 8100 MW, the load duration curve shows that one could expect the load to be above 70 percent of the peak (i.e., above 0.7×8100 MW = 5760 MW) about 40 percent of the year. The minimum load (i.e., load exceeded 100 percent of the time) is about 33 percent of the peak value.

Typically, for U.S. utility systems the minimum annual load is 27 to 33 percent of the peak annual load. Generally, the load level exceeds 90 percent of the peak value 1 to 5 percent of the time, exceeds 80 percent of the peak value 5 to 30 percent of the time, and exceeds 33 to 45 percent of the peak 95 percent of the time. Annual load factors [(average load/peak annual load) $\times$ 100 percent] typically range from 55 to 65 percent.

Related Calculations: Generally, considerable economic savings can be obtained by

TABLE 3 Comparison of Energy Sources and Electrical Generating Systems

Energy source (fuel) and generation system type	*Fuel cost*	*System efficiency*	*Capital cost, $/kW*	*System operation and maintenance (excluding fuel) costs, $/MWh*	*Largest available unit ratings*
Coal-fired steam cycle	Intermediate	High	Very high	Low to medium	Large
Oil-fired steam cycle	Highest	High	High	Lowest	Large
Natural-gas–fired steam cycle	High	High	High	Lowest	Large
Nuclear	Low	Intermediate	Highest	Medium	Largest
Oil-fired combustion turbine	Highest	Low	Lowest	Highest	Smallest
Natural-gas–fired combustion turbine	High	Low	Lowest	Highest	Smallest
Oil-fired combined cycle	Highest	Very high	Intermediate	Medium	Intermediate
Natural-gas–fired combined cycle	High	Very high	Intermediate	Medium	Intermediate
Hydroelectric	Lowest	Highest	Intermediate to highest	Low	Large
Geothermal steam	Low	Lowest	Intermediate	Medium	Intermediate

using higher capital cost, lower operating cost units (such as steam-cycle units) to serve the base load (Fig. 1) and by using lower capital cost, higher operating cost units (such as combustion turbines) to serve the peaking portion of the load. The intermediate load range is generally best served by a combination of base-load, peaking, combined-cycle, and hydroelectric units that have intermediate capital and operating costs and have design

System reliability and availability	*System complexity*	*Fuel availability*	*Cooling water requirements*	*Major environmental impacts*
High	Very high	Best	Large	Particulates, SO_2 and oxides of nitrogen (NO_x) in stack gases; disposal of scrubber sludge and ashes
Very high	High	Fair	Large	SO_2 and NO_x in stack gases; disposal of scrubber sludge
Very high	High	Fair	Large	NO_x in stack gases
High	Highest	Good	Largest	Safety; radioactive waste disposal
Lowest	Moderate	Fair	Smallest	SO_2 and NO_x in stack gases
Lowest	Moderate	Fair	Smallest	NO_x in stack gases
Medium	Moderate	Fair	Moderate	SO_2 and NO_x in stack gases
Medium	Moderate	Fair	Moderate	NO_x in stack gases
Highest if water is available	Lowest	Limited by area	Small	Generally requires construction of a dam
High	Low	Extremely limited by area	Low	H_2S emissions from system

provisions which reliably permit the required load fluctuations and hours per year of operation.

The optimum combination or mix of base-load, intermediate-load, and peaking power-generating units of various sizes involves use of planning procedures and production cost vs. capital cost tradeoff evaluation methods.

	April	August	December	Annual
Load factor	0.667	0.586	0.652	0.623
Min. load / peak load factor	0.371	0.378	0.386	0.330

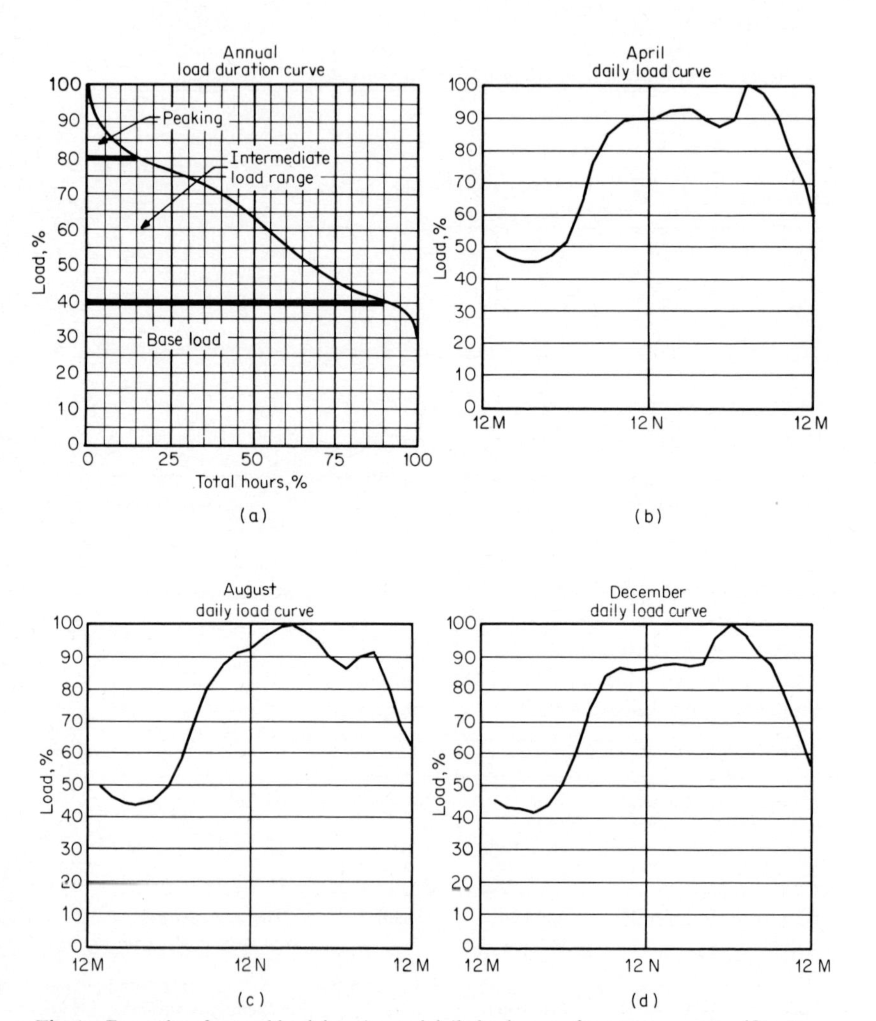

Fig. 1 Examples of annual load duration and daily load curves for a power system. (Courtesy Westinghouse Electric Corp.)

OPTIMUM ELECTRIC-POWER GENERATING UNIT

Determine the qualities of an optimum new electric-power generating unit to be applied to an existing utility system. (Table 4 is a summary of all the necessary steps in making this kind of determination.)

TABLE 4 Steps to Determine the Optimum New Electric-Power Generating Unit

Step 1	Identify all possible energy source (fuel) and electric generation system combination alternatives.
Step 2	Eliminate alternatives which fail to meet system commercial-availability criteria.
Step 3	Eliminate alternatives which fail to meet energy source (fuel) commercial-availability criteria.
Step 4	Eliminate alternatives which fail to meet other functional or site-specific criteria.
Step 5	Eliminate alternatives which are always more costly than other feasible alternatives.
Step 5*a*	Calculate the appropriate annual fixed-charge rate.
Step 5*b*	Calculate fuel costs on a dollars per million Btu basis.
Step 5*c*	Calculate the average net generation unit heat rates.
Step 5*d*	Construct screening curves for each system.
Step 5*e*	Use screening curve results to choose those alternatives to be evaluated further.
Step 5*f*	Construct screening curves for feasible renewable and alternative energy sources and generation systems and compare with alternatives chosen in Step 5*e*.
Step 6	Determine coincident maximum predicted annual loads over the entire planning period.
Step 7	Determine the required planning reserve margin.
Step 8	Evaluate the advantages and disadvantages of smaller and larger generation unit and plant ratings.
Step 8*a*	Consider the economy-of-scale savings associated with larger unit and plant ratings.
Step 8*b*	Consider the operational difficulties associated with unit ratings which are too large.
Step 8*c*	Take into account the range of ratings commercially available for each generation system type.
Step 8*d*	Consider the possibility of jointly owned units.
Step 8*e*	Consider the forecast load growth.
Step 8*f*	Determine the largest unit and plant ratings which can be used in generation expansion plans.
Step 9	Develop alternative generation expansion plans.
Step 10	Compare generation expansion plans on a consistent basis.
Step 11	Determine the optimum generation expansion plan by using an iterative process.
Step 12	Use the optimum generation expansion plan to determine the next new generation units or plants to be installed.
Step 13	Determine the generator ratings for the new generation units to be installed.
Step 14	Determine the optimum plant design.
Step 15	Evaluate tradeoff of annual operation and maintenance costs vs. installed capital costs.
Step 16	Evaluate tradeoffs of thermal efficiency vs. capital costs and/or operation and maintenance costs.
Step 17	Evaluate tradeoff of unit availability (reliability) and installed capital costs and/or operation and maintenance costs.
Step 18	Evaluate tradeoff of unit rating vs. installed capital costs.

Calculation Procedure:

1. Identify Alternatives

As indicated in Table 2, there are over 60 possible combinations of fuel and electric-power generating systems which either have been developed or are in some stage of development. In the development of power-generating expansion plans (to be considered later), it is necessary to evaluate a number of installation sequences with the various combinations of fuel and electric-power generating systems with various ratings. Even with the use of large computer programs, the number of possible alternatives is too large to reasonably evaluate. For this reason, it is necessary to reduce this large number of alternatives to a reasonable and workable number early in the planning process.

2. *Eliminate Alternatives Which Fail to Meet System Commercial Availability Criteria*

Reducing the number of alternatives for further consideration generally begins with elimination of all of those systems which are simply not developed to the stage where they can be considered to be available for installation on a utility system in the required time period. The alternatives which might typically be eliminated for this reason are indicated in Table 5.

3. *Eliminate Alternatives which Fail to Meet Energy-Source Fuel Commercial Availability Criteria*

At this point, the number of alternatives is further reduced by elimination of all of those systems that require fuels which are generally not commercially available in the required quantities. Alternatives which might typically be eliminated for this reason are also indicated in Table 5.

4. *Eliminate Alternatives which Fail to Meet Other Functional or Site-Specific Criteria*

In this step those alternatives from Table 2 are eliminated which, for one reason or another, are not feasible for the particular existing utility power system involved. Such systems might include wind (unless 100- to 200-kW units with a fluctuating and inter-

TABLE 5 Systems Which Might Be Eliminated

Reason for elimination	*Systems eliminated*
Systems are not commercially available for installation on a utility system.	Fuel-cell systems Magnetohydrodynamic systems Thermoelectric systems Thermionic systems Solar photovoltaic or thermal-cycle systems Ocean thermal gradient open steam or closed ammonia cycle systems Ocean-wave hydraulic systems Nuclear-fusion systems
Energy source (fuel) is not commercially available in the required quantities.	Shale oil Tar sands Coal-derived liquids and gases Wood Vegetation Hydrogen Refuse-derived fuels
Systems typically do not satisfy other functional, feasibility, or site-specific criteria.	Wind Geothermal Conventional hydroelectric Tidal hydroelectric

ruptible power output can suffice), geothermal (unless the utility is located in the geyser regions of northern California), conventional hydroelectric (unless the utility is located in a region where elevated water is either available or can feasibly be made available by the construction of a river dam), and tidal hydroelectric (unless the utility is located near one of the few feasible oceanic coastal basin sites).

Table 5 is intended to be somewhat representative of the current technology. It is by no means, however, intended to be all-inclusive or representative for all electric-power generating installation situations. For example, in certain situations the electric-power generating systems which are used extensively today (such as coal and oil) may be similarly eliminated for such reasons as inability to meet government clean-air and/or disposal standards (coal), fuel unavailability for a variety of reasons including government policy (oil or natural gas), lack of a site where a dam can be constructed without excessive ecological and socioeconomic impacts (hydroelectric), or inability to obtain the necessary permits and licenses for a variety of environmental and political reasons (nuclear).

Similarly, even now, it is conceivable that in a specific situation a number of those alternatives which were eliminated such as wind and wood, might be feasible. Also, in the future, several of the generating systems such as solar photovoltaic might become available in the required ratings or might be eliminated because of some other feasibility criterion.

It should be emphasized that for each specific electric-generating-system installation it is necessary to identify those energy alternatives which must be eliminated from further consideration on the basis of criteria which are appropriate for the specific situation under consideration.

5. Eliminate Alternatives Which Are Always More Costly than Other Feasible Alternatives

In this step those remaining fuel and electric-power generating system alternatives from Table 2 are eliminated from further consideration which will not, under any reasonable foreseeable operational criteria, be less costly than other feasible alternatives.

Typically, the elimination of alternatives in this stage is based on a comparison of the total power-generating costs of the various systems, considering both the fixed costs (i.e., capital plus fixed operation and maintenance costs) and the production costs (fuel costs plus variable operation and maintenance costs) for the various systems.

This comparison is generally made by means of *screening curves,* such as those in Figs. 2 through 4. For each combination of fuel and electric-power generating system still under consideration, the annual operation costs per installed kilowatt (dollars per year per kilowatt) is plotted as a function of capacity factor (or equivalent full-load operation hours per year).

ANNUAL CAPACITY FACTOR

Determine the annual capacity factor of a unit rated at 100 MW which produces 550,000 MWh per year.

Calculation Procedure:

1. Compute Annual Capacity Factor as a Percentage

The factor is

$$\left(\frac{550{,}000\ \text{MWh}/100\ \text{MW}}{8760\ \text{h/yr}}\right)100\% = 68.2\%$$

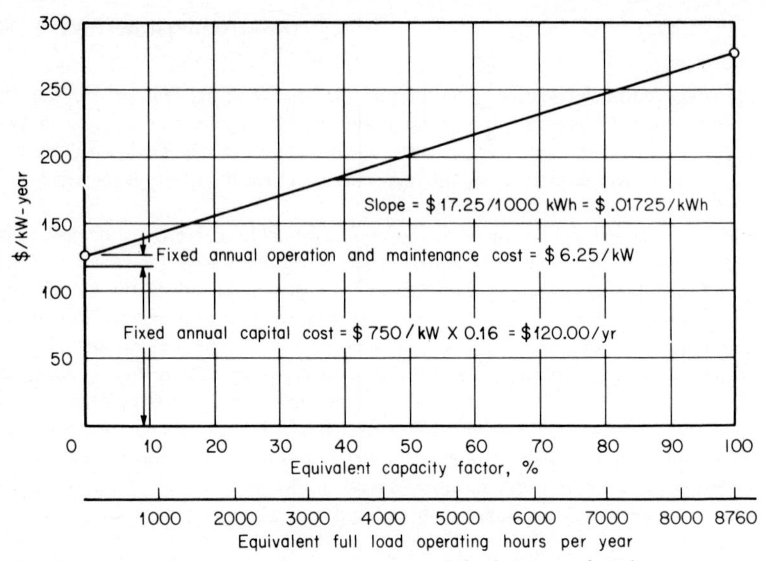

Fig. 2 Construction of a screening curve for a coal-fired steam-cycle unit.

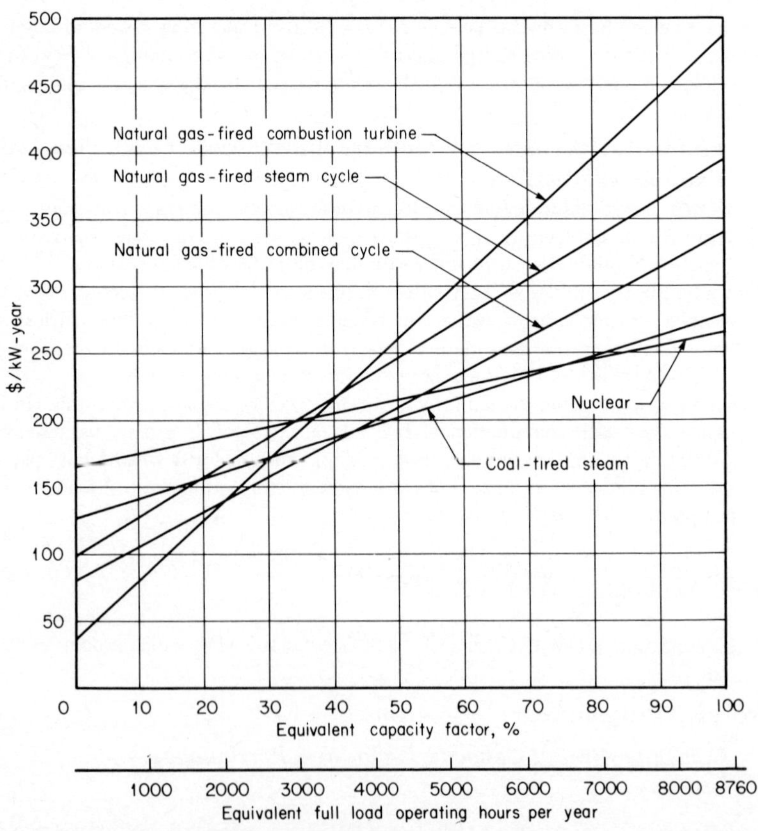

Fig. 3 Screening curves for electric-generation system alternatives based on assumption of availability of natural gas.

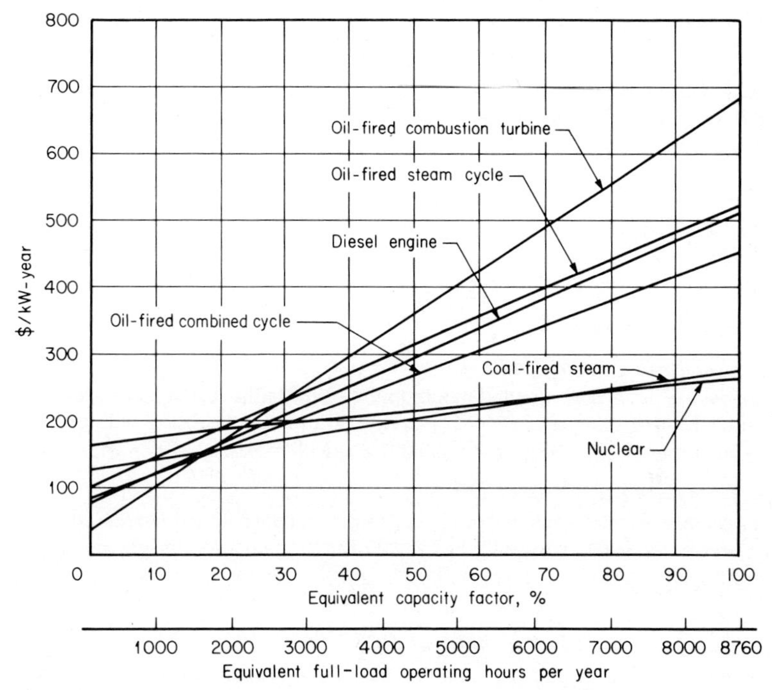

Fig. 4 Screening curves for electric-generation system alternatives based on assumption that natural gas is unavailable.

2. Compute Annual Capacity Factor in Hours per Year

The factor is (68.2/100) (8760 h/yr) = 5550 h/yr.

ANNUAL FIXED-CHARGE RATE

Estimate the annual fixed rate for an investor-owned electric utility.

Calculate Procedure:

1. Examine the Appropriate Factors

As shown in Table 6, the annual fixed-charge rate represents the average, or "levelized," annual carrying charges including interest or return on the installed capital, depre-

TABLE 6 Typical Fixed-Charge Rate for Investor-Owned Electric Utility

Charge	*Rate, %*
Return	7.7
Depreciation	1.4
Taxes	6.5
Insurance	0.4
Total	16.0

ciation or return of the capital, tax expense, and insurance expense associated with the installation of a particular generating unit for the particular utility or company involved.

Related Calculations: Fixed-charge rates for investor-owned utilities generally range from 15 to 20 percent; fixed-charge rates for publicly owned utilities are generally about 5 percent lower.

FUEL COSTS

Calculate fuel costs on a dollars per megajoule (and million Btu) basis.

Calculation Procedure:

1. Compute Cost of Coal

On a dollars per megajoule (dollars per million Btu) basis, coal at \$39.68/tonne (\$36/ton) with a heating value of 27.915 MJ/kg (12,000 Btu/lb), the cost is (\$39.68/tonne)/[(1000 kg/tonne)(27.915 MJ/kg)] = \$0.001421/MJ = \$1.50/million Btu.

2. Compute the Costs of Oil

On a dollars per megajoule basis, oil at \$28 per standard 42-gal barrel (\$0.17612/L) with a heating value of 43.733 MJ/kg (18,800 Btu/lb) and a specific gravity of 0.91, the cost is (\$0.17612/L)/[(43.733 MJ/kg)(0.91 kg/L)] = \$0.004425/MJ = \$4.67/million Btu.

3. Compute the Cost of Natural Gas

On a dollars per megajoule basis, natural gas at \$0.1201/m^3 (\$3.40 per thousand standard cubic feet) with a heating value of 39.115 MJ/m^3 = 1050 Btu/1000 ft^3 costs (\$0.1201/m^3)/(39.115 MJ/m^3) = \$0.00307/MJ = \$3.24/million Btu.

4. Compute Cost of Nuclear Fuel

On a dollars per megajoule basis, nuclear fuel at \$75.36/MWday costs (\$75.36/MWday)/[(1.0 J/MWs)(3600 s/h)(24 h/day)] = \$0.00087/MJ = \$0.92/million Btu.

AVERAGE NET HEAT RATES

A unit requires 158,759 kg/h (350,000 lb/h) of coal with a heating value of 27.915 MJ/kg (12,000 Btu/lb) to produce 420,000 kW output from the generator. In addition, the unit has electric power loads of 20,000 KW from required power-plant auxiliaries, such as boiler feed pumps. Calculate the average net generation unit rate.

Calculation Procedure:

1. Define the Net Heat Rate

The average net heat rate (in Btu/kWh or J/kWh) of an electric-power generating unit is calculated by dividing the total heat input to the system (in units of Btu/h or MJ/h) by the net electric power generated by the plant (in kilowatts), taking into account the boiler, turbine, and generator efficiencies and any auxiliary power requirements.

2. Compute the Total Heat Input to Boiler

Total heat input to boiler equals (158,759 kg/h)(27.915 MJ/kg) = 4.43×10^6 MJ/h = 4200×10^6 Btu/h.

3. Compute the Net Power Output of the Generating Unit
Net generating-unit power output is 420,000 kW − 20,000 kW = 400,000 kW.

4. Determine the Net Heat Rate of the Generating Unit
The net generating unit heat rate is $(4.43 \times 10^6$ MJ/h)/400,000 kW = 11.075 MJ/kWh = 10,500 Btu/kWh.

CONSTRUCTION OF SCREENING CURVE

A screening curve provides a plot of cost per kilowattyear as a function of capacity factor or operating load. An example is the screening curve of Fig. 2 for a coal-fired steam-cycle system, based on the data in Table 7. Assume the total installed capital cost for a 600-MW system is $450 million and the fixed-charge rate is 16 percent. In addition, assume the total fixed operation and maintenance cost is $3,750,000 per year for the unit. Verify the figures given in Fig. 2.

Calculation Procedure:

1. Determine the Fixed Annual Capital Cost
The installed cost per kilowatt is ($450 × 10^6)/600,000 kW = $750/kW. Multiplying by the fixed charge rate, we obtain the fixed annual cost: ($750/kW)(0.16) = $120/kWyr.

2. Compute Fixed Operation and Maintenance Costs
Fixed operation and maintenance cost on a per-kilowatt basis is ($3,750,000/yr)/600,000 kW = $6.25/kWyr.

3. Compute Cost per Year at a Capacity Factor of Zero
The cost in dollars per year per kilowatt at a capacity factor of zero is $126.25/kWyr, which is the sum of the annual fixed capital cost $120/kWyr plus the annual fixed operation and maintenance cost of $6.25/kWyr.

4. Determine the Fuel Cost
Coal at $39.68/tonne with a heating value of 27.915 MJ/kg costs $0.001421/MJ as determined in a previous example. With an average unit heat rate of 11.075 MJ/kWh, the fuel cost for the unit on a dollars per kilowatthour basis is ($0.001421/MJ)(11.075 MJ/kWh) = $0.01575/kWh.

With a levelized variable operation and maintenance cost for the system of $0.00150/kWh (Table 7), the total variable production cost for the coal-fired steam-cycle unit is $0.01725/kWh (i.e., the fuel cost $0.01575/kWh plus the variable operation and maintenance cost of $0.00150/kWh, or $0.01725/kWh). Hence, the total annual fixed and variable costs on a per-kilowatt basis to own and operate a coal-fired steam-cycle system 8760 h per year (100 percent capacity factor) would be $126.25/kWyr + ($0.01725/kWh)(8760 h/yr) = 126.26/kWh + $151.11/kWyr = $277.36/kWyr.

Related Calculation: As indicated in Fig. 2, the screening curve is linear. The y intercept is the sum of the annual fixed capital, operation, and maintenance costs and is a function of the capital cost, fixed-charge rate, and fixed operation and maintenance cost. The slope of the screening curve is the total variable fuel, operation, and maintenance cost for the system (i.e., $0.01725/kWh), and is a function of the fuel cost, heat rate, and variable operation and maintenance costs.

TABLE 7 Data Used for Screening Curves of Figs. 2 Through 4

System	*Total installed capital cost, millions of dollars*	*Unit rating, MW*	*Installed capital cost, \$/kW*	*Annual levelized fixed-charge rate, %*	*Annual fixed capital cost, \$/kWyr*	*Total annual fixed O&M cost, millions of dollars per year*	*Annual fixed O&M costs, \$/kWyr*	*Annual fixed capital, O&M costs, \$/kWyr*
Coal-fired steam cycle	450.0	600	750	16	120.00	3.75	6.25	126.25
Oil-fired steam cycle	360.00	600	600	16	96.00	3.30	5.50	101.50
Natural-gas-fired steam cycle	348.0	600	580	16	92.80	3.00	5.00	97.80
Nuclear	900.0	900	1000	16	160.00	5.13	5.70	165.07
Oil-fired combined cycle	130.5	300	435	18	78.30	1.275	4.25	82.55
Natural-gas-fired combined cycle	126.0	300	420	18	75.60	1.20	4.00	79.60
Oil-fired combustion turbine	8.5	50	170	20	34.00	0.175	3.50	37.50
Natural-gas-fired combustion turbine	8.0	50	160	20	32.00	0.162	3.25	35.25
Diesel engine	3.0	8	375	20	75.00	0.024	3.00	78.00

*\$/t = \$/ton × 1.1023
\$/L = (\$/42-gal barrel) × 0.00629
\$/m^3 = (\$/MCF) × 0.0353

†MJ/kg = (Btu/lb) × 0.002326
MJ/m^3 = (Btu/SCF) × 0.037252

‡MJ/t = (million Btu/ton) × 1163
MJ/m^3 = (million Btu/bbl) × 6636
MJ/m^3 = (million Btu/MCF) × 37.257
MJ/MWday = (million Btu/MWday) × 1055

§\$/MJ = (\$/million Btu) × 0.000948

¶MJ/kWh = (Btu/kWh) × 0.001055
J/kWh = (Btu/kWh) × 1055

Table 7 shows typical data and screening curve parameters, for all those combinations of energy source and electric-power generating system (listed in Table 2) which were not eliminated on the basis of some criterion in Table 5. Table 7 represents those systems which would generally be available today as options for an installation of an electric-power-generating unit.

Fuel costs, $/ standard unit*	*Fuel energy content†*	*Energy content per standard unit,‡ millions of Btu per standard unit*	*Fuel cost,§ dollars per million Btu*	*Average net heat rate for unit,▸ Btu/ kWh*	*Fuel cost, $/ kWh*	*Levelized variable O&M costs, $/ kWh*	*Total variable costs (fuel + variable O&M cost), $/ kWh*
$36/ton	12,000 Btu/lb	24 million Btu/ton	1.50	10,500	0.01575	0.00150	0.01725
$28/bbl	18,800 Btu/lb with 0.91 specific gravity	6 million Btu/barrel	4.67	10,050	0.04693	0.00130	0.04823
$3.40 per 1000 ft^3 (MCF)	1050 Btu per standard cubic foot (SCF)	1.05 million Btu/MCF	3.24	10,050	0.03256	0.00120	0.03376
$75.36 per MWday		81.912 million Btu/MWday	.92	11,500	0.01058	0.00085	0.01143
$28/bbl	18,800 Btu/lb with 0.91 specific gravity	6 million Btu/barrel	4.67	8,300	0.03876	0.00350	0.04226
$3.40/ MCF	1050 Btu/ SCF	1.05 million Btu/MCF	3.24	8,250	0.02673	0.00300	0.02973
$28/bbl	18,800 Btu/lb with 0.91 specific gravity	6 million Btu/barrel	4.67	14,700	0.06865	0.00500	0.07365
$3.40/ MCF	1050 Btu/ SCF	1.05 million Btu/MCF	3.24	14,500	0.04698	0.00450	0.05148
$28/bbl	18,800 Btu/lb with 0.91 specific gravity	6 million Btu/bbl	4.67	10,000	0.04670	0.00300	0.04970

Figure 3 illustrates screening curves plotted for all non-oil-fired systems in Table 7. For capacity factors below 23.3 percent (2039 equivalent full-load operating hours per year), the natural-gas–fired combustion turbine is the least costly alternative. At capacity factors from 23.3 to 42.7 percent, the natural-gas–fired combined-cycle system is the most economical. At capacity factors from 42.7 to 77.4 percent, the coal-fired steam-cycle system provides the lowest total cost, and at capacity factors above 77.4 percent, the nuclear plant offers the most economic advantages. From this it can be concluded that the optimum generating plan for a utility electric power system would consist of some rating and installation sequence combination of those four systems.

Steam-cycle systems, combined-cycle systems, and combustion-turbine systems can generally be fired by either natural gas or oil. As indicated in Table 7, each system firing with oil rather than natural gas generally results in higher annual fixed capital cost and higher fixed and variable operation, maintenance, and fuel costs. Consequently, oil-fired systems usually have both higher total fixed costs and higher total variable costs than natural-gas–fired systems. For this reason, firing with oil instead of natural gas results in higher total costs at all capacity factors.

In addition, as indicated in Table 3, the fact that oil-fired systems generally have more environmental impact than natural-gas systems means that if adequate supplies of natural gas are available, oil-fired steam cycles, combined cycles, and combustion-turbine systems would be eliminated from further consideration. They would never provide any benefits relative to natural-gas–fired systems.

If natural gas were not available, the alternatives would be limited to the non-natural-gas–fired systems of Table 7. The screening curves for these systems are plotted in Fig. 4. From the figure, if natural gas is unavailable, at capacity factors below 16.3 percent oil-fired combustion turbines are the least costly alternative. At capacity factors from 16.3 to 20.0 percent, oil-fired combined-cycle systems are the most economical. At capacity factors from 20.0 to 77.4 percent a coal-fired steam-cycle system provides the lowest total cost, and at capacity factors above 77.4 percent nuclear is the best.

If any of those systems in Table 5 (which were initially eliminated from further consideration on the basis of commercial availability or functional or site-specific criteria) are indeed possibilities for a particular application, a screening curve should also be constructed for those systems. The systems should then be evaluated in the same manner as the systems indicated in Figs. 3 and 4. The screening curves for renewable energy sources such as hydroelectric, solar, wind, etc., are essentially horizontal lines because the fuel, variable operation, and maintenance costs for such systems are negligible.

NONCOINCIDENT AND COINCIDENT MAXIMUM PREDICTED ANNUAL LOADS

For a group of utilities which are developing generating-system expansion plans in common, the combined maximum predicted annual peak loads used in generating-system expansion studies should be the coincident maximum loads (demands) expected in the year under consideration. Any diversity (or noncoincidence) in the peaks of the various utilities in the group should be considered. Such diversity, or noncoincidence, will in general be most significant if all the various utilities in the planning group do not experience a peak demand in the same season.

Assume the planning group consists of four utilities which have expected summer and winter peak loads in the year under consideration as indicated in Table 8. Determine the noncoincident and coincident annual loads.

Calculation Procedure:

1. Analyze the Data in Table 8

Utilities A and D experience the highest annual peak demands in the summer season and utilities B and C experience the highest annual peak demands in the winter season.

2. Compute Noncoincident Demands

Total noncoincident summer maximum demand for the group of utilities is less than the annual noncoincident maximum demand by [1 − (8530 MW/8840 MW)](100 percent) = 3.51 percent. The total noncoincident winter maximum demand is less than

TABLE 8 Calculation of Coincident Maximum Demand for a Group of Four Utilities

	Maximum demand, MW		
	Summer	*Winter*	*Annual*
Utility A	3630	3150	3630
Utility B	2590	2780	2780
Utility C	1780	1900	1900
Utility D	530	410	530
Total noncoincident maximum demand	8530	8240	8840
Seasonal diversity factor	0.9496	0.9648	
Total coincident maximum demand	8100 MW	7950 MW	8100 MW*

*Maximum of summer and winter.

the annual noncoincident maximum demand by [1 − (8240 MW/8840 MW)](100 percent) = 6.79 percent.

3. Compute Coincident Demands

If the seasonal diversity for the group averages 0.9496 in the summer and 0.9648 in the winter, the total coincident maximum demand values used for generation expansion planning in that year would be (8530 MW)(0.9496) = 8100 MW in the summer and (8240 MW)(0.9648) = 7950 MW in the winter.

REQUIRED PLANNING RESERVE MARGIN

All utilities must plan to have a certain amount of reserve generation capacity to supply the needs of their power customers in the event that a portion of the installed generating capacity is unavailable.

Reserve generating capability is also needed to supply any expected growth in the peak needs of electrical utility customers that might exceed the forecast peak demands. In generating-system expansion planning such reserves are generally identified as a percentage of the predicted maximum annual hourly demand for energy.

Compute the reserve capacity for a group of utilities (Table 8) having a predicted maximum hourly demand of 8100 MW.

Calculation Procedure:

1. Determine What Percentage Increases Are Adequate

Lower loss-of-load probabilities are closely related to higher planning reserve margins. Experience and judgment of most utilities and regulators associated with predominantly thermal power systems (as contrasted to hydroelectric systems) has shown that planning reserves of 15 to 25 percent of the predicted annual peak hourly demand are adequate.

2. Calculate Reserve and Installed Capacity

The range of additional reserve capacity is (0.15)(8100) = 1215 MW to (0.25)(8100) = 2025 MW. The total installed capacity is, therefore, 8100 + 1215 = 9315 MW to 8100 + 2025 = 10,125 MW.

Related Calculations: The reliability level of a particular generation expansion plan for a specific utility or group of utilities is generally determined from a loss-of-load probability (LOLP) analysis. Such an analysis determines the probability that the utility, or group of utilities, will lack sufficient installed generation capacity on-line to meet the electrical demand on the power system. This analysis takes into account the typical unavailability, because both planned (maintenance) outages and unplanned (forced) outages, of the various types of electric-power generating units which comprise the utility system.

For generating-system expansion planning a maximum loss-of-load probability value of 1 day in 10 years has traditionally been used as an acceptable level of reliability for an electric-power system. Owing primarily to the rapid escalation of the costs of power-generating-system equipment and to limits in an electrical utility's ability to charge rates which provide for the financing of large construction projects, the trend recently has been to consider higher loss-of-load probabilities as possibly being acceptable.

Planning for generating-system expansion on a group basis generally results in significantly lower installed capacity requirements than individual planning by utilities. For example, collectively the utilities in Table 8 would satisfy a 15 percent reserve requirement with the installation of 9315 MW, whereas individually, on the basis of total annual noncoincident maximum loads, utilities would install a total of 10,166 MW to retain the same reserve margin of 15 percent.

Generating-expansion planning by a group also requires a certain amount of joint planning of the electrical transmission system to ensure that the interconnections between the various utilities in the planning group have sufficient capacity to facilitate the seasonal

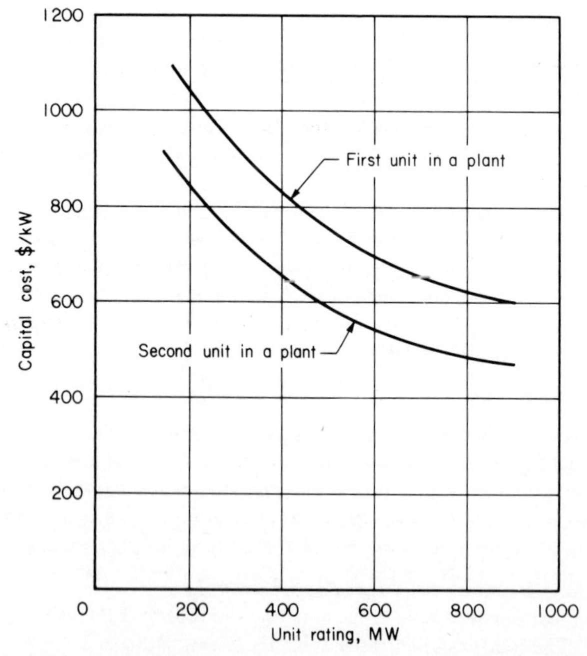

Fig. 5 Typical capital costs vs. unit rating trend for first and second coal-fired steam-cycle units.

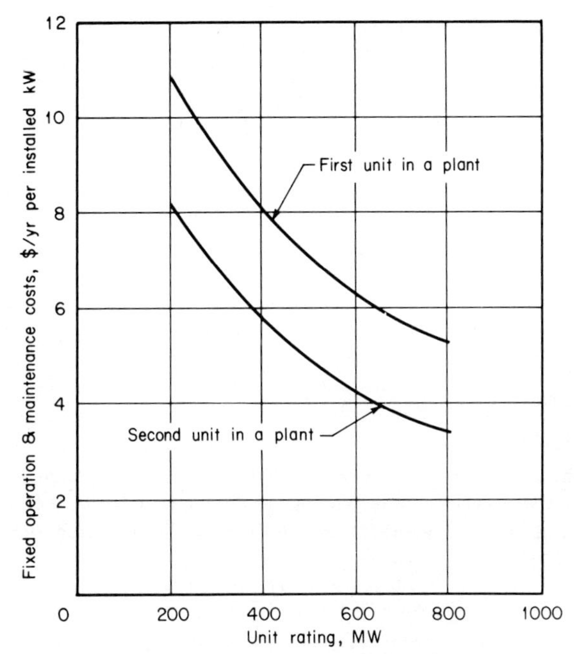

Fig. 6 Typical fixed operation and maintenance costs vs. unit rating trend for first and second coal-fired steam-cycle units.

transfer of power between the utilities. This enables each utility in the planning group to satisfy the applicable reserve criteria at all times of the year.

For all types of electric-power generating units, it is generally the case that smaller unit ratings have higher installed capital costs (on a dollar per kilowatt basis) and higher annual fixed operation and maintenance costs (on a dollar per kilowatthour basis), with the increase in the costs becoming dramatic at the lower range of ratings which are commercially available for that type of electric-power generating unit. Figures 5 and 6 illustrate this for coal-fired steam units. Smaller generating units also generally have somewhat poorer efficiencies (higher heat rates) than the larger units.

The installation of generating units which are too large, however, can cause a utility to experience a number of operational difficulties. These may stem from excessive operation of units at partial loads (where unit heat rates are poorer) or inability to schedule unit maintenance in a manner such that the system will always have enough spinning reserve capacity on-line to supply the required load in the event of an unexpected (forced) outage of the largest generation unit.

In addition, it is not uncommon for large units to have somewhat higher forced-outage rates than smaller units, which implies that with larger units a somewhat larger planning reserve margin might be needed to maintain the same loss-of-load probability.

RATINGS OF COMMERCIALLY AVAILABLE SYSTEMS

The ratings of the various system types indicated in Table 7 are typical of those which are commonly available today. Actually, each of the various systems is commercially available in the range of ratings indicated in Table 9. Evaluate the different systems.

TABLE 9 Commercially Available Unit Ratings

Type	*Configuration**	*Rating range*
Fossil-fired (coal, oil, natural gas) steam turbines	Tandem-compound	20 MW to 800 MW
	Cross-compound	500 MW to 1300 MW
Nuclear steam turbine	Tandem-compound	500 MW to 1300 MW
Combined-cycle systems	Two- or three-shaft	100 MW to 300 MW
Combustion-turbine systems	Single-shaft	< 1 MW to 110 MW
Hydroelectric systems	Single-shaft	< 1 MW to 800 MW
Geothermal systems	Single-shaft	< 20 MW to 135 MW
Diesel systems	Single-shaft	< 1 MW to 20 MW

*See Fig. 7.

Calculation Procedure:

1. Consider Nuclear Units

Nuclear units are available in tandem-compound turbine configurations (Fig. 7*a*) where a high-pressure (HP) turbine, and one to three low-pressure (LP) turbines are on one shaft system driving one generator at 1800 r/min in ratings from 500 MW to 1300 MW.

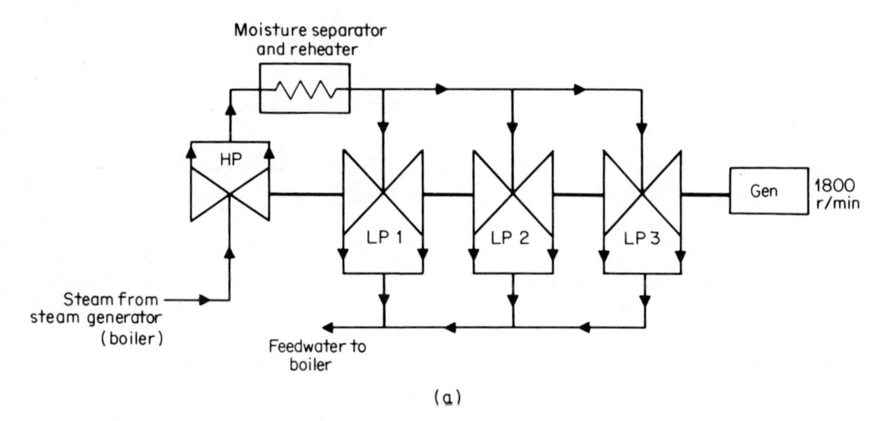

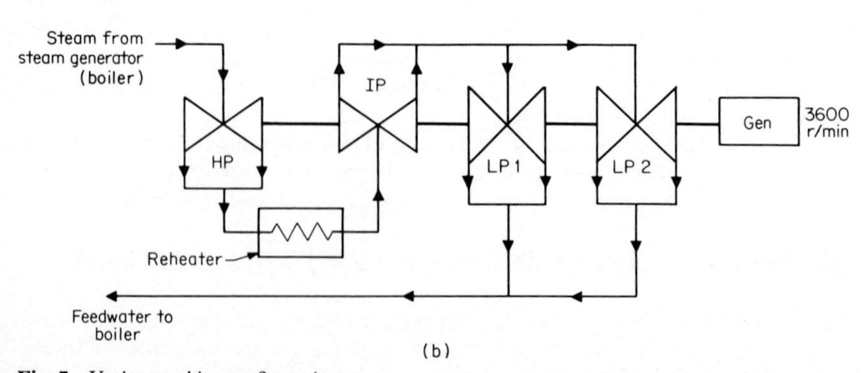

Fig. 7 Various turbine configurations.

2. *Consider Fossil-Fuel Units*

Steam units fired by fossil fuel (coal, oil, or natural gas) are available in tandem-compound 3600-r/min configurations (Fig. 7*b*) up to 800 MW, and in cross-compound turbine configurations (Fig. 7*c*) where an HP and intermediate-pressure (IP) turbine are on one shaft driving a 3600-r/min electric generator. One or more LP turbines on a second shaft system drive an 1800-r/min generator in ratings from 500 to 1300 MW.

Related Calculations: Combined-cycle systems are generally commercially available in ratings from 100 to 300 MW. Although a number of system configurations are available, it is generally the case that the gas- and steam-turbine units in combined-cycle systems drive separate generators, as shown in Fig. 7*d*.

Because nuclear and larger fossil-fired unit ratings are often too large for a particular company or electric utility to assimilate in a single installation, it has become common for

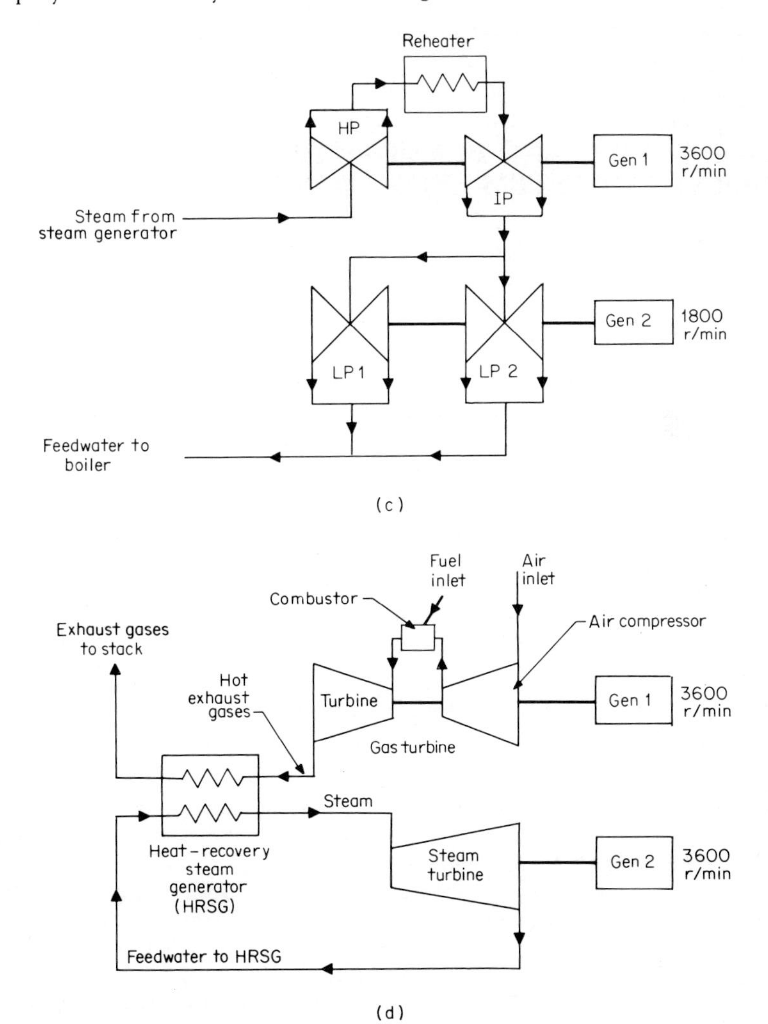

Fig. 7 (*Cont.*)

the smaller and medium-size utilities to install and operate these types of units on a joint, or pool, basis. In this arrangement, one utility has the responsibility for installation and operation of the units for all of the partners. Each of the utilities pays a percentage of all capital and operation costs associated with the unit in accordance with the ownership splits.

Such an arrangement enables all of the owners of the unit to reap the benefits of the lower installed capital costs (dollars per kilowatt) and lower operation costs (dollars per kilowatthour) typical of larger-size units. The operational difficulties associated with having too much of an individual company's total installed capacity in a single generating unit are minimized.

For a number of technical and financial reasons (including excessive fluctuations in reserve margins, uneven cash flows, etc.) utilities find it beneficial to provide for load growth with capacity additions every 1 to 3 years. Hence, the rating of units used in a generating-system expansion plan is to a certain extent related to the forecast growth in the period.

LARGEST UNITS AND PLANT RATINGS USED IN GENERATING-SYSTEM EXPANSION PLANS

The group of utilities in Table 8 is experiencing load growth as shown in Table 10. Determine the largest nuclear and fossil unit ratings allowable through year 15 and beyond.

Calculation Procedure:

1. Select Unit Ratings

Generally the largest unit installed should be 7 to 15 percent of the peak load of the utility group. For this reason, in columns 5 and 7 of Table 10, 900-MW nuclear and 600-MW fossil units are selected through year 15. Beyond year 15, 1100-MW nuclear and 800-MW fossil were chosen.

2. Determine the Ratings for 1 Percent Growth per Year

For lower annual load growth of 1 percent per year instead of 2.1 to 3.2 percent as indicated, financial considerations would probably encourage the utility to install 600-MW nuclear and 300- to 400-MW fossil units instead of the 600- to 1100-MW units.

ALTERNATIVE GENERATING-SYSTEM EXPANSION PLANS

At this point, it is necessary to develop numerous different generating-system expansion plans or strategies. The development of two such plans or strategies is indicated in Table 10. The plans should be based upon the forecast maximum (peak) coincident hourly electrical demand (load) for each year in the planning period for the group of utilities which are planning together. The planning period for such studies is commonly 20 to 40 yr.

If the installed capacity for the group of utilities is initially 9700 MW, columns 5 and 7 might be representative of two of the many generating-system expansion plans or strategies which a planner might develop to provide the required capacity for each year in the planning period. Determine the total installed capacity and percentage reserve for plan B.

Calculation Procedure:

1. Compute the Installed Capacity in Year 6

The total installed capacity in year 6 is 9700 MW initially, plus a 300-MW combined-cycle unit in year 2, plus a 900-MW nuclear unit in year 3, plus a 50-MW natural-gas–fired combustion-turbine unit in year 5, plus a 600-MW coal-fired steam unit in year 6 = 11,550 MW. This exceeds the required installed capacity of 11,209 MW.

2. Compute the Reserve Percentage

The percentage reserve is [(11,550 MW/9,747 MW) − 1.0](100 percent) = 18.5 percent, which exceeds the targeted planning reserve level of 15 percent.

Related Calculations: The excess of the actual reserves in a given generating-system expansion plan over the targeted planning reserve level increases the total cost of the plan, but also to some degree improves the overall reliability level. It therefore needs to be considered in the comparison of generating-system plans.

The generating-system expansion plans developed usually contain only those types of electric-power generating systems which were found in the screening curve analysis to yield minimum total annual cost in some capacity factor range. For example, if natural gas is available in sufficient quantities over the planning period, the types of electric-power generating systems used in the alternative generating-system expansion plans would be limited to nuclear units, coal-fired steam-cycle units, and natural-gas–fired combined-cycle and combustion-turbine units, as indicated in Table 10.

After a number of different generating-system expansion strategies are developed, the plans must be compared on a consistent basis so that the best plan to meet a given reliability index can be determined. The comparison between the various generating-system expansion plans is generally performed by calculating for each plan the production and investment costs over the life of the plan (20 to 40 yr) and then evaluating those costs using discounted revenue requirements (i.e., present worth, present value) techniques (see Sec. 19).

The production costs for each generating-system expansion plan are generally calculated by large computer programs which simulate the dispatching (or loading) of all the units on the entire power system, hourly or weekly, over the entire planning period. These programs generally employ a probabilistic technique to simulate the occasional unavailability of the various units on the power system. In addition, load forecast, economic, and technical data for each existing and new unit on the power system and for the power system as a whole (as indicated in the first four columns of Table 11) are required.

The investment costs for each plan are generally calculated by computer programs which simulate the net cash flows due to the investments in the various plans. Annual book depreciation, taxes, insurance, etc., appropriate for the particular utility involved, are considered. These programs generally require the economic data and corporate financial model data indicated in the last column of Table 11.

To determine the optimum generating-system plan over the planning period, sufficient generating-system expansion plans similar to those indicated in Table 10 must be developed and evaluated so that all of the reasonable combinations of electric-power–generating-system types, ratings, and installation timing sequences are represented. Even with the use of large computer programs, the number of possible alternative plans based on all combinations of plant types, ratings, etc. becomes too cumbersome to evaluate in detail.

TABLE 10 Two Alternative Generation Expansion Plans Developed for a Utility

Column 1	*Column 2*	*Column 3*	*Column 4*
Year	*Forecast annual growth in peak load coincident, %/yr*	*Forecast maximum or peak coincident demand or load, MW*	*Required installed capacity with 15% minimum reserve margin, MW*
0 (current year)		8,100	
1	3.2	8,359	9,613
2	3.2	8,627	9,921
3	3.2	8,903	10,238
4	3.2	9,188	10,566
5	3.2	9,482	10,904
6	2.8	9,747	11,209
7	2.8	10,020	11,523
8	2.8	10,301	11,846
9	2.8	10,589	12,177
10	2.1	10,811	12,433
11	2.1	11,038	12,694
12	2.1	11,270	12,961
13	2.1	11,507	13,233
14	2.5	11,795	13,564
15	2.5	12,089	13,903
16	2.5	12,392	14,250
17	2.5	12,701	14,607
18	2.7	13,044	15,001
19	2.7	13,397	15,406
20	2.7	13,758	15,822
21	2.7	14,130	16,249
22	2.7	14,511	16,688
23	3.0	14,947	17,189
24	3.0	15,395	17,704
25	3.0	15,857	18,235
26	3.0	16,333	18,782
27	3.0	16,823	19,346
28	3.0	17,327	19,926
29	3.0	17,847	20,524
30	3.0	18,382	21,140

Key: N = nuclear steam-cycle unit, C = coal-fired steam-cycle unit, CC = natural-gas–fired combined-cycle unit, and CT = natural-gas–fired combustion turbine.

Column 5	Column 6	Column 7	Column 8
Generating-system expansion plan A		*Generating-system expansion plan B*	
Capacity installation, MW	*Total installed capacity, MW*	*Capacity installation, MW*	*Total installed capacity, MW*
9700	9,700	9700	9,700
——	9,700	——	9,700
600 C	10,300	300 CC	10,000
50 CT	10,350	900 N	10,900
900 N	11,250	——	——
——	11,250	50 CT	10,950
50 CT	11,300	600 C	11,550
300 CC	11,600	50 CT	11,600
600 C	12,220	300 CC	11,900
50 CT	12,250	900 N	12,800
900 N	13,150	——	——
——	13,150	50 CT	12,850
——	13,150	600 C	13,450
300 CT	13,450	——	——
600 C	14,050	300 CC	13,750
——	——	900 N	14,650
1100 N	15,150	——	——
——	15,150	50 CT	14,700
50 CT	15,200	800 C	15,500
300 CC	15,500	300 CC	15,800
800 C	16,300	50 CT	15,850
50 CT	16,350	1100 N	16,950
800 C	17,150	——	——
50 CT	17,200	800 C	17,750
1100 N	18,300	——	——
——	——	1100 N	18,850
800 C	19,100	——	——
1100 N	20,200	800 C	19,650
——	——	300 CC	19,950
800 C	21,000	1100 N	21,050
800 C	21,800	800 C	21,850

TABLE 11 Data Generally Required for Computer Programs to Evaluate Alternative Expansion Plans

Load-forecast data	*Data for each existing unit*	*Data for each new unit*	*General technical data regarding power system*	*Economic data and corporate financial model data*
Generally determined from an analysis of historical load, energy requirement, and weather-sensitivity data using probablistic mathematics	Fuel type	Capital cost and/or levelized carrying charges	Units required in service at all times for system area protection and system integrity	Capital fuel and O&M costs and inflation rates for various units
Future annual load (MW) and system energy requirements (MWh) on a seasonal, monthly, or weekly basis.	Fuel cost	Fuel type	Hydroelictric unit type and data—run of river, pondage, or pumped storage	Carrying charge or fixed-charge rates for various units
Seasonal load variations	Unit incremental heat rates (unit efficiency)	Fuel costs	Minimum fuel allocations (if any)	Discount rate (weighted cost or capital)
Load-peak variance	Unit fuel and start-up	Unit incremental heat rates	Future system load data	Interest rate during construction
Load diversity (for multiple or interconnected power systems)	Unit maximum and minimum rated capacities	Unit fuel and start-up costs	Data for interconnected company's power system or power pool	Planning period (20 to 50 yr)
Sales and purchases to other utilities	Unit availability and reliability data such as partial and full forced outage rates	Unit availability and reliability data such as mature and immature full and partial forced outage rates and scheduled outages	Reliability criteria such as: a spinning reserve or loss-of-load probability (LOLP) operational requirements	Book life, tax life, depreciation rate and method, and salvage value or decommissioning cost for each unit
Seasonally representative load-duration curve shapes	Scheduled outage rates and maintenance schedules	Unit commercial operation dates	Limitations on power system ties and interconnections with the pool and/or other companies	Property and income tax rates
	O&M (fixed, variable and average)	Unit maximum and minimum capacities	Load management	Investment tax credits
	Seasonal derating (if any) and seasonal derating period	Sequence of unit additions	Required licensing and construction lead time for each type of generation unit	Insurance rates
	Sequence of unit retirements (if any)	Operation and maintenance costs (fixed and variable)		
	Minimum downtime and/or dispatching sequence (priority of unit use)	Time required for licensing and construction of each type of unit		

TABLE 12 Time Required to License and Construct Power Plants in the United States

Type	*Years*
Nuclear	8 to 14
Fossil-fired steam	6 to 10
Combined-cycle units	4 to 8
Combustion turbine	3 to 5

It is generally the case, therefore, that generating-system planners use an iterative process to determine the optimum plan.

For example, early in the evaluation process, a smaller number of alternative plans is evaluated. On the basis of a preliminary evaluation, one or more of those plans are modified in one or more ways and reevaluated on a basis consistent with the initial plan to determine if the modifications make the plan less than optimum.

As indicated in Table 12, it takes a number of years to license and construct a new power plant. The initial years of the various alternative generating-system expansion plans represent new power-generating facilities for which the utility is already committed. For this reason, the initial years of all of the alternative generating-system plans are generally the same.

The resulting optimum generating-system expansion plan is generally used, therefore, to determine the nature of next one or two power-generating facilities after the committed units. For example, if plan A of Table 10 is optimum, the utility would already have to be committed to the construction of the 600-MW coal-fired unit in year 2, the 50-MW combustion turbine in year 3, and the 900-MW nuclear unit in year 4. Because of the required lead times for the units in the plan, the optimum plan, therefore, would in essence have determined that licensing and construction must begin shortly for the 50-MW combustion turbine in year 6, the 300-MW combined-cycle unit in year 7, the 600-MW coal-fired unit in year 8, and the 900-MW nuclear unit in year 10.

GENERATOR RATINGS FOR INSTALLED UNITS

After determining the power ratings in MW of the next new generating units, it is necessary to determine the apparent power ratings in MVA of the electric generator for each of those units. For a 0.90 power factor and 600-MW turbine, determine the generator rating.

Calculate Procedure:

1. Compute the Rating

Generator rating in MVA = turbine rating in MW/power factor. Hence, the generator for a 600-MW turbine would be rated at 600 MW/0.90 = 677 MVA.

Related Calculations: The turbine rating in MW used in the above expression may be the rated or guaranteed value, the 5 percent over pressure value (approximately 105 percent of rated), or the maximum calculated value [i.e., 5 percent over rated pressure and valves wide open (109 to 110 percent of rated)] with or without one or more steam-cycle feedwater heaters out of service. This depends upon the manner in which an individual utility operates its plants.

OPTIMUM PLANT DESIGN

At this point, it is necessary to specify the detailed design and configuration of each of the power-generating facilities. Describe a procedure for realizing an optimum plant design.

Calculation Procedure:

1. Choose Design

Consider for example, the 600-MW coal-fired plant required in year 8; a single design (i.e., a single physical configuration and set of rated conditions), must be chosen for each component of the plant. Such components may include coal-handling equipment, boiler, stack-gas cleanup systems, turbine, condenser, boiler feed pump, feedwater heaters, cooling systems, etc. for the plant as a whole.

2. Perform Economic Analyses

In order to determine and specify the optimum plant design for many alternatives, it is necessary for the power-plant designer to repeatedly perform a number of basic economic analyses as the power-plant design is being developed. These analyses, almost without exception, involve one or more of the following tradeoffs:

(a) Operation and maintenance cost vs. capital costs

(b) Thermal efficiency vs. capital costs and/or operation and maintenance (O&M) costs

(c) Unit availability (reliability) vs. capital costs and/or O&M costs

(d) Unit rating vs. capital cost

ANNUAL OPERATION AND MAINTENANCE COSTS VS. INSTALLED CAPITAL COSTS

Evaluate the tradeoffs of annual O&M costs vs. installed capital for units A and B in Table 13.

Calculation Procedure:

1. Examine Initial Capital Costs

Units A and B have the same heat rate [10.550 MJ/kWh (10,000 Btu/kWh)], plant availability (95 percent), and plant rating (600 MW). As a result, the two alternatives would also be expected to have the same capacity factors and annual fuel expense.

Unit B, however, has initial capital costs which are $5 million higher than those of unit A, but has annual O&M costs (excluding fuel) which are $1.5 million less than those of Unit A. An example of such a case would occur if unit B had a more durable, higher capital-cost cooling tower filler material [e.g., polyvinyl chloride (PVC) or concrete] or condenser tubing material (stainless steel or titanium), whereas unit A had lower capital-cost wood cooling-tower filler or carbon steel condenser tubing.

2. Analyze Fixed and Annual Costs

For an 18 percent fixed-charge rate for both alternatives, the annual fixed charges are $900,000 higher for unit B ($81.9 million per year vs. $81.0 million per year). Unit B, however, has annual O&M costs which are $1.5 million lower than those of unit A. Unit B, therefore, would be chosen over unit A because the resulting total annual fixed capital, operation, and maintenance costs (excluding fuel, which is assumed to be the same for

TABLE 13 Evaluation of Annual O&M Costs vs. Installed Capital Costs

Cost component	*Unit A*	*Unit B*
Net unit heat rate	10.55 MJ/kWh (10,000 Btu/kWh)	10.55 MJ/kWh (10,000 Btu/kWh)
Unit availability	95%	95%
Unit rating	600 MW	600 MW
Installed capital cost	$\$450 \times 10^6$	$\$455 \times 10^6$
Levelized or average fixed-charge rate	18.0%	18.0%
Levelized or average annual O&M cost (excluding fuel)	$\$11.2 \times 10^6$/yr	$\$9.7 \times 10^6$/yr
For unit A:		
Annual fixed capital charges $= (\$450 \times 10^6)(18/100)$	=	$\$81.00 \times 10^6$/yr
Annual O&M cost (excluding fuel)	=	$\$11.20 \times 10^6$/yr
Total annual cost used for comparison with unit B	=	$\$92.20 \times 10^6$/yr
For unit B:		
Annual fixed capital charges $= (\$455 \times 10^6)(18/100)$	=	$\$81.90 \times 10^6$/yr
Annual O&M cost (excluding fuel)	=	$\$\ 9.70 \times 10^6$/yr
Total annual cost used for comparison with unit A	=	$\$91.60 \times 10^6$/yr

both alternatives) are lower for unit B by $600,000 per year. In this case, the economic benefits associated with the lower annual operation and maintenance costs for unit B are high enough to offset the higher capital costs.

THERMAL EFFICIENCY VS. INSTALLED CAPITAL AND/ OR ANNUAL OPERATION AND MAINTENANCE COSTS

Table 14 describes two alternative units which have different thermal performance levels but have the same plant availability (reliability) and rating. Unit D has a net heat rate (thermal performance level) that is 0.211 MJ/kWh (200 Btu/kWh) higher (i.e., 2 percent poorer) than that of unit C, but has both installed capital costs and levelized annual O&M costs which are somewhat lower than those of unit C. Determine which unit is a better choice.

Calculation Procedure:

1. Compute the Annual Fixed Charges and Fuel Cost

If the two units have the same capacity factors, unit D with a higher heat rate requires more fuel than unit C. Because unit D has both installed capital costs and levelized annual operation and maintenance costs which are lower than those of unit C, the evaluation problem becomes one of determining whether the cost of the additional fuel required each year for unit D is more or less than the reductions in the annual capital and O&M costs associated with unit D.

TABLE 14 Evaluation of Thermal Efficiency vs. Installed Capital Costs and Annual Costs

Cost components	*Unit C*	*Unit D*
Net unit heat rate	10.550 MJ/kWH (10,000 Btu/kWh)	10.761 MJ/kWh (10,200 Btu/kWh)
Unit availability	95%	95%
Unit rating	600 MW	600 MW
Installed capital cost	$\$450 \times 10^6$	$\$445 \times 10^6$
Levelized or average fixed-charge rate	18.0%	18.0%
Levelized or average annual O&M costs (excluding fuel)	$\$11.2 \times 10^6$/yr	$\$11.1 \times 10^6$/yr
Levelized or average capacity factor	70%	70%
Levelized or average fuel cost over the unit lifetime	$1.50/million Btu ($0.001422/MJ)	$1.50/million Btu ($0.001422/MJ)

For unit C:	
Annual fixed capital charges = ($\$450 \times 10^6$)(18/100)	= $\$ 81.00 \times 10^6$/yr
Annual O&M cost (excluding fuel)	= $\$ 11.20 \times 10^6$/yr
Annual fuel expense = (10.550 MJ/kWh)(600,000 kW)(8760 h/yr)(70/100)($0.001422/MJ) [= (10,000 Btu/kWh)(600,000 kW)(8760 h/yr)(70/100)($\$1.50/10^6$ Btu)]	= $\$ 55.19 \times 10^6$/yr
Total annual cost used for comparison with unit D	= $\$147.39 \times 10^6$/yr
For unit D:	
Annual fixed capital charges = ($\$445 \times 10^6$)(18/100)	= $\$ 80.10 \times 10^6$/yr
Annual O&M cost (excluding fuel)	= $\$ 11.10 \times 10^6$/yr
Annual fuel expense = (10.761 MJ/kWh)(600 000 kW)(8760 h/yr)(70/100)($0.001422/MJ) [= (10,200 Btu/kWh)(600,000 kW)(8760 h/yr)(70/100)($\$1.50/10^6$ Btu)]	= $\$ 56.29 \times 10^6$/yr
Total annual cost used for comparison with unit C	= $\$147.49 \times 10^6$/yr

The simplest method of determining the best alternative is to calculate the total annual fixed charges and fuel costs using the following expressions:

Annual fixed charges (dollars per year) = TICC × FCR/100, where TICC = total installed capital cost for unit C or D (dollars) and FCR = average annual fixed-charge rate (percent/yr).

Annual fuel expense (dollars per year) = HR × rating × 8760 × CF/100 × FC/10^6, where HR = average net heat rate in J/kWh (Btu/kWh), rating = plant rating in kW, CF = average or levelized unit capacity factor in percent, and FC = average or levelized fuel costs over the unit lifetime in dollars per megajoule (dollars per million Btu). The calculated values for these parameters are provided in Table 14.

2. *Make a Comparison*

As shown in Table 14, even though the annual fixed charges for unit C are $900,000 per year higher ($81.0 million per year vs. $80.10 million per year) and the annual O&M costs for unit C are $100,000 per year higher ($11.2 million per year vs. $11.1 million per year), the resulting total annual costs are $100,000 per year lower for unit C ($147.39 million per year vs. $147.49 million per year). This stems from the annual fuel expense for unit C being $1.10 million per year lower than for unit D ($55.19 million per year vs. $56.29 million per year) because the heat rate of unit C is 0.2110 MJ/kWh (200 Btu/kWh) better. In this case, the unit C design should be chosen over that of unit D because the economic benefits associated with the 0.2110 MJ/kWh (200 Btu/kWh) heat-rate improvement more than offsets the higher capital and O&M costs associated with unit C.

Related Calculations: The heat rate for a steam-cycle unit changes significantly with the turbine exhaust pressure (the saturation pressure and temperature provided by the cooling system) as shown in Fig. 8; with the percentage of rated load (amount of partial load operation of the unit) as shown in Fig. 9; with the choice of throttle (gauge) pressure of 12,411 kPa (1800 psig) vs. a gauge pressure measured at 16,548 kPa (2400 psig) vs. 24,132 kPa (3500 psig) as shown in Fig. 9 and Table 15; with throttle and reheater temperature and reheater pressure drop; and with a number of steam-cycle configuration and component performance changes as shown in Table 16. Therefore it is necessary for the power-plant designer to investigate carefully the choice of each of these parameters.

For example, as shown in Fig. 8, a change in cooling-tower performance (such as a change in the cooling-tower dimensions or a change in the rated circulating water flow) which increases the turbine-exhaust saturation temperature from 44.79°C to 48.62°C would cause the turbine-exhaust saturation pressure to rise from an absolute pressure of 9.48 kPa (2.8 inHg) to 11.52 kPa (3.4 inHg) which, as indicated in Fig. 8, would increase the heat-rate factor from 0.9960 to 1.0085, (i.e., a change of 0.0125 or 1.25 percent).

For a net turbine heat rate of 8.440 MJ/kWh (8000 Btu/kWh), this results in an increase of 0.106 MJ/kWh (100 Btu/kWh); i.e., (8.440 MJ/kWh)(0.0125) = 0.106 MJ/kWh.

From Fig. 9, operation of a unit at a gauge pressure of 16,548 kPa/538°C/538°C (2400 psig/1000°F/1000°F) at 70 percent of rated load instead of 90 percent would increase the net turbine heat rate by 0.264 MJ/kWh (250 Btu/kWh) from 8.440 MJ/kWh (8000 Btu/kWh) to 8.704 MJ/kWh (8250 Btu/kWh).

From Table 15 and Fig. 9, a change in the throttle gauge pressure from 16,548 kPa (2400 psig) to 12,411 kPa (1800 psig) would increase the heat rate from 0.160 to 0.179 MJ/kWh (152 to 168 Btu/kWh), i.e., from 1.9 percent to 2.1 percent of 8.440 MJ/kWh (8000 Btu/kWh).

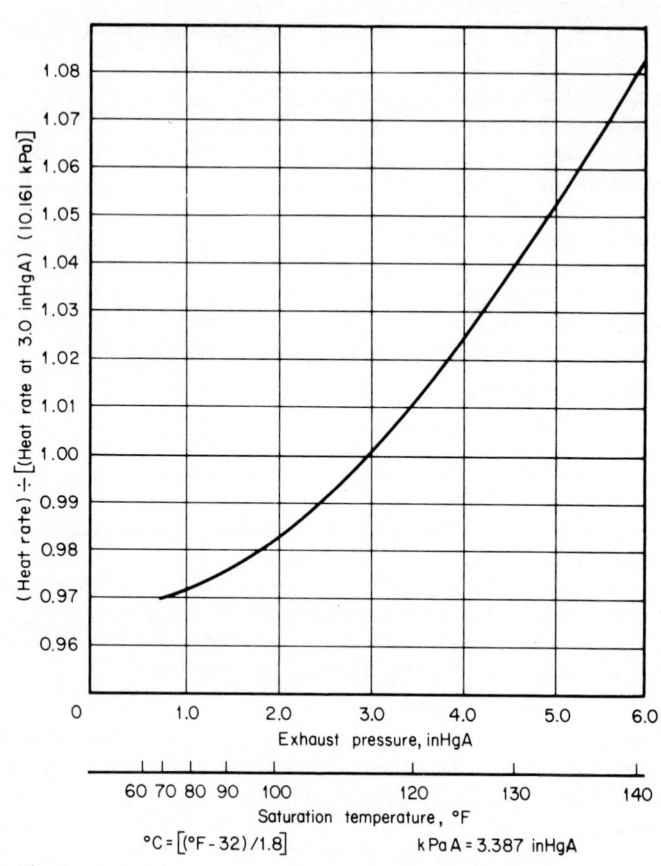

Fig. 8 Typical heat rate vs. exhaust pressure curve for fossil-fired steam-cycle units.

TABLE 15 Effect of Steam Condition Changes on Net Turbine Heat Rates

Steam condition	*% change in net heat rate*			
Throttle pressure, psi	1800	2400	3500	3500
Number of reheats	1	1	1	1
Throttle pressure (change from preceding column)		1.9–2.1	1.8–2.0	1.6–2.0
50°F Δ throttle temperature	0.7	0.7–0.8	0.8–0.9	0.7
50°F Δ first reheat temperature	0.8	0.8	0.8	0.4
50°F Δ second reheat temperature				0.6
One point in % Δ reheated pressure drop	0.1	0.1	0.1	0.1
Heater above reheat point	0.7	0.6	0.5–0.6	

kPa = psi × 6.895

Δ°C = Δ°F/1.8

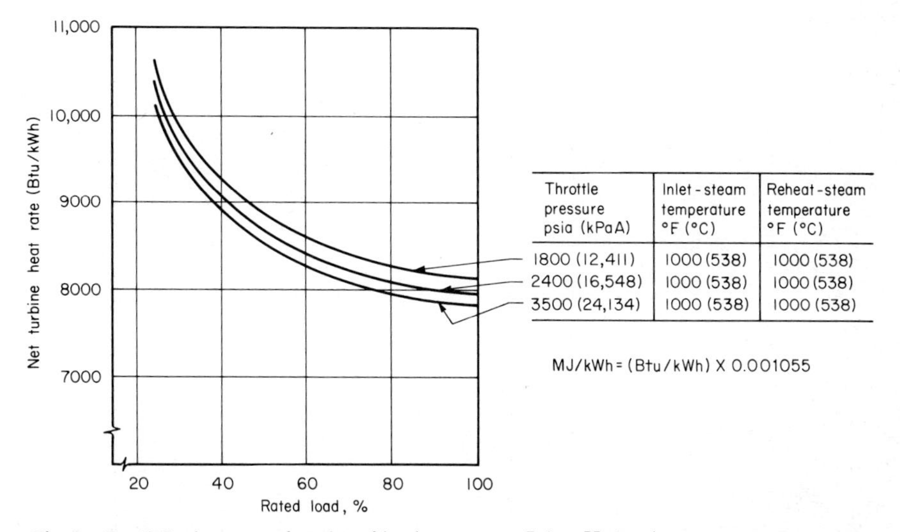

Fig. 9 Fossil-fired steam-cycle unit turbine heat rates at 7.6-cmHgA exhaust pressure vs. percent rated load.

TABLE 16 Effect of Steam-Cycle Changes on Net Turbine Heat Rates

	*Change in net heat rate**	
Cycle configuration	*%*	***Btu/kWh†***
1. Extraction line pressure drops of 3% rather than 5% (constant throttle flow)	−0.14	−11
2. Bottom heater drains flashed to condenser through 15°F drain cooler rather than 10°F‡	+0.01–0.02	+1–2
3. Change deaerator heater to closed cascading type with a 5°F temperature difference (TD) and a 10°F drain cooler	+0.24	+19
4. Make all drain coolers 15°F rather than 10°F	+0.01	+1
5. Reduce demineralized condenser make-up from 3% to 1%	−0.43	−35
6. Make top heater 0°F TD rather than −3° TD (constant throttle flow)	+0.01	+1
7. Make low-pressure heater TDs 3°F rather than 5°F	−0.11	−9
8. Eliminate drain cooler on heater 7	+0.08	+6.1

*+ is poorer

†MJ/kWh = (Btu/kWh) × 0.001055

‡Δ°C = Δ°F/1.8

REPLACEMENT FUEL COST

Table 17 describes the pertinent data for two alternatives which have the same net unit heat rate and rating. Unit F, however, has an average unit availability about 3 percent lower than unit E (92 percent vs. 95 percent). The capacity factor for each unit is 70 percent. Determine the replacement fuel cost.

TABLE 17 Evaluation of Reliability vs. Installed Capital and O&M Costs

	Unit E	*Unit F*
Net unit heat rate	10.550 MJ/kWh) (10,000 Btu/kWh)	10.550 MJ/kWh) (10,000 Btu/kWh)
Unit availability	95%	92%
Unit rating	600 MW	600 MW
Installed capital cost	$\$450 \times 10^6$	$\$440 \times 10^6$
Levelized or average fixed charge rate	18.0%	18.0%
Levelized or average annual O&M cost (excluding fuel)	$\$11.2 \times 10^6$/yr	$\$12.0 \times 10^6$/yr
Desired levelized or average capacity factor	70%	70%
Actual levelized or average capacity factor	70%	67.8%
For unit E:		
Annual fixed capital charges = ($\$450 \times 10^6$)(18/100)		= $\$81.00 \times 10^6$/yr
Annual O&M cost (excluding fuel)		= $\$11.20 \times 10^6$/yr
Total annual cost used for comparison with unit F		= $\$92.20 \times 10^6$/yr
For unit F:		
Annual fixed capital charges = ($\$440 \times 10^6$)(18/100)		= $\$79.20 \times 10^6$/yr
Annual O&M cost (excluding fuel)		= $\$12.00 \times 10^6$/yr
Replacement energy required for unit F as compared with unit E = (600 MW)(8760 h/yr)(70/100)[1 − (92%/95%)] = (600 MW)(8760 h/yr)(70 − 67.789)/100 = 116,210 MWh/yr		
Replacement energy cost penalty for unit F as compared with unit E = (116,210 MWh/yr)($15/MWh)		= $\$\ 1.74 \times 10^6$/yr
Total annual cost used for comparison with unit E		= $\$92.94 \times 10^6$/yr

Calculation Procedure:

1. *Analyze the Problem*

Because the ratings and heat rates are the same, it is convenient, for evaluation purposes, to assume that a utility would attempt to produce the same amount of electric power with either unit throughout the year. However, because of its lower plant availability, unit F would in general be expected to produce about 3 percent less electric power than unit E. As a result, during a total of 3 percent of the year when unit F would not be available, as compared with unit E, the utility would have to either generate additional power or purchase power from a neighboring utility to replace the energy which unit F was unable to produce because of its unavailability.

The difference between the cost of either the purchased or generated replacement power and the cost to generate that power on the unit with the higher plant availability represents a replacement energy cost penalty that must be assessed to the unit with the lower power availability (in this case unit F).

2. *Calculate the Replacement Energy Cost*

The replacement energy cost penalty is generally used to quantify the economic costs associated with changes in plant availability, reliability, or forced outage rates. The replacement energy cost penalty is calculated as follows: replacement energy cost penalty $= \text{RE} \times \text{RECD}$, in dollars per hour, where RE = replacement energy required in MWh/yr and RECD = replacement energy cost differential in dollars per megawatthour.

The value of RECD is determined by $\text{RECD} = \text{REC} - \text{AGC}_{ha}$, where REC = cost to either purchase replacement energy or generate replacement energy on a less efficient or more costly unit, in dollars per megawatthour, and AGC_{ha} = average generation cost of the unit under consideration with the highest (best) availability, in dollars per megawatthour. The average generation cost is calculated as $\text{AGC}_{ha} = \text{HR}_{ha} \times \text{FC}_{ha}/10^6$, where HR_{ha} = heat rate of the highest availability unit under consideration in J/kWh (Btu/kWh) and FC_{ha} = the average or levelized fuel cost of the highest availability unit under consideration in dollars per megajoule (dollars per million Btu).

The replacement energy RE is calculated as follows: $\text{RE} = \text{rating} \times 8760 \times [(\text{DCF}/100)(1 - \text{PA}_{la}/\text{PA}_{ha})]$ where rating = the capacity rating of the units in MW, DCF = desired average or levelized capacity factor for the units in percent, PA_{la} = availability of unit under consideration with lower availability in percent, and PA_{ha} = availability of unit under consideration with higher availability in percent.

Implied in the above equation is the assumption that the actual capacity factor for the unit with lower availability (ACF_{la}) will be lower than for the unit with higher availability as follows: $\text{ACF}_{la} = \text{DCF} \times (\text{PA}_{la}/\text{PA}_{ha})$.

As shown in Table 17, even though the annual fixed charges for unit E were $800,000 per year higher ($81.00 million per year vs. $79.20 million per year), the total resulting annual costs for unit E were $740,000 lower ($92.20 million per year vs. $94.94 million per year) because unit F had a $1.74 million per year replacement energy cost penalty and operation and maintenance costs which were $800,000 per year higher than unit E.

Related Calculations: It generally can be assumed that the replacement energy (either purchased from a neighboring utility or generated on an alternate unit) would cost about $10 to $20 per megawatthour more than energy generated on a new large coal-fired unit. In the example in Table 17 a value of replacement energy cost differential of $15/MWh was used.

TABLE 18 Evaluation of Unit Rating vs. Installed Capital Costs

	Unit G	*Unit H*
Unit rating	610 MW	600 MW
Net unit heat rate	10.550 MJ/kWh (10,000 Btu/kWh)	10.550 MJ/kWh (10,000 Btu/kWh)
Unit availability	95%	95%
Installed capital cost	$\$450 \times 10^6$	$\$448 \times 10^6$
Levelized or average fixed-charge rate	18.0%	18.0%
Levelized or average annual O&M cost (excluding fuel)	$\$11.2 \times 10^6$/yr	$\$11.2 \times 10^6$/yr
Capability penalty rate	$500/kW	$500/kW
For unit G:		
Annual fixed capital charges = ($\$450 \times 10^6$)(18/100)		= $\$81.00 \times 10^6$/yr
Total annual costs used for comparison with unit H		= $\$81.00 \times 10^6$/yr
For unit H:		
Annual fixed capital charges = ($\$448 \times 10^6$)(18/100)		= $\$80.64 \times 10^6$/yr
O&M costs same as alternate G		
Total capability penalty for unit H as compared with unit G = (610 MW − 600 MW)(1000 kW/MW)($500/kW) = $\$5.00 \times 10^6$/yr		
Annual capability penalty = ($\$5.0 \times 10^6$)(18/100)		= $\$0.90 \times 10^6$/yr
Total annual costs used for comparison with unit G		= $\$81.54 \times 10^6$/yr

CAPABILITY PENALTY

Compare the capability (capacity) penalty for units G and H in Table 18. Unit G has a rated capacity which is 10 MW higher than that of unit H.

Calculation Procedure:

1. Analyze the Problem

To achieve an equal reliability level the utility would, in principle, have to replace the 10 MW of capacity not provided by unit H with additional capacity on some other new unit. Therefore, for evaluation purposes, the unit with the smaller rating must be assessed what is called a capability (capacity) penalty to account for the capacity difference. The capability penalty CP is calculated by: $CP = (rating_l - rating_s) \times CPR$, where $rating_l$ and $rating_s$ are the ratings of the larger and smaller units, respectively, in kW, and CPR = capability penalty rate in dollars per kilowatt.

2. Calculate the Capability Penalty

For example, if the units have capital costs of approximately $500/kW, or if the capacity differential between the units is provided by additional capacity on a unit which would cost $500/kW, the capability penalty assessed against unit H (as shown in Table 18) is $5 million total. For an 18 percent fixed-charge rate, this corresponds to $900,000 per year.

Note that the annual operation and maintenance costs are the same for both units. In this case, those costs were not included in the total annual costs used for comparison purposes.

As shown in Table 18, even though alternative H had a capital cost $2 million lower than alternative G, when the capability penalty is taken into account, alternative G would be the economic choice.

Related Calculations: Applying the evaluation techniques summarized in Tables 13 through 18 sequentially makes it possible to evaluate units which fall into more than one (or all) of the categories considered, and thereby to determine the optimum plant design.

Section 9 TRANSMISSION LINES

John S. Wade, Jr., Ph.D.
Associate Professor of Engineering,
The Pennsylvania State University,
The Capitol Campus

REFERENCES *Aluminum Electrical Conductor Handbook,* The Aluminum Association; Barger and Smith—"Impedance and Circulating Current Calculations for the UD Multi-Wire Concentric Neutral Circuits," *IEEE Trans.,* May/June 1972; Carson—"Wave Propagation in Overhead Wires With Ground Return," *Bell Syst Tech J,* October 1926; Elgerd—*Electric Energy Systems*

Theory: An Introduction, McGraw-Hill; Gross—*Power System Analysis,* Wiley; Neuenswander—*Modern Power Systems,* International Textbook; Stevenson—*Elements of Power Systems Analysis,* McGraw-Hill; Westinghouse—*Electrical Transmission and Distribution Reference Book,* Westinghouse Electric.

INTRODUCTION

Newly erected overhead transmission lines are composed of aluminum conductors which, even in modest capacities, are stranded in spiral fashion for flexibility. These are primarily classified as:

AAC: all-aluminum conductors

AAAC: all-aluminum alloy conductors

ACSR: aluminum conductors, steel reinforced

ACAR: aluminum conductors, alloy reinforced

Aluminum conductors are compared in conductivity with the International Annealed Copper Standard (IACS) in Table 1. The table lists the percent conductivity, as well as the temperature coefficient of resistance α expressed per °C above 20°C.

TABLE 1 Comparison of Aluminum and Copper Conductors

Material	*Percent conductivity*	α
Aluminum:		
EC-H19	61.0	0.00403
5005-H19	53.5	0.00354
6201-T81	52.5	0.00347
Copper:		
Hard-drawn	97.0	0.00381

CONDUCTOR RESISTANCE

Calculate the resistance of 1000 ft (304.8 m) of solid round aluminum conductor, type EC-H19 (AWG No. 1) at 20°C and 50°C. The diameter = 0.2893 in (7.35 mm).

Calculation Procedure:

1. Calculate Resistance at 20°C

Use $R = \rho l/A$, where R is the resistance in ohms, ρ is the resistivity in ohm-circular mils per foot ($\Omega \cdot$cmil/ft), l is the length of conductor in feet, and A is the area in circular mils (cmil) which is equal to the square of the conductor diameter given in mils. From Table 1, the percentage of conductivity for EC-H19 is 61 percent that of copper. For IACS copper, $\rho = 10.4\ \Omega \cdot$cmil/ft. Therefore, for aluminum, the resistivity is $10.4/0.61 = 17.05\ \Omega \cdot$cmil/ft. The resistance of the conductor at 20°C is: $R = (17.05)(1000)/289.3^2 = 0.204\ \Omega$.

2. Calculate Resistance at 50°C

Use $R_T = R_{20°}[1 + \alpha(T - 20°)]$ where R_T is the resistance at the new temperature, $R_{20°}$ is the resistance at 20°C, and α is the temperature coefficient of resistance. From Table 1, $\alpha = 0.00403$. Hence $R_{50°} = 0.204[1 + 0.00403(50 - 20)] = 0.229\ \Omega$. .

INDUCTANCE OF SINGLE TRANSMISSION LINE

Using flux linkages, determine the self-inductance L of a single transmission line.

Calculation Procedure:

1. Select Appropriate Equation

Use $L = \lambda/i$, where L is the self-inductance in H/m, λ is the magnetic flux linkage in Wb·turns/m, and i is the current in amperes.

2. Consider Magnetic Flux Linkage

Assume a long, isolated, round conductor with uniform current density. The existing flux linkages include those internal to the conductor partially linking the current and those external to the conductor which links all of the current. These will be calculated separately and then summed, yielding total inductance $L_T = L_{\text{int}} + L_{\text{ext}}$.

3. Apply Ampere's Law

The magnitude of the magnetic flux density **B** in Wb/m^2 produced by a long current filament is $B = \mu i/2\pi r$ where μ = the permeability of the flux medium ($4\pi \times 10^{-7}$ H/m for free space and nonferrous material) and r = radius to **B** from the current center in meters. The direction of **B** is tangential to encirclements of the enclosed current, and clockwise if positive current is directed into this page (right-hand rule). The differential flux linkages per meter external to a conductor of radius a meters are $d\lambda = \mu i\, dr/2\pi r$ Wb·turns/m and $r > a$.

4. Consider Flux Inside Conductor

The calculation of differential flux linkages is complicated by **B**, which is a function of only that part of the current residing inside the circle passing through the measuring point of **B**. The complication is compounded by the reduction in current directly affecting $d\lambda$. Thus, if $\mathbf{B} = (\mu i/2\pi r)(\pi r^2/\pi a^2)$ Wb/m^2, then $d\lambda = (\mu i/2\pi r)(\pi r^2/\pi a^2)^2 = (\mu i/2\pi)(r^3/a^4)$ Wb·turns/m for $r < a$.

5. Calculate the Internal Inductance

Integrating the expression in (4) yields: $\lambda = (\mu i/2\pi)(\frac{1}{4})$, from which:

$$L_{\text{int}} = (10^{-7}/2)\ \text{H/m}.$$

Related Calculations: Calculation of the inductance due to the flux external to a long isolated conductor yields an infinite value, since r varies from a to infinity, but then such an isolated conductor is not possible.

INDUCTANCE OF TWO-WIRE TRANSMISSION LINE

Consider a transmission line consisting of two straight, round conductors (radius a meters), uniformly spaced D meters apart, where $D \gg a$ (Fig. 1). Calculate the inductance of the line.

Calculation Procedure:

1. Consider Flux Linkages

It is realistic to assume uniform and equal but opposite current density in each conductor. The oppositely directed currents, therefore, produce a net total flux linkage of

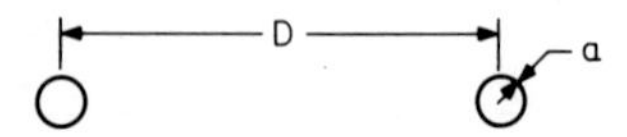

Fig. 1 A two-wire transmission line.

zero because the net current in any cross section of both conductors is zero. This is true for any multiconductor system whose cross-sectional currents add to zero (for example, in a balanced three-phase system).

2. Calculate Inductance of One Conductor

Use $\lambda = (\mu i/2\pi)[\frac{1}{4} + \ln(D/a)]$ Wb·turns/m. Because $L = \lambda/i$, $L = (2 \times 10^{-7})[\frac{1}{4} + \ln(D/a)]$ H/m. Inductance L may be expressed in more compact form by $L = (2 \times 10^{-7}) \ln(D/r')$ where $r' = a \exp(-\frac{1}{4})$ is the *geometric mean radius* GMR. The value of r' is $0.788a$.

3. Calculate Total Inductance L_T

$L_T = 2L = (4 \times 10^{-7}) \ln(D/r')$ H/m.

In more conventional units, $L_T = 1.482 \times \log(D/r')$ mH/mi.

INDUCTIVE REACTANCE OF TWO-WIRE TRANSMISSION LINE

Calculate the inductive reactance of 10 mi (16.1 km) of a two-conductor transmission line (Fig. 1) where $D = 8$ ft (2.44 m) and $a = 0.1$ in (2.54 mm) at a frequency of 60 Hz (377 rad/s).

Calculation Procedure:

1. Calculate the Geometric Mean Radius

The GMR is $r' = (0.7788)(2.54 \times 10^{-3}) = 0.001978$ m.

2. Calculate L_T

$L_T = (4 \times 10^{-7}) \ln(D/r') = (4 \times 10^{-7}) \ln(2.44/0.001978) = 28.5 \times 10^{-7}$ H/m.

3. Calculate Inductive Reactance X_L

$X_L = (377)(28.5 \times 10^{-7} \text{ H/m})(16.1 \times 10^3 \text{ m}) = 17.3\ \Omega$.

Related Calculations: A larger conductor and/or smaller conductor spacing reduces the inductive reactance.

INDUCTANCE OF STRANDED-CONDUCTOR TRANSMISSION LINE

Determine the inductance of a transmission line having six identical round conductors (Fig. 2) arranged so that the currents on each side occupy the conductors equally with uniform current density.

Calculation Procedure:

1. *Calculate Flux of One Conductor*

The flux linkages of conductor 1, which carries one-third of the current, can be deduced from the method established for the two-conductor line by: $\lambda_1 = (\mu/2\pi)(i/3)[\frac{1}{4} + \ln(D_{11'}/a) + \ln(D_{12'}/a) + \ln(D_{13'}/a) - \ln(D_{12}/a) - \ln(D_{13}/a)]$ where $D_{11'}$ is the distance between conductors 1 and 1′ and so on. When terms are collected, the equation becomes: $\lambda_1 = (\mu i/2\pi)\{\ln[(D_{11'}D_{12'}D_{13'})^{1/3}/(r'D_{12}D_{13})^{1/3}]\}$ Wb·turns/m. The flux linkages of the other two conductors carrying current in the same direction are found in a similar manner.

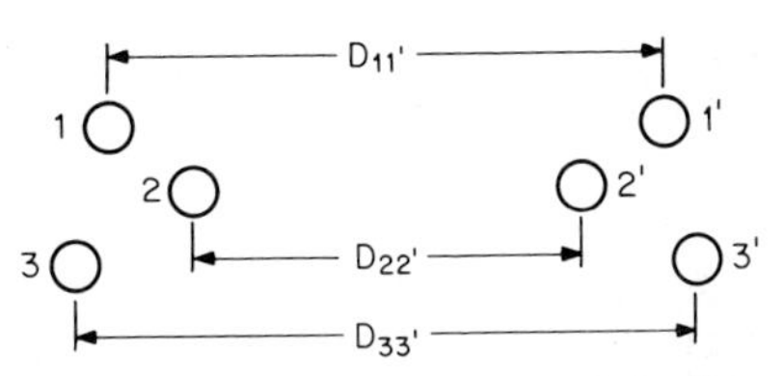

Fig. 2 Stranded-conductor transmission line.

2. *Calculate the Inductances*

The inductances of conductors one, two, and three then become: $L_1 = \lambda_1/(i/3) = (3 \times 2 \times 10^{-7})[\ln(D_{11'}D_{12'}D_{13'})^{1/3}/(r'D_{12}D_{13})^{1/3}]$ H/m, $L_2 = \lambda_2/(i/3) = (3 \times 2 \times 10^{-7})[\ln(D_{21'}D_{22'}D_{23'})^{1/3}/(r'D_{12}D_{23})^{1/3}]$ H/m, and $L_3 = \lambda_3/(i/3) = (3 \times 2 \times 10^{-7})[\ln(D_{31'}D_{32'}D_{33'})^{1/3}/(r'D_{23}D_{13})^{1/3}]$ H/m.

The inductance of the unprimed side (Fig. 2) of the line is one-third the average value because the inductances are in parallel. Then $L_{avg} = (L_1 + L_2 + L_3)/3$ H/m and $L = (L_1 + L_2 + L_3)/9$ H/m.

3. *Determine the Total Inductance L_T*

Combining the sets of equations in Step 2, we obtain: $L_T = (2 \times 10^{-7})\{\ln[(D_{11'}D_{12'}D_{13'})(D_{21'}D_{22'}D_{23'})(D_{31'}D_{32'}D_{33'})]^{1/9}/(r'^3D_{12}^2D_{13}^2D_{23}^2)^{1/9}\}$ H/m.

Related Calculations: To find the inductance of the primed side of the line in Fig. 2, follow the same procedure as above. Again, summing produces the total inductance.

The root of the product in the denominator for the expression of L_T is a GMR. (The root of the product in the numerator is called the geometric mean distance GMD.) Although tabulated values are usually available, one may calculate GMR $= (r'^nD_{12}^2D_{13}^2D_{23}^2 \cdots D_{(n-1)n}^2)^{1/n}$ m where $1/n$ is the reciprocal of the number of strands.

INDUCTANCE OF THREE-PHASE TRANSMISSION LINES

Determine the inductance per phase for a three-phase transmission line consisting of single conductors arranged unsymmetrically (Fig. 3).

Calculation Procedure:

1. *Use Flux-Linkage Method*

The principle of obtaining inductance per phase by using the flux linkages of one conductor is utilized once again. If the line is unsymmetrical and it remains untransposed, the inductance for each phase will not be equal (transposition of a line occurs when phase a, b, and c swap positions periodically). Transposing a transmission line will equalize the inductance per phase. Induc-

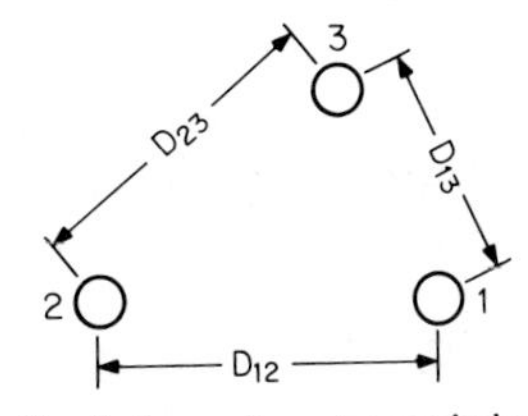

Fig. 3 A three-phase transmission line.

tance, however, varies only slightly when untransposed and it is common practice, in hand calculations, to assume transposition as is done in what follows.

Assume phase a shifts from position 1 to 2, and then to 3; phases b and c also move in the rotation cycle. The average flux linkages for phase a are then given by: $\lambda_a = (2 \times 10^{-7})/(3)[(\frac{1}{4}\mathbf{I_a} + \mathbf{I_a} \ln D_{12}/a + \mathbf{I_a} \ln D_{13}/a + \mathbf{I_b} \ln D_{21}/a + \mathbf{I_c} \ln D_{31}/a) + (\frac{1}{4}\mathbf{I_a} + \mathbf{I_a} \ln D_{21}/a + \mathbf{I_a} \ln D_{23}/a + \mathbf{I_b} \ln D_{32}/a + \mathbf{I_c} \ln D_{12}/a) + (\frac{1}{4}\mathbf{I_a} + \mathbf{I_a} \ln D_{32}/a + \mathbf{I_a} \ln D_{31}/a + \mathbf{I_b} \ln D_{13}/a + \mathbf{I_c} \ln D_{23}/a)]$ where $\mathbf{I_a}$, $\mathbf{I_b}$, and $\mathbf{I_c}$ are the rms phase currents. Then $\mathbf{I_a} + \mathbf{I_b} + \mathbf{I_c} = 0$. Also, $D_{12} = D_{21}$, $D_{23} = D_{32}$, and $D_{13} = D_{31}$. After terms are combined, the average flux linkage becomes: $\lambda_a = (2 \times 10^{-7})(\mathbf{I_a})/(3)[\ln (D_{12}D_{13}D_{23})/a^3 + \frac{3}{4}]$ Wb·turns/m.

2. ***Calculate L***

Using r'^3 and dividing by $\mathbf{I_a}$, we find the inductance per phase is $L_\phi = (2 \times 10^{-7})[\ln (D_{12}D_{13}D_{23})^{1/3}/r']$ H/m.

Related Calculations: If the conductor for each phase is concentrically stranded, the distance between conductors remains the same, but r' is replaced by a tabulated GMR. The inductance in conventional form is given by: $L = 0.7411 \log [(D_{12}D_{13}D_{23})^{1/3}/\text{GMR}]$ mH/mi; D and GMR are given in feet.

PER-PHASE INDUCTIVE REACTANCE

Calculate the per-phase inductive reactance per mile (1600 m) for a three-phase line at 377 rad/s. The conductors are aluminum conductors, steel-reinforced (ACSR) Redwing (Table 2) arranged in a plane as shown in Fig. 4.

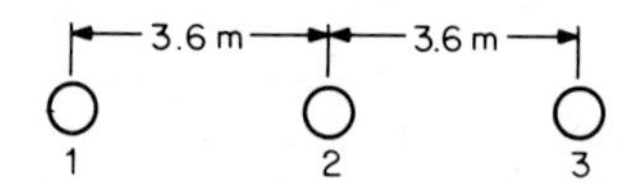

Fig. 4 A plane-spaced transmission line.

Calculation Procedure:

1. ***Calculate L_ϕ***

From Table 2, GMR = 0.0373 ft (0.01 m). Substituting in the equation for per-phase inductance, we find $L_\phi = (2 \times 10^{-7}) \ln (3.6 \times 7.3 \times 3.6)^{1/3}/0.01 = 12.2 \times 10^{-7}$ H/m.

2. ***Calculate Inductive Reactance X_L***

$X_L = 377 \times 12.2 \times 10^{-7}$ H/m $\times$ 1600 m = 0.74 Ω/mi.

INDUCTANCE OF SIX-CONDUCTOR LINE

Calculate the per-phase inductance of the transmission line of Fig. 5 where the conductors are arranged in a double circuit configuration.

Calculation Procedure:

1. ***Use Suitable Expression for L_ϕ***

Use $L_\phi = (2 \times 10^{-7}) \ln (\text{GMD}/\text{GMR}_c)$ H/m where GMR_c is the GMR of a conductor.

Code word	Size, Mcmil	Stranding aluminum/steel	Outside diameter, in	Resistance: DC, Ω/1000 ft at 20°C	Resistance: AC, 60 Hz, Ω/mi at 25°C	GMR, ft	Phase to neutral, 60 Hz, reactance at 1-ft spacing: Inductive Ω/mi, X_a	Phase to neutral, 60 Hz, reactance at 1-ft spacing: Capacitive Ω/mi, X_a
Waxwing	266.8	18/ 1	0.609	0.0646	0.3448	0.0198	0.476	0.1090
Partridge	266.8	26/ 7	0.642	0.0640	0.3452	0.0217	0.465	0.1074
Ostrich	300	26/ 7	0.680	0.0569	0.3070	0.0229	0.458	0.1057
Merlin	336.4	18/ 1	0.684	0.0512	0.2767	0.0222	0.462	0.1055
Linnet	336.4	26/ 7	0.721	0.0507	0.2737	0.0243	0.451	0.1040
Oriole	336.4	30/ 7	0.741	0.0504	0.2719	0.0255	0.445	0.1032
Chickadee	397.5	18/ 1	0.743	0.0433	0.2342	0.0241	0.452	0.1031
Ibis	397.5	26/ 7	0.783	0.0430	0.2323	0.0264	0.441	0.1015
Lark	397.5	30/ 7	0.806	0.0427	0.2306	0.0277	0.435	0.1007
Pelican	477	18/ 1	0.814	0.0361	0.1957	0.0264	0.441	0.1004
Flicker	477	24/ 7	0.846	0.0359	0.1943	0.0284	0.432	0.0992
Hawk	477	26/ 7	0.858	0.0357	0.1931	0.0289	0.430	0.0988
Hen	477	30/ 7	0.883	0.0355	0.1919	0.0304	0.424	0.0980
Osprey	556.5	18/ 1	0.879	0.0309	0.1679	0.0284	0.432	0.0981
Parakeet	556.5	24/ 7	0.914	0.0308	0.1669	0.0306	0.423	0.0969
Dove	556.5	26/ 7	0.927	0.0307	0.1663	0.0314	0.420	0.0965
Eagle	556.5	30/ 7	0.953	0.0305	0.1651	0.0327	0.415	0.0957
Peacock	605	24/ 7	0.953	0.0283	0.1536	0.0319	0.418	0.0957
Squab	605	26/ 7	0.966	0.0282	0.1529	0.0327	0.415	0.0953
Teal	605	30/19	0.994	0.0280	0.1517	0.0341	0.410	0.0944
Rook	636	24/ 7	0.977	0.0269	0.1461	0.0327	0.415	0.0950
Grosbeak	636	26/ 7	0.990	0.0268	0.1454	0.0335	0.412	0.0946
Egret	636	30/19	1.019	0.0267	0.1447	0.0352	0.406	0.0937
Flamingo	666.6	24/ 7	1.000	0.0257	0.1397	0.0335	0.412	0.0943
Crow	715.5	54/ 7	1.051	0.0240	0.1304	0.0349	0.407	0.0932
Starling	715.5	26/ 7	1.081	0.0238	0.1294	0.0355	0.405	0.0948
Redwing	715.5	30/19	1.092	0.0237	0.1287	0.0373	0.399	0.0920

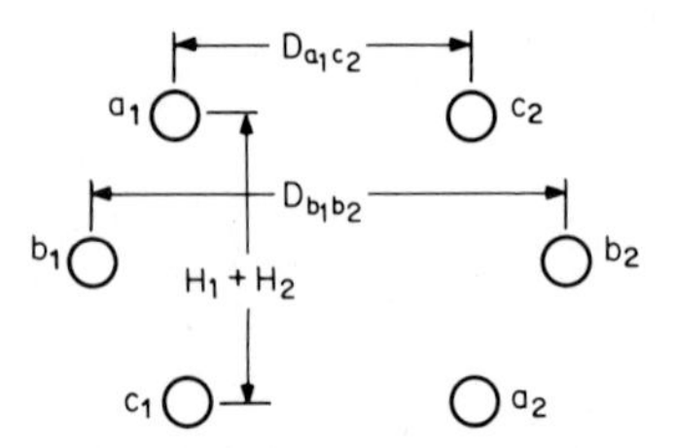

Fig. 5 Line conductors arranged in a double circuit configuration.

2. ***Calculate*** **GMD**

The GMD includes the distances between all the phase combinations. However, the expression for GMD can be reduced to one-half the distances that are represented in the original expression and the root becomes ⅙ rather than ¹⁄₁₂. Thus, GMD = $(D_{a1b1}D_{a1b2}D_{a1c1}D_{a1c2}D_{b1c1}D_{b1c2})^{1/6}$ m.

3. ***Calculate*** **GMR**

Use GMR = $(\text{GMR}_c^3 D_{a1a2}D_{b1b2}D_{c1c2})^{1/6}$ m.

INDUCTIVE REACTANCE OF SIX-CONDUCTOR LINE

Calculate the per-phase inductive reactance of a six-conductor, three-phase line at 377 rad/s consisting of Teal ACSR conductors (Fig. 5). Distance $D_{a1c2} = 4.8$ m, $H_1 = H_2 = 2.4$ m, and $D_{b1b2} = 5.4$ m.

Calculation Procedure:

1. ***Determine*** **GMD**

The necessary dimensions for calculating GMD are: $D_{a1b2} = D_{b1c2} = (2.4^2 + 5.1^2)^{1/2} = 5.64$ m, $D_{a1b1} = D_{b1c1} = (2.4^2 + 0.3^2)^{1/2} = 2.42$ m, $D_{a1c1} = 4.8$ m, and $D_{a1c2} = 4.8$ m. Hence, GMD = $[(2.42^2)(5.64^2)(4.8^2)]^{1/6} = 4.03$ m.

2. ***Determine*** **GMR_c**

From Table 2, for Teal, $\text{GMR}_c = 0.0341$ ft (0.01 m). Then, $D_{a1a2} = D_{c1c2} = (4.8^2 + 4.8^2)^{1/2} = 6.78$ m, $D_{b1b2} = 5.4$ m, and GMR = $[(0.01^3)(6.78^2)(5.4)]^{1/6} = 0.252$ m.

3. ***Calculate Inductive Reactance per Phase***

$X_L = (377)(2 \times 10^{-7})\,[\ln (4.03/0.252)] = 0.209 \times 10^{-3}\ \Omega/\text{m}\ (0.336\ \Omega/\text{mi})$.

INDUCTIVE REACTANCE OF BUNDLED TRANSMISSION LINE

Calculate the inductive reactance per phase at 377 rad/s for the bundled transmission line whose conductors are arranged in a plane shown in Fig. 6. Assume ACSR Crow.

Calculation Procedure:

1. ***Determine*** **GMD**

Assume distances are between bundle centers and transposition of phases. Then, GMD = $[(9^2)(18)]^{1/3} = 11.07$ m. From Table 2, $\text{GMR}_c = 0.034$ ft (0.01 m). The

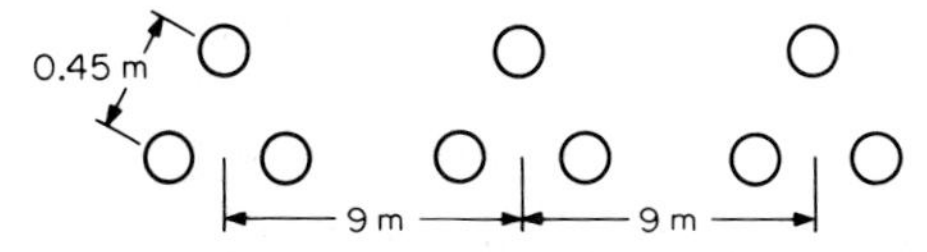

Fig. 6 A bundled transmission line.

GMR should include all conductor spacings from each other in the usual product form with, in this case, three values of GMR_c. Because of redundancy, $GMR = (0.01 \times 0.45^2)^{1/3} = 0.127$ m.

2. ***Calculate Inductive Reactance per Phase***
$X_L = 377 \times 2 \times 10^{-7} \times \ln(11.07/0.127) = 0.337 \times 10^{-3}\ \Omega/\text{m}\ (0.544\ \Omega/\text{mi})$

Related Calculations: For a two-conductor bundle, $GMR = (GMR_c D)^{1/2}$ and for a four-conductor bundle, $GMR = (GMR_c D^3 2^{1/2})^{1/4}$. In each case, D is the distance between adjacent conductors.

At voltage levels above 230 kV, the corona loss surrounding single conductors, even though they are expanded by nonconducting central cores, becomes excessive. Therefore, to reduce the concentration of electric-field intensity, which affects the level of ionization, the radius of a single conductor is artificially increased by arranging several smaller conductors, in what approximates a circular configuration. This idea is depicted in Fig. 7. Other arrangements that prove satisfactory, depending on the voltage level, are shown in Fig. 8.

Fig. 7 A multiconductor configuration replacing a large conductor.

Fig. 8 Practical configurations replacing a large conductor.

A benefit that accrues from bundling conductors of one phase of a line is an increase in GMR. Also, the inductance per phase is reduced, as well as corona ionization loss.

INDUCTIVE REACTANCE DETERMINED BY USING TABLES

Determine the inductive reactance per phase using data in Tables 2 and 3 for ACSR Redwing with the spacing given in Fig. 9.

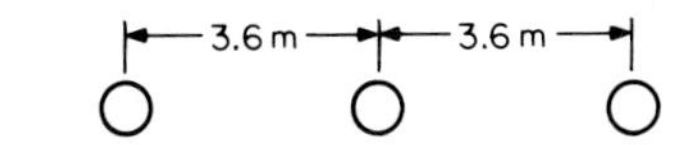

Fig. 9 A three-phase line where conductors are separated by 3.6 m (12 ft).

Calculation Procedure:

1. *Use Appropriate Tables*

The *Aluminum Electrical Conductor Handbook* provides tabulated data that reduce the amount of calculation necessary to find the inductive reactance for either a single- or three-phase line where circuits are neither paralleled or bundled. To determine the reactance by this method, it is convenient to express the inductive reactance by: $X_L = 2\pi f(2 \times 10^{-7}) \ln (1/\text{GMR}_c) + 2\pi f(2 \times 10^{-7}) \ln \text{GMD}$ or $X_L = 2\pi f(0.7411 \times 10^{-3}) \log (1/\text{GMR}_c) + 2\pi f(0.7411 \times 10^{-3}) \log \text{GMD}$. The first term in the latter expression is the reactance at *1-ft spacing* which is tabulated in Table 2 for 60 Hz. The second term is the *spacing component* of inductive reactance tabulated in Table 3 for a frequency of 60 Hz. Conversion from English to SI units, and vice versa, may be required.

2. *Determine* GMD

A distance of 3.6 m = 12 ft. Hence, $\text{GMD} = [(12^2)(24)]^{1/3} = 15.1$ ft (4.53 m).

3. *Determine* X_L

From Tables 2 and 3, $X_L = 0.399 + 0.329 = 0.728\ \Omega/\text{mi}$. In SI units, $X_L = (0.728\ \Omega/\text{mi})(1\ \text{mi}/1.6\ \text{km}) = 0.45\ \Omega/\text{km}$.

EFFECT OF MUTUAL FLUX LINKAGE

Referring to Fig. 10, find the voltage, in V/m, induced in a nearby two-conductor line by the adjacent three-phase transmission line carrying balanced currents having a magnitude of 50 A.

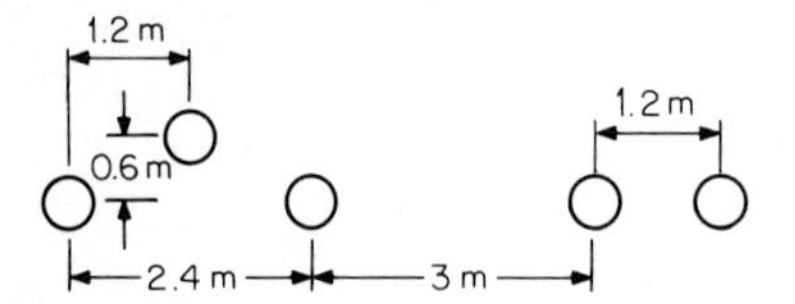

Fig. 10 A three-phase line in proximity to a two-wire line.

Calculation Procedure:

1. *Define Approach*

Flux linkages within the 1.2-m-wide plane of the two-conductor line from each phase of the transmission line shall be summed. Then, from Faraday's law, the derivative of this result will yield the answer.

2. *Calculate Distances from Each Involved Conductor*

$D_{b1} = 2.4 + 3 = 5.4$ m, $D_{b2} = 1.2 + 2.4 + 3 = 6.6$ m, $D_{a1} = 3$ m, $D_{a2} = 3 + 1.2 = 4.2$ m, $D_{c1} = (0.6^2 + 4.2^2)^{1/2} = 4.24$ m, and $D_{c2} = (0.6^2 + 5.4^2)^{1/2} = 5.43$ m.

3. *Calculate Flux Linkages*

$\lambda = \mu(i_a/2\pi) \ln (D_{a2}/D_{a1}) + \mu(i_b/2\pi) \ln (D_{b2}/D_{b1}) + \mu(i_c/2\pi) \ln (D_{c2}/D_{c1})$. This equation is a function of time. Substituting values in the expression and combining terms, we find $\lambda = \sqrt{2}[33.364 \sin \omega t + 20.07 \sin (\omega t - 120°) + 24.74 \sin (\omega t + 120°)] \times 10^{-7}$ Wb·turns/m, where $\omega = 2\pi f$.

4. *Apply Faraday's Law*

The voltage per unit length is $v = d\lambda/dt = \sqrt{2}[33.64\omega \cos \omega t + 20.07\omega \cos (\omega t - 120°) + 24.74\omega \cos (\omega t + 120°)] \times 10^{-7}$ V/m.

5. *Determine* V

Transforming to the frequency domain and rms values, one obtains $\mathbf{V} = (0.424 + j0.143) \times 10^{-3}$ V/m $= (0.68 + j0.23)$ V/mi.

TABLE 3 Separation Component X_d of Inductive Reactance at 60 Hz,* Ohms per Conductor per Mile†

	Separation of conductors											
	Inches											
Feet	***0***	***1***	***2***	***3***	***4***	***5***	***6***	***7***	***8***	***9***	***10***	***11***
0	—	−0.3015	−0.2174	−0.1682	−0.1333	−0.1062	−0.0841	−0.0654	−0.0492	−0.0349	−0.0221	−0.0106
1	0	0.0097	0.0187	0.0271	0.0349	0.0423	0.0492	0.0558	0.0620	0.0679	0.0735	0.0789
2	0.0841	0.0891	0.0938	0.0984	0.1028	0.1071	0.1112	0.1152	0.1190	0.1227	0.1264	0.1299
3	0.1333	0.1366	0.1399	0.1430	0.1461	0.1491	0.1520	0.1549	0.1577	0.1604	0.1631	0.1657
4	0.1682	0.1707	0.1732	0.1756	0.1779	0.1802	0.1825	0.1847	0.1869	0.1891	0.1912	0.1933
5	0.1953	0.1973	0.1993	0.2012	0.2031	0.2050	0.2069	0.2087	0.2105	0.2123	0.2140	0.2157
6	0.2174	0.2191	0.2207	0.2224	0.2240	0.2256	0.2271	0.2287	0.2302	0.2317	0.2332	0.2347
7	0.2361	0.2376	0.2390	0.2404	0.2418	0.2431	0.2445	0.2458	0.2472	0.2485	0.2498	0.2511
	Separation of conductors, ft											
8	0.2523	15	0.3286	22	0.3751	29	0.4086	36	0.4348	43	0.4564	
9	0.2666	16	0.3364	23	0.3805	30	0.4127	37	0.4382	44	0.4592	
10	0.2794	17	0.3438	24	0.3856	31	0.4167	38	0.4414	45	0.4619	
11	0.2910	18	0.3507	25	0.3906	32	0.4205	39	0.4445	56	0.4646	
12	0.3015	19	0.3573	26	0.3953	33	0.4243	40	0.4476	47	0.4672	
13	0.3112	20	0.3635	27	0.3999	34	0.4279	41	0.4506	48	0.4697	
14	0.3202	21	0.3694	28	0.4043	35	0.4314	42	0.4535	49	0.4722	

*From formula: at 60 Hz, $X_d = 0.2794 \log_{10} d$, d = separation in feet.

†From: *Electrical Transmission and Distribution Reference Book,* Westinghouse Electric Corporation, 1950.

INDUCTIVE REACTANCE OF CABLES IN DUCTS OR CONDUIT

Find the reactance per 1000 ft (304.8 m) if three single conductors each having 2-in (5-cm) outside diameters and 750-cmil cross sections are enclosed in a magnetic conduit.

Calculation Procedure:

1. Determine Inductance

Use $L = (2 \times 10^{-7})[\frac{1}{4} + \ln(D/a)]$ H/m. Common practice dictates that the reactance be given in ohms per thousand feet. Hence, $X_L = [0.0153 + 0.1404 \log(D/a)]$.

2. Use Nomogram for Solution

A nomogram based on the preceding equation is provided in Fig. 11. Two factors are used to improve accuracy. The equation for X_L produces a smaller reactance than an open-wire line. If randomly laid in a duct, the value of D is somewhat indeterminate because the outer insulation does not always touch. Therefore, if cables are not clamped on rigid supports, a multiplying factor of 1.2 is used. Further, if confined in a conduit of magnetic material in random lay, a multiplying factor of approximately 1.5 is used. Figure 11 includes a correction table for cables bound together rather than randomly laid. Here, sector refers to a cable whose three conductors approximate 120° sectors.

3. Determine X_L Graphically

Draw a line from 750 MCM to 2-in spacing and read the inductive reactance of 0.038 Ω per thousand feet. Then, $X_L = (1.5)(0.038) = 0.057$ Ω per thousand feet $= 0.19 \times 10^{-3}$ Ω/m.

Related Calculations: For a three-phase cable having concentric stranded conductors with a total of 250 MCM each and a diameter of 0.89 in (2.225 cm), the nomogram yields $X_L = 0.0315$ Ω per thousand feet (10^{-4} Ω/m). If the cable is in a magnetic conduit, the tabulated correction factor is used. Thus, $X_L = (1.149)(0.0315) = 0.0362$ Ω per thousand feet, or 1.2×10^{-4} Ω/m.

CAPACITANCE ASSOCIATED WITH TRANSMISSION LINES

Determine the balanced charging current fed from one end to a 230-kV, three-phase transmission line having a capacitive reactance of 0.2 MΩ·mi/phase (0.32 MΩ·km/phase). The line is 80 mi (128.7 km) long.

Calculation Procedure:

1. Determine Capacitive Reactance

The total capacitive reactance per phase, which is assumed to shunt each phase to ground, is $X_C = 0.32/128.7 = 0.0025$ MΩ.

2. Calculate Charging Current

For the voltage-to-neutral value of $230/\sqrt{3} = 133$ kV, the charging current I_c is $(133 \times 10^3)/(0.0025 \times 10^6) = 53.2$ A.

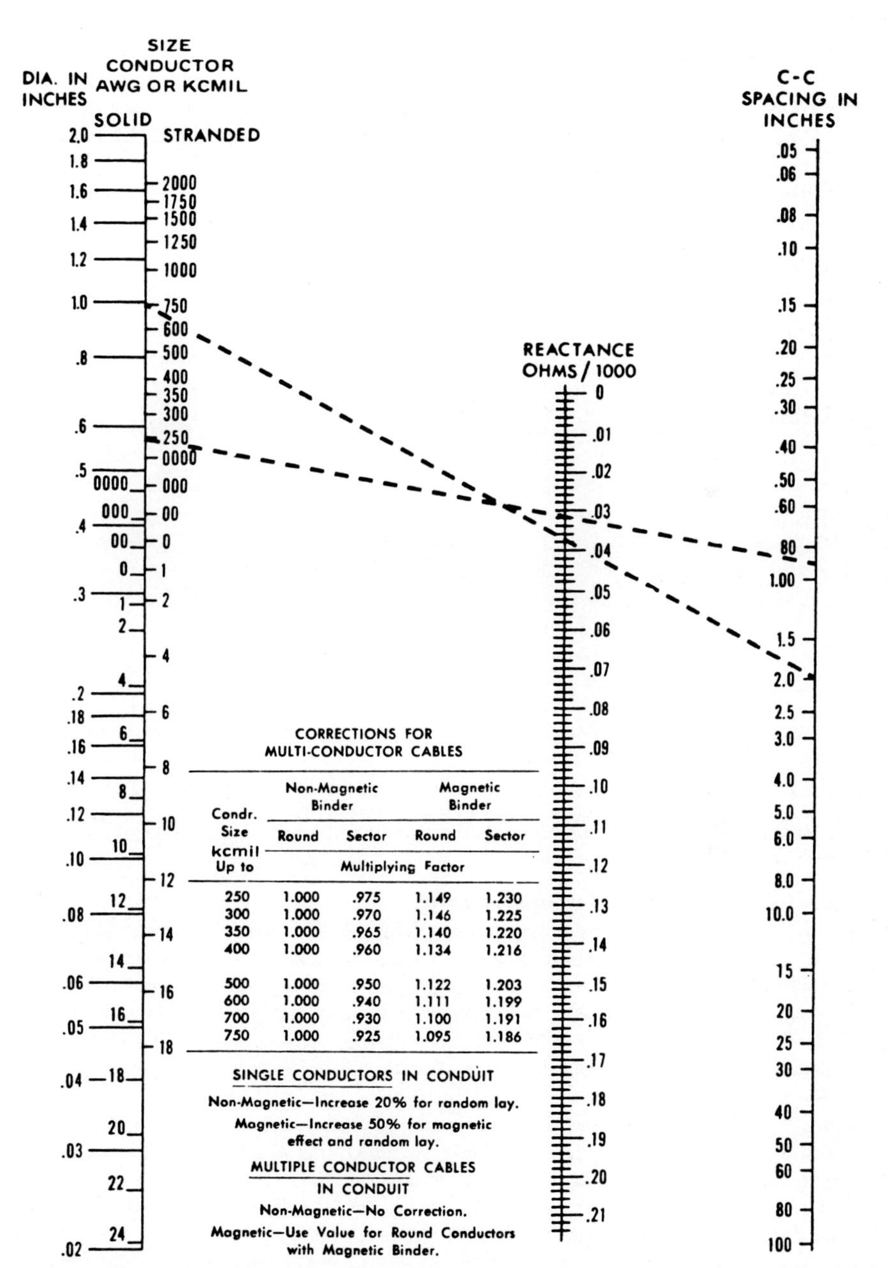

CORRECTIONS FOR MULTI-CONDUCTOR CABLES

Condr. Size kcmil Up to	Non-Magnetic Binder		Magnetic Binder	
	Round	Sector	Round	Sector
	Multiplying Factor			
250	1.000	.975	1.149	1.230
300	1.000	.970	1.146	1.225
350	1.000	.965	1.140	1.220
400	1.000	.960	1.134	1.216
500	1.000	.950	1.122	1.203
600	1.000	.940	1.111	1.199
700	1.000	.930	1.100	1.191
750	1.000	.925	1.095	1.186

SINGLE CONDUCTORS IN CONDUIT

Non-Magnetic—Increase 20% for random lay.

Magnetic—Increase 50% for magnetic effect and random lay.

MULTIPLE CONDUCTOR CABLES IN CONDUIT

Non-Magnetic—No Correction.

Magnetic—Use Value for Round Conductors with Magnetic Binder.

Fig. 11 Nomogram for determining series inductive reactance of insulated conductors to neutral. From: *Aluminum Electrical Conductor Handbook*, (2d ed.) the Aluminum Association, Inc., 1982, page 9-11.

CAPACITANCE OF A TWO-WIRE LINE

Determine the capacitance of a long, round conductor carrying a uniform charge density ρ_L on its outer surface (surplus charge always migrates to the outer surface of any conductor). The conductor is surrounded by an outward (for positive charge) vectorial electric field that appears to radiate from the center of the conductor, although it originates with ρ_L on the surface.

Calculation Procedure:

1. Determine the Electric Potential

The magnitude of the electric field intensity **E** is given by: $E = \rho_L/(2\pi\epsilon r)$ where ϵ is the permittivity. In free space, $\epsilon = 10^{-9}/36\pi$ F/m. For consistency, the distance r from the center of the conductor is in meters and ρ_L is in coulombs per meter. Integration of E yields the electric potential V between points near the conductor (Fig. 12): $V_{ab} =$

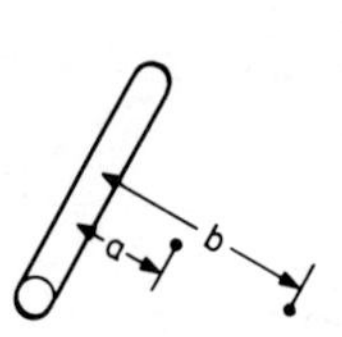

Fig. 12 Long conductor carrying a uniform charge.

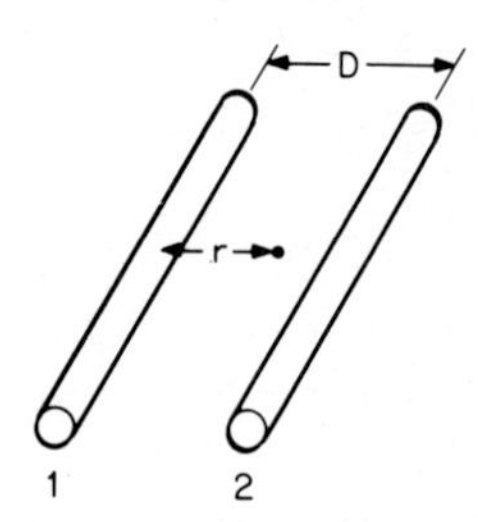

Fig. 13 A charged two-conductor line.

$(\rho_L/2\pi\epsilon) \ln (b/a)$ V. The notation V_{ab} indicates that the voltage is the potential at point a with respect to point b.

2. Consider a Two-Conductor Line

Consider the two conductors as forming a long, parallel conductor system (Fig. 13). Each conductor has an equal but opposite charge, typical of two-wire transmission systems. Further, it is assumed that the charge density per unit area is uniform in each conductor, even though a charge attraction exists between conductors, making it nonuniform. This assumption is completely adequate for open-wire lines for which $D \gg a$.

Because the conductors bear charges of opposite polarities, the electric field at point r in the plane of the conductors between them is $E = (\rho_L/2\pi\epsilon)[1/r - 1/(D - r)]$ V/m, where r is the distance from the center of conductor 1 $(r \geq a)$, and D is the center-to-center spacing between conductors. By integrating E, potential V_{1r} at conductor 1 with respect to point r is obtained: $V_{1r} = (\rho_L/2\pi\epsilon) \ln [r(D - a)/a(D - r)]$ V.

If r extends to conductor 2, and $D \gg a$, the potential of conductor 1 with respect to conductor 2 is: $V_{12} = (\rho_L/\pi\epsilon) \ln (D/a)$ V.

3. Calculate Capacitance

The capacitance per unit length is determined from $C' = q/v$ F, where $q = \rho_L l$ and l is the total line length. Dividing by l yields $C = C'/l$ or $C = \pi\epsilon/\ln (D/a)$ F/m, which is the capacitance between conductors per meter length.

4. *Determine Capacitance to Vertical Plane between Conductors at D/2*

The potential for conductor 1 with respect to this plane (or neutral) is $V_{1n} = (\rho_L/2\pi\epsilon) \ln (D/a)$ V for $D \gg a$. It follows that this potential to the so-called neutral plane is one-half the conductor-to-conductor potential. It is easily shown that the potential from conductor 2 is of the same magnitude. If the neutral were grounded, it would not affect the potentials. The capacitance to neutral is then $C = (2\pi\epsilon)/\ln (D/a)$ F/m or $0.0388/\log (D/a)$ μF/mi.

CAPACITIVE REACTANCE OF TWO-WIRE LINE

Find the capacitive reactance to neutral for a two-conductor transmission line if $D = 8$ ft (2.4 m), $a = 0.25$ in (0.00625 m), and the length of the line is 10 mi (16 km). The frequency is 377 rad/s.

Calculation Procedure:

1. *Calculate Capacitive Reactance*

Recall that $X_C = 1/\omega C$. Substituting for $C = 0.0388/\log (D/a)$, obtain $X_C = 1/[(377)(0.0388)(10)/\log (2.4/0.00625)] = 0.018$ MΩ to neutral.

Related Calculations: This is a large value of shunt impedance and is usually ignored for a line this short. It also follows that the capacitive reactance between conductors is twice the above value.

CAPACITANCE OF THREE-PHASE LINES

Determine the capacitance for a three-phase line.

Calculation Procedure:

1. *Consider Capacitance to Neutral*

The capacitance to neutral of a three-phase transmission line is best established by considering equilateral spacing of the conductors initially. Other spacing which is unsymmetrical is commonly considered using the geometric mean distance in the equilateral case. The error that results is insignificant, especially when consideration is given to the uncertainties of an actual line stemming from line towers and terrain irregularities.

2. *Determine Phase Voltages*

As shown in Fig. 14, $\mathbf{V_{an}}$, the potential of phase a with respect to the center of the triangle (the neutral), can be found by superimposing the potentials from all phases along the dimension from phase a to the center. The net charge is zero for any cross section, as it was for the two-wire line. Also, $D \gg a$. Superscripts to identify the three-phase potentials along the dimension for a to b are necessary. Thus $|\mathbf{V}^{\mathbf{a}}_{\mathbf{an}}| =$

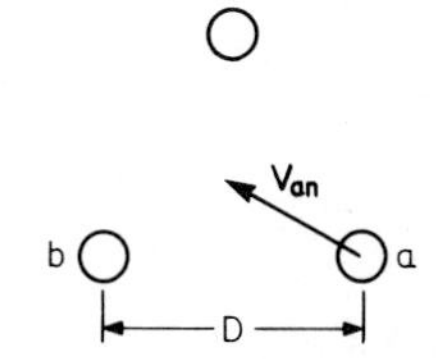

Fig. 14 Equilateral spacing of a three-phase line. a denotes the conductor radii.

$(\rho_{La}/2\pi\epsilon) \ln [(D/\sqrt{3})/D]$ due to phase a, $|\mathbf{V}_{an}^{b}| = (\rho_{Lb}/2\pi\epsilon) \ln [(D/\sqrt{3})/D]$ due to phase b, and $|\mathbf{V}_{an}^{c}| = (\rho_{Lc}/2\pi\epsilon) \ln [D/\sqrt{3})/D]$ due to phase c.

The sum of the above three equations yields $\mathbf{V}_{an}$. Also $\rho_{La} + \rho_{Lb} + \rho_{Lc} = 0$. Thus $|\mathbf{V}_{an}| = (\rho_{La})/2\pi\epsilon) \ln (D/a)$ V. This equation has the same form as the equation for the potential to neutral of a two-wire line. The phase-to-neutral potentials of the other phases differ only in phase angle.

3. Determine Capacitance to Neutral

Dividing ρ_{La} by $|\mathbf{V}_{an}|$, we obtain $C = (2\pi\epsilon)/\ln (D/a)$ F/m $= 0.0388/\log (D/a)$ μF/mi.

CAPACITIVE REACTANCE OF THREE-PHASE LINES

Find the capacitive reactance to neutral of a three-phase line at 377 rad/s (Fig. 15). The conductors are ACSR Waxwing and the line is 60 mi (96.6 km) long.

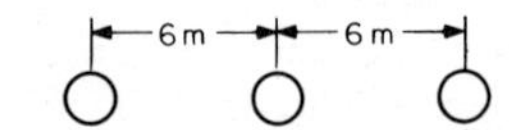

Fig. 15 A three-phase line where conductors are separated by 6 m (20 ft).

Calculation Procedure:

1. Calculate Capacitance

From Table 2, the external diameter of Waxwing is 0.609 in (0.015 m). Even though the conductors are not in equilateral spacing, use of GMD produces a sufficiently accurate capacitance to neutral. Therefore, $\text{GMD} = [(6^2)(12)]^{1/3} = 7.54$ m and $a = 0.015/2 = 0.0075$ m. Hence, $C = 0.0388/\log (7.54/0.0075) = 0.0129$ μF/mi or 0.008 μF/km.

2. Calculate Capacitive Reactance

$X_C = 1/(377)(0.008 \times 10^{-6})(96.6) = 0.0034$ MΩ to neutral

CHARGING CURRENT AND REACTIVE POWER

Determine the input charging current and the charging apparent power in megavars for the previous example if the line-to-line voltage is 230 kV.

Calculation Procedure:

1. Calculate Charge Current I_c

$I_c = (V/\sqrt{3})/X_C$ where V is the line-to-line voltage. Then, $I_c = (230 \times 10^3)/(\sqrt{3})(0.0034 \times 10^6) = 39.06$ A.

2. Calculate Reactive Power Q

$Q = \sqrt{3}VI = (\sqrt{3})(230 \times 10^3)(39.06) = 15.56$ Mvar for the three-phase line.

TRANSMISSION-LINE MODELS

Short transmission lines [up to 80 km (50 mi)], are represented by their series impedance consisting of the line resistance R_L and inductive reactance X_L. In cases where R_L is less than 10 percent of X_L, R_L is sometimes ignored.

The model for medium lines [up to 320 km (200 mi)] is represented in Fig. 16, where the line capacitance C_L is considered. Expressions for admittance $\mathbf{Y_L}$ and impedance $\mathbf{Z_L}$ are: $\mathbf{Y_L} = j\omega C_L/2$ and $\mathbf{Z_L} = R_L + jX_L$. The voltage $\mathbf{V_S}$ and current $\mathbf{I_S}$ for the sending end of the line are: $\mathbf{V_S} = (\mathbf{V_R Y_L}/2 + \mathbf{I_R})\mathbf{Z_L} + \mathbf{V_R}$ and $\mathbf{I_S} = \mathbf{V_S Y_L}/2 + \mathbf{V_R Y_L}/2 + \mathbf{I_R}$, where $\mathbf{V_R}$ is the receiving-end voltage and $\mathbf{I_R}$ is the receiving-end current. The above equations may be written as: $\mathbf{V_S} = \mathbf{AV_R} + \mathbf{BI_R}$ and $\mathbf{I_S} = \mathbf{CV_R} + \mathbf{DI_R}$, where $\mathbf{A} = \mathbf{D} = \mathbf{Z_L Y_L}/2 + 1$, $\mathbf{B} = \mathbf{Z_L}$, and $\mathbf{C} = \mathbf{Y_L} + \mathbf{Z_L Y_L^2}/4$.

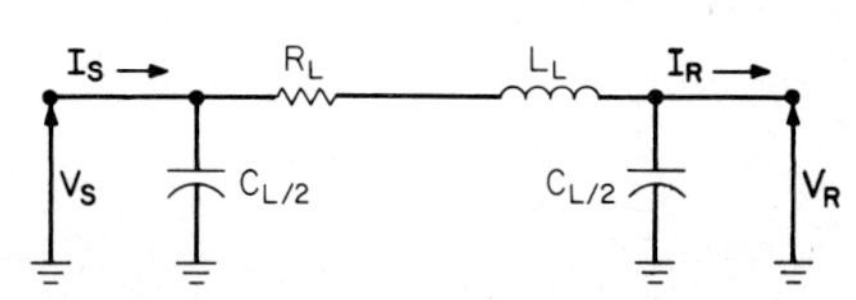

Fig. 16 A model of a single-phase, medium-length line.

For long transmission lines, $\mathbf{V_S} = (\mathbf{V_R} + \mathbf{I_R Z_c})e^{\gamma x}/2 + (\mathbf{V_R} - \mathbf{I_R Z_c})e^{-\gamma x}/2$ and $\mathbf{I_S} = (\mathbf{V_R}/\mathbf{Z_c} + \mathbf{I_R})e^{\gamma x}/2 + (\mathbf{V_R}/\mathbf{Z_c} - \mathbf{I_R})e^{-\gamma x}/2$, where the characteristic impedance $\mathbf{Z_c} = \sqrt{\mathbf{Z}/\mathbf{Y}}\ \Omega$ and the propagation constant $\gamma = \sqrt{\mathbf{ZY}}$ per kilometer (or per mile).

MEDIUM TRANSMISSION LINES

Calculate the sending-end voltage and current for a 320-km (200-mi) transmission line. The receiving-end line-to-line voltage is 230 kV and the current is 200 A, at a power factor of 0.8 lagging. The line parameters per kilometer are: $R = 0.2\ \Omega$, $L = 2$ mH, $C = 0.01\ \mu$F, and $f = 60$ Hz.

Calculation Procedure:

1. *Determine* $\mathbf{Y_L}$ *and* $\mathbf{Z_L}$

$\mathbf{Y_L} = j754$ microsiemens (μS) and $\mathbf{Z_L} = 40 + j150.8\ \Omega$.

2. *Calculate* A, B, C, *and* D

$\mathbf{A} = \mathbf{D} = \mathbf{Y_L Z_L}/2 + 1 = j754(40 + j150.8)/2 + 1 = 0.887\underline{/1.9°}$, $\mathbf{B} = \mathbf{Z_L} = 40 + j150.8 = 156\underline{/75.14°}$ and $\mathbf{C} = \mathbf{Y_L} + \mathbf{Z_L Y_L^2}/4 = j754 + (40 + j150.8)(j754)^2/4 = 732\underline{/90°}$.

3. *Calculate* $\mathbf{V_S}$ *and* $\mathbf{I_S}$

The receiving-end phase voltage is $230/\sqrt{3} = 132.8$ kV. Hence $\mathbf{V_S} = (0.887\underline{/1.9°})(132.8 \times 10^3\underline{/0°}) + (156\underline{/75.14°})(200\underline{/-36.9°}) = 148\underline{/5.3°}$ kV and $\mathbf{I_S} = (732 \times 10^{-6}\underline{/90°})(132.8 \times 10^3\underline{/0°}) + (0.887\underline{/1.9°})(200\underline{/-36.9°}) = 142\underline{/-4°}$ A. The magnitude of the sending-end line voltage is 262 V.

Related Calcuations: In case the sending-end values $\mathbf{V_S}$ and $\mathbf{I_S}$ are known, the receiving-end voltage and current may be calculated by: $\mathbf{V_R} = \mathbf{DV_S} - \mathbf{BI_S}$ and $\mathbf{I_R} = -\mathbf{CV}_S + \mathbf{AI_S}$.

LONG TRANSMISSION LINES

Recompute the medium-line model example (320 km long) with the same values of potential, current, and line parameters using the long-line model.

Calculation Procedure:

1. *Calculate* Z *and* Y

Because $R = 0.2\ \Omega/\text{km}$ and $L = 2\ \text{mH/km}$, the series impedance per unit length is $\mathbf{Z} = 0.2 + j0.754 = 0.78\angle 75.1°\ \Omega/\text{km}$. Since $C = 0.01\ \mu\text{F/km}$, the shunt admittance per kilometer is $\mathbf{Y} = j3.77\ \text{S/m}$.

2. *Calculate* $\mathbf{Z_c}$

The characteristic (surge) impedance is $\mathbf{Z_c} = [0.78\angle 75.1°/(3.77 \times 10^{-6}\angle 90°)]^{1/2} = 455\angle -7.45°\ \Omega$. If the resistance is less than one-tenth the inductive reactance per unit length, the characteristic impedance approaches a real number.

3. *Calculate the Propagation Constant* γ

$\gamma = [(0.78\angle 75.1°)(3.77 \times 10^{-6}\angle 90°)]^{1/2} = 1.72 \times 10^{-3}\angle 82.55°$. A small resistance results in a value approaching an imaginary number. To be useful, γ must be in rectangular form. Thus, $\gamma = 0.223 \times 10^{-3} + j1.71 \times 10^{-3}$.

The real part of γ is the *attenuation factor* α. $\alpha = 0.223 \times 10^{-3}$ nepers/km. The imaginary part is the *phase-shift constant* β. $\beta = 1.71 \times 10^{-3}$ rad/km.

4. *Calculate* $\mathbf{V_S}$ *and* $\mathbf{I_S}$

The per-phase receiving voltage to neutral is 132.8 kV. Substituting the above values in the equations for $\mathbf{V_S}$ and $\mathbf{I_S}$ yields: $\mathbf{V_S} = [(132.8 \times 10^3\angle 0°)/2][\exp\,(0.223 \times 10^{-3})(200)][\exp\,(j1.71 \times 10^{-3})(200)] + [(200\angle -36.9°)(455\angle -7.45°)/2][\exp\,(0.223 \times 10^{-3})(200)][\exp\,(j1.71 \times 10^{-3})(200)] + [(132.8 \times 10^3\angle 0°)/2][\exp\,(-0.223 \times 10^{-3})(200)][\exp\,(-j1.71 \times 10^{-3})(200)] - [(200\angle -36.9°)(455\angle -7.45°/2][\exp\,(-0.223 \times 10^{-3})(200)][\exp\,(-j1.71 \times 10^{-3})(200)]$ and $\mathbf{I_S} = \{[(132.8 \times 10^3\angle 0°/(455\angle -7.45°)]/2\}[\exp\,(0.223 \times 10^{-3})(200)][\exp\,(j1.71 \times 10^{-3})(200)] + (200\angle -36.9°/2)[\exp\,(0.223 \times 10^{-3})(200)][\exp\,(j1.71 \times 10^{-3})(200)] - \{[(132.8 \times 10^3\angle 0°)/455\angle -7.45°]/2\}\,[\exp\,(-0.223 \times 10^{-3})(200)]\,[\exp\,(j1.71 \times 10^{-3})(200)] + (200\angle -36.9°/2)[\exp\,(-0.223 \times 10^{-3})(200)][\exp\,(-j1.71 \times 10^{-3})(200)]$. When terms are combined, $\mathbf{V_S} = 150.8\angle 8.06°$ kV (phase to neutral) and $\mathbf{I_S} = 152.6\angle -4.52°$ A (line). The magnitude of the line voltage at the input is 261.2 kV. These results correspond to the results of the medium-line model of the previous example with little difference for a 320-km line.

Related Calculations: Equations for $\mathbf{V_S}$ and $\mathbf{I_S}$ are most easily solved by a suitable Fortran algorithm. With such a program, the value of x can be incremented outward from the receiving end to display the behavior of the potential and current throughout the line. Such a program is suitable for lines of any length.

The first term in each equation for $\mathbf{V_S}$ and $\mathbf{I_S}$ may be viewed as representing a traveling wave from the source to the load end of the line. If x is made zero, the wave is incident at the receiving end. The second term in each equation represents a wave reflected from the load back toward the source. If x is made zero, the value of this wave is found at the receiving end. The sum of the two terms at the receiving end should be 132.8 kV and 200 A in magnitude.

If the impedance at the load end is equal to the characteristic (surge) impedance $\mathbf{Z_c}$, the reflected terms (second terms in equations for $\mathbf{V_S}$ and $\mathbf{I_S}$) are zero. The line is said to

TABLE 4 Separation Component X_d of Capacitive Reactance at 60 Hz,* Megohm-Miles per Conductor†

	Separation of conductors											
	Inches											
Feet	*0*	*1*	*2*	*3*	*4*	*5*	*6*	*7*	*8*	*9*	*10*	*11*
0	—	−0.0737	−0.0532	−0.0411	−0.0326	−0.0260	−0.0206	−0.0160	−0.0120	−0.0085	−0.0054	−0.0026
1	0	0.0024	0.0046	0.0066	0.0085	0.0103	0.0120	0.0136	0.0152	0.0166	0.0180	0.0193
2	0.0206	0.0218	0.0229	0.0241	0.0251	0.0262	0.0272	0.0282	0.0291	0.0300	0.0309	0.0318
3	0.0326	0.0334	0.0342	0.0350	0.0357	0.0365	0.0372	0.0379	0.0385	0.0392	0.0399	0.0405
4	0.0411	0.0417	0.0423	0.0429	0.0435	0.0441	0.0446	0.0452	0.0457	0.0462	0.0467	0.0473
5	0.0478	0.0482	0.0487	0.0492	0.0497	0.0501	0.0506	0.0510	0.0515	0.0519	0.0523	0.0527
6	0.0532	0.0536	0.0540	0.0544	0.0548	0.0552	0.0555	0.0559	0.0563	0.0567	0.0570	0.0574
7	0.0577	0.0581	0.0584	0.0588	0.0591	0.0594	0.0598	0.0601	0.0604	0.0608	0.0611	0.0614
	Separation of conductors, ft											
8	0.0617	15	0.0803	22	0.0917	29	0.0999	36	0.1063	43	0.1116	
9	0.0652	16	0.0823	23	0.0930	30	0.1009	37	0.1071	44	0.1123	
10	0.0683	17	0.0841	24	0.0943	31	0.1019	38	0.1079	45	0.1129	
11	0.0711	18	0.0858	25	0.0955	32	0.1028	39	0.1087	46	0.1136	
12	0.0737	19	0.0874	26	0.0967	33	0.1037	40	0.1094	47	0.1142	
13	0.0761	20	0.0889	27	0.0978	34	0.1046	41	0.1102	48	0.1149	
14	0.0783	21	0.0903	28	0.0989	35	0.1055	42	0.1109	49	0.1155	

*From formula: for 60 Hz, $X_d = 0.06831 \log_{10} d$, d = separation in feet.

†From: *Electrical Transmission and Distribution Reference Book,* Westinghouse Electric Corporation, 1950.

be matched to the load. This is hardly possible in a transmission line, but is achieved at much higher (e.g., radio) frequencies. This eliminates the so-called standing waves stemming from the summation of terms in the equations for $\mathbf{V_S}$ and $\mathbf{I_S}$. Such quantities as standing-wave ratio SWR and reflection coefficient σ are easily calculated, but are beyond the needs of power studies.

COMPLEX POWER

Determine the complex power **S** at both ends of the 320-km (200-mi) transmission line using the results of the preceding example.

Calculation Procedure:

1. Calculate S at Receiving End

Use $\mathbf{S} = 3\mathbf{VI}^*$, where **V** is the phase-to-neutral voltage and $\mathbf{I}^*$ is the complex conjugate of the line current under balanced conditions. Then, at the receiving end, $\mathbf{S} = (3)(132.8 \times 10^3\angle 0°)(200\angle 36.9°) = 63{,}719$ kW $+ j47{,}841$ kvar.

2. Calculate S at Sending End

At the sending end, $\mathbf{S} = (3)(150.8 \times 10^3\angle 8.06°)(152.6\angle 4.52°) = 67.379$ kW $+ j15{,}036$ kvar.

Related Calculations: The transmission line must, in this case, be furnishing some of the receiving end's requirements for reactive power from the supply of stored charge, because the apparent power input in kvar is less than the output. Power loss of the line may be determined by $Q = 47{,}841 - 15{,}036 = 32{,}905$ kvar made up by stored line charge and $P = 67{,}379 - 63{,}719 = 3660$ kW line-resistance losses.

SURGE IMPEDANCE LOADING

A convenient method for comparing the capability of transmission lines to support energy flow (but not accounting for resistance-loss restrictions) is through the use of *surge impedance loading* SIL. If the line is assumed terminated in its own surge impedance value as a load (preferably a real number), then a hypothetical power capability is obtained which can be compared with other lines.

Compare two 230-kV lines for their power capability if $Z_{c1} = 500\ \Omega$ for line 1 and $Z_{c2} = 400\ \Omega$ for line 2.

Calculation Procedure:

1. Determine Expression for SIL

If Z_c is considered as a load, the load current I_L may be expressed by $I_L = V_L/\sqrt{3}Z_c$ kA, where V_L is the magnitude of the line-to-line voltage in kilovolts. Then, SIL $= \sqrt{3}V_L I_L = V_L^2/Z_c$.

2. Calculate SIL Values

$\text{SIL}_1 = 230^2/500 = 106$ MW and $\text{SIL}_2 = 230^2/400 = 118$ MW. Line 2 has greater power capability than line 1.

Section 10 ELECTRIC-POWER NETWORKS

Lawrence J. Hollander, P.E.
Associate Dean, Cooper Union

REFERENCES Slemon and Straughen—*Electric Machines,* Addison-Wesley; Matsch—*Electromagnetic and Electromechanical Machines,* Harper & Row; Stein and Hunt—*Electric Power System Components,* Van Nostrand Reinhold; Nasar and Unnewehr—*Electromechanics and Electric Machines,* Wiley; Zaborszky and Rittenhouse—*Electric Power Transmission,* Rensselaer Polytechnic Institute Bookstore; Fitzgerald and Kingsley—*Electric Machinery,* McGraw-Hill; Beeman—*Industrial Power Systems Handbook,* McGraw-Hill; Greenwood—*Electrical Power Systems Engineering: Problems and Solutions,* McGraw-Hill; Guile and Paterson—*Electrical Power Systems,* Pergamon; Pansini—*Basic Electrical Power Transmission,* Hayden; Stevenson—*Elements of Power*

System Analysis, McGraw-Hill; Anderson—*Analysis of Faulted Power Systems,* Iowa State University Press; Sullivan—*Power System Planning,* McGraw-Hill; IEEE—*Recommended Practice for Grounding of Industrial and Commercial Power Systems;* IEEE—*Recommended Practice for Electric Power Systems in Commercial Buildings;* Kosow—*Electric Machinery and Transformers,* Prentice-Hall; Siskind—*Electrical Machines: Direct and Alternating Currents,* McGraw Hill.

POWER SYSTEM REPRESENTATION: GENERATORS, MOTORS, TRANSFORMERS, AND LINES

The following components comprise a simplified version of a power system, listed in sequential physical order from the generator location to the load: (1) two steam-electric generators, each at 13.2 kV, (2) two step-up transformers, 13.2/66 kV, (3) sending-end, high-voltage bus at 66 kV, (4) one long transmission line at 66 kV, (5) receiving-end bus at 66 kV, (6) a second 66-kV transmission line with a center-tap bus, (7) step-down transformer at receiving-end bus, 66/12 kV, supplying four 12-kV motors in parallel, and (8) a step-down transformer, 66/7.2 kV, off the center-tap bus, supplying a 7.2-kV motor. Draw a one-line diagram for the three-phase, 60-Hz system, including appropriate oil circuit breakers (OCBs).

Calculation Procedure:

1. Identify the Appropriate Symbols

For electric-power networks an appropriate selection of graphic symbols is shown in Fig. 1.

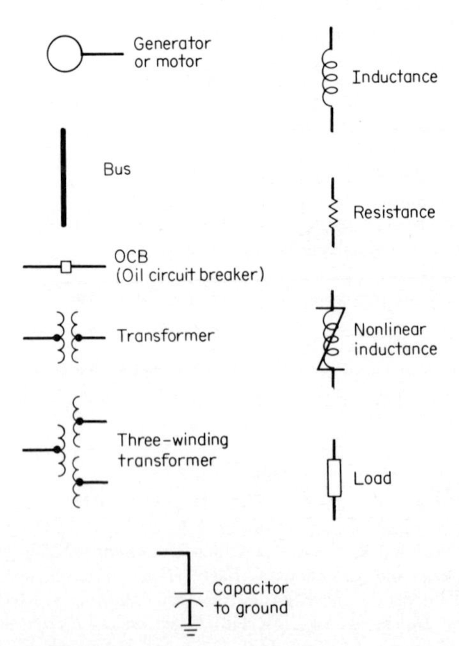

Fig. 1 Common power symbols used in one-line diagrams.

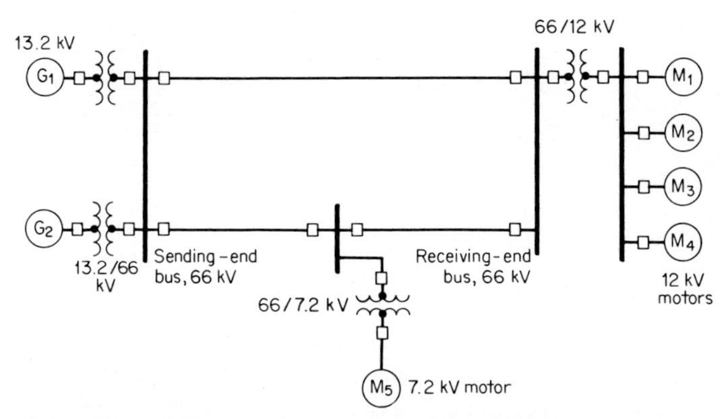

Fig. 2 Three-phase power system represented by one-line diagram.

2. *Draw the Required System*

The system described in the problem is shown in Fig. 2. The oil circuit breakers are added at the appropriate points for proper isolation of equipment.

Related Calculations: It is the general procedure to use one-line diagrams for representing three-phase systems. When analysis is done using symmetrical components, different diagrams may be drawn that will represent the electric circuitry for positive, negative, and zero-sequence components. Additionally, it is often necessary to identify the grounding connection, or whether the device is wye- or delta-connected. This type of notation is shown in Fig. 3.

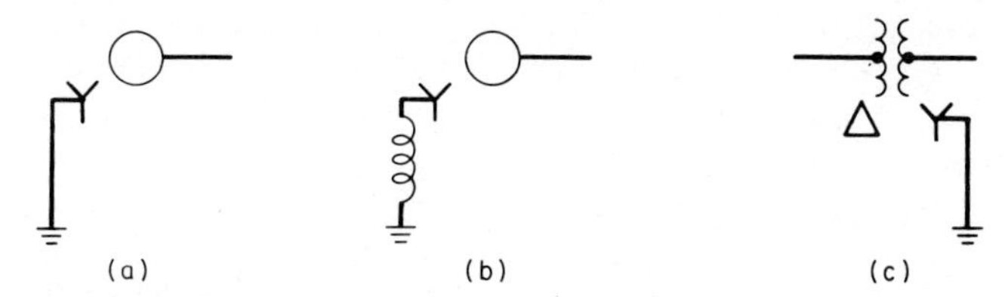

Fig. 3 Identification for wye-connected generator or motor. (*a*) Solidly grounded. (*b*) Grounded through an inductance. (*c*) The transformer is identified as being delta-wye, with the wye-side solidly grounded.

PER-UNIT METHOD OF SOLVING THREE-PHASE PROBLEMS

For the system shown in Fig. 4, draw the electric circuit or reactance diagram, with all reactances marked in per-unit (p.u.) values, and find the generator terminal voltage assuming both motors operating at 12 kV, three-quarters load, and unit power factor.

Calculation Procedure:

1. *Establish Base Voltage through the System*

By observation of the magnitude of the components in the system, a base value of apparent power S is chosen; it should be of the general magnitude of the components, and

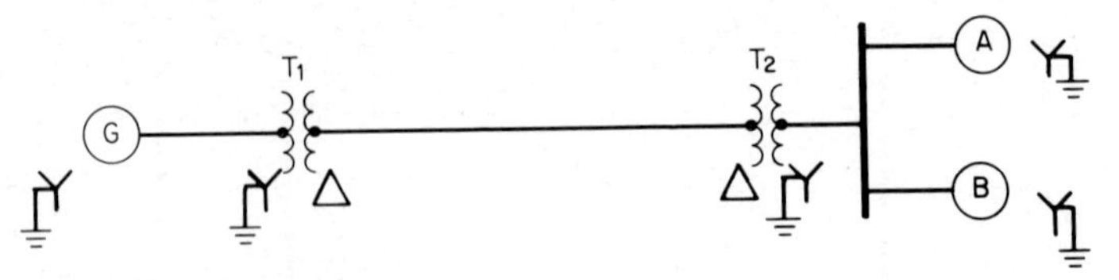

Fig. 4 One-line diagram of electric-power system supplying motor loads. The table below gives specifications.

Generator	*Transformers (each)*	*Motor A*	*Motor B*	*Transmission line*
13.8 kV	25,000 kVA	15,000 kVA	10,000 kVA	$X = 65\ \Omega$
25,000 kVA	13.2/69 kV	13.0 kV	13.0 kV	
Three phase	$X_L = 11\%$	$X'' = 15\%$	$X'' = 15\%$	
$X'' = 15\%$				

the choice is arbitrary. In this problem, 25,000 kVA is chosen as the base S, and simultaneously, at the generator end 13.8 kV is selected as a base voltage V_{base}.

The base voltage of the transmission line is then determined by the turns ratio of the connecting transformer: (13.8 kV)(69 kV/13.2 kV) = 72.136 kV. The base voltage of the motors is determined likewise but with the 72.136-kV value: thus, (72.136 kV)(13.2 kV/69 kV) = 13.8 kV. The selected base S value remains constant throughout the system, but the base voltage is 13.8 kV at the generator and at the motors, and 72.136 kV on the transmission line.

2. *Calculate the Generator Reactance*

No calculation is necessary for correcting the value of the generator reactance because it is given as 0.15 p.u. (15 percent), based on 25,000 kVA and 13.8 kV. If a different S base were used in this problem, then a correction would be necessary as shown for the transmission line, motors, and transformers.

3. *Calculate the Transformer Reactance*

It is necessary to make a correction when the transformer nameplate reactance is used because the calculated operation is at a different voltage, 13.8 kV/72.136 kV instead of 13.2 kV/69 kV. Use the equation for correction: per-unit reactance = (nameplate per-unit reactance)(base kVA/nameplate kVA)(nameplate kV/base kV)^2 = (0.11)(25,000/25,000)(13.2/13.8)^2 = 0.101 p.u. This applies to each transformer.

4. *Calculate the Transmission-Line Reactance*

Use the equation: $X_{\text{per unit}}$ = (ohms reactance)(base kVA)/(1000)(base kV)^2 = (65)(25,000)/(1000)(72.1)^2 = 0.313 p.u.

5. *Calculate the Reactance of the Motors*

Corrections need to be made in the nameplate ratings of both motors because of differences of ratings in kVA and kV as compared with those selected for calculations in this problem. Use the correcting equation from Step 3, above. For motor A, X''_A = (0.15 per unit)(25,000 kVA/15,000 kVA)(13.0 kV/13.8 kV)^2 = 0.222 p.u. For motor B, similarly, X''_B = (0.15 p.u.)(25,000 kVA/10,000 kVA)(13.0 kV/13.8 kV)^2 = 0.333 p.u.

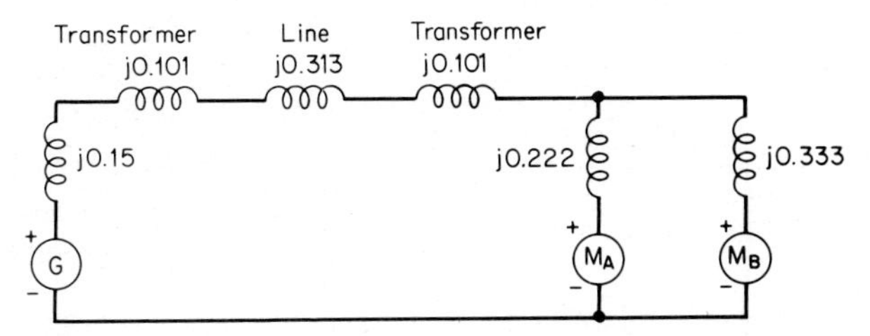

Fig. 5 One-line reactance circuit diagram (reactances shown on a per unit basis).

6. *Draw the Reactance Diagram*

The completed reactance diagram is shown in Fig. 5.

7. *Calculate Operating Conditions of the Motors*

If the motors are operating at 12 kV, this represents 12 kV/13.8 kV = 0.87 per-unit voltage. At unity power factor the load is given as three-quarters or 0.75 per unit. Thus, expressed per unit, the combined motor current is obtained by using the equation $I_{\text{per unit}}$ = per-unit power/per-unit voltage = 0.75/0.87 = $0.862\angle 0°$ per unit.

8. *Calculate the Generator Terminal Voltage*

The voltage at the generator terminals, $\mathbf{V}_G = \mathbf{V}_{\text{motor}}$ + drop through transformers and transmission line = $0.87\angle 0° + 0.862\angle 0°(j0.101 + j0.313 + j0.101) = 0.87 + j0.444 = 0.977\angle 27.03°$ per unit. In order to obtain the actual voltage, multiply the per-unit voltage by the base voltage at the generator. Thus, $\mathbf{V}_g = (0.977\angle 27.03°)(13.8 \text{ kV}) = 13.48\angle 27.03°$ kV.

Related Calculations: In the solution of these problems the selection of base voltage and apparent power are arbitrary. However, the base voltage in each section of the circuit must be related in accordance with transformer turns ratios. The base impedance may be calculated from the equation Z_{base} = (base kV)^2(1000)/(base kVA). For the transmission-line section in this problem, $Z_{\text{base}} = (7.136)^2(1000)/(25{,}000) = 208.1\ \Omega$; thus the per-unit reactance of the transmission line equals (actual ohms)/(base ohms) = 65/208.1 = 0.313 per unit.

PER-UNIT BASES FOR SHORT-CIRCUIT CALCULATIONS

For the system shown in Fig. 6, assuming S bases of 30,000 and 75,000 kVA, respectively, calculate the through impedance in ohms between the generator and the output terminals of the transformer.

Calculation Procedure:

1. *Correct Generator 1 Impedance (Reactance)*

Use the equation for changing S base: per-unit reactance = (nameplate per-unit reactance)(base kVA/nameplate kVA) = $X'' = (0.20)(30{,}000/40{,}000) = 0.15$ p.u. on a 30,000-kVA base. Similarly, $X'' = (0.20)(75{,}000/40{,}000) = 0.375$ p.u. on a 75,000-kVA base.

2. *Correct Generator 2 Impedance (Reactance)*

Use the equation as in Step 1, above: $X'' = (0.25)(30{,}000/30{,}000) = 0.25$ p.u. on a 30,000-kVA base. $X'' = (0.25)(75{,}000/30{,}000) = 0.625$ p.u. on a 75,000-kVA base.

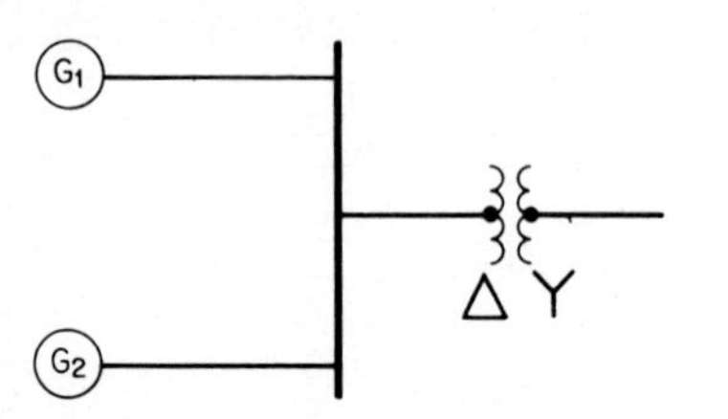

Fig. 6 System of paralleled generators and transformer with calculations demonstrating different apparent-power bases.

Generator 1	*Generator 2*	*Transformer*
40,000 kVA	30,000 kVA	75,000 kVA
13.2 kV	13.2 kV	13.2 kV (delta)
$X'' = 0.20$ per unit	$X'' = 0.25$ per unit	66 kV (wye)
		$X = 0.10$ per unit

3. *Correct the Transformer Impedance (Reactance)*

Use the same equation as for the generator corrections: $X = (0.1)(30{,}000/75{,}000) = 0.04$ p.u. on a 30,000-kVA base; it is 0.10 on a 75,000-kVA base (Fig. 7).

4. *Calculate the Through Impedance per Unit*

In either case the through impedance is equal to the parallel combination of generators 1 and 2, plus the series impedance of the transformer. For the base of 30,000 kVA, $jX_{total} = (j0.15)(j0.25)/(j0.15 + j0.25) + j0.04 = j0.134$ p.u. For the base of 75,000 kVA,

$$jX_{total} = (j0.375)(j0.625)/(j0.375 + j0.625) + j0.10 = j0.334 \text{ p.u.}$$

5. *Convert Impedances (Reactances) to Ohms*

The base impedance (reactance) in ohms is (1000)(base kV)^2/(base kVA). Thus, for the case of the 30,000-kVA base, the base impedance is $(1000)(13.2)^2/30{,}000 = 5.808\ \Omega$. The actual impedance in ohms of the given circuit is equal to (per-unit impedance)(base impedance) $= (j0.134)(5.808) = 0.777\ \Omega$, referred to the low-voltage side of the transformer.

The base impedance (reactance) in ohms for the case of the 75,000-kVA base is $(1000)(13.2)^2/75{,}000 = 2.32\ \Omega$. The actual impedance in ohms of the given circuit is equal to $(j0.334)(2.32) = 0.777\ \Omega$, referred to the low-voltage side of the transformer.

6. *Compare Results of Different S Bases*

It is seen that either *S* base, selected arbitrarily, yields the same answer in ohms. When working on a per-unit basis, the same per-unit values hold true for either side of the transformers; in actual ohms, amperes, or volts, turns ratio adjustments are necessary

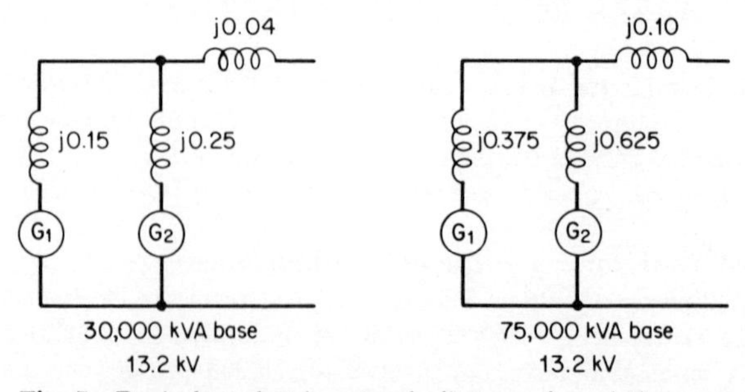

Fig. 7 Equivalent electric network diagrams for solution on a 30,000-kVA base and on a 75,000-kVA base.

depending on the reference side for which calculations are being made. In this problem the total reactance is 0.777 Ω, referred to the low-voltage side of the transformers; it is $(0.777)(66\ \text{kV}/13.2\ \text{kV})^2 = 19.425\ \Omega$, referred to the high-voltage side. Expressed per unit it is 0.134 (on a 30,000-kVA base) or 2.32 (on a 75,000-kVA base) on either side of the transformer.

Related Calculations: This problem illustrates the arbitrary selection of the *S* base, provided the value selected is used consistently throughout the circuit. Likewise, the *S* base may be selected in one part of the circuit, but in all other parts of the circuit the *S*-base values must be related in accordance with transformer turns ratios. In both cases, a little experience and general observation of the given information will suggest appropriate selections of per-unit bases that will yield comfortable numbers with which to work.

CHANGING THE BASE OF PER-UNIT QUANTITIES

The reactance of a forced-oil-cooled 345-kV/69-kV transformer is given as 22 percent; it has a nameplate capacity of 450 kVA. Calculations for a short-circuit study are being made using bases of 765 kV and 1 MVA. Determine (1) the transformer reactance on the study bases and (2) the *S* base for which the reactance will be 10 percent on a 345-kV base.

Calculation Procedure:

1. Convert the Nameplate Reactance to Study Bases

The study bases are 765 kV and 1000 kVA. Let subscript 1 denote the given nameplate conditions and subscript 2 denote the new or revised conditions. Use the equation: X_2 p.u. $= (X_1\ \text{p.u.})[(\text{base kVA}_2)/(\text{base kVA}_1)][(\text{base kV}_1)/(\text{base kV}_2)]^2$. Thus, $X_2 = (0.22)[(1000)/(450)]\,[(345)/(765)]^2 = 0.099$ p.u.

2. Calculation of Alternative S Base

Assume it is desired that the per-unit reactance of the transformer be 0.10 on a 345-kV base; use the same equation as in the first step, and solve for base kVA_2. Thus, base $\text{kVA}_2 = (X_2/X_1)(\text{base kVA}_1)[(\text{base kV}_2)/(\text{base kV}_1)]^2 = [(0.10)/(0.22)](450)[(345)/(345)]^2 = 204.5$ kVA.

Related Calculations: The equation used in this problem may be used to convert any impedance, resistance, or reactance expressed per unit (or percent) from one set of kV, kVA bases to any other set of kV, kVA bases.

WYE-DELTA AND DELTA-WYE CONVERSIONS

A one-line portion of an electrical network diagram is shown in Fig. 8. Using wye-delta and/or delta-wye conversions, reduce the network to a single reactance.

Calculation Procedure:

1. Convert Wye to Delta

The network reduction may be started at almost any point. One starting point is to take the wye-formation of reactances *a*, *b*, and *c* and convert to a delta. Refer to Fig. 9 for the appropriate equation. Thus, substituting *X*'s for *Z*'s in the equations yields: $X_{ab} = (X_aX_b + X_bX_c + X_cX_a)/X_c = [(j0.30)(j0.15) + (j0.15)(j0.45) +$

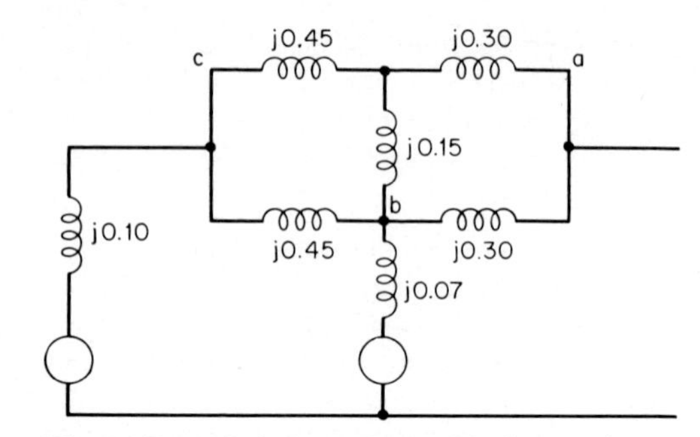

Fig. 8 Portion of electrical network diagram (reactances shown on a per-unit basis).

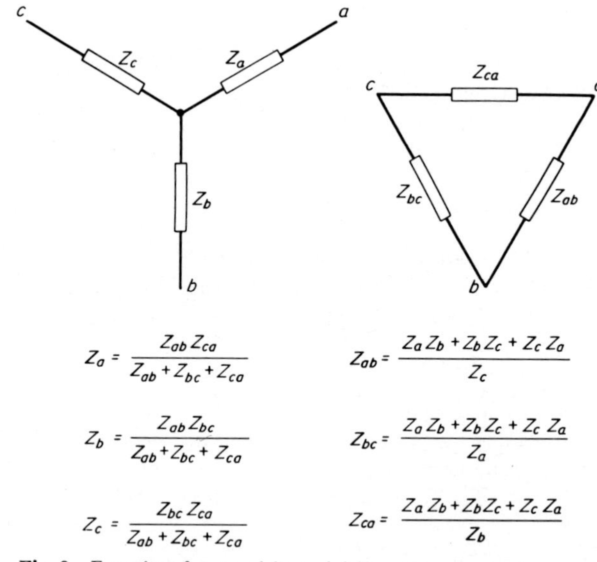

Fig. 9 Equations for wye-delta and delta-wye conversions.

$(j0.45)(j0.30)]/(j0.45) = [j^2 0.045 + j^2 0.0675 + j^2 0.135]/j0.45 = (j^2 0.2475)/j0.45 = j0.55$ p.u. (Fig. 10).

$X_{bc} = (X_a X_b + X_b X_c + X_c X_a)/X_a = j^2 0.2475/j0.30 = j0.819$ p.u. and $X_{ca} = (X_a X_b + X_b X_c + X_c X_a)/X_b = j^2 0.2475/j0.15 = j1.65$ p.u.

2. *Combine Parallel Reactances*

It is noted that after the first wye-delta conversion, points a and b are connected by two reactances in parallel, yielding a combined reactance of $(j0.55)(j0.30)/(j0.55 + j0.30) = j0.194$ p.u. Points b and c are connected by two reactances in parallel, yielding a combined reactance of $(j0.819)(j0.45)/(j0.819 + j0.45) = j0.290$ p.u.

3. *Convert Delta to Wye*

Points a, b, and c now form a new delta connection that will be converted to a wye. $X_a = X_{ab} X_{ca}/(X_{ab} + X_{bc} + X_{ca}) = (j0.194)(j1.65)/(j0.194 + j0.290 +$

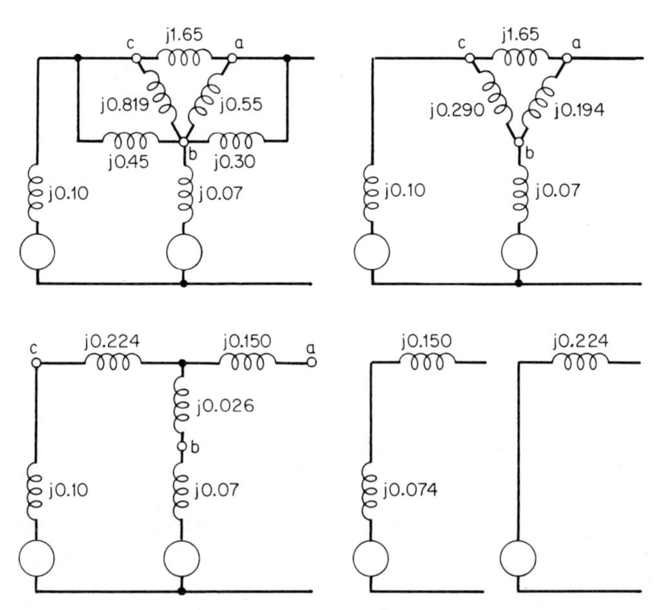

Fig. 10 Network reduction of power circuit.

$j1.65) = j0.150$ p.u. $X_b = X_{ab}X_{bc}/(X_{ab} + X_{bc} + X_{ca}) = (j0.194)(j0.290)/(j0.194 + j0.290 + j1.65) = j0.026$ p.u. And $X_c = X_{bc}X_{ca}/(X_{ab} + X_{bc} + X_{ca}) = (j0.290)(j1.65)/(j0.194 + j0.290 + j1.65) = j0.224$ p.u.

4. Combine Reactances in Generator Leads

In each generator lead there are now two reactances in series. In the *c* branch the two reactances total $j0.10 + j0.224 = j0.324$ per unit. In the *b* branch the total series reactance is $j0.07 + j0.026 = j0.096$ p.u. These may be paralleled, leaving one equivalent generator and one reactance, $(j0.324)(j0.096)/(j0.324 + j0.096) = j0.074$ p.u.

5. Combine Remaining Reactances

The equivalent generator reactance is now added to the reactance in branch *a*, yielding $j0.074 + j0.150 = j0.224$ p.u., as illustrated in Fig. 10.

Related Calculations: This problem illustrates the delta-wye and wye-delta conversion used for network reductions. Such conversions find many applications in network problems.

PER-UNIT REACTANCES OF THREE-WINDING TRANSFORMERS

For the three-phase, 60-Hz system shown in Fig. 11, the leakage reactances of the three-winding transformer are: $X_{ps} = 0.08$ per unit at 50 MVA and 13.2 kV, $X_{pt} = 0.07$ p.u. at 50 MVA and 13.2 kV, and $X_{st} = 0.20$ p.u. at 20 MVA and 2.2 kV, where subscripts p, s, and t refer to primary, secondary, and tertiary windings. Using a base of

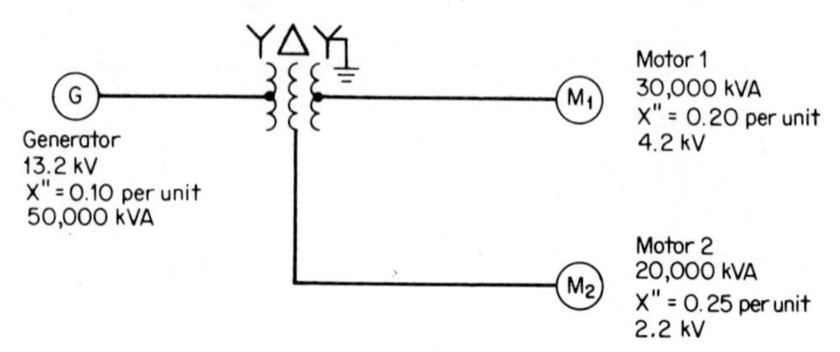

Three-winding transformer

Primary	Secondary	Tertiary
13.2 kV	4.2 kV	2.2 kV
50 MVA	30 MVA	20 MVA

Fig. 11 Three-winding transformer interconnecting motor loads to generator.

50,000 kVA and 13.2 kV, calculate the various reactances in the circuit and draw the simplified-circuit network.

Calculation Procedure:

1. Correct the Generator Reactance

The generator reactance is given as 0.10 p.u. on the base of 50,000 kVA and 13.2 kV. No correction is needed.

2. Correct the Three-Winding Transformer Reactances

X_{ps} and X_{pt} are given as 0.08 and 0.07 p.u., respectively, on the base of 50,000 kVA and 13.2 kV. No correction is needed. However, X_{st} is given as 0.20 p.u. at 20,000 kVA and 2.2 kV; it must be corrected to a base of 50,000 kVA and 13.2 kV. Use the equation $X_2 = X_1(\text{base } S_2/\text{base } S_1)(V_{\text{base }1}/V_{\text{base }2})^2 = (0.20)\ [(50{,}000)/(20{,}000)]\ [(2.2)/(13.2)]^2 = 0.014$ p.u; this is the corrected value of X_{st}.

3. Calculate the Three-Phase, Three-Winding Transformer Reactances as an Equivalent Wye

Use the equation $X_p = \frac{1}{2}(X_{ps} + X_{pt} - X_{st}) = \frac{1}{2}(0.08 + 0.07 - 0.014) = 0.068$ p.u., $X_s = \frac{1}{2}(X_{ps} + X_{st} - X_{pt}) = \frac{1}{2}(0.08 + 0.014 - 0.07) = 0.012$ p.u., and $X_t = \frac{1}{2}(X_{pt} + X_{st} - X_{ps}) = \frac{1}{2}(0.07 + 0.014 - 0.08) = 0.002$ p.u.

4. Correct the Motor Reactances

X'' of motor load 1 is given as 0.02 on a base of 30,000 kVA and 4.2 kV. The corrected value is $(0.20)(50{,}000/30{,}000)(4.2/13.2)^2 = 0.034$ p.u.

X'' of motor load 2 is given as 0.25 on a base of 20,000 kVA and 2.2 kV. The corrected value is determined similarly: $(0.25)(50{,}000/20{,}000)(2.2/13.2)^2 = 0.017$ p.u.

5. Draw the Simplified-Circuit Network

See Fig. 12.

Related Calculations: Three-winding transformers are helpful in suppressing third-harmonic currents; the tertiary winding is connected in closed delta for that purpose. The third-harmonic currents develop from the exciting current. This problem illustrates the handling of the reactances of three-winding transformers.

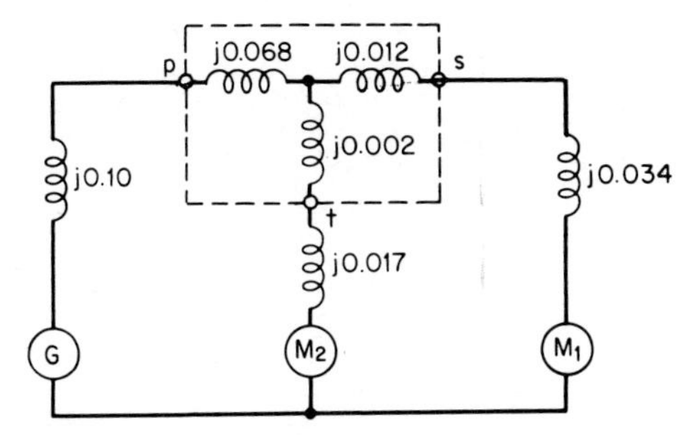

Fig. 12 Three-winding transformer equivalent circuit.

CALCULATION OF COMPLEX POWER $P + jQ$

At a certain point in the solution of a balanced three-phase problem, the current per phase is equal to $5.0\angle -37°$ A and the line voltage is $69\angle 0°$ kV. With a per-unit base of 1000 kVA and 69 kV, determine the complex power.

Calculation Procedure:

1. Change Voltage to a per-Unit Basis

The voltage is given as $69\angle 0°$ kV. Since the base voltage is 72 kV, the per-unit voltage $= 69/72 = 0.96\angle 0°$ p.u.

2. Change Current to a per-Unit Basis

Use the equation: $S = \sqrt{3}V_L I_L$; thus, $I_L = S/\sqrt{3}V_L$. For the base conditions specified, $I_L = 1{,}000{,}000\ \text{VA}/(\sqrt{3})(72{,}000\ \text{V}) = 8.02$ A. The given current is $5\angle -37°$ A/ $8.02 = 0.623\angle -37°$ per unit.

3. Calculate Complex Power

Use the equation for complex power: $\mathbf{S} = P + jQ = \mathbf{VI}^*$ where the asterisk indicates the conjugate value of the current. $\mathbf{S} = (0.96\angle 0°)(0.623\angle +37°) = 0.598\angle 37°$ $= 0.478 + j0.360$ p.u. The real power $P = 0.478$ p.u., or $(0.478)(1000\ \text{kVA}) = 478$ kW. The reactive power $Q = 0.360$ p.u., or $(0.360)(1000\ \text{kVA}) = 360$ kvar. $P + jQ$ $= 478\ \text{kW} + j360$ kvar, or $0.478 + j0.360$ p.u.

Related Calculations: This problem illustrates the finding of complex power using the conjugate of the current and per-unit notation. Alternatively, the real power may be obtained from the equation: $P = \sqrt{3}V_L I_L \cos\theta = (\sqrt{3})(69{,}000\ \text{V})(5.0\ \text{A}) \cos 37° =$ 478 kW. Reactive power is $Q = \sqrt{3}V_L I_L \sin\theta = (\sqrt{3})(69{,}000\ \text{V})(5.0\ \text{A}) \sin 37° =$ 360 kvar.

CHECKING VOLTAGE PHASE SEQUENCE WITH LAMPS

Given the three leads of a 120-V, three-phase, 60-Hz system. Two identical lamps and an inductance are connected arbitrarily as shown in Fig. 13. Determine the phase sequence of the unknown phases for the condition that one lamp glows brighter than the other.

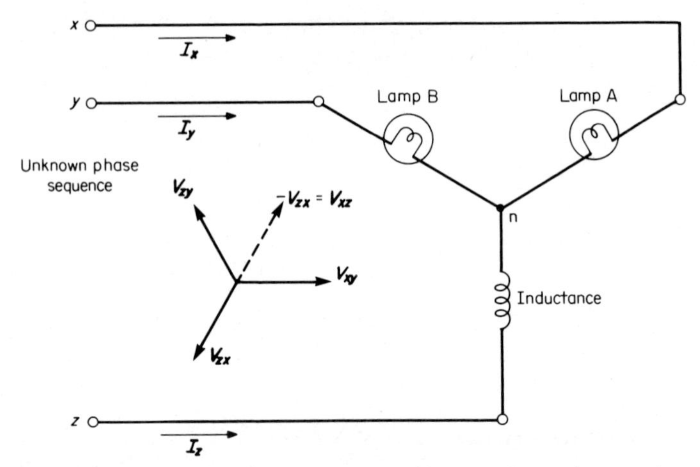

Fig. 13 Determination of phase sequence with two resistance lamps and an inductance.

Calculation Procedure:

1. Assume per-Unit Values of Lamps and Inductance

For whatever kVA, kV bases may be assumed, let each lamp be a pure resistance of 1.0 p.u., and let the inductance be without resistance, or $1.0\angle 90°$ p.u. That is, $R_A = R_B = 1.0\angle 0°$ per unit, and $X = 1.0\angle 90°$ per unit. Practically, the inductive angle will not be 90°, but slightly less because of resistance; this solution depends upon the inductance having a high ratio of X_L/R.

2. Assume a Phase Rotation

Assume that the phase rotation is *xy-zx-yz*, or that $\mathbf{V}_{xy} = 1.0\angle 0°$, $\mathbf{V}_{zx} = 1.0\angle -120°$, and $\mathbf{V}_{yz} = 1.0\angle +120°$ p.u.

3. Develop Voltage Equations for Lamps

Three equations may be written for the circuit: (1) $\mathbf{I}_x + \mathbf{I}_y + \mathbf{I}_z = 0$; (2) $\mathbf{V}_{zx} - R_A\mathbf{I}_x + X\mathbf{I}_z = 0$; and (3) $\mathbf{V}_{zy} - R_B\mathbf{I}_y + X\mathbf{I}_z = 0$. Substitute the assumed values for $\mathbf{V}_{zx}$, $\mathbf{V}_{zy}$ $(= -\mathbf{V}_{yz})$, R_A, and R_B. Equation (1) remains unchanged; $\mathbf{I}_x + \mathbf{I}_y + \mathbf{I}_z = 0$. Equation (2) becomes $1\angle -120° - \mathbf{I}_x + 1\angle 90°\ \mathbf{I}_z = 0$, or $-\mathbf{I}_x + 1\angle 90°\ \mathbf{I}_z = -(1\angle -120°) = 1\angle 60°$. Equation (3) becomes $-(1\angle 120°) - \mathbf{I}_y + 1\angle 90° \mathbf{I}_z = 0$, or $-\mathbf{I}_y + 1\angle 90°\ \mathbf{I}_z = 1\angle 120°$.

4. Determine the Voltage across Lamp A

The voltage across lamp A is proportional to the current through it, namely I_x. The solution of the simultaneous equations for I_x yields

$$I_x = \frac{\begin{vmatrix} 0 & 1 & 1 \\ 1\angle 60° & 0 & j \\ 1\angle 120° & -1 & j \end{vmatrix}}{\begin{vmatrix} 1 & 1 & 1 \\ -1 & 0 & j \\ 0 & -1 & j \end{vmatrix}} = \frac{-0.50 - j1.866}{1 + j2} = 0.863\angle 191.6° \text{ p.u.}$$

5. Determine the Voltage across Lamp B

The voltage across lamp B is proportional to the current through it, namely $\mathbf{I}_y$. The solution of the simultaneous equations for I_y *yields*

$$\mathbf{I}_y = \frac{\begin{vmatrix} 1 & 0 & 1 \\ -1 & 1\angle 60^\circ & j \\ 0 & 1\angle 120^\circ & j \end{vmatrix}}{1 + j2} = \frac{0.5 + j0.134}{1 + j2} = 0.232\angle -48.4^\circ \text{ p.u.}$$

6. Confirm the Voltage Sequence

From the previous two steps it is determined that the current through lamp A is greater than through lamp B. Therefore, lamp A is brighter than lamp B, indicating that the phase sequence is as was assumed, *xy-zx-yz*.

7. Let the Phase Sequence Be Reversed

The reversed phase sequence is *xy-yz-zx*. In this case the assumed voltages become: $\mathbf{V}_{xy} = 1.0\angle 0^\circ$, $\mathbf{V}_{zx} = 1.0\angle 120^\circ$, and $\mathbf{V}_{yz} = 1.0\angle -120^\circ$.

8. Determine the Voltage across Lamp A for Reversed Sequence

The voltage across lamp A, by solution of simultaneous equations, is proportional to

$$\mathbf{I}_x = \frac{\begin{vmatrix} 0 & 1 & 1 \\ 1\angle -60^\circ & 0 & j \\ 1\angle -120^\circ & -1 & j \end{vmatrix}}{1 + j2} = \frac{-0.5 - j0.134}{1 + j2} = 0.232\angle 131.5^\circ \text{ p.u.}$$

9. Determine the Voltage across Lamp B for Reversed Sequence

The voltage across lamp B, by solution of simultaneous equations, is proportional to

$$\mathbf{I}_y = \frac{\begin{vmatrix} 1 & 0 & 1 \\ -1 & 1\angle -60^\circ & j \\ 0 & 1\angle -120^\circ & j \end{vmatrix}}{1 + j2} = \frac{0.5 + j1.866}{1 + j2} = 0.863\angle 11.6^\circ \text{ p.u.}$$

10. Confirm the Voltage Sequence

Thus, for the connection of Fig. 13, lamp B will glow brighter than lamp A for the voltage sequence pattern *xy-yz-zx*.

Related Calculations: This problem indicates one method of checking phase sequence of voltages. Another method is based on a similar analysis and uses a combination of a voltmeter, a capacitor, and an inductance. These two methods rely on the use of an unbalanced load impedance.

TOTAL POWER IN BALANCED THREE-PHASE SYSTEM

A three-phase balanced 440-V, 60-Hz system has a wye-connected per-phase load impedance of $22\angle 37^\circ\ \Omega$. Determine the total power absorbed by the load.

Calculation Procedure:

1. *Determine the Line-to-Neutral Voltage*

The given voltage class, 440, is presumed to be the line-to-line voltage. The line-to-neutral voltage is $440\ \text{V}/\sqrt{3} = 254\ \text{V}$.

2. *Determine the Current per Phase*

The current per phase for a wye connection is the line current $= \mathbf{V}_{\text{phase}}/\mathbf{Z}_{\text{phase}} = 254\angle 0°\ \text{V}/22\angle 37°\ \Omega = 11.547\angle -37°\ \text{A}$.

3. *Calculate the Power per Phase*

The power per phase $P_{\text{phase}} = V_{\text{phase}} I_{\text{phase}} \cos\theta = (254\angle 0°\ \text{V})(11.547\ \text{A}) \cos 37° = 2342.3\ \text{W}$.

4. *Calculate the Total Power*

The total power is $P_{\text{total}} = 3P_{\text{phase}} = (3)(2342.3\ \text{W}) = 7027\ \text{W}$. Alternatively, $P_{\text{total}} = \sqrt{3} V_L I_L \cos\theta = (\sqrt{3})(440\ \text{V})(11.545\ \text{A}) \cos 37° = 7027\ \text{W}$.

Related Calculations: This problem illustrates two approaches to finding the total power in a three-phase circuit. It is assumed herein that the system is balanced. For unbalanced systems, symmetrical components are used and the analysis is done separately for the zero-sequence, positive-sequence, and negative-sequence networks.

DIVISION OF LOAD BETWEEN TRANSFORMERS IN PARALLEL

Two single-phase transformers are connected in parallel on both the high- and low-voltage sides and have the following characteristics: transformer A, 100 kVA, 2300/120 V, 0.006-Ω resistance, and 0.025-Ω leakage reactance referred to the low-voltage side; transformer B, 150 kVA, 2300/115 V, 0.004-Ω resistance, and 0.015-Ω leakage reactance referred to the low-voltage side. A 125-kW load at 0.85 power factor, lagging, is connected to the low-voltage side, with 125 V at the terminal of the transformers. Determine the primary voltage and the current supplied by each transformer.

Calculation Procedure:

1. *Determine the Admittance Y_A of Transformer A*

The impedance referred to the low-voltage side is given as $0.006 + j0.025\ \Omega$. Convert the impedance from rectangular form to polar form ($0.0257\angle 76.5°$), and take the reciprocal to change the impedance into an admittance ($38.90\angle -76.5°$); in rectangular form this is equal to $9.08 - j37.82$ S.

2. *Determine the Admittance Y_B of Transformer B*

The impedance referred to the low-voltage side is given as $0.004 + j0.015\ \Omega$. Convert the impedance from rectangular form to polar form ($0.0155\angle 75.07°$) and take the recip rocal to change the impedance into an admittance ($64.42\angle -75.07°$); in rectangular form this is equal to $16.60 - j62.24$ S.

3. *Determine the Total Admittance of the Paralleled Transformers*

The total admittance $\mathbf{Y}_{\text{total}}$ is the sum of the admittance of transformer A and that of transformer B, namely $25.68 - j100.06$ S, as referred to the low-voltage side of the transformers. In polar form this is $103.3\angle -75.6°$.

4. Determine the Total Current

For the total current, use the equation $P/(V \cos \theta) = 125{,}000 \text{ W}/(125 \text{ V})(0.85) = 1176$ A.

5. Calculate the Primary Voltage

Assume that the secondary voltage is the reference phasor, $125 + j0$ V. The load current lags the secondary voltage by an angle whose cosine is 0.85, the power-factor angle. The load current in polar form is $1176\angle{-31.79°}$ and in rectangular form is $999.6 - j619.5$ A. The primary voltage is obtained from the equation $\mathbf{V}_1 = a_{\text{total}}\mathbf{V}_2 + (a_{\text{total}}\mathbf{I}_1/\mathbf{Y}_{\text{total}})$ where a_{total} = turns ratio for the two transformers in parallel. This same equation may be developed into another form: $\mathbf{V}_1 = (\mathbf{V}_2\mathbf{Y}_{\text{total}} + \mathbf{I}_1)/(\mathbf{Y}_{\text{total}}/a_{\text{total}})$, where $Y_{\text{total}}/a_{\text{total}} = Y_A/a_A + Y_B/a_B$, a_A is the turns ratio of transformer A (i.e., $a_A = 2300/120 = 19.17$), and a_B is the turns ratio of transformer B (i.e., $a_B = 2300/115 = 20.0$). $Y_{\text{total}}/a_{\text{total}} = (9.08 - j37.82)/19.17 + (16.60 - j62.24)/20 = 0.47 - j1.97 + 0.83 - j3.11 = 1.30 - j5.08 = 5.24\angle{-75.65°}$. Thus, $\mathbf{V}_1 = [(125)(25.68 - j100.06) + 999.6 - j619.5]/5.24\angle{-75.65°} = 13{,}787\angle{-72.2°}/5.24\angle{-75.65°} = 2631\angle{3.45°}$ V (primary-side voltage of two transformers in parallel).

6. Calculate the Division of Load between Transformers

First, calculate the current through transformer A using the equation:

$$\mathbf{I}_A = \frac{\mathbf{I}_{\text{total}} + \mathbf{V}_1[(\mathbf{Y}_{\text{total}}/a_{\text{total}}) - (\mathbf{Y}_{\text{total}}/a_{\text{total}})]}{\mathbf{Z}_A\mathbf{Y}_{\text{total}}}$$

$$= \frac{(999.6 - j619.5) + 2631\angle{3.45°}\{[(25.68 - j100.06)/19.17] - (1.30 + j5.08)\}}{(0.0257\angle{76.5°})(103.3\angle{-75.6°})}$$

$$= 1494\angle{-41.02°}/(0.0257\angle{76.5°})(103.3\angle{75.6°})$$

$$= 562.6\angle{41.92°} = 418.7 - j375.9 \text{ A}$$

Similarly, the current through transformer B is calculated as

$$\mathbf{I}_B = \frac{\mathbf{I}_{\text{total}} + \mathbf{V}_1[(\mathbf{Y}_{\text{total}}/a_B) - (\mathbf{Y}_{\text{total}}/a_{\text{total}})]}{\mathbf{Z}_B\mathbf{Y}_{\text{total}}}$$

$$= 643.9\angle{-22.97°} = 593 - j251 \text{ A}$$

The portion of load carried by transformer A = (562.6 A)(0.120 kV) = 67.5 kVA = 67.5 percent of its rating of 100 kVA. The portion of load carried by transformer B = (643.9 A)(0.115 kV) = 74.0 kVA, or (74.0 kVA)(100 percent)/150 kVA = 49.3 percent of its rating of 150 kVA.

Related Calculations: This problem illustrates the approach to calculating the division of load between two single-phase transformers. The method is applicable to three-phase transformers under balanced conditions, and may be extended to three-phase transformers under unbalanced conditions by applying the theory of symmetrical components.

PHASE SHIFT IN WYE-DELTA TRANSFORMER BANKS

A three-phase, 300-kVA, 2300/23,900-V, 60-Hz transformer is connected wye-delta as shown in Fig. 14. The transformer supplies a load of 280 kVA at a power factor of 0.9

lagging. The supply voltages (line to neutral) on the high-voltage side are $\mathbf{V_{AN}} = 13{,}800\underline{/0°}$, $\mathbf{V_{BN}} = 13{,}800\underline{/-120°}$, and $\mathbf{V_{CN}} = 13{,}800\underline{/+120°}$. Find the phasor voltages and currents for the transformer.

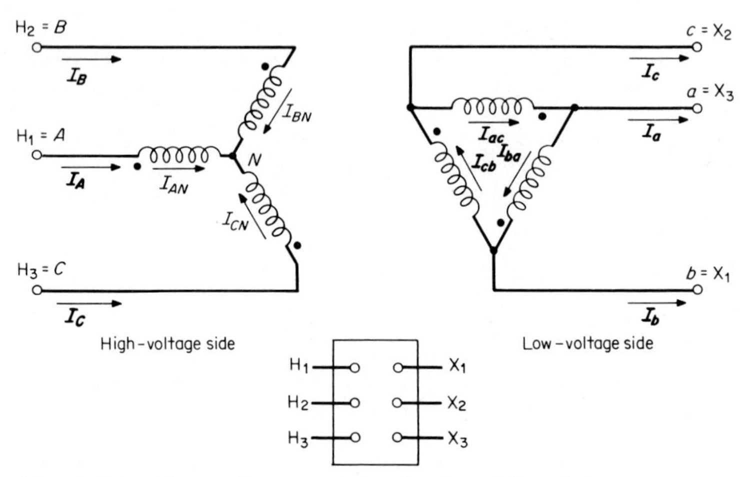

Fig. 14 Wye-delta transformer connections for phase-shift analysis.

Calculation Procedure:

1. Determine the Turns Ratio

The turns ratio of windings that are drawn in parallel (linked magnetically) is the same for each pair of windings in a three-phase transformer. In this case, the turns ratio a = 13,800 V/2300 V = 6.

2. Determine the Low-Voltage-Side Phasor Voltages

On the delta side the voltage $\mathbf{V_{ab}} = \mathbf{V_{AN}}/a = 13{,}800\underline{/0°}/6 = 2300\underline{/0°}$ V, $\mathbf{V_{bc}} = \mathbf{V_{BN}}/a = 13{,}800\underline{/-120°}/6 = 2300\underline{/-120°}$ V, and $\mathbf{V_{ca}} = \mathbf{V_{CN}}/a = 13{,}800\underline{/+120°}/6 = 13{,}800\underline{/+120°}$ V.

3. Determine the Line-to-Line Voltages on the High-Voltage Side

The supply voltages were given on the basis of the line-to-neutral voltages in each phase. The line-to-line voltages are obtained from phasor additions: $\mathbf{V_{AB}} = \mathbf{V_{AN}} - \mathbf{V_{BN}} = 13{,}800\underline{/0°} - 13{,}800\underline{/-120°} = 13{,}800 + 6900 + j11{,}951 = 20{,}700 + j11{,}951 = 23{,}900\underline{/30°}$ V, $\mathbf{V_{BC}} = \mathbf{V_{BN}} - \mathbf{V_{CN}} = 13{,}800\underline{/-120°} - 13{,}800\underline{+120°} = -6900 - j11{,}951 + 6900 - j11{,}951 = -j23{,}900 = 23{,}900\underline{/-90°}$ V, and $\mathbf{V_{CA}} = \mathbf{V_{CN}} - \mathbf{V_{AN}} = 13{,}800\underline{/+120°} - 13{,}800\underline{/0°} = -6900 + j11{,}951 - 13{,}800 = -20{,}000 + j11{,}951 = 23{,}800\underline{/150°}$ V. The high-voltage line-to-line voltages lead the low-voltage line-to-line voltages by 30°; that is the convention for both wye-delta and delta-wye connections.

4. Determine the Load Current on the High-Voltage Side

Use the equation: $I_L = \text{VA}/3V_{\text{phase}}$. The magnitude of the current on the high-voltage side is $I_{AN} = I_{BN} = I_{CN} = 300{,}000\ \text{VA}/(3)(13{,}800\ \text{V}) = 7.25$ A. The power-factor angle $= \cos^{-1} 0.9 = 25.84°$. With a lagging power-factor angle the current lags the respective voltage by that angle. $\mathbf{I_{AN}} = 7.25\angle -25.84°$ A, $\mathbf{I_{BN}} = 7.25\angle -120° - 25.84° = 7.25\angle -145.84°$ A, and $\mathbf{I_{CN}} = 7.25\angle +120° - 25.84° = 7.25\angle 94.16°$ A.

5. Determine the Load Current on the Low-Voltage Side

$\mathbf{I_{ab}} = a\mathbf{I_{AN}}$, $\mathbf{I_{bc}} = a\mathbf{I_{BN}}$, and $\mathbf{I_{ca}} = a\mathbf{I_{CN}}$. Thus, $\mathbf{I_{ab}} = (6)(7.25\angle -25.84°) = 43.5\angle -25.84°$ A, $\mathbf{I_{bc}} = (6)(7.25\angle -145.84°) = 43.5\angle -145.84°$ A, and $\mathbf{I_{ca}} = (6)(7.25\angle 94.16°) = 43.5\angle 94.16°$ A. The line currents may be obtained by phasor addition. Thus, $\mathbf{I_a} = \mathbf{I_{ac}} - \mathbf{I_{ba}} = \mathbf{I_{ab}} - \mathbf{I_{ca}} = 43.5\angle -25.84° - 43.5\angle 94.16° = 39.15 - j18.96 + 3.16 - j43.4 = 42.31 - j62.36 = 75.4\angle -55.84°$ A. $\mathbf{I_b} = \mathbf{I_{ba}} - \mathbf{I_{cb}} = \mathbf{I_{bc}} - \mathbf{I_{ab}} = 43.5\angle -145.84° - 43.5\angle -25.84° = -35.99 - j24.43 - 39.15 + j18.96 = -75.14 - j5.47 = 75.4\angle -175.84°$ A. $\mathbf{I_c} = \mathbf{I_{cb}} - \mathbf{I_{ac}} = \mathbf{I_{ca}} - \mathbf{I_{bc}} = 43.5\angle 94.16° - 43.5\angle -145.84° = -3.16 + j43.4 + 35.99 + j24.43 = 32.83 + j67.83 = 75.4\angle 64.16°$ A.

6. Compare Line Currents on High- and Low-Voltage Sides

Note that the line currents on the low-voltage side (delta) lag the line currents on the high-voltage side (wye) by 30° in each phase, respectively.

High-voltage side	**Low-voltage side**
$\mathbf{I_A} = \mathbf{I_{AN}} = 7.25\angle -25.84°$ A	$\mathbf{I_a} = 75.4\angle 55.84°$ A
$\mathbf{I_B} = \mathbf{I_{BN}} = 7.25\angle -145.84°$ A	$\mathbf{I_b} = 75.4\angle -175.84°$ A
$\mathbf{I_C} = \mathbf{I_{CN}} = 7.25\angle 94.16°$ A	$\mathbf{I_c} = 75.4\angle 64.16°$ A

Related Calculations: This problem illustrates the method of calculating the phase shift through transformer banks that are connected wye-delta or delta-wye. For the standard connection shown, the voltage and current on the high-voltage side will lead the respective quantities on the low-voltage side by 30°; it makes no difference whether the connection is wye-delta or delta-wye.

CALCULATION OF POWER, APPARENT POWER, REACTIVE POWER, AND POWER FACTOR

For the equivalent circuit shown in Fig. 15, generator A is supplying 4000 W at 440 V and 0.90 power factor, lagging. The motor load is drawing 9500 W at 0.85 power factor, lagging. Determine the apparent power in VA, the reactive power in vars, the power in W, and power factor of each generator and of the motor load.

Calculation Procedure:

1. Determine the Current from Generator A

Use the equation: $P = EI\cos\theta$ and solve for the current. $I_A = P_A/E_A\cos\theta_A = 4000\ \text{W}/(440\ \text{V})(0.90) = 10.1$ A.

2. Determine the Reactive Power and Apparent Power Values for Generator A

To find the apparent power, divide the power in watts by the power factor: VA = W/pf = 4000 W/0.90 = 4444.4 VA. The reactive power is W tan ($\cos^{-1}$ pf) = (4000 W) tan ($\cos^{-1}$ 0.90) = 1937.3 vars.

3. Calculate the Losses in Z_A

The power loss in watts in Z_A is $I_A^2 R_A$ = (10.1)²(1.5 Ω) = 153 W. The loss in vars in Z_A is $I_A^2 X_A$ = (10.1)²(1.5 Ω) = 153 vars. The quantities of power and reactive power delivered to the motor from generator A are 4000 − 153 = 3847 W, and 1937.3 − 153 = 1784.3 vars, respectively.

4. Calculate the Requirements of the Motor Load

At the motor, the apparent power = W/pf = 9500 W/0.85 = 11,176.5 VA. The reactive power = W tan ($\cos^{-1}$ pf) = (9500 W) tan ($\cos^{-1}$ 0.85) = 5887.6 vars.

5. Calculate the Motor Requirements to Be Supplied by Generator B

The power (in watts) needed from generator B = 9500 − 3847 = 5653 W. The reactive power (in vars) needed from generator B = $\sqrt{5653^2 + 4{,}103.3^2}$= 6985.2 vars.

6. Calculate the Current from Generator B

First calculate the voltage at the load as represented by the voltage drop from generator A. $V_{\text{load}} = \text{VA}_{\text{genA}}/I_A = \sqrt{3847^2 + 1784.3^2}/10.1$ = 419.9 V. Therefore, I_B = 6985.2 VA/419.9 V = 16.64 A.

7. Calculate the Losses in Z_B

The loss in watts in $Z_B = I_B^2 R_B$ = (16.64)²(0.6) = 166.1 W. The loss in vars in Z_B = $I_B^2 X_B$ = (16.64)²(0.7) = 193.8 vars. Thus, generator B must supply 5653 + 166.1 = 5819.1 A, 4103.3 + 193.8 = 4297.1 vars, and $\sqrt{(5819.1)^2 + (4297.1)^2}$ = 7233.7 VA.

8. Summarize the Results

	Apparent power, kVa	*Reactive power, kvar*	*Power, kW*
Generator A	4.444	1.937	4.000
Generator B	7.234	4.297	5.819
Motor load	11.177	5.888	9.500
Losses:			
Z_A	0.216	0.153	0.153
Z_B	0.255	0.193	0.166

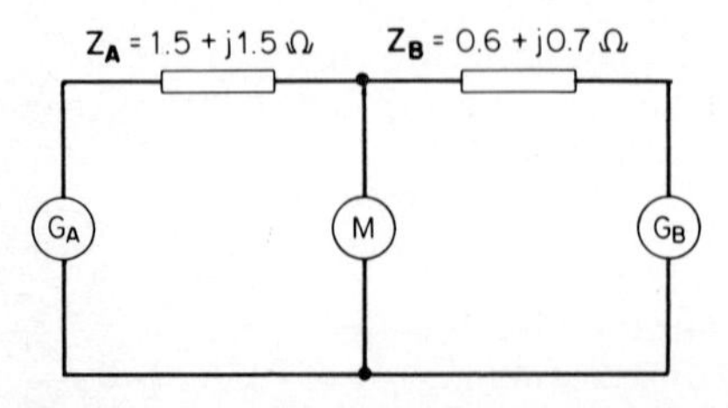

Fig. 15 Equivalent circuit of motor load fed by two generators.

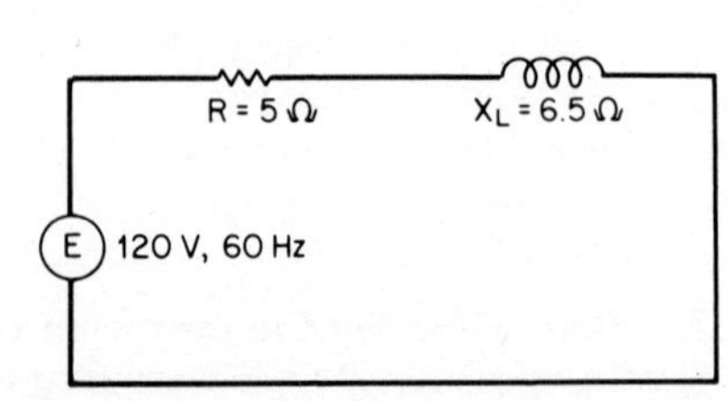

Fig. 16 Series power circuit used to illustrate real, reactive, and apparent power.

It is noted that the sum of the kW and kvars values generated equals the sum of the kW and kvar values of the motor load and losses in Z_A and Z_B, respectively.

Related Calculations: This problem illustrates the relationship of kVA, kvar, and kW values and power factor. For the three-phase case the factor $\sqrt{3}$ must be included. For example, $P = \sqrt{3}EI \cos \theta$ and vars $= \sqrt{3}EI \sin \theta$.

THE POWER DIAGRAM: REAL AND APPARENT POWER

A series circuit, as shown in Fig. 16, has a resistance of 5 Ω and a reactance $X_L = 6.5$ Ω. The power source is 120 V, 60 Hz. Determine the real power, the reactive power, the apparent power, and the power factor; construct the power diagram.

Calculation Procedure:

1. ***Determine the Impedance of the Circuit***

The total impedance of the circuit is $\mathbf{Z_T} = R + jX_L = 5 + j6.5 = 8.2\angle 52.43^\circ\ \Omega$.

2. ***Determine the Phasor Current***

The phasor current is determined from the equation: $\mathbf{I} = \mathbf{E}/\mathbf{Z_T} = 120\angle 0^\circ\ \text{V}/8.2\angle 52.43^\circ\ \Omega = 14.63\angle -52.43^\circ\ \text{A} = 8.92 - j11.60$ A.

3. ***Calculate the Real Power***

Use the equation: $P = EI \cos \theta = (120)(14.63) \cos 52.43^\circ = 1070.0$ W where $\cos \theta =$ power factor $= 0.61$.

4. ***Calculate the Reactive Power***

Use the equation: vars $= EI \sin \theta = (120)(14.63) \sin 52.43^\circ = 1391.5$ vars. The $\sin \theta$ is sometimes known as the reactive factor; in this case $\sin \theta = \sin 52.43^\circ = 0.793$.

5. ***Calculate the Apparent Power***

The apparent power is $EI = (120)(14.63) = 1755.6$ VA. Also, it may be calculated from the equation: real power + j(reactive power) = 1070.0 W + j1391.5 vars $= 1755.6\angle 52.43^\circ$ VA.

6. ***Construct the Power Diagram***

The power diagram is shown in Fig. 17. Also, it is evident that: power factor = real power/apparent power = 1,070.0 W/1755.6 VA = 0.61.

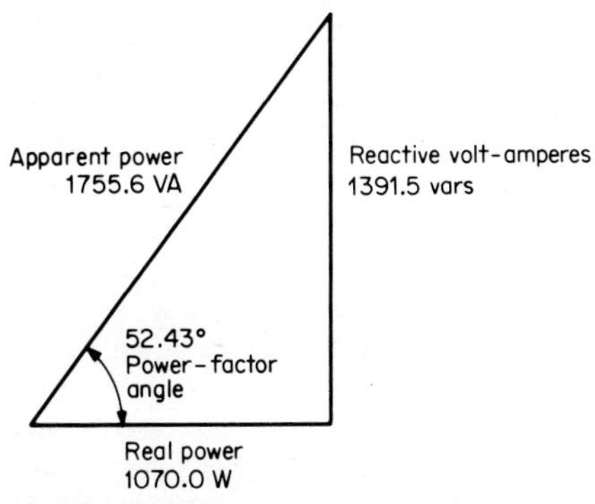

Fig. 17 The power diagram (or power triangle).

Related Calculations: The procedure is similar for a three-phase circuit where the appropriate equation for real power is $P = \sqrt{3}E_{line}I_{line} \cos \theta$. The apparent power is $VA = \sqrt{3}E_{line}I_{line}$, and the reactive power is vars $= \sqrt{3}E_{line}I_{line} \sin \theta$.

STATIC CAPACITORS USED TO IMPROVE POWER FACTOR

A manufacturing company has several three-phase motors that draw a combined load of 12 kVA, 0.60 power factor, lagging, from 220-V mains. It is desired to improve the power factor to 0.85 lagging. Calculate the line current before and after adding the capacitors, and determine the reactive power rating of the capacitors that are added.

Calculation Procedure:

1. Calculate the Line Current before Adding the Capacitors

Use the equation: $P = \sqrt{3}E_{line}I_{line} \cos \theta$, solving for I_{line}. $I_{line} = P/\sqrt{3}E_{line} \cos \theta = 7.2 \text{ kW}/(\sqrt{3})(220 \text{ V})(0.6) = 31.5 \text{ A}$.

2. Calculate the Line Current after Adding the Capacitors

Use the same equation as in the first step. $I_{line} = P/\sqrt{3}E_{line} \cos \theta = 7.2 \text{ kW}/(3)(220 \text{ V})(0.85) = 22.2 \text{ A}$. Note that the improved power factor has resulted in less current; 31.5 A is required at 0.60 power factor, and 22.2 A is required at 0.85 power factor.

3. Calculate the Rating in kvar of Capacitors

The reactive power in kvar needed in the circuit before the capacitors are added is determined by multiplying the apparent power in kVA by the sine of the power-factor angle: $(12 \text{ kVA}) \sin (\cos^{-1} 0.60) = 9.6$ kvar. The real power = kVA × power factor = (12 kVA)(0.60) = 7.2 kW; this remains constant before and after the capacitor bank is added (see Fig. 18). The rating of the capacitor bank is the difference in kvar between the before and after cases, namely, 9.6 kvar − 4.5 kvar at 220 V (three-phase).

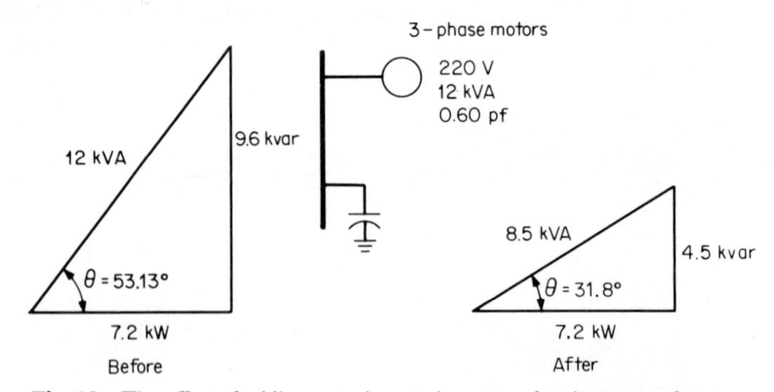

Fig. 18 The effect of adding capacitors to improve a lagging power factor.

Related Calculations: Lagging power factors may be improved by the addition of sources of leading reactive power; these sources are usually capacitor banks and sometimes synchronous capacitors (i.e., synchronous motors on the line that are operated in the overexcited state). In either case, leading reactive power is supplied at the installation, so that the current and losses are reduced over long transmission or distribution lines.

THREE-PHASE SYNCHRONOUS MOTOR USED TO CORRECT POWER FACTOR

A three-phase synchronous motor is rated 2200 V, 100 kVA, 60 Hz. It is being operated to improve the power factor in a factory; overexcitation results in its operating at a leading power factor of 0.75 at 75 kW full load. Determine (1) the reactive power in kvar delivered by the motor, (2) draw the power triangle, and (3) determine the power factor needed for the motor to deliver 50 kvar with a 75-kW load.

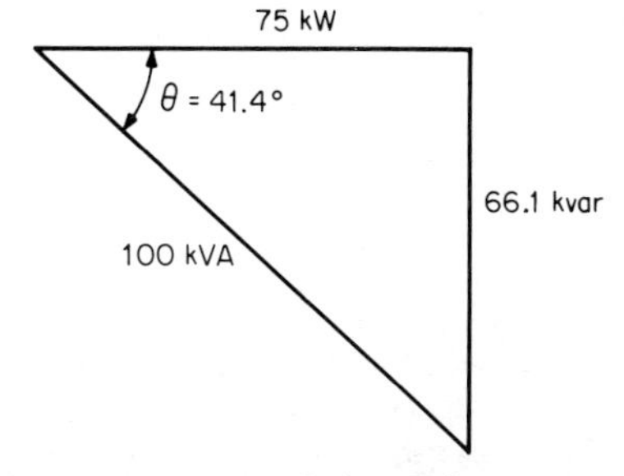

Fig. 19 Power triangle for synchronous motor operating at a leading power factor.

Calculation Procedure:

1. Calculate the Power-Factor Angle

The power-factor angle is determined from $\theta = \cos^{-1}$ (power factor) $= \cos^{-1} 0.75 = 41.4°$, leading.

2. Calculate the Number of kvar *Delivered*

Reactive power $=$ kVA $\times \sin\theta = (\text{kW}/\cos\theta)\sin\theta =$ kW $\tan\theta = 75 \tan 41.4° = 66.1$ kvar. The power triangle is shown in Fig. 19.

3. Calculate the Power Factor for Delivering 50 kvar

If the load is 75 kW and the reactive power is 50 kvar, $\theta = \tan^{-1}$ (reactive power/load kW) $= \tan^{-1}$ (50 kvar/75 kW) $= 33.7°$. The power factor $= \cos 33.7° = 0.832$ leading.

Related Calculations: Synchronous motors operated in the overexcited mode deliver reactive power instead of absorbing it. In this manner they act as synchronous capacitors and are useful in improving power factor. In a factory having many induction motors (a lagging power-factor load), it is often desirable to have a few machines act as synchronous capacitors to improve the overall power factor.

POWER CALCULATION OF TWO-WINDING TRANSFORMER CONNECTED AS AN AUTOTRANSFORMER

A single-phase transformer is rated 440/220 V, 5 kVA, at 60 Hz. Calculate the capacity in kVA of the transformer if it is connected as an autotransformer to deliver 660 V on the load side with a supply voltage of 440 V.

Calculation Procedure:

1. Draw the Connection Diagram

See Fig. 20 for the connection diagram.

2. Calculate Rated Current in the Windings

The rated current in the H_1-H_2 winding is $I_H = 5{,}000$ VA/440 V $= 11.36$ A. The rated current in the X_1-X_2 winding is $I_X = 5000$ VA/220 V $= 22.73$ A.

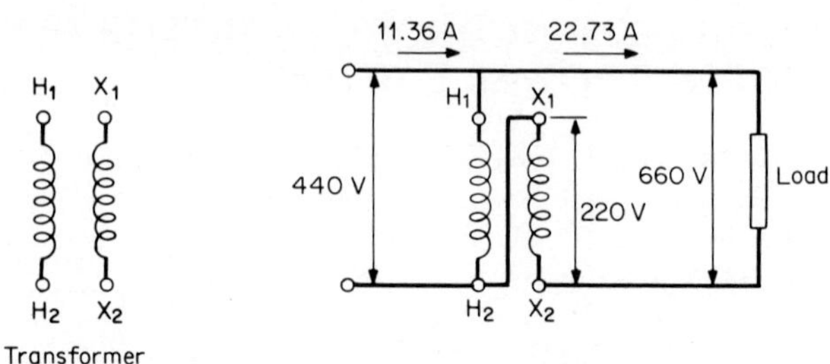

Fig. 20 Single-phase transformer connected as autotransformer.

3. Calculate the Apparent Power Delivered to the Load

The voltage across the load is 660 V and the current through it is 22.73 A. Thus, the load receives (660 V)(22.73 A) = 15,000 VA. For a 2:1 ratio two-winding transformer connected as an autotransformer, the delivered apparent power is tripled.

Related Calculations: It can be shown that for autotransformer connection of single-phase transformers the capacity increases with a certain relationship to the turns ratio. For example, for a 1:1-ratio two-winding transformer the autotransformer connection produces a 1:2-ratio and the capacity is doubled; for a 2:1-ratio two-winding transformer the autotransformer connection becomes a 2:3-ratio transformer and the capacity is tripled; for a 3:1-ratio two-winding transformer the autotransformer connection becomes a 3:4-ratio transformer and the capacity increases 4 times. A serious consideration in using the autotransformer connection between two transformer windings is the solid electrical connection between them that may necessitate special insulation requirements.

TWO-WATTMETER METHOD FOR DETERMINING THE POWER OF A THREE-PHASE LOAD

In the circuit shown in Fig. 21, $\mathbf{E}_{ab} = 220\underline{/0°}$ V, $\mathbf{E}_{bc} = 220\underline{/-120°}$ V, and $\mathbf{E}_{ca} = 220\underline{/+120°}$ V. Determine the reading of each wattmeter and show that the sum of W_1 and W_2 = the total power in the load.

Calculation Procedure:

1. Calculate $\mathbf{I}_a$

Use the relation: $\mathbf{I}_a = \mathbf{I}_{ac} - \mathbf{I}_{ba}$, where the delta load currents are each equal to the voltage divided by the load impedance. $\mathbf{E}_{ac} = -\mathbf{E}_{ca} = 220\underline{/-60°}$, and $\mathbf{E}_{ba} = -\mathbf{E}_{ab} = 220\underline{/180°}$. $\mathbf{I}_a = (220\underline{/-60°})/(12 + j14) - (220\underline{/180°})/(16 + j10) = 11.96\underline{/-109.4°} - 11.64\underline{/148°} = -3.97 - j11.28 + 9.87 - j6.17 = 5.9 - j17.45 = 18.42\underline{/-71.32°}$ A.

2. Calculate $\mathbf{I}_c$

Use the relation: $\mathbf{I}_c = \mathbf{I}_{cb} - \mathbf{I}_{ac} = 220\underline{/60°}/(8 - j8) - 220\underline{/-60°}/(12 + j14) = 19.47\underline{/105°} - 11.96\underline{/-109.4°} = -5.04 + j18.81 + 3.97 + j11.28 = -1.07 + j30.09 = 30.11\underline{/92.04°}$ A.

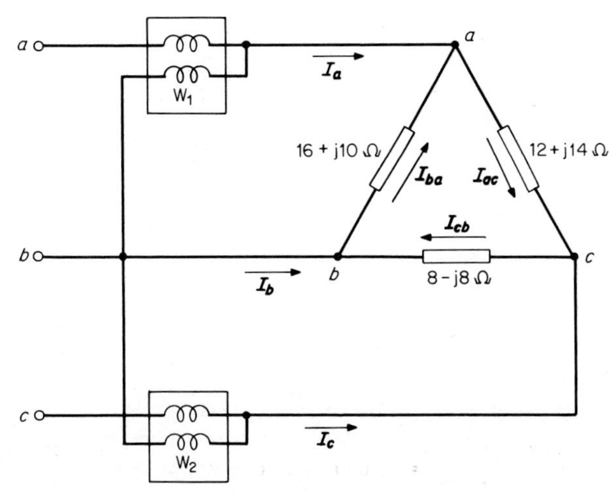

Fig. 21 Two-wattmeter connection for measuring three-phase power.

3. Calculate Power Reading of Wattmeter 1

Use the equation: $W_1 = \mathbf{E_{ab}I_a} \cos \theta = (220 \text{ V})(18.42 \text{ A}) \cos (0° + 71.32°) =$ 1297.9 W.

4. Calculate Power Reading of Wattmeter 2

Use the equation: $W_2 = \mathbf{E_{cb}I_c} \cos \theta = (220 \text{ V})(30.11 \text{ A}) \cos (60° - 92.04°) =$ 5615.2 W.

5. Calculate the Total Power of the Load

The total power $= W_1 + W_2 = 1297.9 + 5615.2 = 6913.1$ W.

6. Calculate the Power Loss in Each Load Resistance

Use the equation: $P = I^2R$. For resistance in delta-side ac, $R = 12\ \Omega$; $P = (11.96)^2(12) = 1716.5$ W. For resistance in delta-side ab, $R = 16\ \Omega$; $P = (11.64)^2(16) = 2167.8$ W. The total power absorbed by the load $= 1716.5 + 3032.6 + 2167.8 =$ 6916.9 W, which compares with the sum of W_1 and W_2 (the slight difference results from rounding off the numbers throughout the problem solution).

Related Calculations: When the two-wattmeter method is used for measuring power in a three-phase circuit, care must be given to the proper connections of the voltage and current coils. The common potential lead must be connected to the phase not being used for the current-coil connections.

OPEN-DELTA TRANSFORMER OPERATION

A new electric facility is being installed wherein the loading is light at this time. The economics indicate that two single-phase transformers be connected in open-delta, and that the future load be handled by installing the third transformer to close the delta, whenever the load growth reaches the necessary level. Each single-phase transformer is rated 2200/220 V, 200 kVA, 60 Hz.

If for the open-delta connection (using two transformers) each transformer is fully loaded, and for the closed-delta connection (using three transformers), each transformer is fully loaded, what is the ratio of three-phase loading of the open delta as compared with the closed delta? Refer to Fig. 22.

Calculation Procedure:

1. Calculate the Line Currents for Closed Delta

Use the three-phase equation: $\text{kVA} = \sqrt{3}E_{\text{line}}I_{\text{line}}$ solving for I_{line}. On the high-voltage side, $I_{\text{line}} = 600{,}000\ \text{VA}/(\sqrt{3})(2200\ \text{V}) = 157.5\ \text{A}$. Similarly, on the low-voltage side, $I_{\text{line}} = 600{,}000\ \text{VA}/(\sqrt{3})(220\ \text{V}) = 1575\ \text{A}$.

2. Calculate the Transformer Currents for the Closed Delta

The transformer currents $= I_{\text{line}}/\sqrt{3} = 157.5/\sqrt{3} = 90.9$ A on the high-voltage side, or 909 A on the low-voltage side. Another method of determining these currents is to divide the transformer kVA value by the voltage; thus [200 kVA (per transformer)]/2.2 kV = 90.9 A on the high-voltage side, and [200 kVA (per transformer)]/0.22 kV = 909 A on the low-voltage side.

3. Calculate the Transformer Current for the Open-Delta

In the open-delta case the magnitude of the line currents must be the same as that of the transformer currents. Again, if each transformer is carrying 200 kVA, the line currents on the high-voltage side must be 90.9 A and on the low-voltage side, 909 A. That is, the line currents are reduced from 157.5 A to 90.9 A, and 1575 A is reduced to 909 A, respectively.

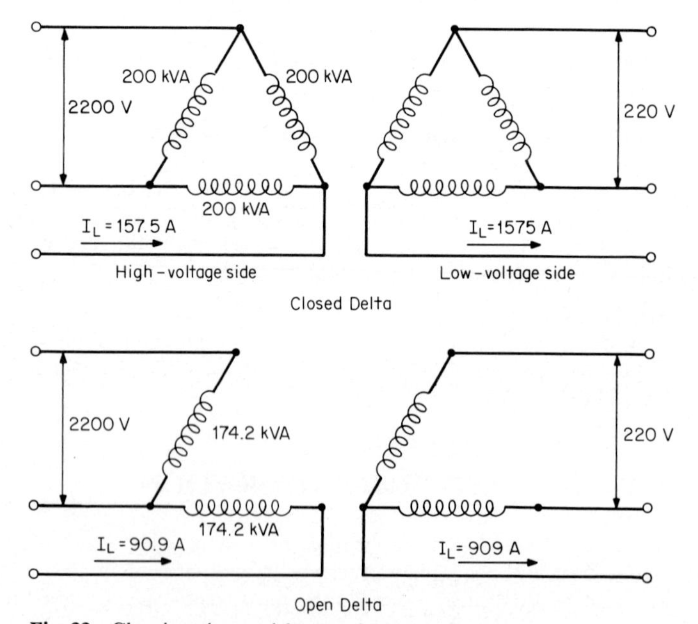

Fig. 22 Closed- and open-delta transformer connections.

4. Calculate the Apparent Power of the Open Delta
The three-phase apparent power of the open delta = $\sqrt{3}E_LI_L$ = ($\sqrt{3}$)(2200 V)(90.9 A) = 346.4 kVA. A comparison may be made with the 600 kVA loading of the closed delta; thus, (346.4 kVA)(100 percent)/600 kVA = 57.7 percent. The 346.4 kVA is carried by only two transformers (174.2 kVA each). Thus, each transformer in the open-delta case carries only (174.2 kVA)(100 percent)/200 kVA = 87.1 percent of the loading for the closed-delta case.

Related Calculations: Calculations involving the open-delta connection are often required, sometimes for load growth planning as in this problem, and sometimes for contingency planning. For the open delta, the flow-through three-phase kVA value is reduced to 57.7 percent of the closed-delta value. The individual transformer loading is reduced to 87.1 percent on the two transformers that form the open delta as compared with 100 percent for each of three transformers in a closed delta. The basic assumptions in these calculations are a balanced symmetrical system of voltages and currents and negligible transformer impedances.

REAL AND REACTIVE POWER OF A THREE-PHASE MOTOR IN PARALLEL WITH A BALANCED-DELTA LOAD

In the system shown in Fig. 23, the balanced-delta load has an impedance on each side of 12 − j10 Ω. The three-phase induction motor is rated 230 V, 60 Hz, 8 kVA at 0.72 power factor, lagging. It is wye-connected. Determine (1) the line current, (2) the power factor, and (3) the power requirements of the combined load.

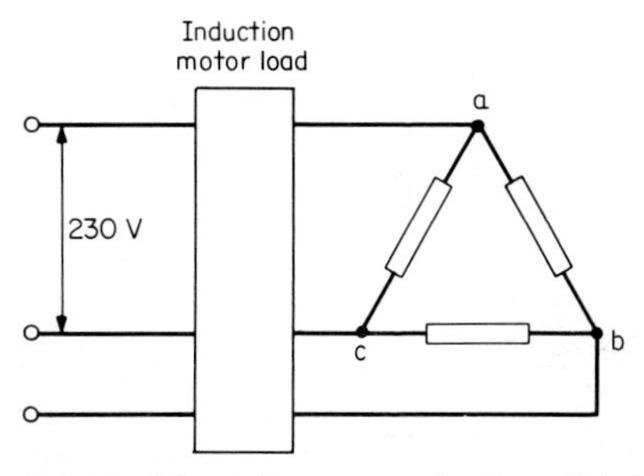

Fig. 23 Three-phase motor load paralleled with a balanced-delta load.

Calculation Procedure:

1. Calculate the Motor Current
The motor is wye-connected; thus, the line current is equal to the phase current. Use the equation: $I_{\text{line}} = \text{kVA}/\sqrt{3}\text{kV}_{\text{line}}$ = 8 kVA/($\sqrt{3}$)(0.230 kV) = 20.08 A.

2. Calculate the Equivalent Impedance of the Motor
Use the equation: $(V_{\text{line}}/\sqrt{3})/I_{\text{phase}} = Z_{\text{phase}}$ = (230/$\sqrt{3}$)/20.08 = 6.61 Ω. The angle $\theta = \cos^{-1}$ (power factor) = $\cos^{-1}$ 0.72 = 43.95°. $R_{\text{motor}} = Z \cos\theta$ = 6.61 cos 43.95° = 4.76 Ω. $X_{\text{motor}} = Z \sin\theta$ = 6.61 sin 43.95° = 4.59 Ω. Thus, $\mathbf{Z}_{\text{motor}}$ = 4.76 + j4.59 Ω per phase = $6.61\angle 43.95°$ Ω per phase.

3. Convert Balanced-Delta Load to Equivalent Wye
Use the equations: $\mathbf{Z}_a = (\mathbf{Z}_{ab}\mathbf{Z}_{ca})/(\mathbf{Z}_{ab} + \mathbf{Z}_{bc} + \mathbf{Z}_{ca}) = (12 - j10)^2/(3)(12 - j10)$ = ⅓(12 − j10) = 4.00 − j3.33 = $5.21\angle -39.8°$ Ω.

4. Combine the Two Balanced-Wye Loads
The motor load of 4.76 + j4.59 Ω per phase is to be paralleled with the load of 4.00 − j3.33 Ω per phase. Use the equation: $\mathbf{Z}_{\text{total}} = \mathbf{Z}_{\text{motor}}\mathbf{Z}_{\text{load}}/(\mathbf{Z}_{\text{motor}} + \mathbf{Z}_{\text{load}})$ = (6.61

$\underline{/43.95°})(5.21\underline{/-39.8°})/(4.76 + j4.59 + 4.00 - j4.33) = 3.93\underline{/2.45°}$ Ω per phase (equivalent-wye connection).

5. Calculate the Total Line Current and Power Factor

Use the equation: $I_{total} = (V_{line}/\sqrt{3})/Z_{total} = (230\text{ V}/\sqrt{3})/3.93\ \Omega = 33.8$ A. The power factor = cos θ = cos 2.45° = 0.999 (essentially unity power factor).

6. Calculate the Power Requirements

Use the equation: $W_{total} = 3I^2_{total}R_{total} = (3)(33.8\text{ A})^2(3.93\ \Omega) = 13{,}469$ W. The apparent power = $\sqrt{3}V_{line}I_{line} = (\sqrt{3})(230\text{ V})(33.8\text{ A}) = 13{,}465$ VA. The apparent power is approximately equal to the real power, the difference being a result of rounding off the numbers throughout the solution of the problem; the power factor is unity.

Related Calculations: This problem illustrates one technique in handling paralleled loads. There are other approaches to this problem. For example, the current could be obtained for each load separately and then added to obtain a combined, or total, current.

Section 11 LOAD-FLOW STUDIES

John S. Wade, Jr., Ph.D.
Associate Professor of Engineering
The Pennsylvania State University,
The Capitol Campus

REFERENCES: Elgerd—*Electric Energy Systems Theory: An Introduction,* McGraw-Hill; Gross—*Power System Analysis,* Wiley; Neuenswander—*Modern Power Systems,* International Textbook; Stevenson—*Elements of Power Systems Analysis,* McGraw-Hill.

INTRODUCTORY TRANSMISSION-LINE STUDY

The sending-end bus of a transmission line has a line-to-line voltage of 110 kV and delivers $\mathbf{S} = 20 + j5$ MVA toward the receiving end. As is often the case, the series resistance of the line is small compared with the inductive reactance and is neglected; then, $\mathbf{Z} = j120\ \Omega$. Using the per-unit (p.u.) system, solve for the voltage at the receiving end and draw a phasor diagram depicting the voltage relationships.

Calculation Procedure:

1. Convert Data to Per-Unit Values

Per-unit values rely on a voltage-base value (usually nominal) and an apparent-power base (guided by equipment capacity). Choose 110 kV for the base voltage and arbitrarily choose a power base of 100 MVA. Therefore, the sending-end line voltage is 1 p.u. and the sending-end apparent power is $\mathbf{S} = (20 + j5)/100 = 0.2 + j0.05$ p.u.

The p.u. impedance is found by dividing the ohmic line impedance by the base impedance, which is $|Z_{base}| = (110\text{ kV})^2/100\text{ MVA} = 121\ \Omega$. Then the per-unit impedance for the transmission line is $|\mathbf{Z}| = 120/121 = 0.99$ p.u. per phase.

2. *Calculate the Receiving-End Voltage*

Because the three-phase system is balanced, the single-phase model of Fig. 1 may be used. The sending-end voltage $\mathbf{V}_1$ is 1 p.u.; the sending-end phase voltage, therefore, is $\mathbf{V}_1/\sqrt{3}$. If the phase voltage is treated this way, the sending-end complex power is divided by 3: $\mathbf{S} = 0.2/3 + j0.05/3$. (It will be shown later that this step is unnecessary.)

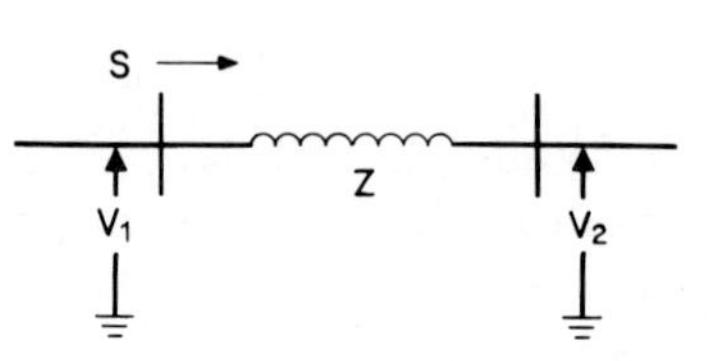

Fig. 1 A single-phase model of a balanced three-phase system.

The current in the transmission line must be $\mathbf{I} = (\mathbf{V}_1/\sqrt{3} - \mathbf{V}_2/\sqrt{3})/\mathbf{Z}$. With terms rearranged, $\mathbf{V}_2/\sqrt{3} = \mathbf{V}_1/\sqrt{3} - \mathbf{IZ}$. The per-unit MVA value on the basis of a single phase is $\mathbf{S}/3 = (\mathbf{V}_1/\sqrt{3})\mathbf{I}^* = (P + jQ)/3$ out of the sending bus. Conjugation of both sides yields $[(\mathbf{V}_1/\sqrt{3})\mathbf{I}^*]^* = (P + jQ)^*/3$. This is rearranged to give $\mathbf{I} = [(P - jQ)/3]/(\mathbf{V}_1^*/\sqrt{3})$. Because $\mathbf{V}_1$ can be made the reference with a zero angle, $\mathbf{V}_1 = \mathbf{V}_1^* = V_1$.

The equation for $\mathbf{V}_2/\sqrt{3}$ at the receiving end becomes $\mathbf{V}_2/\sqrt{3} = V_1/\sqrt{3} - [(P - jQ)/3]\mathbf{Z}/(V_1/\sqrt{3})$. However, multiplying through by $\sqrt{3}$ yields a simplification; that is, $\mathbf{V}_2 = V_1 - (P - jQ)\mathbf{Z}/V_1$. Note that this equation now employs three-phase values that could have been used at the outset. More important, for illustrative purposes, it is rewritten as $\mathbf{V}_2 = V_1 - Q(X/V_1) - jP(X/V_1)$ since $\mathbf{Z} = jX$ p.u.

Substituting known values into the equation yields $\mathbf{V}_2 = 1.0 - (0.05)(0.99/1) - j(0.2)(0.99/1) = 0.95 - j0.198 = 0.97\underline{/-11.8^\circ}$. When multiplied by the line voltage, $\mathbf{V}_2 = 106.7\underline{/-11.8^\circ}$ kV.

3. *Draw Phasor Diagram*

The phasor diagram of Fig. 2 illustrates that the receiving-end voltage $\mathbf{V}_2$ lags $\mathbf{V}_1$, a condition which always indicates the direction of real-power flow. The two potentials could have the same magnitude and power would still flow *if* the receiving-end potential lags the sending-end value.

Related Calculations: Term $P(X/V_1)$ is determined by the variable P. It can have a negative j direction and therefore greatly affects the angle between the ends of the line; however, its effect on the magnitude of $\mathbf{V}_2$ is small. If line resistance is a factor, this observation still holds.

Term $Q(X/V_1)$ has a negative real value in opposition to $\mathbf{V}_1$. Therefore, the variable Q (in vars) reduces $\mathbf{V}_2$. This is detrimental. It results in line power loss when resistance also is considered.

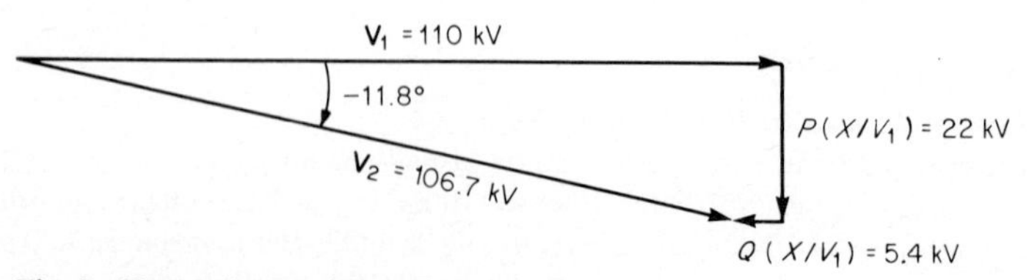

Fig. 2 Phasor diagram for single-phase model.

The model of Fig. 1 should indicate that Q required at the receiving end be injected there by capacitors or synchronous machines. Also, line inductive reactance should be minimized where possible.

The inherent design of a useful transmission line usually has a small resistance per phase in proportion to the unavoidable series reactance. Line analysis, however, often produces useful results where resistance is ignored.

AN EXHAUSTIVE SINGLE-TRANSMISSION-LINE STUDY

A 230-kV, three-phase transmission line is connected between sending bus 1 and receiving bus 2. The line has a series impedance of $15 + j180\ \Omega$/phase. The parallel admittance is $j0.00005$ S. The two buses have injected complex power from the generating system and each also furnishes loads in direct fashion. The values are as follows:

Bus 1	*Bus 2*
Load $\mathbf{S}_{1L} = -(40 + j10)$ MVA	Load $\mathbf{S}_{2L} = -(80 + j25)$ MVA
Generated power $\mathbf{S}_{1G} = (100 + j30)$ MVA	Generated power $\mathbf{S}_{2G}$ to be found

A voltage of 1 p.u. is maintained at bus 1. Find the voltage at bus 2 assuming the power flow is from bus 1 to 2. Also find the generated power required at bus 2 and the line loss.

Calculation Procedure:

1. Convert to per-Unit Values

The nominal potential at bus 1 (230 kV) is chosen as the base voltage and 100 MVA is arbitrarily selected as the complex power base. Dividing the complex power by the chosen system base yields:

Bus 1	*Bus 2*
Load $\mathbf{S}_{1L} = -(0.4 + j0.1)$	Load $\mathbf{S}_{2L} = -(0.8 + j0.25)$
Generated power $\mathbf{S}_{1G} = (1 + j0.3)$	Generated power $\mathbf{S}_{2G}$ to be found

The magnitude of the base impedance can always be found: $|\mathbf{Z}|_{\text{base}} = (230\ \text{kV})^2/100\ \text{MVA} = 529\ \Omega$. Then, the per-unit line impedance is $(15 + j180)/529 = (0.028 + j0.34)$. The per-unit parallel admittance is $(j0.00005)(529) = j0.0265$.

2. Draw Circuit Model

Modern power analysis utilizes voltage and complex power in many calculations. Current $\mathbf{I}$ is implicit, especially in digital computations for load-flow studies. Figure 3 is a model of a transmission line. The bus generation is represented by a circle and the outflow of power to a load by an arrow. The model includes a pi network and, as customary, one-half the per-unit parallel admittance is placed at either end of the line where it injects reactive power into the adjacent bus.

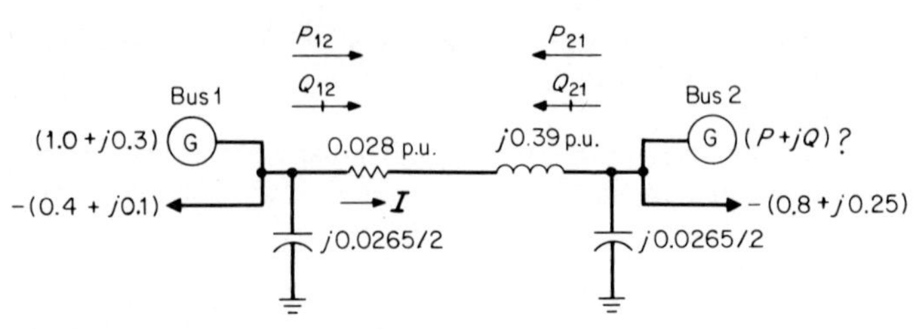

Fig. 3 Model of a two-bus system.

3. *Derive Model Equations*

The current in the model transmission line of Fig. 3 is determined from the phase-to-ground voltage: $\mathbf{I} = (\mathbf{V}_1 - \mathbf{V}_2)/\mathbf{Z}$. Because $\mathbf{S}_{12} = \mathbf{V}_1\mathbf{I}_1^*$, $\mathbf{S}_{12} = \mathbf{V}_1(\mathbf{V}_1^* - \mathbf{V}_2^*)/\mathbf{Z}^*$. If it is assumed in the beginning that $\mathbf{S}_{12} = P_{12} + jQ_{12}$ is a positive complex power (outflow from bus 1), then, as has been shown, $\mathbf{V}_2$ lags $\mathbf{V}_1$ by an angle whose symbol is δ (delta). Multiplying through by $\mathbf{V}_1$ in the equations yields $\mathbf{S}_{12} = (|\mathbf{V}_1|^2 - |\mathbf{V}_1||\mathbf{V}_2| \cos \delta - j|\mathbf{V}_1||\mathbf{V}_2| \sin \delta)/\mathbf{Z}^*$.

If, at this point, it is assumed that $\mathbf{S}_{12}$ is a three-phase quantity and $\mathbf{V}_1$ as well as $\mathbf{V}_2$ are line potentials instead of phase voltages as originally assumed, the single-phase equation above is rewritten to accommodate the change as $\mathbf{S}_{12}/3 = (|\mathbf{V}_1|^2/3 - |\mathbf{V}_1|/\sqrt{3}|\mathbf{V}_2|/\sqrt{3} \cos \delta - j|\mathbf{V}_1|/\sqrt{3}|\mathbf{V}_2|/\sqrt{3} \sin \delta)/\mathbf{Z}^*$. Notice that the value 3 may be canceled so that the equation reduces to an expression for $\mathbf{S}_{12}$ in terms of line potentials $|\mathbf{V}_1|$ and $|\mathbf{V}_2|$.

Multiplying numerator and denominator by $(\mathbf{Z}^*)^*$ yields $\mathbf{S}_{12} = (\mathbf{R}|\mathbf{V}_1|^2 - R|\mathbf{V}_1||\mathbf{V}_2| \cos \delta + X|\mathbf{V}_1||\mathbf{V}_2| \sin \delta)/(R^2 + X^2) + j(X|\mathbf{V}_1|^2 - X|\mathbf{V}_1||\mathbf{V}_2| \cos \delta - R|\mathbf{V}_1||\mathbf{V}_2| \sin \delta)/(R^2 + X^2)$. The real term is P_{12} and the imaginary term is Q_{12}.

Similarly, for $\mathbf{S}_{21} = \mathbf{V}_2(-\mathbf{I}^*)$ the outflow of bus 2 is $\mathbf{S}_{21} = (R|\mathbf{V}_2|^2 - R|\mathbf{V}_1||\mathbf{V}_2| \cos \delta - X|\mathbf{V}_1||\mathbf{V}_2| \sin \delta)/(R^2 + X^2) + j(X|\mathbf{V}_2|^2 - X|\mathbf{V}_1||\mathbf{V}_2| \cos \delta + R|\mathbf{V}_1||\mathbf{V}_2| \sin \delta)/(R^2 + X^2)$. In this expression the real term is P_{21} and the imaginary term is Q_{21}.

The above four equations are nonlinear because a product of potentials appears in each term. There are seven variables: $|\mathbf{V}_1|$, $|\mathbf{V}_2|$, δ, P_{12}, P_{21}, Q_{12}, and Q_{21}. Therefore, the model must have three of these as known values in order to solve the equations. (*Note:* **I** is implicitly represented.)

4. *Obtain Solution*

The net per-unit power, which is known at bus 1, must be delivered to the transmission line and so becomes $\mathbf{S}_{12}$. There is no other possible exchange; therefore $\mathbf{S}_{12} = (1 + j0.3) - (0.4 + j0.1) + j|\mathbf{V}_1|^2(0.0265/2)$. The last term is the reactive power generated by half the line's parallel admittance. Because $|\mathbf{V}_1| = 1$ p.u., $\mathbf{S}_{12} = (0.6 + j0.213)$.

The problem is solved when all seven variables are known. Voltage $|\mathbf{V}_2|$ and δ can be found from $\mathbf{S}_{12} = P_{12} + jQ_{12}$. Recall that $P_{12} = (R|\mathbf{V}_1|^2 - R|\mathbf{V}_1||\mathbf{V}_2| \cos \delta + X|\mathbf{V}_1||\mathbf{V}_2| \sin \delta)/(R^2 + X^2)$ and $Q_{12} = (X|\mathbf{V}_1|^2 - X|\mathbf{V}_1||\mathbf{V}_2| \cos \delta - R|\mathbf{V}_1||\mathbf{V}_2| \sin \delta)/(R^2 + X^2)$. Substituting values in these equations, where $|\mathbf{V}_1| = 1$ p.u., we obtain $0.6 = (0.028|1|^2 - 0.028|1|\mathbf{V}_2 \cos \delta + 0.34|1|\mathbf{V}_2 \sin \delta)/(0.028^2 + 0.34^2)$ and $0.213 = (0.34|1|^2 - 0.34|1|V_2 \cos \delta - 0.028|1|\mathbf{V}_2 \sin \delta)/(0.028^2 + 0.34^2)$. The unknowns are on the right side and can be isolated by consolidation and division of one equation by the other to eliminate $|\mathbf{V}_2|$. Performing the algebra yields $\tan\delta = \sin\delta/\cos\delta = 0.217$ and $\delta = 12.3°$. The value for $|\mathbf{V}_2|$ can be found from $0.0418 = (-0.028\cos\delta + 0.34\sin\delta)|\mathbf{V}_2|$.

Then $|\mathbf{V}_2| = 0.93\angle -12.3°$ p.u.

The equations for $\mathbf{S}_{21}$ determine the load flow toward bus 1 from bus 2. Because P_{12} is positive, it should be expected that, given no-fault balanced conditions, P_{21} would be negative. This signifies power flow from the transmission line into bus 2. However, no such preconceived notion should be entertained regarding Q_{21}. The reactive-power flow from bus 1 may not be enough to magnetize the line and ensure that no reactive-power flow from bus 1 reaches bus 2. In this event, Q_{21} would be positive to augment the magnetization.

Substituting in the appropriate equations, we obtain $P_{21} = (0.028)|0.93|^2 - 0.028|1||0.93|\cos 12.3° - 0.34|1||0.93|\sin 12.7°)/(0.028^2 + 0.34^2) = -0.589$ p.u. MW (input to bus 2). Similarly, solving for Q_{21}, we have $Q_{21} = -0.08$ p.u. Mvar (input to bus 2).

The line losses can now be calculated because all seven variables have been identified. In this case, line magnetization has been produced from bus 1 only. $\mathbf{S}_{loss} = (P_{12} + jQ_{12}) + (P_{21} + jQ_{21}) = (0.6 - 0.589) + j(0.213 - 0.08)$. Thus $\mathbf{S}_{loss} = (0.011 + j0.133) = 0.133\angle 85.3°$. This should be in proportion to $\mathbf{Z}$ because $\mathbf{S}_{loss} = |\mathbf{I}|^2\mathbf{Z}$. The value of $\mathbf{Z}$ is: $\mathbf{Z} = (15 + j180) = 181\angle 85.2°$. The polar angles of $\mathbf{S}_{loss}$ and $\mathbf{Z}$ are within the tolerance afforded by a pocket calculator.

The per-unit generation at bus 2 is found by taking the difference between the bus 2 load and the inflow $\mathbf{S}_{21}$ of the line plus the contribution of the parallel admittance. Then, $\mathbf{S}_{2G} = \mathbf{S}_{2L} - (\mathbf{S}_{21} + \mathbf{S}_{2C}) = (0.8 + j0.25) - [0.589 + j0.08 + j|0.93|^2(0.0265/2)] = 0.211 + j0.159$.

5. ***Obtain Final Values***

The potentials are expressed in kilovolts when the per-unit values are multiplied by the line-voltage base. The multiplication yields $|\mathbf{V}_1| = 230$ kV and $|\mathbf{V}_2| = (0.93)(230$ kV$) = 214$ kV. The complex powers of interest are multiplied by the base of 100 MVA, yielding $\mathbf{S}_{12} = 60 + j21.3$ MVA, $\mathbf{S}_{21} = -(58.9 + j8)$ MVA, $\mathbf{S}_{loss} = 1.1 + j13.3$ MVA, and $\mathbf{S}_{2G} = 21.1 + j15.9$ MVA.

Related Calculations: The difference angle δ is a relative one. Therefore its value may be applied to either the voltage at bus 1 or that of bus 2. If one is interested in calculating the line current $\mathbf{I}$ the basic equation is: $\mathbf{I} = (\mathbf{V}_1 - \mathbf{V}_2/\mathbf{Z}$ where $\mathbf{V}_1$ and $\mathbf{V}_2$ are the phase voltages. Substituting the known values, one finds $\mathbf{I} = [230\angle 0°/\sqrt{3} - (214/\sqrt{3})\angle -12.3°]/(15 + j180) = 0.16\angle -19.9°$ kA. As a check, $\mathbf{S}_{12} = \sqrt{3}\,\mathbf{V}_1\mathbf{I}^*$ may be used to solve for $\mathbf{I}$. From this equation, $\mathbf{I} = 0.16\angle -19.5°$. A small difference exists because of rounding off with a pocket calculator.

The previous calculations represent a load-flow study for a single transmission line. Even at this simple level, a digital computer is useful in reducing computation time and improving accuracy. Problems involving more than two buses generally require a digital computer for their solution.

A SHORT TRANSMISSION LINE

Problems involving short transmission lines can usually be solved with acceptable accuracy if the phase-to-ground capacitance is ignored and the reactance-to-resistance ratio of the series impedance is large. The line then becomes a single, lumped inductive reactance.

Assuming that the resistance and parallel admittance can be ignored, solve the previous problem.

Calculation Procedure:

1. Convert to Per-Unit Values

The per-unit values from the previous problem are unchanged except for the omitted line parameters. The per-unit line impedance now becomes $X = j0.34$ p.u.

2. Draw Circuit Model

An approximate model is provided in Fig. 4.

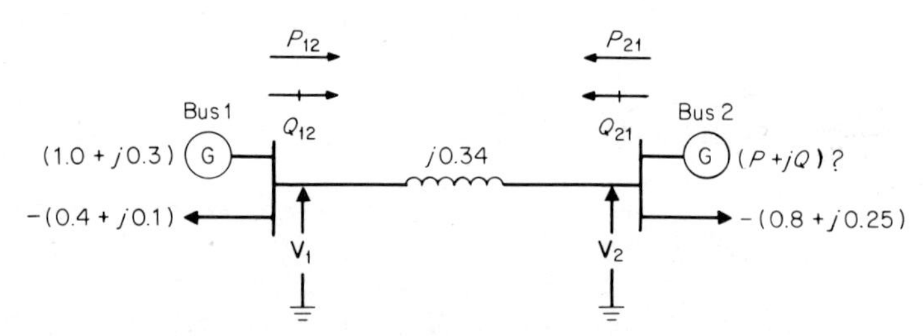

Fig. 4 Approximate model of a short transmission line.

3. Derive Model Equations

The equations of the previous problem are written with $R = 0$ for $\mathbf{S_{12}}$ and $\mathbf{S_{21}}$. This results in $P_{12} = |\mathbf{V_1}||\mathbf{V_2}| \sin \delta/X$; $Q_{12} = |\mathbf{V_1}|^2/X - |\mathbf{V_1}||\mathbf{V_2}| \cos \delta/X$; $P_{21} = -|\mathbf{V_1}||\mathbf{V_2}| \sin \delta/X$; $Q_{21} = |\mathbf{V_2}|^2/X - |\mathbf{V_1}||\mathbf{V_2}| \cos \delta/X$.

4. Obtain Solution

The net bus power is not the same as in the previous problem because line capacitance is ignored. Therefore, $\mathbf{S_{12}} = (1 + j0.3) - (0.4 + j0.1) = 0.6 + j0.2$. Substitution of values in the abbreviated equations yields $P_{12} = 0.6 = |1||\mathbf{V_2}| \sin (\delta/0.34)$ and $Q_{12} = 0.2 = |1|^2/0.34 - |1||\mathbf{V_2}| \cos \delta/0.34$. This reduces to $0.204 = |\mathbf{V_2}| \sin \delta$ and $0.932 = |\mathbf{V_2}| \cos \delta$. Dividing the first equation by the second equation yields $\tan \delta = \sin \delta/\cos \delta = 0.204/0.932 = 0.219$; therefore, $\delta = 12.4°$.

Using the equation for P_{12} to solve for $|\mathbf{V_2}|$ yields $|\mathbf{V_2}| = (0.6)(0.34)/\sin 12.4° = 0.95\underline{/-12.4°}$ (for $\mathbf{V_1}$ taken as the angle reference). Comparison of $\mathbf{V_2}$ with the more accurate values previously calculated indicates a magnitude difference of 0.02 p.u. and a phase angle nearly the same.

Solving for $\mathbf{S_{21}}$, $P_{21} = -|1||0.95| \sin 12.4°/0.34 = -0.6$ and $Q_{21} = |0.95|^2/0.34 - |1||0.95| \cos 12.4°/0.34 = -0.075$. Then, $\mathbf{S_{21}} = -(0.6 + j0.075)$ p.u. MVA. Note that $|P_{12}| = |P_{21}|$ because the line resistance has been ignored. (Power flowing out of bus 1 must enter bus 2.) Complex power loss is $\mathbf{S_{loss}} = j(0.2 - 0.075) = j0.125$ p.u. MVA; this value is 6 percent of the result obtained in the previous problem.

$\mathbf{S_{2G}} = (0.8 + j0.25) - (0.6 + j0.075) = (0.2 + j0.18)$ p.u. This value is the difference between the output and injected input from the transmission line. The results obtained approximate the values found in the previous problem.

5. Calculate the Line Current

Use $\mathbf{I}^* = \mathbf{S}_{12}/\sqrt{3}\mathbf{V}_1 = (60 + j20)/(\sqrt{3})(230) = 0.159\angle -18.4°$ kA.

6. Calculate the Power Loss in Line

Even though the resistance was not considered, an approximation to the line loss is still possible. From $P_{loss} = 3|\mathbf{I}|^2R$, where $R = 15\ \Omega$ from the previous problem, $P_{loss} = (3)(0.159)^2(15) = 1.14$ MW. This is very close to the value of 1.1 MW found in the previous problem.

Related Calculations: The equation $P_{12} = |\mathbf{V}_1||\mathbf{V}_2| \sin \delta/X$ is significant because it is indicative of a transmission line's ultimate capacity for carrying a load. Burn-down considerations aside (burn-down is a resistive phenomenon), inspection of the equation indicates that an upper limit must be reached as $\delta = 90°$, similar to pullout torque of a synchronous machine. Also, the increase of power-handling capability is realized with the increase of potential, which is squared in the equation. For approximately the same reactance, a 500-kV line can carry 52 times as much power as a 69-kV line.

A GENERAL LOAD-FLOW STUDY

Determine what variables and parameters are involved in a system consisting of more than two buses and an energy loop configuration. What is meant by bus power? What does the solution of a load-flow study satisfy?

Calculation Procedure:

1. Draw Circuit Model

Figure 5 shows, as an example, a three-bus system model having a complexity that requires a minimum of digital computer calculations.

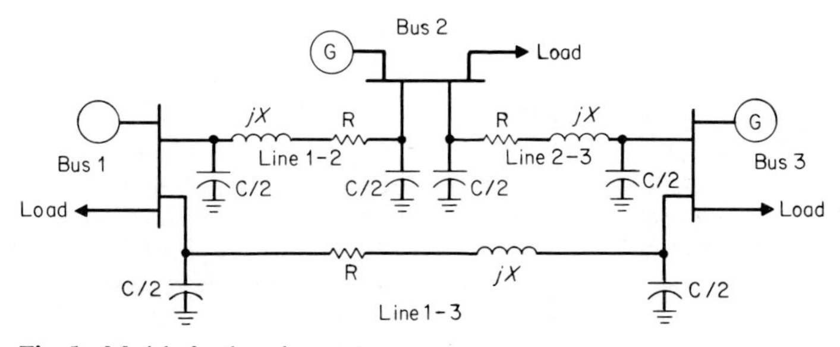

Fig. 5 Model of a three-bus system.

2. Define Parameters

The parameters to be considered are specified in the model of Fig. 5. Because a computer solution will be employed, resistance is usually retained in the problem, although capacitance to ground is often ignored for short lines [less than 16 km (10 mi)]. The reactance might include the leakage reactance of transformers between buses. Transformer winding resistance, however, is comparatively small and often neglected. All parameters are expressed on a per-unit basis according to the potential at which they are found and the arbitrary selection of the system base apparent power.

3. *Define Variables*

Complex power-line flow and the potentials at all buses, as well as bus power, will constitute the sought-after variables in a solution of a load-flow study. The entire solution is on a per-unit basis.

Bus power is the difference between the positive generated complex power injected into the bus, which may include banks of reactive-power-furnishing capacitors, and the negative complex load power leaving the system. A bus may also simply be a tie point between intersecting transmission lines and, therefore, have a bus power that is zero. It could also be a load bus with no generation and have a negative complex bus power. There can be many combinations that arise in a large system.

To begin a load-flow study, two of the variables mentioned above must be known at each bus. The four possible variables are $|\mathbf{V_i}|$, δ_i, and the two parts of the bus power, P_i and Q_i. The loads on the buses are always known and represent system demand to be satisfied by total generation. (The generation at the buses is partly known, as explained below.)

Specifically, the system is divided into three types of buses. They are:

(a) The swing (slack) bus where $|\mathbf{V_i}|$ and δ_i are known and act as the reference system potential. The generation at this bus stands ready to make up system losses and the complex bus power is part of the solution.

(b) The control bus where $|\mathbf{V_j}|$ and P_j are known quantities. The voltage magnitude is selected to prop up system potential in its area of influence. The imaginary part of the complex power Q_j is part of the solution because the program must determine the injected reactive power necessary to achieve the desired $|\mathbf{V_j}|$. The angle δ_j is also an outcome of the solution.

(c) The load bus where the known complex bus power is negative. The result will be the values of $|\mathbf{V_k}|$ and δ_k.

4. *Enumerate Results of Study*

The results of a load-flow study of a three-bus system, or something more complicated, are enumerated as: (*a*) the complex potential at each bus; (*b*) the complex bus power; (*c*) the complex line-flow power; (*d*) line losses; and (*e*) reactive power injection. In more sophisticated programs there are usually results for (*f*) overload status of line; (*g*) location of taps on tap-changing transformers; (*h*) power-system interchanges; (*i*) cost of system operation for a given mode; (*j*) format identifying plant and substation by name.

Related Calculations: Selection of the roles each bus will play is a combination of engineering judgment and the nature of the bus (is generated power or reactive power available for injection?). A system with many buses will usually have scattered control buses and important load buses, but only one swing or reference bus. Figure 5 depicts a system so small that control buses would not be necessary. If the swing bus were designated as bus 1, its fixed voltage would adequately influence the voltage level at the adjacent buses.

NODAL EQUATIONS AS A BASIS FOR LOAD-FLOW STUDIES

Determine the basic nodal equations for a load-flow study.

Calculation Procedure:

1. Draw Circuit Model

The circuit model is provided in Fig. 6 with current injection at each node and all the appropriate parameters in place.

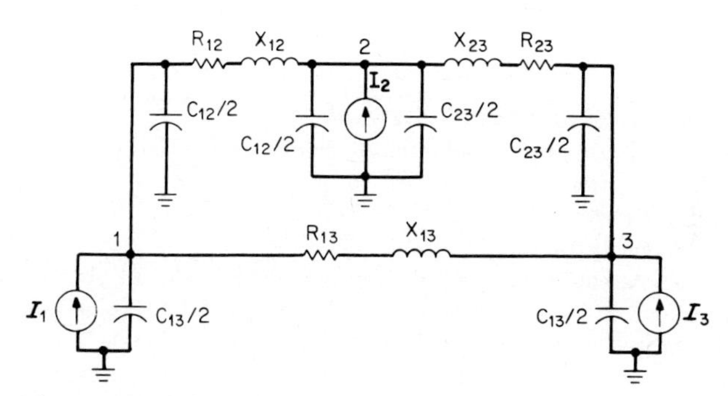

Fig. 6 Circuit model containing current injection.

2. Determine Equations for Bus Potentials

The matrix form of the three equations necessary to solve for the three bus potentials is:

$$\begin{bmatrix} \mathbf{Y}_{11} & \mathbf{Y}_{12} & \mathbf{Y}_{13} \\ \mathbf{Y}_{21} & \mathbf{Y}_{22} & \mathbf{Y}_{23} \\ \mathbf{Y}_{31} & \mathbf{Y}_{32} & \mathbf{Y}_{33} \end{bmatrix} \begin{bmatrix} \mathbf{V}_1 \\ \mathbf{V}_2 \\ \mathbf{V}_3 \end{bmatrix} = \begin{bmatrix} \mathbf{I}_1 \\ \mathbf{I}_2 \\ \mathbf{I}_3 \end{bmatrix}$$

where $\mathbf{Y}_{11} = 1/(R_{12} + jX_{12}) + 1/(R_{13} + jX_{13}) + j\omega C_{12}/2 + j\omega C_{13}/2$, $\mathbf{Y}_{22} = 1/(R_{12} + jX_{12}) + 1/(R_{23} + jX_{23}) + j\omega C_{12}/2 + j\omega C_{23}/2$, $\mathbf{Y}_{33} = 1/(R_{13} + jX_{13}) + 1/(R_{23} + jX_{23}) + j\omega C_{13}/2 + j\omega C_{23}/2$, $\mathbf{Y}_{12} = \mathbf{Y}_{21} = -1/(R_{12} + jX_{12})$, $\mathbf{Y}_{13} = \mathbf{Y}_{31} = -1/(R_{13} + jX_{13})$, and $\mathbf{Y}_{23} = \mathbf{Y}_{32} = -1/(R_{23} + jX_{23})$.

The currents are eliminated as explicit variables through $\mathbf{S} = \mathbf{VI}^*$, where $\mathbf{S}$ is the bus power. Thus, $\mathbf{I} = \mathbf{S}^*/\mathbf{V}^* = (P - jQ)/\mathbf{V}^*$. This can be applied to $\mathbf{I}_1$, $\mathbf{I}_2$, and $\mathbf{I}_3$ in the previous equations. For example, the equation for bus 1 can be written as $\mathbf{Y}_{11}\mathbf{V}_1 = (P_i - jQ)/\mathbf{V}_1^* - \mathbf{Y}_{12}\mathbf{V}_2 - \mathbf{Y}_{13}\mathbf{V}_3$. Then, the potential for bus 1 is: $\mathbf{V}_1 = (1/\mathbf{Y}_{11})[(P_1 - jQ_1)/\mathbf{V}_1^* - \mathbf{Y}_{12}\mathbf{V}_2 - \mathbf{Y}_{13}\mathbf{V}_3]$. Similarly, for buses 2 and 3, we obtain: $\mathbf{V}_2 = (1/\mathbf{Y}_{22})[(P_2 - jQ_2)/\mathbf{V}_2^* - \mathbf{Y}_{21}\mathbf{V}_1 - \mathbf{Y}_{23}\mathbf{V}_3]$ and $\mathbf{V}_3 = (1/\mathbf{Y}_{33})[(P_3 - jQ_3)/\mathbf{V}_3^* - \mathbf{Y}_{31}\mathbf{V}_1 - \mathbf{Y}_{32}\mathbf{V}_2]$.

Related Calculations: For larger systems containing more buses, the basic equation is

$$\mathbf{V}_i = \frac{1}{\mathbf{Y}_{ii}} \left[\frac{(P_i - jQ_i)}{\mathbf{V}_i^* - \sum_{i \neq j} \mathbf{Y}_{ij}\mathbf{V}_j} \right]$$

It is observed that this equation becomes a rearranged version of the equations for the preceding two-bus problems if subscript j is limited to a value of 2. However, a form of $\mathbf{V}_i$ is on both sides of the equation so that, in solving for $\mathbf{V}_i$, an iterative technique using a first guess is involved.

SOLUTION TECHNIQUE FOR A MULTIBUS SYSTEM

Describe a solution method for a multibus load-flow study.

Calculation Procedure:

1. Employ the Gauss-Siedel Method

This method uses successive substitutions to determine each bus potential after reasonable guesses for the potential at each bus (except one) have been made. One bus has its potential fixed at the desired magnitude and approximate angle. This bus is the swing (slack) bus mentioned earlier.

In the system's simplest form, control buses may not be required. As an example, consider Fig. 5. If bus 1 is designated the swing bus, buses 2 and 3 become load buses where $P_i + jQ_i$, the bus power, is negative. If desired, the swing bus might have a fixed potential of $\mathbf{V}_1 = 1\angle 0°$. A convenient initial value for the potential of the load buses would be the same.

2. Develop Solution

The solution begins with the equation: $\mathbf{V}_2 = [(P_2 - jQ_2)/\mathbf{V}_2^* - \mathbf{Y}_{12}\mathbf{V}_1 - \mathbf{Y}_{23}\mathbf{V}_3]/\mathbf{Y}_{22}$. The values on the right side of the equation are either known or have been assumed ($\mathbf{V}_3 = 1\angle 0°$). When the corrected value of $\mathbf{V}_2$ is determined, it is substituted in the right side of the equation for $\mathbf{V}_3$ so that $\mathbf{V}_3 = [(P_3 - jQ_3)/\mathbf{V}_3^* - \mathbf{Y}_{12}\mathbf{V}_1 - \mathbf{Y}_{23}\mathbf{V}_2]/\mathbf{Y}_{33}$. This then yields a corrected value of $\mathbf{V}_3$ for substitution back in the equation for $\mathbf{V}_2$.

Keep in mind that an equation for the potential at the swing bus is never used because $\mathbf{V}_1$ is preselected. However, continued use of $\mathbf{V}_1$, when it is multiplied by the appropriate coupling admittance, stabilizes the convergence procedure in the repeated use of the other equations. Voltages $\mathbf{V}_2$ and $\mathbf{V}_3$ will continually change value as use of the potential equations is rotated until the accuracy dictated by the computer program is reached for them.

After determination of the unknown potentials, the complex power flows may be found by using the equations discussed earlier, that is, $\mathbf{S}_{12}$, $\mathbf{S}_{21}$, $\mathbf{S}_{13}$, and $\mathbf{S}_{31}$ and the attendant complex power losses which are readily available.

Another important quantity yet to be identified is the complex power of the swing bus. This is $\mathbf{S}_1 = (P_{12} + jQ_{12}) + (P_{13} + jQ_{13})$. If there is a known load at the swing bus the complex bus power cannot be considered to be equal to the total generated power found there. Instead, $\mathbf{S}_{1G} = \mathbf{S}_1 - \mathbf{S}_{\text{load}}$. (Recall that $\mathbf{S}_{\text{load}}$ is regarded as negative.)

Related Calculations: The Gauss-Siedel iterative method cannot be accomplished readily without programmed digital computations. The repetitive operations and the extent of numerical information demand the speed, accuracy, and large memory of a digital computer.

A DIGITAL PROGRAM APPLIED TO A FOUR-BUS SYSTEM

Figure 7 is a diagram of a four-bus system that will be solved on a digital computer. The system consists of a swing bus where the complex potential is known; a control bus where the magnitudes of the potential and the real power are known; and two load buses where the complex negative load power is known. (Of course, the line parameters also are available.) Develop a computer program for the problem.

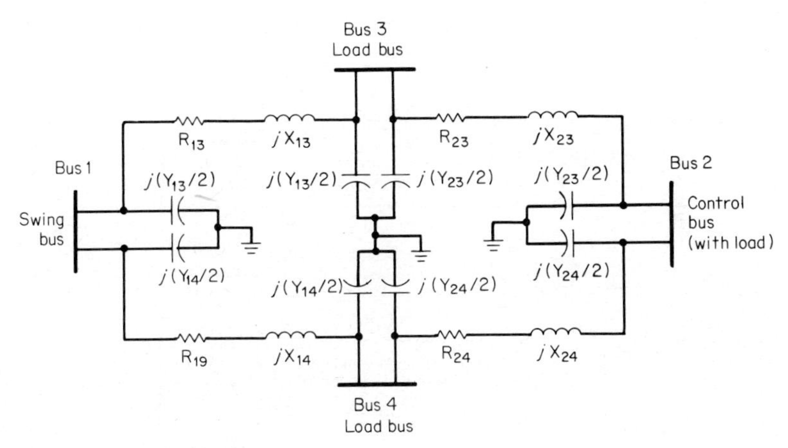

Fig. 7 Model of a four-bus system.

Calculation Procedure:

1. Analyze the Problem

The loop system's voltage is maintained at a reasonable level at the load buses. This stems from it being propped up by the invariant swing-bus potential at bus 1 and a magnitude of potential at bus 2 which is maintained by injecting into bus 1 the necessary reactive power from a bank of capacitors or a synchronous machine. This results in a net positive value of Q_2 if the potential tends to be low. (Q_2 can be positive even though the fixed power P_2 flows out of the system and is therefore negative.)

A control bus requires that its value of Q be ascertained before the fixed magnitude of potential is determined. The necessary equation is a simple rearrangement of the nodal equation discussed earlier, using the imaginary part only. This results in

$$Q_i = -\mathrm{Im}\left[\mathbf{Y}_{ii}|\mathbf{V}_i|^2 - \mathbf{V}_i^* \sum_{i\neq j} \mathbf{V}_j\mathbf{Y}_{ij}\right]$$

The potential, whose magnitude is controlled by division by the absolute value, becomes

$$\frac{\mathbf{V}_i}{|\mathbf{V}_i|} = \frac{(P_i - jQ_i)/\mathbf{V}_i^* + \sum_{i\neq j} \mathbf{V}_j\mathbf{Y}_{ij}}{\mathbf{Y}_{ii}}$$

This is applied to control bus 2 by substituting the subscript 2 for i. The division by $|\mathbf{V}_i|$ results in a magnitude of 1 p.u. Applying a constant could result in other magnitudes.

2. *Develop Flowchart*

The flowchart is provided in Fig. 8.

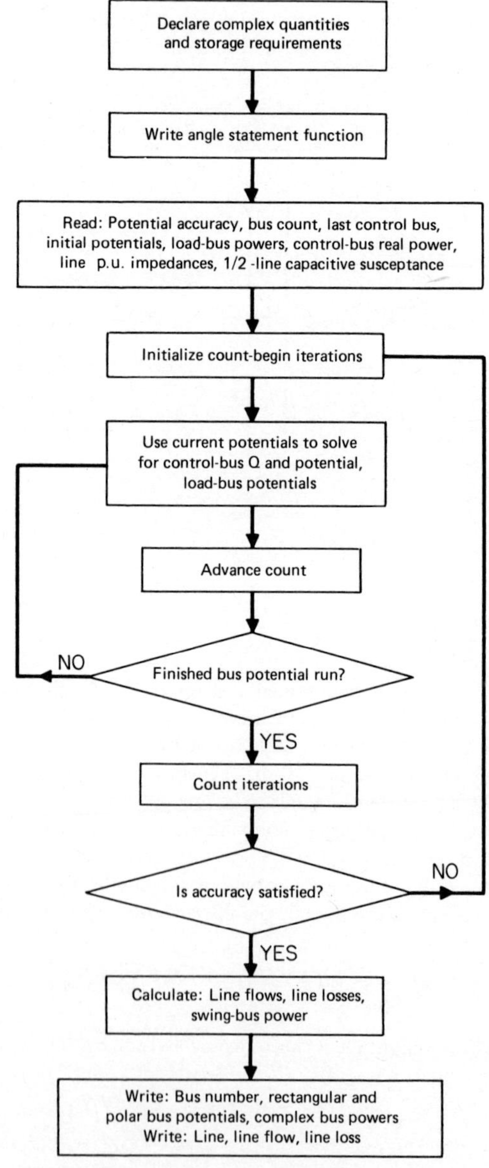

Fig. 8 Flowchart for a four-bus system.

3. Determine the Data for Program

The data for the four-bus problem as consecutively entered in the Fortran program and identifier names for the data follow:

Accuracy of the expected potentials: VE = 0.0001

Number of buses: NBUS = 4

Last control bus: NCONT = 2

Swing-bus voltage and first guess: $V(I) = 1\angle 0^\circ$

Load bus powers: $S(3) = -1.0 - j0.5$, $S(4) = -1.3 - j0.6$

Control bus power in per-unit megawatts: $P(2) = -0.8$

Line impedances:

$\mathbf{Z}(1, 3) = 0.01 + j0.12$, $\mathbf{Z}(1, 4) = 0.005 + j0.09$,

$\mathbf{Z}(2, 3) = 0.008 + j0.1$, $\mathbf{Z}(2, 4) = 0.009 + j0.12$

One-half the capacitive susceptance of each line:

$\mathbf{YC}(1, 3) = 0.0 + j0.02$, $\mathbf{YC}(1, 4) = 0.0 + j0.01$,

$\mathbf{YC}(2, 3) = 0.0 + j0.008$, $\mathbf{YC}(2, 4) = 0.0 + j0.009$

4. Run Program

The results are provided in Fig. 9 and the listed program is given in Fig. 10.

PROGRAM CONVERGED IN 12 ITERATIONS

	VOLTS				COMPLEX BUS POWER	
BUS	REAL	IMAG	MAGN	ANGLE	REAL	IMAG
1	1.000	0.0	1.000	0.0	3.142	0.815
2	0.977	-0.215	1.000	-12.426	-0.800	0.813
3	0.947	-0.166	0.961	-9.928	-1.000	-0.500
4	0.945	-0.154	0.957	-9.228	-1.300	-0.600

	LINE FLOWS		LINE LOSSES	
LINE	REAL	IMAG	REAL	IMAG
1-2	0.0	0.0	0.0	0.0
1-3	1.408	0.306	0.021	0.212
1-4	1.734	0.507	0.016	0.276
2-1	0.0	0.0	0.0	0.0
2-3	-0.384	0.422	0.003	0.018
2-4	-0.415	0.391	0.003	0.023
3-1	-1.387	-0.096	0.021	0.212
3-2	0.387	-0.404	0.003	0.018
3-4	0.0	0.0	0.0	0.0
4-1	-1.718	-0.231	0.016	0.276
4-2	0.418	-0.369	0.003	0.023
4-3	0.0	0.0	0.0	0.0

Fig. 9 Printout of results for a four-bus system.

```
,*** PSU-WATFIV *** 19-5 (03/06/80-- 836)

           C  LOAD FLOW PROGRAM WITH ONE CONTROL BUS AND TWO LOAD BUSES
           C
           C  DECLARE AND ORGANIZE INPUT DATA
   1            COMPLEX*8 Z(10,10),        Y(10,10),S(10),V(10),SUM(10),VPAST(10),
               1SL(10,10),SLOSS(10,10),VP,YC(10,10)
   2            DIMENSION P(10),Q(10)
           C  WRITE A DUMMY STATEMENT FUNCTION
   3            AV(VP)= ATAN2(AIMAG(VP),REAL(VP))*57.29578
           C READ DATA
   4            READ,VE
   5            READ,NBUS,NCONT
   6            READ,(V(I),I=1,NBUS)
   7            KK= NCONT+1
   8            READ,(S(I),I=KK,NBUS)
   9            IF(NCONT.EQ.1)GO TO 11
  10            READ,(P(I),I=2,NCONT)
           C  READ LINE IMPEDANCES, Z(I,J)=0, WHERE I=J
  11         11 DO 1 I=1,NBUS
  12            READ,(Z(I,J),J=1,NBUS)
  13          1 CONTINUE
  14            DO 6 I=1,NBUS
  15            READ,(YC(I,J),J=1,NBUS)
  16          6 CONTINUE
           C  FIND THE LINE ADMITTANCES FROM 1/Z(I,J)
  17            DO 2 I=1,NBUS
  18            DO 2 J=1,NBUS
  19            IF(CABS(Z(I,J)).NE.0.0) GO TO 3
  20            Y(I,J)=(0.0,0.0)
  21            GO TO 2
  22          3 Y(I,J)=1./Z(I,J)
  23          2 CONTINUE
           C  FIND NODAL ADMITTANCE AT EACH BUS
  24            DO 4 I=1,NBUS
  25            Y(I,I)=(0.0,0.0)
  26            DO 4 J=1,NBUS
  27            IF(I.EQ.J) GO TO 4
  28            Y(I,I)=Y(I,I)+Y(I,J)+YC(I,J)
  29          4 CONTINUE
           C  BEGIN CALCULATIONS FOR BUS POTENTIALS
  30            L=0
  31          5 DO 20 M=2,NBUS
  32            DO 10 K=2,NBUS
  33            SUM(K)=(0.0,0.0)
  34            DO 10 I=1,NBUS
  35            IF(K.EQ.I) GO TO 10
  36            SUM(K)=SUM(K)+V(I)*Y(I,K)
  37         10 CONTINUE
  38            VPAST(M)=V(M)
  39            IF(M.GT.NCONT) GO TO 25
  40            Q(M)=-AIMAG(Y(M,M)*(CABS(VPAST(M)))**2-CONJG(VPAST(M))*SUM(M))
  41            S(M)=CMPLX(P(M),Q(M))
  42            V(M)=(CONJG(S(M)/VPAST(M))+SUM(M))/Y(M,M)
  43            V(M)=V(M)/CABS(V(M))
  44            GO TO 20
  45         25 V(M)=(CONJG(S(M)/VPAST(M))+SUM(M))/Y(M,M)
  46         20 CONTINUE
  47            DO 30 N=2,NBUS
  48            IF(CABS(V(N)-VPAST(N)).GT.VE) GO TO 40
```

```
49          30 CONTINUE
50             GO TO 50
51          40 L=L+1
52             GO TO 5
        C   FOR BUSES DIRECTLY CONNECTED CALCULATE LINE FLOWS TO INCLUDE THE
        C   EFFECTS OF CAPACITIVE SUSCEPTANCE.
53          50 DO 60 I=1,NBUS
54             DO 60 J=1,NBUS
55             IF(I.EQ.J.OR.CABS(Y(I,J)).EQ.0.0) GO TO 65
56             SL(I,J)=V(I)*CONJG((V(I)-V(J))*Y(I,J)+V(I)*YC(I,J))
57             GO TO 60
58          65 SL(I,J)=(0.0,0.0)
59          60 CONTINUE
        C   CALCULATE SWING BUS POWER
60             S(1)=(0.0,0.0)
61             DO 70 J=1,NBUS
62             S(1)=S(1)+SL(1,J)
63          70 CONTINUE
        C   CALCULATE LINE LOSS
64             DO 80 I=1,NBUS
65             DO 80 J=1,NBUS
66             IF(I.EQ.J) GO TO 90
67             SLOSS(I,J)=SL(I,J)+SL(J,I)
68             GO TO 80
69          90 SLOSS(I,J)=(0.0,0.0)
70          80 CONTINUE
71             PRINT,'PROGRAM CONVERGED IN',L,'ITERATIONS'
72             PRINT, ' '
73             WRITE(6,400)
74         400 FORMAT(' ',23X,'VOLTS',18X,'COMPLEX BUS POWER'//)
75             WRITE(6,500)
76         500 FORMAT(' ',7X,'BUS',4X,'REAL',4X,'IMAG',4X,'MAGN',3X,'ANGLE',8X
              1,'REAL',4X,'IMAG'//)
77             PRINT, ' '
78             DO 200 I=1,NBUS
79             WRITE(6,600) I,V(I),CABS(V(I)),AV(V(I)),S(I)
80         600 FORMAT(' ',9X,I1,4F8.3,4X,2F8.3//)
81         200 CONTINUE
82             WRITE(6,700)
83         700 FORMAT(' ',16X,'LINE FLOWS',5X,'LINE LOSSES'//)
84             WRITE(6,800)
85         800 FORMAT(' ',6X,'LINE',4X,'REAL',4X,'IMAG',4X,'REAL',4X,'IMAG'//)
86             DO 300 I=1,NBUS
87             DO 300 J=1,NBUS
88             IF(I.EQ.J) GO TO 300
89             WRITE(6,900)I,J,SL(I,J),SLOSS(I,J)
90         900 FORMAT(' ',7X,I1,'-',I1,6F8.3)
91         300 CONTINUE
92             STOP
93             END

STATEMENTS EXECUTED=     2179

CORE USAGE        OBJECT CODE=    6896 BYTES,ARRAY AREA=    4400 BYTES,TOTAL AREA AVAILABLE=  198656  BYTES

DIAGNOSTICS        NUMBER OF ERRORS=       0, NUMBER OF WARNINGS=       0, NUMBER OF EXTENSIONS=       2

,COMPILE TIME=    0.07 SEC,EXECUTION TIME=     0.04 SEC,                                 WATFIV - JUN 1977 V1L6
```

Fig. 10 Program listing for a four-bus system.

Related Calculations: For brevity, the listed Fortran program lacks several routines that would improve its usefulness. These are:

1. Prevention of divergence of the iterative portion of the program by abortion after a preset number of iterations and the printing of a termination message.
2. Because there is a limited amount of reactive power that can be injected at a control bus, the program should reflect this by changing the status of a bus to a load bus when a limit has been reached.
3. The input data should be described and reported.

Section 12 POWER-SYSTEM CONTROL

John S. Wade, Jr., Ph.D.
Associate Professor of Engineering,
The Pennsylvania State University,
The Capitol Campus

REFERENCES Elgerd—*Electric Energy Systems Theory: An Introduction,* McGraw-Hill; Gross—*Power System Analysis,* Wiley; NERC-OC Publications—*National Energy Reliability Council—Operating Committee;* Neuenswander—*Modern Power Systems,* International Textbook; Stevenson—*Elements of Power Systems Analysis,* McGraw-Hill.

INTRODUCTION

The National Electric Reliability Council—Operating Committee (NERC-OC) is responsible for the promulgation of rules for its member companies so that power flow to consumers is reliable, especially on a regional basis. The "control area" is the basic unit recognized by NERC-OC. It may consist of a single large private company, a government-operated system such as the Tennessee Valley Authority (TVA), or several investor-owned companies banded together in a power pool. The distinguishing feature of control areas is a single control center entrusted with the authority to operate the system within its area.

The primary responsibility of a control area is to match its load with its own power generation. However, through interties with neighboring control areas, it is also prepared for mutual aid and scheduled sales of energy. Such mutual aid in emergencies may take the form of the flow of energy to a control area not directly adjacent to it. Even normally scheduled sales of energy may find intervening control areas between buyer and seller.

Frequency measured in hertz (Hz) is the direct indication of the status of intertied control areas. In North America, the normal frequency is 60 Hz. However, fluctuations in demand against generation cause all the intertied control areas to witness the same variation in frequency (normally no more than ± 0.1 Hz). The frequency falls below 60 Hz when the demand outweighs the generation. Conversely, when the generation is more than necessary, the frequency rises.

In this dynamic system, constant regulation is needed to maintain the frequency within narrow limits. A single control area may be responsible for excess demand. In that case, the interties will show a net flow of energy above scheduled sales. This net inflow is called the area control error (ACE). The dispatcher will take steps within one minute to increase generation and return the ACE to zero (an NERC-OC rule). The dispatcher has ten minutes to accomplish this. An outflow requiring reduced generation in some other case may also be observed.

Because consumer demand is ever-changing, time lags are present in matching generation to this dynamic situation. An increase in demand is met by three inherent properties of the power system:

1. An expenditure of kinetic energy, present in the rotating mass of the generators, as they decelerate (the frequency falls).
2. A general reduction in load that is frequency-sensitive (largely motors). This can amount to a 1 percent load reduction per 1 percent frequency reduction.
3. The automatic regulation of selected generating plants to increase the energy output.

It is important to note that the entire system of interconnected control areas slows down its rotating mass of generators at the same rate. This results in a uniform drop in frequency, while supplying energy to the excess demand.

CHARACTERISTICS OF AN UNREGULATED, ISOLATED CONTROL AREA

Determine the drop in frequency of an unregulated, isolated control area because of a sudden load demand. The control area has no interties and no automatic regulation. The dispatcher is absent. The kinetic energy of the rotating mass, before the disturbance, is 25,000 MWs. The load demand increases from 5000 MW to 5200 MW as electric-arc furnaces come on-line.

Calculation Procedure:

1. Calculate the Kinetic Energy

Kinetic energy K of a group of alternators is proportional to their mutual frequency squared. Therefore, $K_i/f_i^2 = K/f^2$, where subscript i indicates the initial value, and no subscript indicates ongoing values. The frequency f may be written as $f = f_i + \delta f$. The subsequent kinetic energy K is found to be $K = K_i(f_i + \delta f)^2/f_i^2$. Expanding and dropping the second-order term δf^2 yields $K = K_i(1 + 2\delta f/f_i)$. This is now included in the dynamic equation $d[K_i(1 + 2\delta f/f_i)]/dt + LC(\delta f) + \text{demand} = 0$, or

$(2K_i/f_i)[d(\delta f)/dt] + LC(\delta f) + \text{demand} = 0$. Here each term is in megawatts, and the change in frequency δf is the dependent variable. Term LC represents the system load change with frequency, and demand is the 200-MW increase in load.

$LC = (5000)(1 \text{ percent})/(60)(1 \text{ percent})$, or $LC = 83.3$ MW/Hz. Since $2K_i/f_i = 833.3$, the first-order differential equation becomes $833.3d(\delta f)/dt + 83.3(\delta f) = -200$. This reduces to $d(\delta f)/dt + 0.1\delta f = -0.24$.

2. *Solve for δf*

From Laplace transforms with zero initial conditions, the transform is $s\,\Delta F + 0.1\,\Delta f = -0.24/s$. Thus, $\Delta F = -0.24/s(s + 0.1)$. From a table of inverse transforms, in the time domain, $\delta f = 2.4(e^{-0.1t} - 1)$ Hz.

The time constant is $\tau = 1/0.1 = 10$ s. Therefore, in about 5 time constants ($t = 50$ s), the change in frequency is -2.4 Hz. That means that the frequency has a final value of 57.6 Hz.

At the above reduced frequency, the load's sensitivity to frequency produces a load reduction. However, 57.6 Hz is not acceptable for operation of steam turbines that are tuned for 60 Hz. An intolerable situation has therefore developed that will lead to equipment damage. Interties with other systems to create more rotating mass would yield an improved result.

3. *Calculate Initial Rate of Change in Frequency*

The initial rate of change in frequency is $d(\delta f)/dt\,|_{t=0} = -0.24$ Hz/s. The actual load carried at 57.6 Hz is $5200 - 83.3(2.4) \cong 5000$ MW.

Related Calculations: The example illustrates that a control area with limited rotating mass could have severe frequency changes without the mutual aid of interties with other control areas. If 10 times the kinetic energy used here was available by interties and also 10 times the load, the equation for δf would now be $\delta f = 0.24(e^{-0.1t} - 1)$. The total drop in frequency would now be $(60 - 0.24) = 59.76$ Hz and is now within tolerable limits. It should be noted that the time constant ($\tau = 10$ s) has not changed, and 50 seconds are still necessary to obtain 59.76 Hz.

POWER-SYSTEM REGULATION

Develop a mathematical model for a power system.

Calculation Procedure:

1. *Consider a Time-Dependent Model*

In Sec. 11, the load-flow study required a static model whose variables were considered independent of time and whose parameters were declared constant. This model was used to obtain an instantaneous result, even though the system load flow must change with time to meet varying consumer demands.

A time-dependent model that accounts for load changes may be applied to an on-line digital computer to furnish remote signals to generating stations for the ongoing process of matching generator output to demand. Faster changing systems, owing to faults, require a different model considered in Sec. 16.

2. *Develop the System Transfer Function*

A transfer function may be realized from the Laplace transform found in the previous example. For brevity, let $2K_i/f_i = u_i$. Then, $u_i\,d(\delta f)/dt + LC(\delta f) = -P_D$ where

P_D is the load demand. When the equation is transformed to the s domain, we get $s\,\Delta F + (LC/u_i)\,\Delta F = -\Delta P_D/u_i$. Then, $\Delta F/\Delta P_D = -1/[u_i(s + LC/u_i)]$, which is a transfer function in the s domain as shown in the block diagram, Fig. 1.

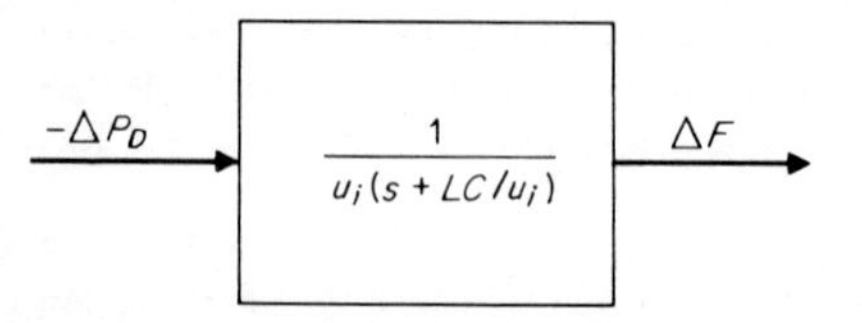

Fig. 1 Block diagram of an open-loop system.

The input in Fig. 1, $-\Delta P_D$, when multiplied by the operator in the block (the transfer function), results in the output ΔF. This system with no feedback elements is an open-loop system. As demonstrated in the previous single-control-area example, there is but a single time constant ($\tau = u_i/LC$). The output lags the input because of this time constant. Fig. 2 compares the time relationship of $-\delta P_D$ to δf.

3. *Add a Controller*

A circle is added to Fig. 1 as a symbol for a comparator (summer) so that the system might be improved by the addition of compensating signals (Fig. 3). If an incremental

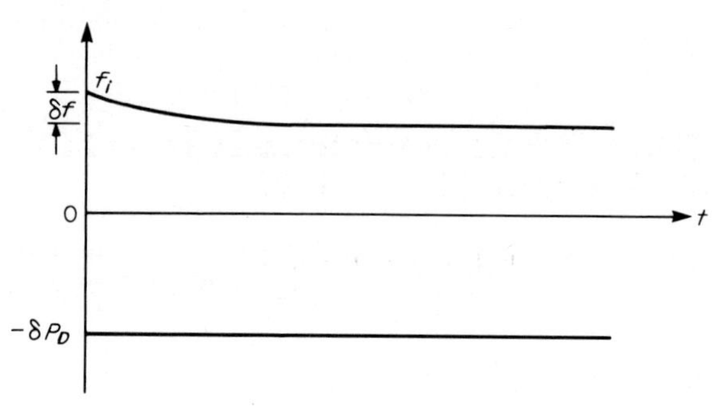

Fig. 2 Time relationship between δf and $-\Delta P_D$.

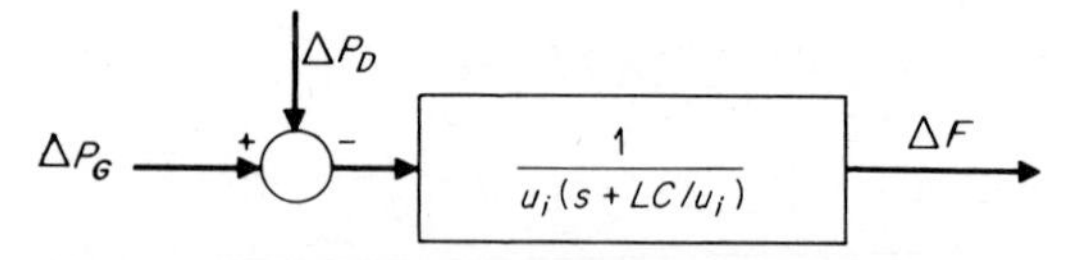

Fig. 3 Addition of a compensating signal ΔP_G.

increase in power generation, ΔP_G, could be applied to the comparator at the same instant and with the same magnitude as the incremental increase in demand, ΔP_D, then the net input would be zero. Consequently, the frequency would remain constant ($\delta f = 0$).

The time-domain expression of this situation is $u_i\, d(\delta f)/dt + LC\delta f = \Delta P_G - \Delta P_D = 0$. The differential equation has now no excitation (forcing function), and because there were no initial conditions, there can be no output ($\delta f = 0$). This ideal situation cannot be realized since the change in generation, ΔP_G, does not anticipate the change in load, and a time lag results.

4. *Add Feedback*

In order to produce any generation to balance a change in load demand, the energy-producing plants must be made aware of the falling (or rising) frequency. This is especially true of generating stations whose output can be varied quickly, such as hydroelectric, pumped-storage, combustion-turbine, and the fastest of coal- or oil-fired plants. Rates of change of megawattseconds per minute are typical.

Before automatic regulation, the central dispatcher made use of voice communication to inform a plant to pick up generation to match demand. In doing so, the dispatcher was closing the loop from ΔF to ΔP_G and became part of the feedback control. Today, modern control centers use signals from on-line computers to automatically control throttles, fuel flow, and governors, producing faster plant energy response.

Figure 4 depicts a control system in which the status of frequency change is fed back to the power-producing elements. This is called a closed-loop system. A transfer function

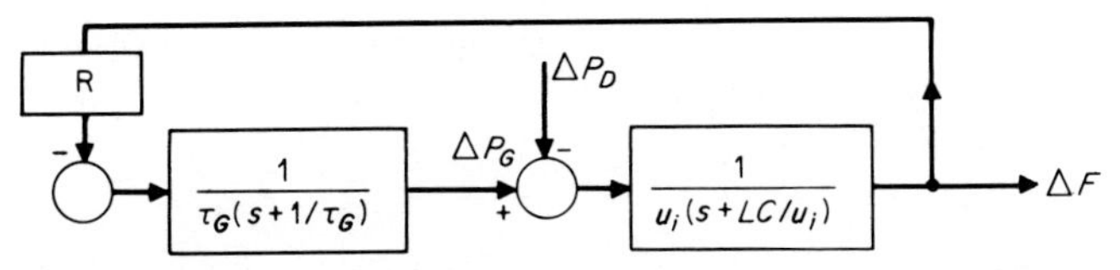

Fig. 4 The addition of feedback.

representing the time delay in plant response is a necessary addition. A function block R, representing a regulating element, is included (at the left) in the diagram.

CHARACTERISTICS OF A PROPORTIONATELY CONTROLLED FEEDBACK SYSTEM

Determine the transfer function $\Delta F/\Delta P_D$ for Fig. 4. Use the final-value theorem with the values of the previous open-loop example to obtain the final value of δf when $\delta P_D = 200u(t)$ MW, a step function.

Calculation Procedure:

1. Determine $\Delta F/\Delta P_D$

As is usual when linearity is assumed, the Laplace transform (s domain) is utilized. From Fig. 4, $\Delta F = -\dfrac{\Delta P_D - \Delta P_G}{u_i(s + LC/u_i)}$, and the feedback branch is

$$\Delta P_G = -\Delta F\left(\frac{R}{\tau_G(s + 1/\tau_G)}\right)$$

Combining these equations to eliminate ΔP_G yields:

$$\frac{\Delta F}{\Delta P_D} = -\frac{\dfrac{1}{u_i(s + LC/u_i)}}{1 + \dfrac{R}{u_i\tau_G\left(s + \dfrac{1}{\tau_G}\right)\left(s + \dfrac{LC}{u_i}\right)}}$$

This follows the well-known form for any negative-feedback system: $C/\text{ref} = G/(1 + GH)$. In the present case ΔF is the controlled variable C; $-\Delta P_D$ is the reference input, ref; $1/u_i(s + LC/u_i)$ is the feed-forward plant G; and $R/\tau_G(s + 1/\tau_G)$ is the feedback element H.

2. *Apply Final-Value Theorem*

The final-value theorem determines δf after the transient behavior has died out. The formula is

$$\delta f|_{t \to \infty} = \lim_{s \to 0} sF(s)$$

where $F(s) = (\Delta F/\Delta P_D)\,\Delta P_D$ in this case. Because $\mathcal{L}(\delta P_D) = \Delta P_D = 200/s$, a step function, the final-value theorem becomes

$$\delta f|_{t \to \infty} = \lim_{s \to 0} \frac{\dfrac{-200}{u_i(s + LC/u_i)}}{1 + \dfrac{R}{\tau_G u_i(s + LC/u_i)(s + 1/\tau_G)}}$$

If s approaches zero, $\delta f|_{t \to \infty} = (-200/LC)/(1 + R/LC)$. This can be written $\delta f|_{t \to \infty} = -200/(LC + R)$. If $LC = 83.3$ MW/Hz and the regulation $R = 0$, then $\delta f|_{t \to \infty} = -2.4$ Hz, as in the previous open-loop calculations. However, if R in megawatts per hertz is not zero, the final value of δf will become less significant as R increases. Notice that δf cannot become zero.

Related Calculations: The use of a regulating factor, or gain, in the feedback branch, here known as R, is called proportional-feedback control. It does not completely restore the frequency to 60 Hz, but does reduce the deviation.

When R represents the total regulation factor for a system, then $LC + R$ is known as the area frequency-response characteristic (AFRC). Because of the composite nature of $LC + R$, members of NERC-OC determine AFRC experimentally as outlined in the NERC-OC manual. Member groups use megawatts per 0.1 Hz as the measure of AFRC.

PROPORTIONAL PLUS INTEGRAL CONTROL

How can a system be completely restored to normal frequency? Use the final-value theorem to show that the addition of a parallel integrating branch will bring a system back to 60 Hz after a load disturbance ΔP_D.

Calculation Procedure:

1. *Modify System*

In order to restore a system's normal frequency, the control system of Fig. 4 is modified to include an integrating element fed back in parallel with the proportional control R. Fig. 5 indicates the arrangement.

2. *Determine $\Delta F/\Delta P_D$*

The equation for a negative-feedback system is $C/\text{ref} = G/(1 + GH)$. This time, $H = (K_I/s + R)$. Therefore, the equation is

$$\frac{\Delta F}{\Delta P_D} = -\frac{1/[u_i(s + LC/u_i)]}{1 + \dfrac{R + K_I/s}{\tau_G u_i(s + LC/u_i)(s + 1/\tau_G)}}$$

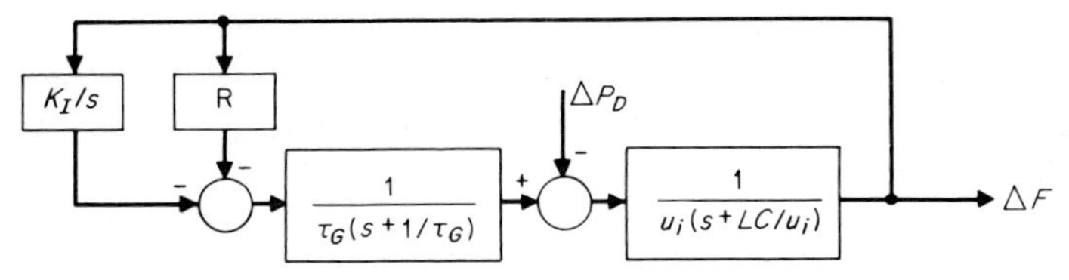

Fig. 5 Introducing integral control.

This reduces to

$$\frac{\Delta F}{\Delta P_D} = -\frac{s\tau_G[s + (1/\tau_G)]}{su_i\tau_G[s + (LC/u_i)][s + (1/\tau_G)] + (sR + K_I)}$$

3. Apply Final-Value Theorem

Again, $\Delta P_D = 200/s$, but the final-value theorem produces

$$\delta f|_{t\to\infty} = \lim_{s\to 0} \frac{-200s\tau_G[s + (1/\tau_G)]}{u_i\tau_G[s + (LC/u_i)][s + (1/\tau_G)] + (sR + K_I)}$$

As s approaches 0, $\delta f|_{t\to\infty} = 0$. There is no final deviation from 60 Hz with proportional-plus-integral control. Therefore this scheme is representative of the usual individually controlled power system.

AVERAGE FREQUENCY

With load demand constantly changing, and consequent frequency deviations, how can the system frequency be made to average 60 Hz? Is the maintenance of a 60-Hz average important?

Calculation Procedure:

1. Consider Role of Feedback Control

Any deviation from normal frequency may be countered by the action of a proportional-plus-integral-feedback system. However, there is a time lag because of the failure of the system to act instantaneously. It is possible then for the average value of frequency to maintain itself above or below 60 Hz if not periodically checked and corrected.

2. Apply Frequency Averaging

Control areas are interconnected, and, except for phase-angle deviation as considered in a later section, they will act synchronously with the same frequency. NERC-OC has provided that a single control area in an interconnected system will direct frequency correction. By integration of the frequency deviation about the normal frequency of 60 Hz, the time error in seconds is established. When the time error exceeds 2 s, an adjustment is sought. After all control areas have been notified, a frequency offset of ± 0.02 Hz is applied to the entire system until the error has been reduced to less than 0.5 s.

It should be obvious that the system generator's rotational speed is changed to produce this offset. This correction may be applied for 5 h at a time.

Related Calculations: By periodically correcting the frequency, electric timekeeping instruments maintain satisfactory accuracy.

SYSTEM OF CONTROL AREAS

Why are control areas intertied? How are interties mathematically realized? What are the specific responsibilities of individual control areas?

Calculation Procedure:

1. Consider Use of Interties

Figure 5 represents one control area operating without interconnection to others. In this condition, frequency averaging would be difficult, and there could be no mutual aid or scheduled sales between this control area and others.

Control areas are intertied by transmission lines so that they can take advantage of the spinning mass of large numbers of generators and differences in incremental costs between units that have different degrees of loading (time zones can be a factor). Power flows through these lines on a short-term basis if a control area suffers an emergency stemming from equipment breakdown or, on a long-term basis, if there is a mutually beneficial sale of energy between corporate units in each of two control areas. Where a scheduled sale exists, the buyer's generation does not satisfy the demand. This causes the phase angle at the buyer's end of the intertie to lag that of the seller. Therefore, as noted in Sec. 11, the energy flows toward the buyer.

Occasionally, this energy reaches the buyer by a circuitous route as a result of the peculiar nature of the voltage level and phase. For instance, Ontario has been known to furnish energy through New York, Pennsylvania, and Ohio for a sale in Michigan. This sometimes occurs even though there are direct interties between Michigan and Ontario. Hence, the intervening control areas along the route register an inflow at one border and an outflow at another on their energy metering. Compensatory settlements for the use of control area transmission facilities, line loss, and other costs are made between all parties involved in the transfer.

2. Develop Mathematical Model of Interties

In Sec. 11 it was seen that transmission-line power from bus i to bus j is $P_{ij} = [|\mathbf{V}_i|(|\mathbf{V}_j|/X)] \sin \phi_{ij}$, provided the line resistance can be considered negligible compared with line inductive reactance X, and where ϕ_{ij} is the angular difference that bus j lags bus i in voltage phase angle. Therefore, power flows out of bus i toward bus j. When the magnitudes of $\mathbf{V}_i$ and $\mathbf{V}_j$ are fixed, $P_{max} = (|\mathbf{V}_i||\mathbf{V}_j|)/X$.

The derivative of P_{ij} with respect to ϕ is $dP_{ij}/d\phi = P_{max} \cos \phi_{ij}$. This is called the *synchronizing coefficient*. This coefficient is useful in indicating whether a tie line (or any transmission line) is strong or weak.

If ϕ_{ij} must be a large angle to support sizable power flow, the synchronizing coefficient is small and the tie line is weak. If an attempt is made to produce an angle ϕ_{ij} greater than 90° by excessive power flow, the intertied control areas will lose synchronization. At 90°, the synchronizing coefficient is zero.

For the case where there is an intertie between control areas, the Laplace-transformed ΔP_{ij} becomes an input to the system of Fig. 5. Because the control system is a dynamic representation, ΔP_{ij} is an incremental change in system loading and affects frequency deviation only when the tie-line phase angle is changing.

It has been established that $dP_{ij}/d\phi = P_{max} \cos \phi_{ij}$ so that an incremental treatment in the time domain yields $\delta P_{ij} = P_{max}[\cos(\phi_i - \phi_j)](\delta\phi_i - \delta\phi_j)$. Here, ϕ_{ij} becomes $\phi_i - \phi_j$, as if an angle of reference elsewhere in the system is used. Incremental change $\delta\phi$

becomes $\delta\phi_i - \delta\phi_j$ and, if this exists, a slight change in rotational frequency similar to that found for a load-changing synchronous machine must take place.

Time is necessary to produce an angular difference so that $\delta\phi_i - \delta\phi_j = 2\pi\int(\delta f_i - \delta f_j)\,dt$. Then a change in intertie power between control areas becomes $\delta P_{ij} = P_{max}[\cos(\phi_i - \phi_j)]2\pi\int(\delta f_i - \delta f_j)\,dt$. Transforming to the s domain from the time domain yields: $\Delta P_{ij} = P_{max}[\cos(\phi_i - \phi_j)](2\pi)(\Delta F_i/s - \Delta F_{ij}j/s)$. For all interties n for a particular control area,

$$\Delta P_{ij} = 2\pi \sum_{j}^{n} P_{max} \cos(\phi_i - \phi_j)[(\Delta F_i/s) - \Delta F_j/s].$$

Here, the subscript for the control area in particular has been dropped; ΔP_{tie} has then become another input to affect the frequency deviation ΔF independently or with ΔP_D (Fig. 6).

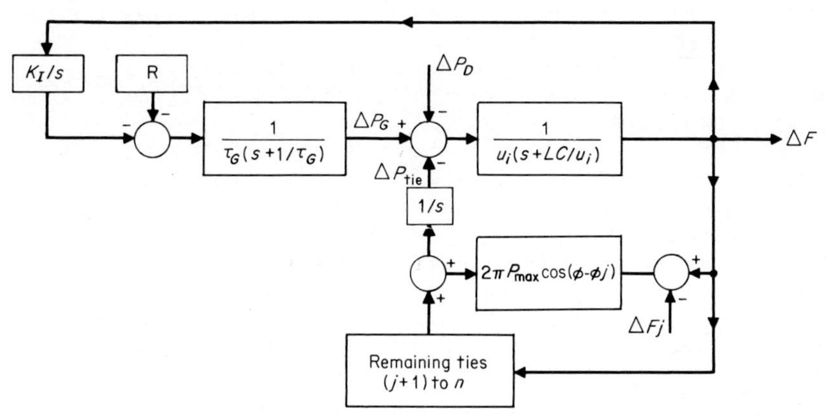

Fig. 6 Control system with interties.

This system can be solved for ΔP_D input or any ΔP_{tie} by a digital or analog computer. Only one composite control area is indicated in the feedback system of Fig. 6. In practice, however, all control areas must be represented in the computer simulation. This is beyond the scope of this handbook.

Related Calculations: NERC-OC assigns to each control area the responsibility of determining the limit of assistance it is able to contribute to controlling system frequency about the normal value (60 Hz). This limit is called "frequency bias" and it is given in megawatts per 0.1 Hz. The AFRC is used as a guide in establishing each control area's capability to aid in restoring system frequency. Therefore, the AFRC is recommended by NERC-OC when the control area does not have its own empirical formula.

As a rule of thumb, on-line operating generation should have the reserve capacity to replace the largest generating unit in the control area. In order not to burden other control areas, when the largest generating unit malfunctions and is lost to the control area, 10 minutes is considered sufficient time for replacing it with what is called "spinning reserve." However, during that interval other control areas will call upon their "frequency bias" to supply energy to the stricken one. In this way, frequency deviation is limited and the entire system benefits.

Section 13 SHORT-CIRCUIT COMPUTATIONS

Lawrence J. Hollander, P.E.
Associate Dean, Cooper Union

REFERENCES Fitzgerald and Kingsley—*Electric Machinery,* McGraw-Hill; Hubert—*Preventive Maintenance of Electrical Equipment,* McGraw-Hill; Kosow—*Electric Machinery and Transformers,* Prentice-Hall; Siskind—*Electrical Machines: Direct and Alternating Currents;* McGraw-Hill; Lawrence and Richards—*Principles of Alternating-Current Machinery,* McGraw-Hill; Beeman—*Industrial Power Systems Handbook,* McGraw-Hill; Greenwood—*Electrical Transients in Power Systems,* Wiley; Knable—*Electrical Power Systems Engineering: Problems and Solutions,* McGraw-Hill; Freeman—*Electric Power Transmission and Distribution,* Intl. Ideas; Guile and Paterson—*Electrical Power Systems,* Pergamon; Pansini—*Basic Electrical Power Transmission,* Hayden; Stevenson—*Elements of Power System Analysis,* McGraw-Hill; Knight—*Power Systems Engineering and Mathematics,* Pergamon; Anderson—*Analysis of Faulted Power Systems,* Iowa State University Press; Gibbs—*Transformer Principles and Practice,* McGraw-Hill; Ragaller—*Current Interruption in High-Voltage Networks,* Plenum Press; Sullivan—*Power System Planning,*

McGraw-Hill; IEEE—*Recommended Practice for Grounding of Industrial and Commercial Power Systems;* IEEE—*Recommended Practice for Electric Power Systems in Commercial Buildings.*

TRANSFORMER REGULATION DETERMINED FROM SHORT-CIRCUIT TEST

A single-phase transformer has the following nameplate data: 2300/220 V, 60 Hz, 5 kVA. A short-circuit test (low-voltage winding short-circuited) requires 66 V on the high-voltage winding to produce rated full-load current; 90 W is measured on the input. Determine the transformer's percent regulation for a load of rated current and a power factor of 0.80, lagging.

Calculation Procedure:

1. Compute the Rated Full-Load Current (High-Voltage Side)

For a single-phase ac circuit use the relation $\text{kVA} = VI/1000$, where kVA = apparent power in kilovolt-amperes, V = potential in volts, and I = current in amperes. Rearranging the equation yields $I = 1000\ \text{kVA}/V = (1000)(5)/2300 = 2.17$ A.

2. Compute the Circuit Power Factor for the Short-Circuit Test

Use the relation for power factor: $\text{pf} = W/\text{VA}$, where W is real power in watts and VA is apparent power in volt-amperes (volts × amperes); $\text{pf} = 90/(66)(2.17) = 0.628$. The power factor is the cosine of the angle between voltage and current; $\cos^{-1} 0.628 = \theta = 51.1°$ lagging. This is the pf angle of the transformer's internal impedance.

3. Compute the Circuit Power-Factor Angle for the Operating Condition

As in Step 2 use the relation: $\cos^{-1} 0.80 = \theta = 36.9°$ lagging. This is the pf angle of the load served by the transformer.

4. Compute the Transformer Output Voltage for Serving an 0.80-pf Load at Rated Full-Load Current

Use the relation for IR drop in the transformer: $V_{IR} = V_{sc} \cos (\theta_{ii} - \theta_{load}) = 66 \cos (51.1° - 36.9°) = 66 \cos 14.2° = 64.0$ V, where V_{sc} = short-circuit test voltage, θ_{ii} = internal-impedance phase angle, and θ_{load} = load phase angle. Refer to Fig. 1. Use the relation for IX drop in the transformer: $V_{IX} = V_{sc} \sin (\theta_{ii} - \theta_{load}) = 66 \sin (51.1° - 36.9°) = 66 \sin 14.2° = 16.2$ V. Thus, $V^2_{input} = (V_{output} + V_{IR})^2 + V^2_{IX} = 2300^2 = (V_{output} + 64)^2 + 16.2^2$. Solution of the equation for V_{output} yields 2236 V.

5. Compute the Transformer Regulation

Use the relation: percent regulation $= [(V_{input} - V_{output})/V_{input}](100\ \text{percent}) = [(2300 - 2236)/2300](100\ \text{percent}) = 2.78$ percent.

Related Calculations: The general method presented here is valid for calculating transformer regulation with both leading and lagging loads.

TERMINAL VOLTAGE OF SINGLE-PHASE TRANSFORMER AT FULL LOAD

A single-phase transformer has the following nameplate data: 2300/440 V, 60 Hz, 10 kVA. The ohmic constants of the equivalent circuit are primary resistance referred to the primary $r_1 = 6.1$, secondary resistance referred to the secondary $r_2 = 0.18$, primary

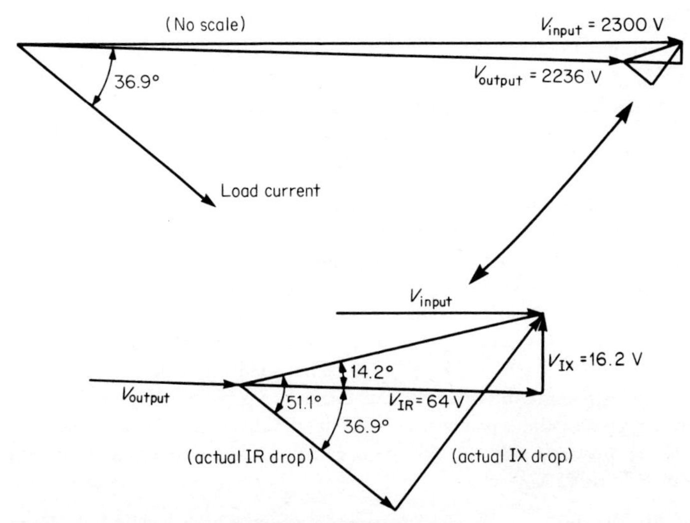

Fig. 1 Transformer internal impedance-drop triangle. In accordance with trigonometric relations, the actual *IR* and *IX* drops are changed to values more easily handled mathematically.

leakage reactance referred to the primary $x_1 = 13.1$, and secondary leakage reactance referred to the secondary $x_2 = 0.52$. With the primary supply voltage set at nameplate rating, 2300 V, calculate the secondary terminal voltage at full load, 0.80 pf, lagging.

Calculation Procedure:

1. Compute the Turns Ratio

The turns ratio a is determined from the equation a = primary voltage/secondary voltage = 2300/440 = 5.23.

2. Compute the Total Resistance of the Transformer Referred to the Primary

The equation for the total series resistance of the simplified equivalent circuit (see Fig. 2) referred to the primary is $R = r_1 + a^2r_2 = 6.1 + (5.23)^2(0.18) = 11.02\ \Omega$.

3. Compute the Total Leakage Reactance of the Transformer Referred to the Primary

The equation for the total leakage reactance of the simplified equivalent circuit (see Fig. 2) referred to the primary is $X = x_1 + a^2x_2 = 13.1 + (5.23)^2(0.52) = 27.32\ \Omega$.

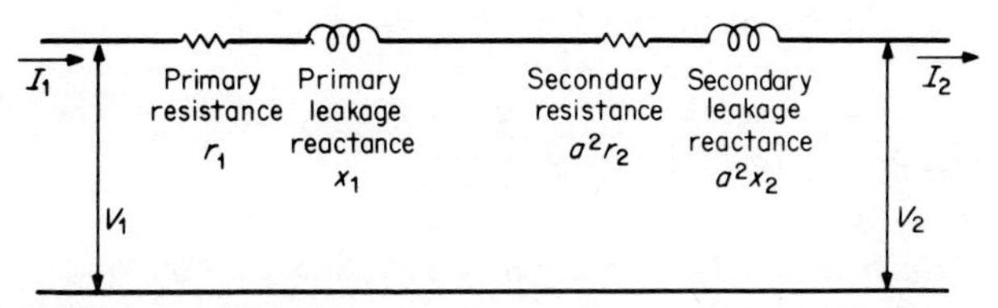

Fig. 2 Transformer approximate equivalent circuit referred to the primary side.

4. Compute the Load Current

The approximate or simplified equivalent circuit is referred to the primary side of the transformer. Use the equation: $I_1 = I_2$ = kVA rating/kV rating = 10/2.3 = 4.35 A, and assume this to be the reference phasor. Thus, in phasor notation, $\mathbf{I_1} = \mathbf{I_2} = 4.35 + j0$. It is given that the load current lags the output voltage $\mathbf{V_2}$ by a pf angle of $\cos^{-1} 0.80 = \theta = 36.87°$.

5. Compute the Output Voltage

Use the equation: $\mathbf{V_1} = \mathbf{V_2} + \mathbf{I_2}(R + jX)$, where all quantities are referred to the primary side of the transformer. With $\mathbf{I_2}$ as the reference phasor, $\mathbf{V_2}$ is written as $(0.8 + j0.6)V_2$. Thus, $|\mathbf{V_1}| = (0.8 + j0.6)V_2 + 4.35(11.02 + j27.32) = 2300$. This is rewritten in the form $2300^2 = (0.8V_2 + 47.94)^2 + (0.6V_2 + 118.84)^2$; by further rearrangement of this equation, V_2 may be determined from the solution for a quadratic equation, $x = [-b \pm (b^2 - 4ac)^{1/2}]/2a$. The quadratic equation is: $0 = V_2^2 + 219.3V_2 - 5{,}273{,}578.8$; the solution in primary terms yields $V_2 = 2189.4$ V.

The actual value of the secondary voltage (load voltage) is computed by dividing V_2 (referred to the primary side) by the turns ratio a. Thus, the actual value of $V_2 = 2189.4/5.23 = 418.62$ V.

Related Calculations: The general method presented here is valid for calculating the terminal voltage for both leading and lagging loads, and may be used for three-phase transformers having balanced loads. For unbalanced loads, the same analysis may be based on the concept of symmetrical components.

VOLTAGE AND CURRENT IN BALANCED THREE-PHASE CIRCUITS

The line-to-line voltage of a balanced three-phase circuit is 346.5 V. Assuming the reference phasor $\mathbf{V_{ab}}$ to be at 0°, determine all the voltages and the currents in a load that is wye-connected and has an impedance $\mathbf{Z_L} = 12\angle 25°\ \Omega$ in each leg of the wye. If the same loads were connected in delta, what would be the currents in the lines and in the legs of the delta?

Calculation Procedure:

1. Draw the Phasor Diagram of Voltages

Assuming a phase sequence of a, b, c, the phasor diagram of voltages is as shown in Fig. 3.

2. Identify All Voltages in Polar Form

From the phasor diagram for a balanced system, the polar form of the voltages may be written: $\mathbf{V_{an}} = 200\angle 330°$ V, $\mathbf{V_{bn}} = 200\angle 210°$ V, $\mathbf{V_{cn}} = 200\angle 90°$ V, $\mathbf{V_{ab}}$ (reference) $= 346.5\angle 0°$ V, $\mathbf{V_{bc}} = 346.5\angle 240°$ V, $\mathbf{V_{ca}} = 346.5\angle 120°$ V.

3. Calculate the Currents for the Wye Connection

Calculate the current in the load for each branch of the wye connection; these currents must lag the respective voltages by the power-factor angle of the load. Thus, $\mathbf{I_{an}}$ must lag $\mathbf{V_{an}}$ by 25°; $\mathbf{I_{an}} = \mathbf{V_{an}}/\mathbf{Z_L} = 200\angle 330°\ \text{V}/12\angle 25°\ \Omega = 16.7\angle 305°$ A. In a similar manner $\mathbf{I_{bn}} = 16.7\angle 185°$ A and $\mathbf{I_{cn}} = 16.7\angle 65°$ A.

4. Calculate Internal Delta Currents for Delta Connection

For the delta-connected load, the voltage across each leg (side) of the delta is the line-to-line voltage rather than the line-to-neutral voltage (see Fig. 4). $\mathbf{I_{ab}} = \mathbf{V_{ab}}/\mathbf{Z_L} = 346.5$

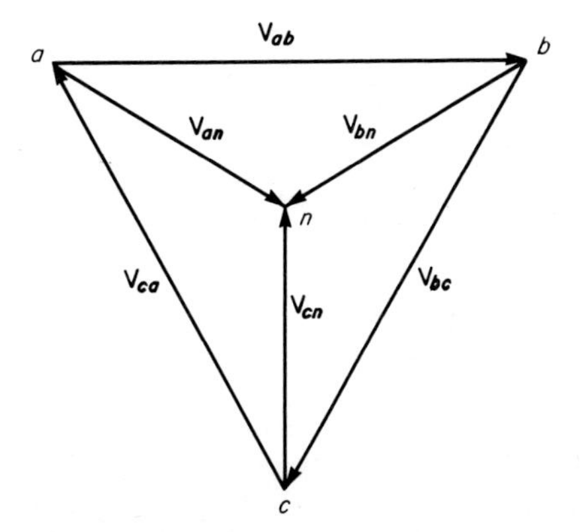

Fig. 3 Phasor diagram of voltages based on a phase sequence of a, b, c.

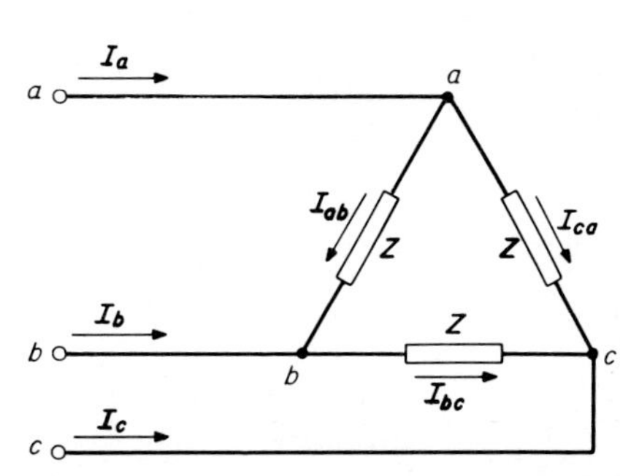

Fig. 4 Circuit diagram of a three-phase delta-connected load.

$\angle 0°$ V/12$\angle 25°$ V = 28.9$\angle -25°$ A. In a similar manner, $\mathbf{I}_{bc}$ = 28.9$\angle 215°$ A and $\mathbf{I}_{ca}$ = 28.9$\angle 95°$ A.

5. Calculate the Line Currents for the Delta Connection

Calculate the line currents for the delta-connected load using the relation: the sum of the currents into a node must equal the sum of the currents leaving a node. At node *a*, $\mathbf{I}_a + \mathbf{I}_{ca} = \mathbf{I}_{ab}$, or $\mathbf{I}_a = \mathbf{I}_{ab} - \mathbf{I}_{ca}$ = 28.9$\angle -25°$ − 28.9$\angle 95°$ = 26.2 − j12.2 + 2.5 − j28.8 = 28.7 − j41.0 = 50.0$\angle -55°$ A. Similarly, $\mathbf{I}_b$ = 50.0$\angle 185°$ A and $\mathbf{I}_c$ = 50.0$\angle 65°$ A. In this case, for the delta-connected load, the line currents are $\sqrt{3}$ greater than the phase currents within the delta and lag them by 30°.

Related Calculations: This procedure may be used for all balanced loads having any combinations of resistance, inductance, and capacitance. Further, it is applicable to both wye- and delta-connected loads. Through the use of symmetrical components, unbalanced loads may be handled in a similar manner. Symmetrical components are most helpful for making the calculations of balanced loads with unbalanced voltages applied to them.

THREE-PHASE SHORT-CIRCUIT CALCULATIONS

Three alternators are connected in parallel on the low-voltage side of a wye-wye three-phase transformer as shown in Fig. 5. Assume that the voltage on the high-voltage side of the transformer is adjusted to 132 kV, the transformer is unloaded, and no currents are flowing among the alternators. If a three-phase short circuit occurs on the high-voltage side of the transformer, compute the subtransient current in each alternator.

Calculation Procedure:

1. Select the Base Quantities for per-Unit Calculations

The selection of base quantities is arbitrary, but usually these quantities are selected to minimize the number of conversions from one set of base quantities to another. In this example, the following base quantities are chosen: 50,000 kVA or 50 MVA, 13.2 kV (low-voltage side), and 138 kV (high-voltage side).

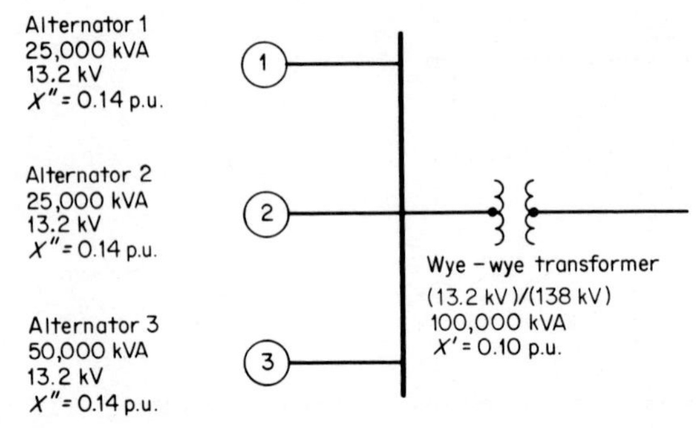

Fig. 5 Paralleled alternators connected to a wye-wye transformer.

2. Convert per-Unit Reactances to Selected Base Quantities

Alternators 1 and 2 are equal in all respects; the per-unit reactance of each is corrected in accordance with the equation: $X_{\text{new base}}'' = X_{\text{old base}}'' \text{MVA}_{\text{new base}} \text{kV}^2_{\text{old base}} / \text{MVA}_{\text{old base}} \text{kV}^2_{\text{new base}} = (0.14)(50)(13.2)^2/(25)(13.2)^2 = 0.28$ per unit. Because both the old and new base voltages in this equation are 13.2 kV, the base voltage has no effect on the calculation in this example; this is not always the case.

The correction for the per-unit reactance of alternator 3 is done in the same manner except that in this case the per-unit reactance that is given is already on the selected base of 50,000 kVA and 13.2 kV. The correction of the per-unit reactance of the transformer is done similarly: $X_{\text{new base}} = (0.10)(50)/100 = 0.05$ per unit.

3. Calculate the Internal Voltage of the Alternators

The internal voltage of the alternators must be calculated according to the actual voltage conditions on the system, just prior to the fault. Because the alternators are unloaded (in this problem), there is no voltage drop; the internal voltage is the same as the voltage on the high-voltage side of the transformer when the transformer ratio is taken into account. $E_{\text{alt 1}} = E_{\text{alt 2}} =$ (actual high-side voltage)/(base high-side voltage) $= 132$ kV/138 kV $= 0.957$ per unit. For alternator 3 the calculation is the same, yielding 0.957 per unit for the internal generated voltage.

4. Calculate the Subtransient Current in the Short Circuit

The subtransient current in the short circuit is determined from the equation: $\mathbf{I}'' = \mathbf{E}_{\text{alt}}/\text{reactance to fault} = 0.957/(j0.07 + j0.05) = 0.957/j0.12 = -j7.98$ per unit (see Fig. 6).

5. Calculate the Voltage on the Low-Voltage Side of the Transformer

The voltage on the low-voltage side of the transformer is equal to the voltage rise from the short-circuit location (zero voltage) through the transformer reactance, $j0.05$ per unit. $\mathbf{V}_{\text{low side}} = \mathbf{I}''\mathbf{X}_{\text{trans}} = (-j7.98)(j0.5) = 0.399$ per unit.

6. Calculate Individual Alternator Currents

The general equation for individual alternator currents is $\mathbf{I}_{\text{alt}}'' =$ (voltage drop across alternator reactance)/(alternator reactance). Thus, $\mathbf{I}_{\text{alt}}''\ 1 = (0.957 - 0.399)/j0.28 = -j1.99$ per unit. $\mathbf{I}_{\text{alt}}''\ 2$ is the same as that of alternator 1. $\mathbf{I}_{\text{alt}}''\ 3 = (0.957 - 0.399)/j0.14 = -j3.99$ per unit.

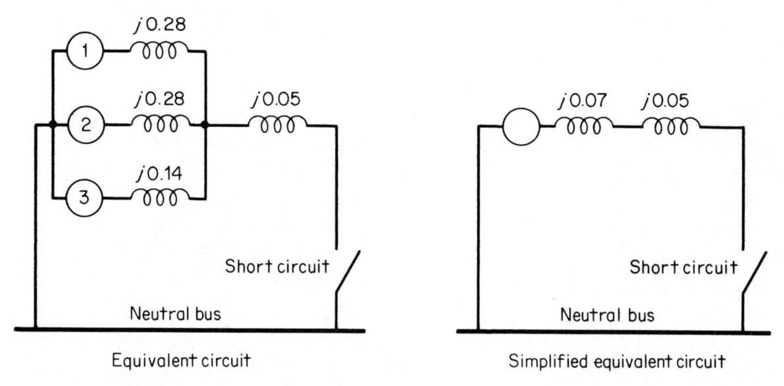

Fig. 6 Development of equivalent circuit where all reactances are on a per-unit basis.

7. *Convert per-Unit Currents to Amperes*

The base current = base kVA/($\sqrt{3}$)(base kV) = 50,000/($\sqrt{3}$)(13.2) = 2187 A. Thus, $\mathbf{I}_{alt}''$ 1 = $\mathbf{I}_{alt}''$ 2 = (2187)($-j$1.99) = 4352$\angle -90°$ A. $\mathbf{I}_{alt}''$ 3 = (2187)($-j$3.99) = 8726$\angle -90°$ A.

Related Calculations: These calculations have yielded subtransient currents because subtransient reactances were used for the alternators. For calculations of transient currents or synchronous currents, the respective reactances must be transient or synchronous, as the case may be.

SUBTRANSIENT, TRANSIENT, AND SYNCHRONOUS SHORT-CIRCUIT CURRENTS

A pumped-storage facility using hydroelectric generators is connected into a much larger 60-Hz system through a single 138-kV transmission line, as shown in Fig. 7. The transformer leakage reactances are each 0.08 per unit; the larger system is assumed to be an infinite bus, and the inductive reactance of the transmission line is 0.55 per unit. The 50,000-kVA, 13.8-kV pumped-storage generating station is taken as the base of the given per-unit values. A solid three-phase short circuit occurs on the transmission line adjacent to the sending-end circuit breaker. Before the short circuit the receiving-end bus was at 100 percent value, unity power factor, and the hydroelectric generators were 75 percent

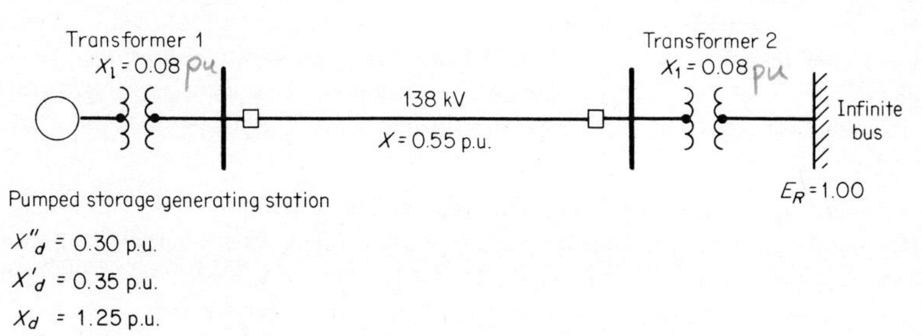

Fig. 7 One-line diagram of a generating and transmission system.

loaded, on the basis of kVA rating. Determine the subtransient, transient, and synchronous short-circuit currents.

Calculation Procedure:

1. Compute the Prefault Voltage behind the Subtransient Reactance

In this first calculation the subtransient short-circuit current is determined; the whole series of calculations will be repeated for determining the transient short-circuit current, and repeated again for determining the synchronous short-circuit current. The total impedance from the internal voltage of the hydroelectric generators to the infinite bus for the subtransient calculation is $X''_d + X_l$ of transformer 1 + X of the transmission line + X_l of transformer 2, or $0.30 + 0.08 + 0.55 + 0.08 = 1.01$ per unit. The receiving-end voltage $\mathbf{E_R} = 1.00 + j0$, and the current from the hydroelectric generators is 0.75 per unit (i.e., 75 percent loaded, on the basis of kVA rating). Thus, $\mathbf{E_{int}}'' = 1.00 + j0 + (0.75)(j1.01) = 1.00 + j0.76 = 1.26\angle 37.2°$ per-unit voltage, where $\mathbf{E_{int}}''$ is the subtransient internal voltage.

2. Compute the Subtransient Current to the Fault

The impedance from the internal voltage of the hydroelectric generators to the fault location $= X_{gf} = X''_d + X_{l1} = 0.30 + 0.08 = 0.38$ per unit, where X_{gf} is the generator-to-fault impedance. The subtransient current from the hydroelectric generators to the fault $= E_{int}''/X_{gf} = 1.26/0.38 = 3.32$ per unit. The subtransient current from the infinite bus to the fault $= E_R/X_{bf} = 1.00/(0.08 + 0.55) = 1.00/0.63 = 1.59$ per unit, where X_{bf} is the impedance from the infinite bus to the fault. The total subtransient short-circuit current at the fault $=$ 3.32 from the hydroelectric generators + 1.59 from the infinite bus $=$ 4.91 per unit.

3. Compute the Effect of the Maximum DC Component Offset

The maximum possible dc offset is taken to be $\sqrt{2}$ times the symmetrical wave, and the value of the total offset wave is the short-circuit current $I_{sc} = \sqrt{I_n^2 + I_w^2}$, where I_n is the current with dc offset neglected and I_w is the current with dc offset. From the hydroelectric generators, current (with maximum dc component) is $3.32\sqrt{2}$, and from the infinite bus, current is $1.59\sqrt{2}$, for a total of $4.91\sqrt{2} = 6.94$ per unit. The greatest rms value of $I_{sc} = \sqrt{4.91^2 + 6.94^2} = 8.5$ per unit.

4. Convert the per-Unit Current to Amperes

The base current $= \sqrt{3}(50{,}000)/138 = 627.6$ A. Therefore, $I_{sc} = (8.5$ per unit$)(627.6) = 5335$ A, the subtransient current at the fault.

5. Compute the Prefault Voltage behind the Transient Reactance

Use the procedure as for the subtransient case; the total impedance from the internal voltage of the hydroelectric generators to the infinite bus for the transient calculation is $X'_d + X_l$ of transformer 1 + X of the transmission + X_l of transformer 2, or $0.35 + 0.08 + 0.55 + 0.08 = 1.06$ per unit. The transient internal voltage of the hydroelectric generators $\mathbf{E}'_{int} = 1.00 + j0 + 0.75(j1.06) = 1.00 + j0.80 = 1.28\angle 38.7°$ per-unit voltage.

6. Calculate the Transient Current to the Fault

The total transient reactance from the internal voltage of the hydroelectric generators to the fault location, $X'_{gf} = X'_d + X_l$ of transformer 1 $= 0.35 + 0.08 = 0.43$ per unit. The transient current from the hydroelectric generators to the fault is $I'_g = E'_{int}/X'_{gf} = 1.28/0.43 = 2.98$ per unit (only the magnitude is considered, not the angle).

The transient current from the infinite bus to the fault, $I'_b = E_R/X'_{bf} = 1.00/(0.08 + 0.55) = 1.00/0.63 = 1.59$ per unit. Thus, the total transient short-circuit current at the fault location $I'_t = 2.98$ (from the hydroelectric generators) + 1.59 (from the infinite bus) = 4.57 per unit.

7. *Convert the per-Unit Current to Amperes*

As before, the base current is 627.6 A; the transient current at the fault location is $I_{sc} = (4.57 \text{ per unit})(627.6 \text{ A}) = 2868$ A.

8. *Calculate the Synchronous Current*

Use the same procedure as for subtransient and transient current but substitute the synchronous reactance of the generator rather than the subtransient or transient reactance. $\mathbf{E_{int}} = 1.00 + j0 + 0.75(j1.96) = 1.00 + j1.47 = 1.78\underline{/55.8°}$ per-unit voltage. $X_{gf} = X_d + X_{t1} = 1.25 + 0.08 = 1.33$ per unit. $I_g = E_{int}/X_{gf} = 1.78/1.33 = 1.34$ per unit (only the magnitude is considered, not the angle). $I_b = E_R/X_{bf} = 1.00(0.08 + 0.55) = 1.00/0.63 = 1.59$ per unit. The total synchronous short-circuit current at the fault location $I_t = 1.33$ (from the hydroelectric generators) + 1.59 (from the infinite bus) = 2.92 per unit. Thus, as before, $I_t = (2.92 \text{ per unit})(627.6 \text{ A}) = 1833$ A.

Related Calculations: For all situations, the calculations for subtransient, transient, and synchronous currents are done with the respective reactances of the generator. The subtransient reactance is the smallest of the three and yields the largest short-circuit current. In addition, the dc offset is used for the subtransient condition, usually considered to be within the first three cycles. Synchronous conditions may be considered to prevail after the first 60 cycles (1 s).

POWER IN UNBALANCED THREE-PHASE CIRCUITS

A balanced three-phase distribution system has 240 V between phases; a 20-Ω resistive load is connected from phase b to phase c; phase a is open. Using symmetrical components, calculate the power delivered to the resistor. Actually, the resistor represents an unbalanced load on a balanced three-phase system. See Fig. 8.

Calculation Procedure:

1. *Determine the Voltages at the Load*

In order to determine the voltages of the system at the point of load, let the voltage between phases b and c be the reference, and assume a sequence of rotation of a, b, c. By inspection write $\mathbf{E_{cb}} = 240\underline{/0°}$, $\mathbf{E_{ba}} = 240\underline{/120°}$, $\mathbf{E_{ac}} = 240\underline{/240°}$, or $\mathbf{E_a} = (240/\sqrt{3})\underline{/90°}$, $\mathbf{E_b} = (240/\sqrt{3})\underline{/-30°}$, and $\mathbf{E_c} = (240/\sqrt{3})\underline{/210°}$, where the phase voltage $240/\sqrt{3} = 138.6$ V.

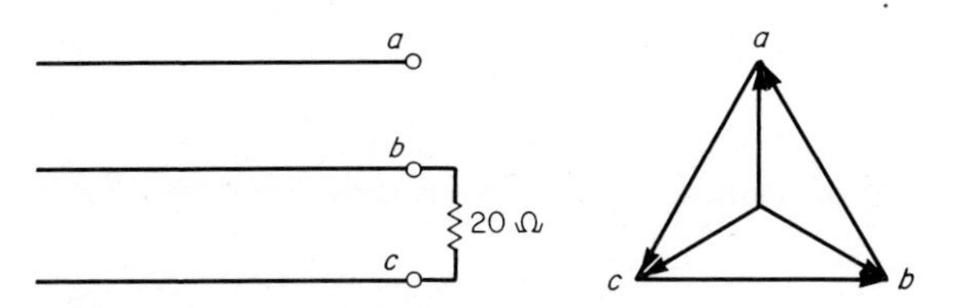

Fig. 8 Unbalanced resistor load on a balanced voltage system.

2. Determine the Corresponding Branch Currents

$\mathbf{I}_{cb} = \mathbf{E}_{cb}/\mathbf{Z} = (240\angle 0°)/(20\angle 0°) = 12\angle 0°$ A, and both $\mathbf{I}_{ba}$ and $\mathbf{I}_{ac} = 0$, because of the open circuit from phase a to phase b and from phase a to phase c.

3. Prepare to Calculate the Power Delivered

Assuming that in an unbalanced three-phase circuit the total power is the sum of the powers represented by the separate phase-sequence-component products, use the relation for total power $P_t = 3E_{a1}I_{a1} \cos \theta_1 + 3E_{a2}I_{a2} \cos \theta_2 + 3E_{a0}I_{a0} \cos \theta_0$. In this problem the positive-sequence voltage is the only voltage to be considered because the input voltages are balanced; negative- and zero-sequence voltages are not present. Consequently, to solve the power equation it is not necessary to calculate I_{a2} or I_{a0} because those terms will go to zero; only I_{a1} need be calculated. The positive-sequence component of the voltage, thus, is the same as the phase voltage, $E_a = E_{a1} = 240/\sqrt{3} = 138.6\angle 90°$. Before proceeding further, calculate the positive sequence component of the current.

4. Calculate the Positive Sequence Component of Current

Use the equation: $\mathbf{I}_{a1} = (\mathbf{I}_a + a\mathbf{I}_b + a^2\mathbf{I}_c)/3$, where in this problem $\mathbf{I}_a = 0$ and $\mathbf{I}_b = -\mathbf{I}_c = 12\angle 0°$. $\mathbf{I}_{a1} = (0 + 12\angle 120° + 12\angle 180° + 240°)/3 = (12\angle 120° + 12\angle 60°)/3 = (-6 + j10.4 + 6 + j10.4)/3 = j20.8/3 = j6.93 = 6.93\angle 90°$ A.

5. Calculate the Power Delivered

Use the equation: $P_t = 3E_{a1}I_{a1} \cos \theta = (3)(138.6)(6.93) \cos (90° - 90°) = 2880$ W.

Related Calculations: It should be recognized that the answer to this problem is obtained immediately from the expression for power in a resistance, $P = E^2/R = 240^2/20 = 2880$ W. However, the more powerful concept of symmetrical components is used in this case to demonstrate the procedure for more complex situations. The general solution shown is most appropriate for cases where there are not only positive-sequence components, but also negative- and/or zero-sequence components. It is important to recognize that the total power is the sum of the powers represented by the separate phase-sequence-component products.

DETERMINATION OF PHASE-SEQUENCE COMPONENTS

A set of unbalanced line currents in a three-phase, four-wire system is as follows: $\mathbf{I}_a = -j12$, $\mathbf{I}_b = -16 + j10$, and $\mathbf{I}_c = 14$. Find the positive-, negative-, and zero-sequence components of the current.

Calculation Procedure:

1. Change Cartesian Form of Currents to Polar Form

Use the standard trigonometric functions (sin, cos, and tan) to convert the given line currents into polar form: $\mathbf{I}_a = -j12 = 12\angle -90°$, $\mathbf{I}_b = -16 + j10 = 18.9\angle 148°$, $\mathbf{I}_c = 14\angle 0°$.

2. Calculate Positive-Sequence Components

Use the equation for positive-sequence components of current: $\mathbf{I}_{a1} = (\mathbf{I}_a + a\mathbf{I}_b + a^2\mathbf{I}_c)/3 = (12\angle -90° + 18.9\angle 148° + 120° + 14\angle 240°)/3 = (0 - j12 - 0.66 - j18.89 - 7.0 - j12.12)/3 = (-7.66 - j43.01)/3 = 14.56\angle 258.9°$. Thus, the a-phase

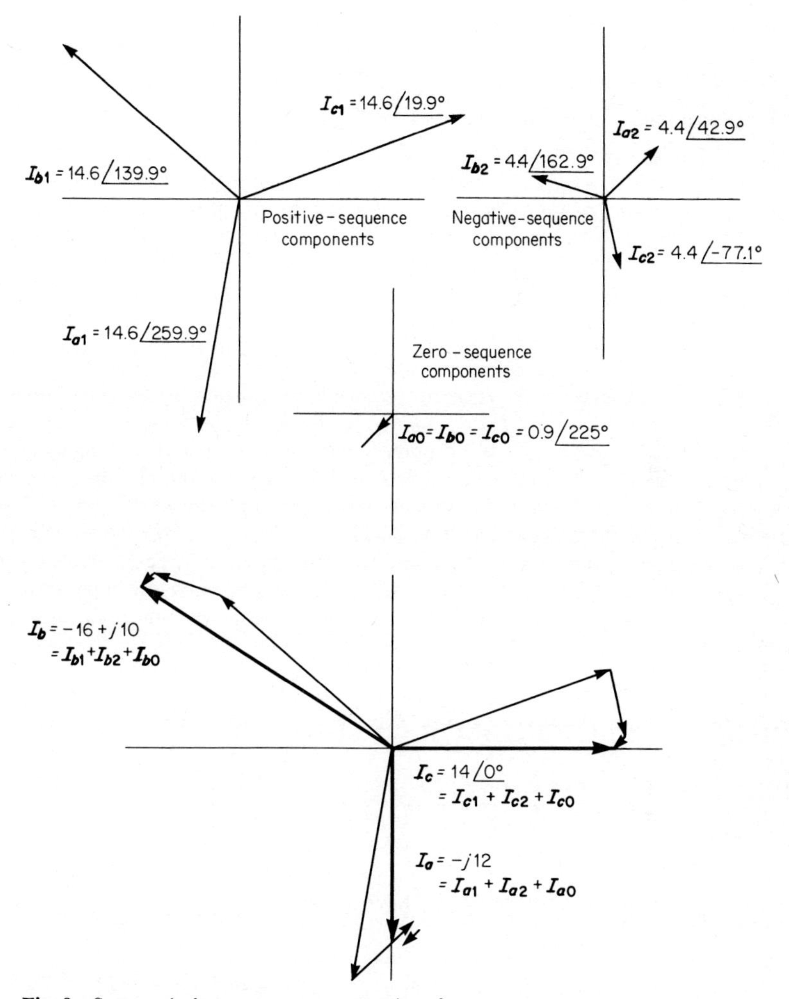

Fig. 9 Symmetrical component representation of currents.

positive-sequence component $\mathbf{I}_{a1} = 14.56\underline{/259.9°}$, the b-phase positive-sequence component $\mathbf{I}_{b1} = 14.56\underline{/259.9° - 120°} = 14.56\underline{/139.9°}$, and the c-phase positive-sequence component $\mathbf{I}_{c1} = 14.56\underline{/259.9° + 120°} = 14.56\underline{/19.9°}$. See Fig. 9.

3. *Calculate Negative-Sequence Components*

Use the equation for negative-sequence components of current: $\mathbf{I}_{a2} = (\mathbf{I}_a + a^2\mathbf{I}_b + a\mathbf{I}_c)/3 = (12\underline{/-90°} + 18.9\underline{/148° + 240°} + 14\underline{/120°})/3 = 4.41\underline{/42.9°}$. Thus, the a-phase negative sequence component $\mathbf{I}_{a2} = 4.41\underline{/42.9°}$, the b-phase negative sequence component $\mathbf{I}_{b2} = 4.41\underline{/42.9° + 120°} = 4.41\underline{/162.9°}$, and the c-phase negative sequence component $\mathbf{I}_{c2} = 4.41\underline{/42.9° + 240°} = 4.41\underline{/-77.1°}$. See Fig. 9.

4. Calculate the Zero-Sequence Components

Use the equation for zero-sequence components of current: $\mathbf{I}_{a0} = (\mathbf{I}_a + \mathbf{I}_b + \mathbf{I}_c)/3 = (-j12 - 16 + j10 + 14)/3 = 0.94\angle 225°$. The zero-sequence component for each phase is the same: $\mathbf{I}_{a0} = \mathbf{I}_{b0} = \mathbf{I}_{c0} = 0.94\angle 225°$. See Fig. 9.

5. Calculate the Phase Currents

This is merely a final check, and demonstrates the procedure for determining the phase currents if the sequence components are known; the calculation may be done graphically as shown in Fig. 9, or mathematically. The mathematical solution follows. The a-phase current $\mathbf{I}_a = \mathbf{I}_{a0} + \mathbf{I}_{a1} + \mathbf{I}_{a2} = 0.94\angle 225° + 14.56\angle 259.9° + 4.41\angle 42.9° = -j12$. The b-phase current $\mathbf{I}_b = \mathbf{I}_{a0} + a^2\mathbf{I}_{a1} + a\mathbf{I}_{a2} = \mathbf{I}_{b0} + \mathbf{I}_{b1} + \mathbf{I}_{b2} = 0.94\angle 225° + 14.56\angle 139.9° + 4.41\angle 162.9° = -16 + j10$. The c-phase current $\mathbf{I}_c = \mathbf{I}_{a0} + a\mathbf{I}_{a1} + a^2\mathbf{I}_{a2} = \mathbf{I}_{c0} + \mathbf{I}_{c1} + \mathbf{I}_{c2} = 0.94\angle 225° + 14.56\angle 19.9° + 4.41\angle -77.1° = 14 + j0$.

Related Calculations: The procedure shown here is applicable for determining the nine symmetrical components, $\mathbf{I}_{a0}$, $\mathbf{I}_{b0}$, $\mathbf{I}_{c0}$, $\mathbf{I}_{a1}$, $\mathbf{I}_{b1}$, $\mathbf{I}_{c1}$, $\mathbf{I}_{a2}$, $\mathbf{I}_{b2}$, and $\mathbf{I}_{c2}$, when one is given unbalanced phase currents. By substitution of voltages for currents, the same equation forms may be used for determining the nine symmetrical components of unbalanced phase voltages. On the other hand, if the symmetrical components are known, the equations used in Step 5 give the unbalanced phase voltages. In many instances, one or more of the negative- or zero-sequence components may not exist (that is, may equal zero). For perfectly balanced three-phase systems with balanced voltages, currents, impedances, and loads, there will be no negative- and zero-sequence components; only positive-sequence components will exist.

PROPERTIES OF PHASOR OPERATORS j AND a

Determine the value of each of the following relations and express the answer in polar form: ja, $1 + a + a^2$, $a + a^2$, $a^2 + ja + ja^2 3$.

Calculation Procedure:

1. Determine the Value of the Operators

A phasor on which a operates is rotated 120° counterclockwise (positive or forward direction), and there is no change in magnitude. See Fig. 10. A phasor on which j operates is rotated by 90° counterclockwise. Although $-j$ signifies rotation of $-90°$ because that position is exactly 180° out of phase with $+j$, the similar situation is not true for $-a$ as compared with $+a$. If $+a = 1\angle 120°$, the position of 180° reversal is at $-60°$; therefore, $-a = 1\angle -60°$.

2. Determine the Value of ja

Use the relations: $j = 1\angle 90°$ and $a = 1\angle 120°$. Thus, $ja = 1\angle 210°$.

3. Determine the Value of $1 + a + a^2$

This is a very common expression occurring with symmetrical components. It represents three balanced phasors of magnitude 1, each displaced 120° from the other. The sum is equal to $1\angle 0° + 1\angle 120° + 1\angle 240° = 0$.

4. Determine the Value of $a + a^2$

Use the relations: $a = 1\angle 120°$ and $a^2 = 1\angle 240°$. Change each to cartesian form: $a = 1\angle 120° = -0.5 + j0.866$, $a^2 = 1\angle 240° = -0.5 - j0.866$. Therefore, $a + a^2$

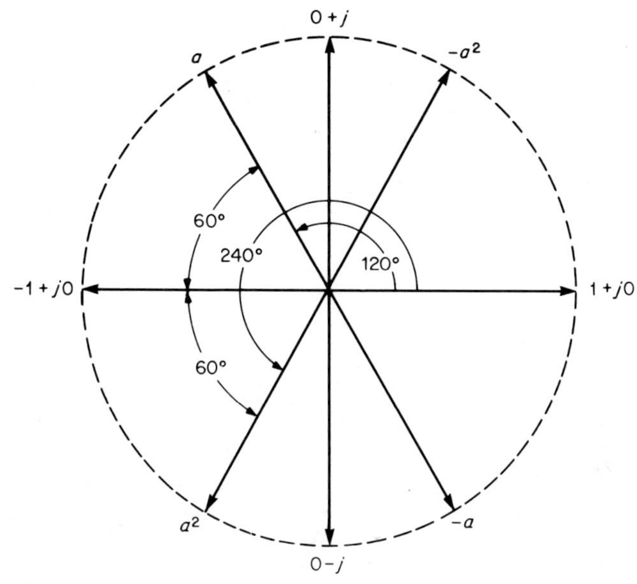

Fig. 10 Properties of phasor operator a.

$= -0.5 + j0.866 - 0.5 - j0.866 = -1.0 + j0 = 1\angle 180° = (1\angle 90°)(1\angle 90°) = j^2$.

5. Determine the Value of $a^2 + ja + ja^2 3$

Use the relation: $a^2 = 1\angle 240° = -0.5 - j0.866$, $ja = (1\angle 90°)(1\angle 120°) = 1\angle 210° = -0.866 - j0.5$, and $ja^2 3 = (1\angle 90°)(1\angle 240°)(3) = 3\angle 330° = 2.6 - j1.5$. The sum is $-0.5 - j0.866 - 0.866 - j0.5 + 2.6 - j1.5 = 1.234 - j2.866 = 3.12\angle -66.7°$. See Fig. 11.

Related Calculations: The j operator is very common in all power calculations; the a operator is used in all calculations involving symmetrical components.

COMPLEX POWER CALCULATED WITH SYMMETRICAL COMPONENTS

The resolution of a set of three-phase unbalanced voltages into symmetrical components yields the following: $\mathbf{V_{a1}} = 150\angle 0°$ V, $\mathbf{V_{a2}} = 75\angle 30°$ V, $\mathbf{V_{a0}} = 10\angle -20°$ V. The component currents are $\mathbf{I_{a1}} = 12\angle 18°$ A, $\mathbf{I_{a2}} = 6\angle 30°$ A, and $\mathbf{I_{a0}} = 12\angle 200°$ A. Determine the complex power represented by these voltages and currents.

Calculation Procedure:

1. Calculate Zero-Sequence Complex Power

Use the relation: zero-sequence complex power $\mathbf{S_0} = P_0 + jQ_0 = 3V_{a0}I^*_{a0}$, where the asterisk signifies the conjugate of the quantity (i.e., if $I_{a0} = 12\angle 200°$, the conjugate of the quantity $\mathbf{I^*_{a0}} = 12\angle -200°$). The zero-sequence complex power is $3\mathbf{V_{a0}}\mathbf{I^*_{a0}} = (3)(10\angle -20°)(12\angle -200°) = 360\angle -220° = (-275.8 + j231.4)$ VA $= -275.8$ W $+ j231.4$ vars.

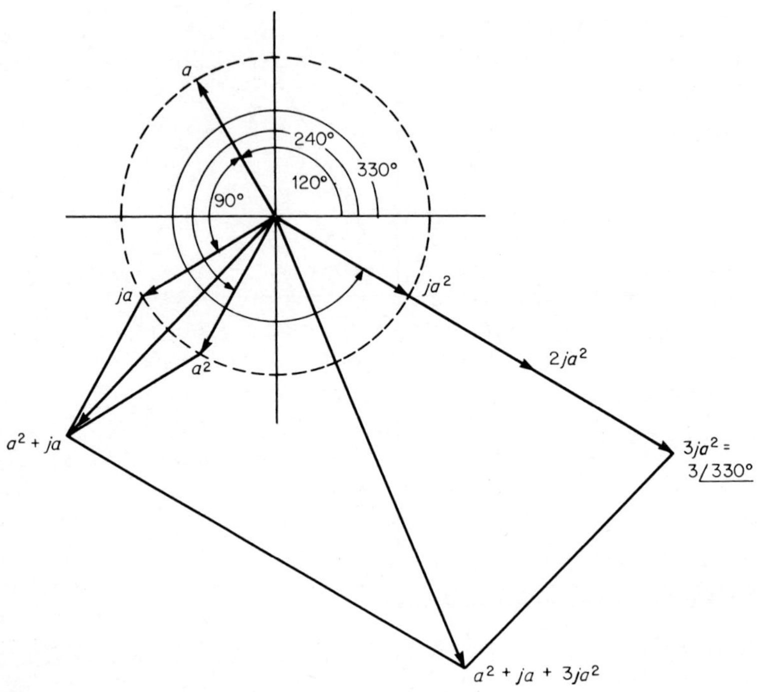

Fig. 11 Phasor representation of $a^2 + ja + 3ja^2$.

2. *Calculate Positive-Sequence Complex Power*

Use the relation: positive-sequence complex power $\mathbf{S}_1 = P_1 + jQ_1 = 3\mathbf{V}_{a1}\mathbf{I}^*_{a1} = (3)(150\angle 0°)(12\angle -18°) = 5400\angle -18° = 5135.7 - j1668.7$ VA $= 5135.7$ W $- j1688.7$ vars.

3. *Calculate Negative-Sequence Complex Power*

Use the relation: negative-sequence complex power $\mathbf{S}_2 = P_2 + jQ_2 = 3\mathbf{V}_{a2}\mathbf{I}^*_{a2} = (3)(75\angle 30°)(6\angle -30°) = 1350\angle 0°$ VA $= 1350$ W $+ j0$ vars.

4. *Calculate Total Complex Power*

Use the relation: $\mathbf{S}_t = P_t + jQ_t = 3\mathbf{V}_{a0}\mathbf{I}^*_{a0} + 3\mathbf{V}_{a1}\mathbf{I}^*_{a1} + 3\mathbf{V}_{a2}\mathbf{I}^*_{a2} = -275.8 + 5135.7 + 1350$ W $+ j(231.4 - 1668.7)$ vars $= 6209.9$ W $- j1437.3$ vars.

5. *Alternative Solution: Compute the Phase Voltage*

As an alternative solution and as a check, compute the phase voltages and currents: $\mathbf{V}_a = \mathbf{V}_{a0} + \mathbf{V}_{a1} + \mathbf{V}_{a2} = 10\angle -20° + 150\angle 0° + 75\angle 30° = 224.4 + j34.1 = 226.9\angle 8.6°$. In this calculation the intervening mathematical steps, wherein the polar form of the phasors are converted to cartesian form, are not shown. Similarly, $\mathbf{V}_b = \mathbf{V}_{b0} + \mathbf{V}_{b1} + \mathbf{V}_{b2} = \mathbf{V}_{a0} + a^2\mathbf{V}_{a1} + a\mathbf{V}_{a2} = 10\angle -20° + 150\angle 240° + 75\angle 150° = 161.9\angle 216.3°$. $\mathbf{V}_c = \mathbf{V}_{c0} + \mathbf{V}_{c1} + \mathbf{V}_{c2} = \mathbf{V}_{a0} + a\mathbf{V}_{a1} + a^2\mathbf{V}_{a2} = 10\angle -20° + 150\angle 120° + 75\angle 270° = 83.4\angle 141.9°$.

6. Compute the Phase Currents

Use the same relations as for phase voltages: $\mathbf{I_a} = \mathbf{I_{a0}} + \mathbf{I_{a1}} + \mathbf{I_{a2}} = 12\angle 200^\circ + 12\angle 18^\circ + 6\angle 30^\circ = 5.9\angle 26.0$ and $\mathbf{I_a^*} = 5.9\angle -26.0^\circ$. $\mathbf{I_b} = \mathbf{I_{b0}} + \mathbf{I_{b1}} + \mathbf{I_{b2}} = \mathbf{I_{a0}} + a^2\mathbf{I_{a1}} + a\mathbf{I_{a2}} = 12\angle 200^\circ + 12\angle 258^\circ + 6\angle 150^\circ = 22.9\angle 214.1^\circ$ and $\mathbf{I_b^*} = 22.9\angle -214.1^\circ$. $\mathbf{I_c} = \mathbf{I_{c0}} + \mathbf{I_{c1}} + \mathbf{I_{c2}} = \mathbf{I_{a0}} + a\mathbf{I_{a1}} + a^2\mathbf{I_{a2}} = 12\angle 200^\circ + 12\angle 138^\circ + 6\angle 270^\circ = 20.3\angle 185.9^\circ$ and $\mathbf{I_c^*} = 20.3\angle -185.9^\circ$.

7. Compute the Complex Power

Use the relation: $\mathbf{S_t} = \mathbf{V_a I_a^*} + \mathbf{V_b I_b^*} + \mathbf{V_c I_c^*} = (226.9\angle 8.6^\circ)(5.9\angle 26.0^\circ) + (161.9\angle 216.3^\circ)(22.9\angle -214.1^\circ) + (83.4\angle 141.9^\circ)(20.3\angle -185.9^\circ)$. Completing the mathematical solution of this equation will yield the same result (6210 W $- j$1436 vars) as did the solution using the symmetrical components.

Related Calculations: This problem illustrates two methods of finding the complex power $P + jQ$; namely, (1) by symmetrical components and (2) by unbalanced phase components. In either case, the complex power is obtained by summation of the products of the respective phasor voltages by the conjugate phasor currents.

IMPEDANCES AND REACTANCES TO DIFFERENT SEQUENCES

A salient-pole generator is connected to a system having a reactance of 9.0 per unit, as shown in Fig. 12; the system base values are 15,000 kVA and 13.2 V. Draw the positive-, negative-, and zero-sequence diagrams for a three-phase short circuit at the load.

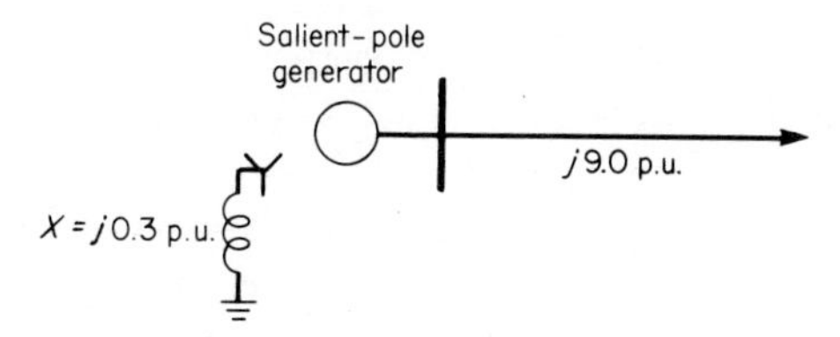

Fig. 12 Representation of salient-pole generator connected to a system.

Calculation Procedure:

1. Assign Impedance and/or Reactance Values to the Generator

When impedance and/or reactance values are not given for parts of a system, it is necessary to make estimates. The literature of the power industry contains extensive listings of typical values of reactance; in most cases the resistance is neglected and only the reactance is used. Typical values for generator reactance are: subtransient, $X''_d = 0.10$ per unit; transient, $X'_d = 0.20$ per unit; synchronous, $X_d = 1.20$ per unit (each of these being positive-sequence values).

The negative-sequence reactance can vary between 0.10 per unit for large two-pole turbine generators to 0.50 for salient-pole machines; the zero-sequence reactance can vary similarly from 0.03 to 0.20 per unit.

2. Draw the Positive-Sequence Diagram

The generated voltage, being of positive sequence a, b, c, is shown in the positive-sequence diagram only. The grounding reactance of the generator does not appear in the positive-sequence diagram. See Fig. 13.

3. Draw the Negative-Sequence Diagram

No generated voltage appears in the negative-sequence diagram; neither is there shown the reactance of the grounding device. See Fig. 13.

4. Draw the Zero-Sequence Diagram

No generated voltage appears in the zero-sequence diagram, but it should be noted that the reactance of the grounding device is multiplied by three.

Related Calculations: If component reactances are not known, it is possible to make estimates of these values in the per-unit system, on the basis of typical values found in handbook tables and manufacturers' literature. For short-circuit calculations, estimated values of reactance will give satisfactory results.

LINE-TO-LINE SHORT-CIRCUIT CALCULATIONS

An ungrounded wye-connected generator having a subtransient reactance $X''_d = 0.12$ per unit, a negative-sequence reactance $X_2 = 0.15$ per unit, and a zero-sequence reactance $X_0 = 0.05$ per unit is faulted at its terminals with a line-to-line short circuit. Determine (1) the line-to-line subtransient short-circuit current and (2) the ratio of that current with respect to three-phase short-circuit current. The generator is rated 10 MW, 13.8 kV, and operates at 60 Hz.

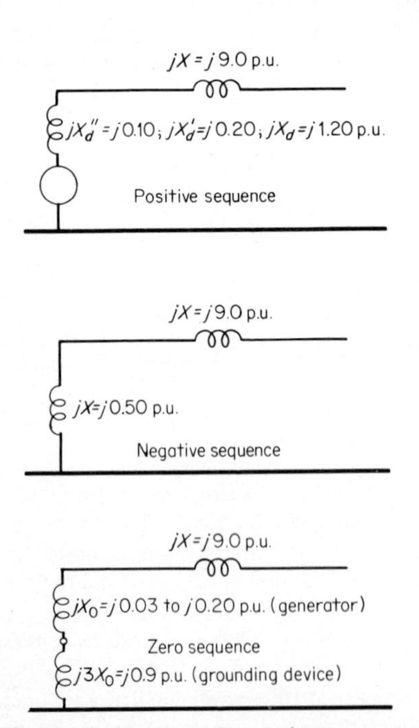

Fig. 13 Positive-, negative-, and zero-sequence diagrams; approximate and typical values of generator reactances are included.

Calculation Procedure:

1. Draw the Sequence Networks

It is necessary to draw only the positive- and negative-sequence diagrams because zero-sequence currents are not involved with line-to-line short circuits. See Fig. 14.

2. Connect the Sequence Networks for Line-to-Line Fault

For a line-to-line fault the positive and negative sequence networks are connected in parallel as shown in Fig. 15.

3. Calculate Line-to-Line Fault Current

Use the equation for phase-a positive-sequence current: $\mathbf{I}_{a1} = \mathbf{E}/\mathbf{Z} = 1\underline{/0°}/$

$(j0.12 + j0.15) = 1\underline{/0^\circ}/j0.27 = -j3.70$ per unit. The phase-a negative-sequence current $\mathbf{I}_{a2} = -\mathbf{I}_{a1} = +j3.70$ per unit. The phase-a fault current $\mathbf{I}_a = \mathbf{I}_{a0} + \mathbf{I}_{a1} + \mathbf{I}_{a2} = 0 - j3.70 + j3.70 = 0$. The phase-b fault current $\mathbf{I}_b = \mathbf{I}_{a0} + a^2\mathbf{I}_{a1} + a\mathbf{I}_{a2} = 0 + 3.70\underline{/-90^\circ + 240^\circ} + 3.70\underline{/90^\circ + 120^\circ} = -3.20 + j1.85 - 3.20 - j1.85 = -6.40$ per unit. Similarly, the phase-c fault current $\mathbf{I}_c = \mathbf{I}_{a0} + a\mathbf{I}_{a1} + a^2\mathbf{I}_{a2} = 0 + 3.70\underline{/-90^\circ + 120^\circ} + 3.7\underline{/90^\circ + 240^\circ} = 6.40$ per unit. Thus, it is calculated that $\mathbf{I}_b = -\mathbf{I}_c = -6.40$ per unit.

4. *Convert the per-Unit Current to Amperes*

First determine the base current using the equation: power $= \sqrt{3}V_{\text{line}}I_{\text{line}} \cos\theta$, or $I_{\text{line}} = \text{power}/\sqrt{3}V_{\text{line}} \cos\theta$. Power/$\cos\theta$ = volt-amperes. Thus, volt-amperes/$\sqrt{3}V_{\text{line}}$ = 10,000 kVA/$(\sqrt{3})$(13.8 kV) = 418.4 A (base current). Therefore, the fault-current magnitude in phase a and phase b = (418.4 A)(6.40 per unit) = 2678 A.

5. *Calculate the Three-Phase Short-Circuit Current*

For the three-phase case, only the positive-sequence network diagram is used. $\mathbf{I}_a = \mathbf{E}/\mathbf{Z} = 1\underline{/0^\circ}/j0.12 = -j8.33$ per unit. Converting this to amperes yields (magnitude only) I_a = (8.33 per unit)(418.4 A) = 3485 A.

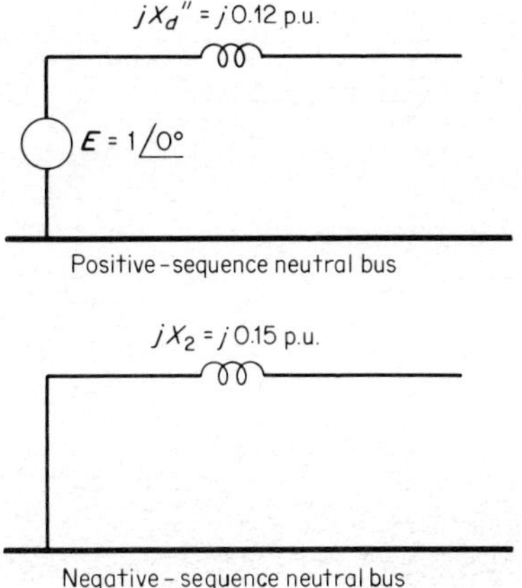

Fig. 14 Positive- and negative-sequence network diagrams.

6. *Calculate Ratio of Short-Circuit Currents*

The ratio of line-to-line short-circuit current with respect to three-phase short-circuit (magnitudes only) is 2678 A/3485 A = 0.768. The same calculation may be done with the per-unit values, namely, 6.40 per unit/8.33 per unit = 0.768.

Related Calculations: In order to calculate the line-to-line short-circuit currents it is necessary to establish the positive- and negative-sequence network diagrams; these two networks are connected in parallel and the calculation of sequence components of current proceeds from that point. It matters not how extensive the networks, as long as each network may be reduced to its simplest form. In the usual case with a number of generators, the generated voltages are paralleled in the positive-sequence network.

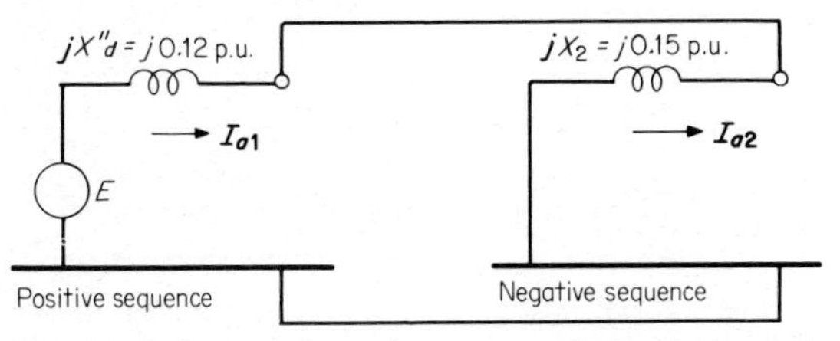

Fig. 15 Positive- and negative-sequence diagrams connected for a line-to-line fault.

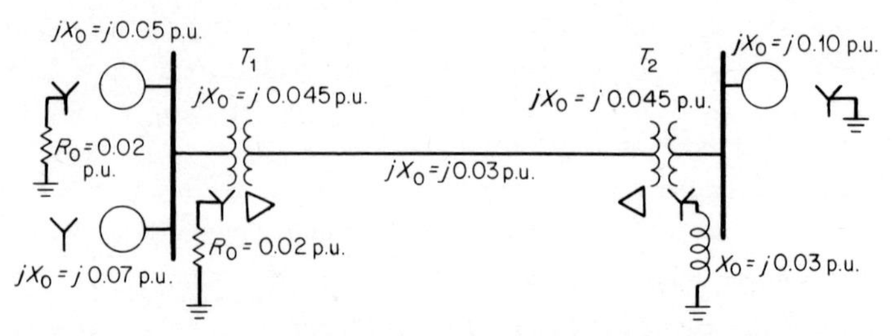

Fig. 16 Zero-sequence quantities shown for a three-phase system.

IMPEDANCE TO ZERO SEQUENCE FOR GENERATORS, TRANSFORMERS, AND TRANSMISSION LINES

A balanced three-phase system is shown in Fig. 16. Draw the zero-sequence network.

Calculation Procedure:

1. Determine the Zero-Sequence Treatment of Transformers

The zero-sequence equivalent circuit for transformers is shown in Fig. 17. It should be noted that zero-sequence current cannot flow in the secondary of the transformer if it does not flow in the primary (provided the transformer itself is not at fault).

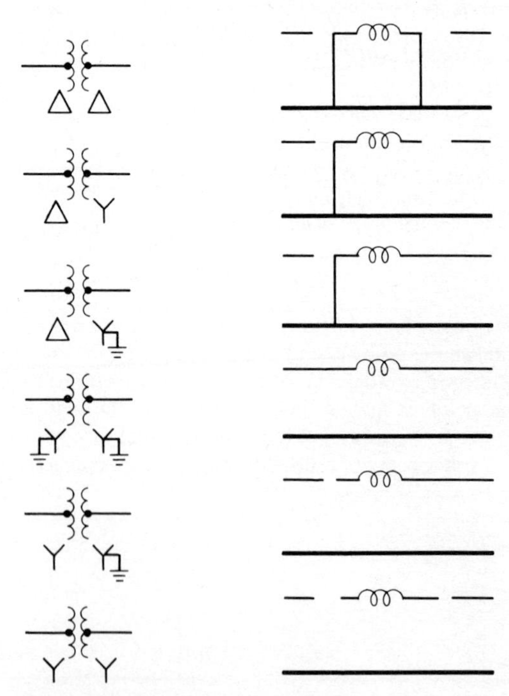

Fig. 17 Zero-sequence equivalent circuits for transformers.

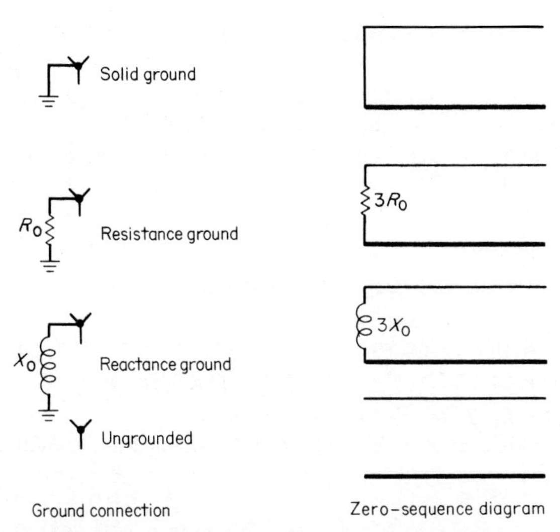

Fig. 18 Zero-sequence diagrams for ground connections.

2. *Determine Zero-Sequence Treatment of Grounding Devices*

Use the relation that the reactance of grounding devices appears in the zero-sequence network diagrams at 3 times the actual value. See Fig. 18.

3. *Draw the Complete Zero-Sequence Network Diagram*

Note that since each end of the transmission line is connected to delta-connected transformers, the line is isolated from zero-sequence currents in the network diagram. See Fig. 19.

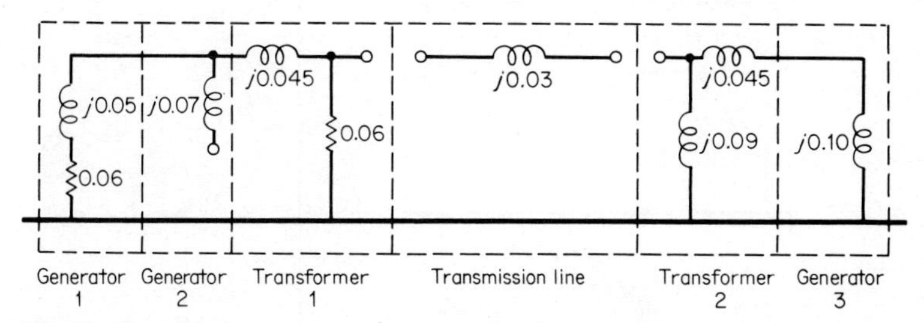

Fig. 19 Complete zero-sequence network diagram where all reactances are on a per-unit basis.

Related Calculations: This problem illustrates a number of different possibilities that exist in deriving paths for zero-sequence currents; most of the situations that occur are demonstrated. Here again, once the sequence network is drawn, it may be simplified by combining elements.

LINE-TO-GROUND SHORT-CIRCUIT CALCULATIONS

A 13.2-kV, 30,000-kVA generator has positive-, negative-, and zero-sequence reactances of 0.12, 0.12, and 0.08 per unit, respectively. The generator neutral is grounded through a reactance of 0.03 per unit. For the given reactance determine the line-to-line voltages and short-circuit currents when a single line-to-ground fault occurs at the generator terminals. It may be assumed that the generator was unloaded before the fault.

Calculation Procedure:

1. *Draw the Sequence Network Diagram*

The internal generator voltage before the fault is equal to the terminal voltage, because the generator at that time is unloaded; it is equal to $\mathbf{E_g} = 1.0 + j0$ per unit. The sequence network diagram for a single line-to-ground fault is shown in Fig. 20.

2. *Calculate the Total Series Impedance*

The total series impedance $= \mathbf{Z_1} + \mathbf{Z_2} + \mathbf{Z_0}$. In the neutral or ground connection, if it exists, the impedance in the zero-sequence network will be 3 times the actual value. Thus, $\mathbf{Z_1} + \mathbf{Z_2} + \mathbf{Z_0} = j0.12 + j0.12 + j0.08 + (3)(j0.03) = j0.41$ per unit.

3. *Calculate the Positive-, Negative-, and Zero-Sequence Components*

The positive-sequence current component $\mathbf{I_{a1}} = \mathbf{E_g}/(\mathbf{Z_1} + \mathbf{Z_2} + \mathbf{Z_0}) = (1.0 + j0)/j0.41 = -j2.44$ per unit. For the single line-to-ground fault, $\mathbf{I_{a1}} = \mathbf{I_{a2}} = \mathbf{I_{a0}}$; thus $\mathbf{I_{a2}} = -j2.44$ and $\mathbf{I_{a0}} = -j2.44$ per unit.

4. *Calculate the Base Current*

The base current $= \text{kVA}_{\text{base}}/(\sqrt{3}\text{kV}_{\text{base}}) = 30{,}000/(\sqrt{3})(13.2) = 1312$ A.

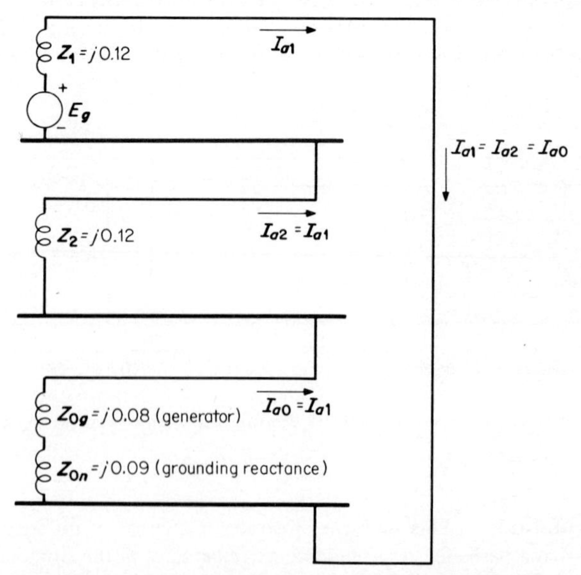

Fig. 20 Sequence network representation of a single line-to-ground fault of unloaded generator; phase a is grounded.

5. *Calculate the Phase Currents*

The phase a current = $\mathbf{I}_{a1} + \mathbf{I}_{a2} + \mathbf{I}_{a0} = 3\mathbf{I}_{a1} = (3)(-j2.44) = -j7.32$ per unit. In SI units, the current in phase a is (7.32 per unit)(1312 A) = 9604 A. When only the magnitude is of interest, the $-j$ may be neglected in the final step.

The phase b current = $a^2\mathbf{I}_{a1} + \mathbf{a}\mathbf{I}_{a2} + \mathbf{I}_{a0} = a^2\mathbf{I}_{a1} + a\mathbf{I}_{a1} + \mathbf{I}_{a1} = \mathbf{I}_{a1}(a^2 + a + 1) = 0$. Similarly, the phase c current = $a\mathbf{I}_{a1} + a^2\mathbf{I}_{a2} + \mathbf{I}_{a0} = \mathbf{a}\mathbf{I}_{a1} + a^2\mathbf{I}_{a1} + \mathbf{I}_{a1} = \mathbf{I}_{a1}(a + a^2 + 1) = 0$. Because only phase a is shorted to ground at the fault, phases b and c are open-circuited and carry no current.

6. *Calculate the Sequence Voltage Components*

With phase a as the reference point at the fault, $\mathbf{V}_{a1} = \mathbf{E}_g - \mathbf{I}_{a1}\mathbf{Z}_1 = 1.0 - (-j2.44)(j0.12) = 1.0 - 0.293 = 0.707$ per unit. $\mathbf{V}_{a2} = -I_{a2}\mathbf{Z}_2 = -(-j2.44)(j0.12) = -0.293$ per unit. $\mathbf{V}_{a0} = -\mathbf{I}_{a0}\mathbf{Z}_0 = -(-j2.44)(j0.08 + j0.09) = -0.415$ per unit.

7. *Convert the Sequence Voltage Components to Phase Voltages*

$\mathbf{V}_a = \mathbf{V}_{a1} + \mathbf{V}_{a2} + \mathbf{V}_{a0} = 0.707 - 0.293 - 0.415 = 0$. $\mathbf{V}_b = a^2\mathbf{V}_{a1} + a\mathbf{V}_{a2} + \mathbf{V}_{a0} = 0.707\underline{/240^\circ} - 0.293\underline{/120^\circ} - 0.415 = -0.622 - j0.866$ per unit. $\mathbf{V}_c = a\mathbf{V}_{a1} = a^2\mathbf{V}_{a2} + \mathbf{V}_{a0} = 0.707\underline{/120^\circ} - 0.293\underline{/240^\circ} - 0.415 = -0.622 + j0.866$ per unit.

8. *Convert the Phase Voltages to Line Voltages*

$\mathbf{V}_{ab} = \mathbf{V}_a - \mathbf{V}_b = 0 - (-0.622 - j0.866) = 0.622 + j0.866 = 1.07\underline{/54.3^\circ}$ per unit. $\mathbf{V}_{bc} = \mathbf{V}_b - \mathbf{V}_c = -0.622 - j0.866 - (-0.622 + j0.866) = -j1.732 = 1.732\underline{/270^\circ}$ per unit. $\mathbf{V}_{ca} = \mathbf{V}_c - \mathbf{V}_a = 0.622 + j0.866 = 1.07\underline{/125.7^\circ}$ per unit.

9. *Convert the Line Voltages to SI Units*

In this problem the generator voltage *per phase*, $\mathbf{E}_g$, was assumed to be 1.0 per unit. Therefore, 1.0 per-unit voltage = 13.2 kV/$\sqrt{3}$ = 7.62 kV. The line voltages in SI units become $\mathbf{V}_{ab} = (1.07\underline{/54.3^\circ})(7.62\text{ kV}) = 8.15\underline{/54.3^\circ}$ kV, $\mathbf{V}_{bc} = (1.732\underline{/270^\circ})(7.62\text{ kV}) = 13.2\underline{/270^\circ}$ kV, and $\mathbf{V}_{ca} = (1.07\underline{/125.7^\circ})(7.62\text{ kV}) = 8.15\underline{/125.7^\circ}$ kV.

10. *Draw the Voltage Phasor Diagram*

See Fig. 21.

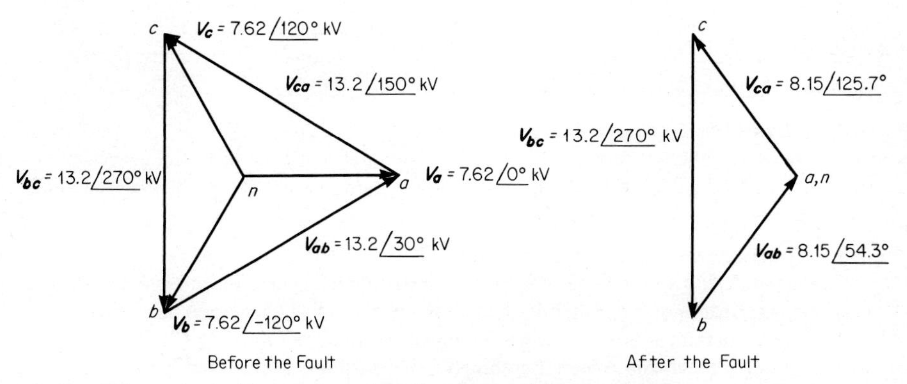

Fig. 21 Voltage phasor diagrams for a single line-to-ground fault on phase a.

Related Calculations: The procedure shown for a single line-to-ground fault is applicable to any other type of fault provided the sequence networks are connected in the proper manner. For example, for a line-to-line fault, the positive-sequence network and the negative-sequence network are connected in parallel without the zero-sequence network.

SUBTRANSIENT-CURRENT CONTRIBUTION FROM MOTORS; CIRCUIT-BREAKER SELECTION

Consider the system shown in Fig. 22, wherein a generator (20,000 kVA, 13.2 kV, $X''_d = 0.14$ per unit) is supplying two large induction motors (each being 7500 kVA, 6.9 kV, $X''_d = 0.16$ per unit). The three-phase step-down transformer is rated 20,000 kVA, 13.2/

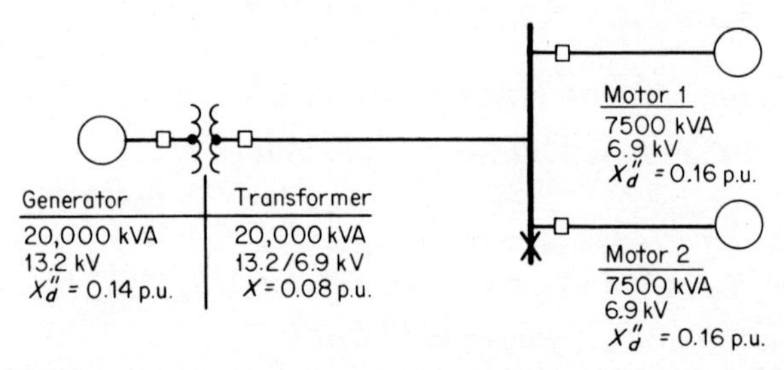

Fig. 22 One-line diagram of a generator supplying two motors.

6.9 kV, and its leakage reactance is 0.08 per unit. A three-phase short circuit occurs on the bus. Find the subtransient fault current and the symmetrical short-circuit interrupting current.

Calculation Procedure:

1. Convert Subtransient Motor Reactances to Generator Base kVA

To convert the motor reactance from a 7500-kVA base to a 20,000-kVA base, use the relation: $X''_d = (0.16)(20{,}000/7500) = 0.427$ per unit. Thus, the reactance of each motor is $j0.427$ per unit, and the combined reactance of the two motors connected in parallel is $j0.427/2 = j0.214$ per unit.

2. Draw the Network Diagram

The network diagram is shown in Fig. 23. The total reactance from the generator to the bus is $j0.14 + j0.08 = j0.22$ per unit, and the total reactance from the motors to the bus is $j0.214$ per unit.

3. Reduce the Network Diagram

The network diagram may be reduced by paralleling the generator and motor voltages (1.00 per unit each) and paralleling the reactance path of (1) generator to faulted bus, and (2) motor to faulted bus, yielding a combined reactance of $(j0.22)(j0.214)/(j0.22 + j0.214) = j0.108$ per unit. The reduced network diagram is shown in Fig. 24.

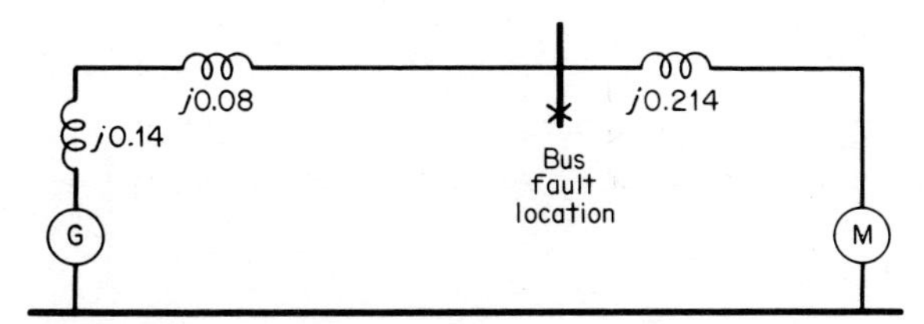

Fig. 23 Equivalent network diagram of a generator supplying a motor load.

4. *Calculate Subtransient Fault Current*

The subtransient symmetrical short-circuit current = $V/jX = 1.00/j0.108 = -j9.22$ per unit. In terms of amperes on the 6.9-kV bus, the base current = 20,000 kVA/($\sqrt{3}$)(6.9 kV) = 1674 A. Thus the subtransient fault current = $(-j9.22)(1674)$ = 15,429 A (the $-j$ is usually ignored at this point because only the magnitude is of interest).

5. *Calculate Symmetrical Short-Circuit Interrupting Current*

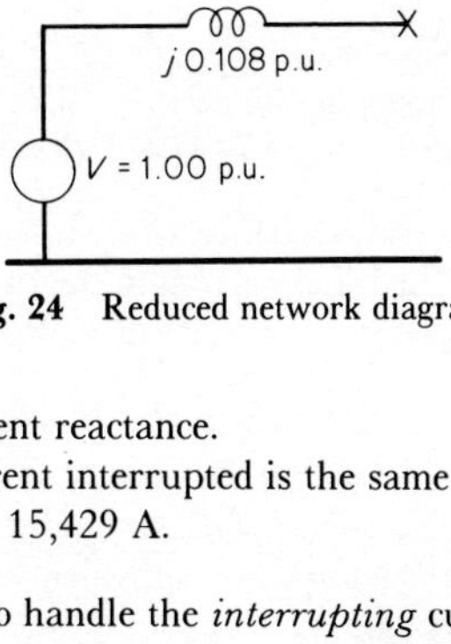

The interrupting current is related to the speed of operation of the circuit breaker, and includes the contribution of current from the motors at the time of current interruption. Were the motors of the synchronous type, the reactance that would be used in the network diagram would be 1.5 times subtransient reactance; this, in effect, represents an approximate transient reactance.

Fig. 24 Reduced network diagram.

For induction motors, the symmetrical short-circuit current interrupted is the same as the subtransient symmetrical short-circuit current, namely 15,429 A.

Related Calculations: The breaker must be selected to handle the *interrupting* current indicated by the calculation of the symmetrical short-circuit current. It is necessary to consult manufacturers' literature for selecting an appropriate breaker on the basis of the symmetrical current that can be interrupted.

INDUCTION-MOTOR INRUSH CURRENT

A three-phase, 240-V, wye-connected, 15-hp, six-pole, 60-Hz, wound-rotor induction motor has the following equivalent-circuit constants referred to the stator: $r_1 = 0.30$, $r_2 = 0.15$, $x_1 = 0.45$, $x_2 = 0.25$, and $x_\phi = 15.5\ \Omega$ per phase. At all loads the core losses, friction, and windage are 500 W. Compare the inrush (starting) current with the load current at 3 percent slip, assuming that the rotor windings are short-circuited.

Calculation Procedure:

1. *Draw the Equivalent Circuit*

The equivalent circuit is shown in Fig. 25.

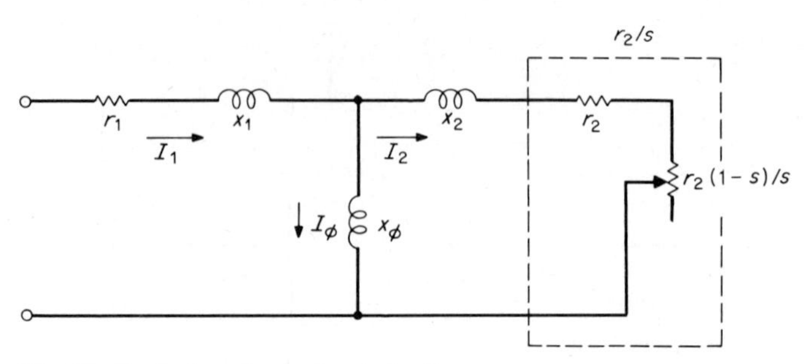

Fig. 25 Equivalent circuit of an induction motor.

2. *Calculate the Total Impedance of the Equivalent Circuit for a Slip of 3 Percent*

The secondary or rotor impedance (referred to the stator) is $r_2/s + jx_2 = 0.15/0.03 + j0.25 = 5 + j0.25\ \Omega$ per phase. This impedance in parallel with jx_ϕ is $(5 + j0.25)(j15.5)/(5 + j0.25 + j15.5) = 4.4 + j4.7\ \Omega$ per phase. The total impedance of the equivalent circuit is $4.4 + j4.7 + r_1 + jx_1 = 4.4 + j4.7 + 3.0 + j0.45 = 4.7 + j5.15 = 6.97\angle 47.62°\ \Omega$ per phase.

3. *Calculate the Stator (Input) Current for a Slip of 3 Percent*

The input current at the running condition of 3 percent slip is $I_1 = V/Z = 240\ \text{V}/(\sqrt{3})(6.97\ \Omega) = 19.88$ A per phase at a power factor of $\cos 47.62° = 0.674$.

4. *Calculate the Total Impedance of the Equivalent Circuit for a Slip of 100 Percent*

The slip at starting condition is 100 percent. This value is used to calculate the total impedance of the equivalent circuit. Thus, the secondary or rotor impedance referred to the stator is $r_2/s + jx_2 = 0.15/1.0 + j0.25 = 0.15 + j0.25\ \Omega$ per phase. This impedance in parallel with jx_ϕ is $(0.15 + j0.25)(j15.5)/(0.15 + j0.25 + j15.5) = 0.145 + j0.248\ \Omega$ per phase. The total impedance of the equivalent circuit is $0.145 + j0.248 + r_1 + jx_1 = 0.145 + j0.248 + 0.30 + j0.45 = 0.445 + j0.698 = 0.83\angle 57.48°\ \Omega$ per phase.

5. *Calculate the Starting Current*

The input current at the condition of start (slip = 100 percent) is $I_1 = V/Z_{\text{start}} = 240\ \text{V}/(\sqrt{3})(0.83\ \Omega) = 166.9$ A per phase at a power factor of $\cos 57.48° = 0.538$.

6. *Compare Starting (Inrush Current) to Running Current at 3 Percent Slip*

The ratio of starting current to running current (at 3 percent slip) is 166.9 A/19.88 A = 8.4.

Related Calculations: The procedure shown is applicable for all values of slip, and for for both squirrel-cage and wound-rotor induction motors. Although the assumption in this problem was that the rotor winding was short-circuited, other conditions may be calculated when the rotor circuit external resistance or reactance is known.

INDUCTION-MOTOR SHORT-CIRCUIT CURRENT

A 1000-hp, 2200-V, 25-Hz wye-connected 12-pole wound-rotor induction motor has a full-load efficiency of 94.5 percent and a power factor of 92 percent. Referred to the stator, the constants of the machine in ohms per phase are: $r_1 = 0.102$, $r_2 = 0.104$, $x_1 = 0.32$, $x_2 = 0.32$, $x_\phi = 16.9$. Determine the motor short-circuit current at the time of occurrence, assuming that immediately before the fault, full-load rated conditions existed.

Calculation Procedure:

1. *Draw the Equivalent Circuit Diagram*

For running condition with the rotor winding short-circuited, the circuit is as shown in Fig. 25.

2. *Calculate the Prefault Stator Current*

$I_{\text{stator}} = I_1(1000 \text{ hp})(746 \text{ W/hp})/(0.92)(0.945)(\sqrt{3})(2200 \text{ V}) = 225.2$ A. The angle equals $\cos^{-1} 0.92$, or 23.1°.

3. *Calculate the Motor Transient Reactance*

The motor transient reactance is determined by neglecting the rotor resistance and using the relation $x_1' = x_1 + x_\phi x_2/(x_\phi + x_2)$. Thus, $x_1' = 0.32 + (16.9)(0.32)/(16.9 + 0.32) = 0.634\ \Omega$ per phase.

4. *Calculate the Voltage behind the Transient Reactance*

Refer to Fig. 26. $\mathbf{E}_1' = \mathbf{V}_1 - (r_1 + jx_1')\mathbf{I}_1 = 2200/\sqrt{3} - (0.102 + j0.634)(225.2\angle{-23.1°}) = 1270.2 - (0.642\angle{80.86°})(225.2\angle{-23.1°}) = 1270.2 - 144.6\angle{57.76°} = 1270.2 - 77.2 - j122.3 = 1193 - j122.3 = 1199.3\angle{-5.85°}$.

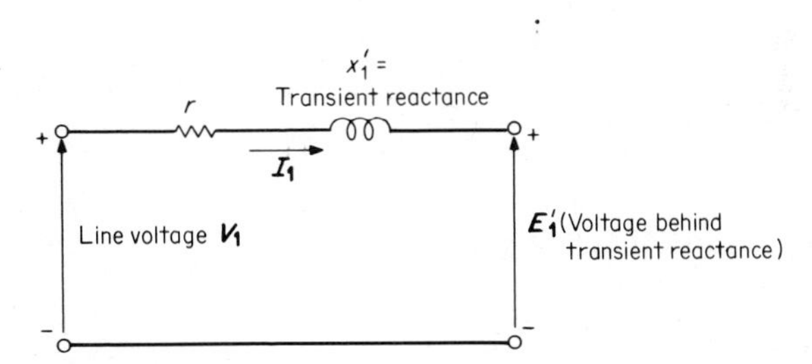

Fig. 26 Transient-reactance diagram of an induction motor.

5. *Calculate the Short-Circuit Current at the Instant of Occurrence*

The initial short-circuit current is considered to be equal to the voltage behind the transient reactance divided by the transient reactance, or $1199.3 \text{ V}/0.634\ \Omega = 1891.6$ A. This is the rms initial short-circuit current per phase from the motor.

Related Calculations: This procedure is used to calculate the initial short-circuit current as dependent upon the voltage behind a calculated transient reactance. This initial short-circuit current decays very rapidly. The calculation is based on the machine running at rated conditions with very small slip.

BUS VOLTAGES CALCULATED BY MATRIX EQUATION AND INVERSION

The three sequence components of bus voltage for a given bus are given as $V_0 = -0.105$, $V_1 = 0.953$, and $V_2 = -0.230$ per unit. Determine the three phase voltages and the three sequence components of the phase voltages.

Calculation Procedure:

1. Write the Phase-Voltage Equations

The three separate equations will be written first in order to show the relation to the matrix equation: $\mathbf{V}_a = \mathbf{V}_0 + \mathbf{V}_1 + \mathbf{V}_2 = -0.105 + 0.953 - 0.230$, $\mathbf{V}_b = \mathbf{V}_0 + a^2\mathbf{V}_1 + a\mathbf{V}_2 = -0.105 + 0.953\underline{/240°} - 0.230\underline{/120°}$, and $\mathbf{V}_c = \mathbf{V}_0 + a\mathbf{V}_1 + a^2\mathbf{V}_2 = -0.105 + 0.953\underline{/120°} - 0.230\underline{/240°}$.

2. Write the Matrix Equation

In matrix form the individual equations become

$$\begin{bmatrix} \mathbf{V}_a \\ \mathbf{V}_b \\ \mathbf{V}_c \end{bmatrix} = \begin{bmatrix} 1 & 1 & 1 \\ 1 & a^2 & a \\ 1 & a & a^2 \end{bmatrix} \begin{bmatrix} -0.105 \\ +0.953 \\ -0.230 \end{bmatrix}$$

3. Solve for $\mathbf{V}_a$

$\mathbf{V}_a = -0.105 + 0.953 - 0.230 = 0.618$.

4. Solve for $\mathbf{V}_b$

$\mathbf{V}_b = -0.105 + 0.953\underline{/240°} - 0.230\underline{/120°} = -0.4665 - j1.0243 = 1.1255\underline{/245.5°}$.

5. Solve for $\mathbf{V}_c$

$\mathbf{V}_c = -0.105 + 0.953\underline{/120°} - 0.230\underline{/240°} = -0.4665 + j1.0245 = 1.1255\underline{/114.5°}$.

6. Write the Full Matrix Equation

In matrix form the full equation becomes:

$$\begin{bmatrix} \mathbf{V}_a \\ \mathbf{V}_b \\ \mathbf{V}_c \end{bmatrix} = \begin{bmatrix} 1 & 1 & 1 \\ 1 & a^2 & a \\ 1 & a & a^2 \end{bmatrix} \begin{bmatrix} -0.105 \\ +0.953 \\ -0.230 \end{bmatrix} = \begin{bmatrix} 0.618\underline{/0°} \\ 1.1255\underline{/245.5°} \\ 1.1255\underline{/114.5°} \end{bmatrix}$$

$\mathbf{V}_{abc} = [\mathbf{T}]\mathbf{V}_{012}$, where

$$[\mathbf{T}] = \begin{bmatrix} 1 & 1 & 1 \\ 1 & a^2 & a \\ 1 & a & a^2 \end{bmatrix}$$

7. Write the Matrix Equation to Determine Sequence Components

$\mathbf{V}_{012} = \frac{1}{3}[\mathbf{T}]^{-1}\mathbf{V}_{abc}$, where

$$[\mathbf{T}]^1 = \begin{bmatrix} 1 & 1 & 1 \\ 1 & a & a^2 \\ 1 & a^2 & a \end{bmatrix}$$

$$\begin{bmatrix} \mathbf{V}_0 \\ \mathbf{V}_1 \\ \mathbf{V}_2 \end{bmatrix} = \frac{1}{3} \begin{bmatrix} 1 & 1 & 1 \\ 1 & a & a^2 \\ 1 & a^2 & a \end{bmatrix} \begin{bmatrix} 0.618\angle 0° \\ 1.126\angle 245.5° \\ 1.126\angle 114.5° \end{bmatrix}$$

The solution yields: $V_0 = -0.105$, $V_1 = 0.953$, and $V_2 = -0.230$ per unit.

Related Calculations: The matrix style of writing and solving equations is most useful in all types of three-phase problems. It is applicable particularly for problems wherein symmetrical components are used.

POWER FLOW THROUGH A TRANSMISSION LINE; ABCD CONSTANTS

The **ABCD** constants of a three-phase transmission line (nominal pi circuit) are $\mathbf{A} = 0.950 + j0.021 = 0.950\angle 1.27°$, $\mathbf{B} = 21.0 + j90.0 = 92.4\angle 76.87°\ \Omega$, $\mathbf{C} = 0.0006\angle 90°$ S, and $\mathbf{D} = \mathbf{A}$. Find the steady-state stability limit of the line if both the sending and receiving voltages are held to 138 kV: (1) with the **ABCD** constants as given, (2) with the shunt admittances neglected, and (3) with both the series resistance and the shunt admittances neglected. Refer to Fig. 27.

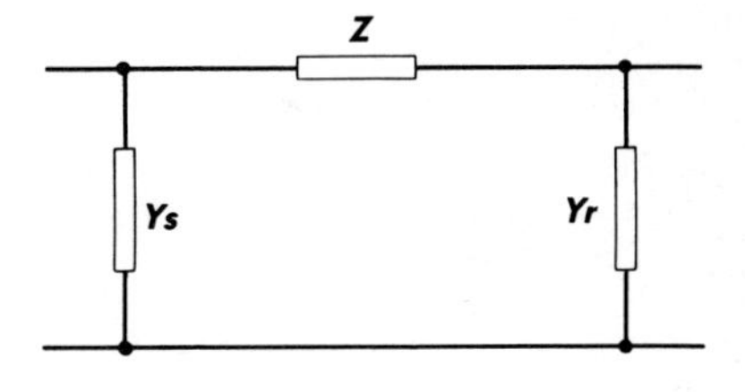

$\mathbf{Y}_s$ = Sending-end shunt admittance in siemens

$\mathbf{Y}_r$ = Receiving-end shunt admittance in siemens

$\mathbf{Z}$ = Series impedance in ohms

$\mathbf{A} = 1 + \mathbf{Y}_r\mathbf{Z}$ *(unitless)*; $\mathbf{A}\angle\alpha$

$\mathbf{B} = \mathbf{Z}$ *(ohms)*; $\mathbf{B}\angle\beta$

$\mathbf{C} = \mathbf{Y}_s + \mathbf{Y}_r + \mathbf{Z}\mathbf{Y}_s\mathbf{Y}_r$ *(siemens)*

$\mathbf{D} = 1 + \mathbf{Y}_s\mathbf{Z}$ *(unitless)*

Fig. 27 **ABCD** constants for a nominal pi circuit. $\mathbf{Y}_s$ = sending-end shunt admittance in siemens; $\mathbf{Y}_r$ = receiving-end shunt admittance in siemens; $\mathbf{Z}$ = series impedance in Ω; $\mathbf{A} = 1 + \mathbf{Y}_r\mathbf{Z}$ (unitless), $\mathbf{A}\angle\alpha$; $\mathbf{B} = \mathbf{Z}$ in Ω, $\mathbf{B}\angle\beta$; $\mathbf{C} = \mathbf{Y}_s + \mathbf{Y}_r + \mathbf{Z}\mathbf{Y}_s\mathbf{Y}_r$ in siemens; $\mathbf{D} = 1 + \mathbf{Y}_s\mathbf{Z}$ (unitless).

Calculation Procedure:

1. Calculate the Steady-State Stability Limit for Nominal pi Circuit

The equation for steady-state stability limit is $P_{max} = |\mathbf{V}_s||\mathbf{V}_r|/|\mathbf{B}| - (|\mathbf{A}||\mathbf{V}_r|^2/|\mathbf{B}|)\cos(\beta - \alpha)$, where $|\mathbf{V}_s|$ = magnitude of sending-end voltage = 138 kV and $|\mathbf{V}_r|$ = magnitude of receiving-end voltage = 138 kV. $P_{max} = (138)(138)(10^6)/92.4 - [(0.950)(138)^2(10^6)/92.4]\cos(76.87° - 1.27°) = 206.1 \times 10^6 - 48.7 \times 10^6 = 157.4$ MW.

2. Calculate the Steady-State Stability Limit with Series Impedance Only

The shunt admittances at the sending end ($\mathbf{Y}_s$) and at the receiving end ($\mathbf{Y}_r$) are both equal to zero; therefore, in the equation $\mathbf{A} = 1 + \mathbf{Y}_r\mathbf{Z}$, the second term on the right is equal to zero, and **A** = 1. Similarly, **D** = 1. **B** remains unchanged at $92.4\angle 76.87°$, and **C** = 0. The steady-state stability limit is determined from the same equation used in step 1. $P_{max} = |\mathbf{V}_s||\mathbf{V}_r|/|\mathbf{B}| - (|\mathbf{A}||\mathbf{V}_r|^2/|\mathbf{B}|)\cos(\beta - \alpha) = (138)(138)(10^6)/92.4 - [(1)(138)^2(10^6)/92.4]\cos 76.87° = 206.1 \times 10^6 - 46.8 \times 10^6 = 159.3$ MW.

3. *Calculate the Steady-State Stability Limit with Series Reactance Only*

Both the shunt admittances and the series resistance are neglected by letting them equal zero. Therefore, **A** = 1, **B** = $j90 = 90\angle 90°$, **C** = 0, and **D** = 1. Again, the same equation is used to determine the steady-state stability limit: $P_{max} = (138)(138)(10^6)/90.0 - [(1)(138)^2(10^6)/90.0]\cos(90° - 0°) = 211.6 - 0 = 211.6$ MW.

Related Calculations: The equation for maximum power represents the steady-state stability limit and may be used for all power transmission problems where **ABCD** constants are known or can be determined from the circuit parameters.

Section 14 SYSTEM GROUNDING

H. W. Beaty
A. Bruning
Electric Power Research Institute

REFERENCES *IEEE Standard Dictionary of Electrical and Electronics Terms,* IEEE; *IEEE Recommended Practice for Grounding of Industrial and Commercial Power Systems;* Wright—*Calculation of Resistances to Ground,* AIEE; Westinghouse Electric Co.—*Electrical Transmission and Distribution Reference Book;* Carson—"Wave Propagation in Overhead Wires with Ground Return," *Bell System Technical Journal;* Lewis—*Transmission of Electrical Power, vol. II: Unbalances and System Disturbances,* Illinois Institute of Technology; Electric Power Research Institute, *Computer Program for Determination of Earth Potentials Due to Faults on Loss of Concentric Neutral on URD Cable;* National Fire Protection Assn., *National Electrical Code,* NFPA; Gross, Chitnis, and Stratton—*Grounding Grids for High Voltage Stations,* AIEE; National Bureau of Standards—Technical Bulletin no. 108; Lewis, Allen, and Wang—*Circuit Constants for Concentric-Neutral Underground Distribution Cables on a Phase Basis,* IEEE; Electric Power Research Institute—*Graphical and Tabular Results of Computer Simulation of Faulted URD Cables, vols. 1 and 2;* James G. Biddle Co.—*Getting Down-to-Earth—Manual on Earth-Resistance Testing for the Practical Man;* IEEE—*Recommended Guide for Measuring Ground Resistance and Potential Gradients in the Earth;* Reynolds, Ironside, Silcocks, and Williams—*A New Instrument for Measuring Ground Impedances,* James G. Biddle Co.; Tagg—*Earth Resistance,* George Newes Ltd.

SELECTION OF GROUNDING SYSTEM

Determine what factors are significant in the selection of a grounding system.

Calculation Procedure:

1. Consider Grounding Impedance

The different levels of grounding impedance are:

a. Solidly grounded: No intentional grounding impedance.

b. Effectively grounded: $R_0 \leq X_1, X_0 \leq 3X_1$, where R is the system fault resistance and X is the system fault reactance. (Subscripts 1, 2, and 0 refer to positive-, negative, and zero-sequence symmetrical components, respectively.)

c. Reactance grounded: $X_0 \leq 10X_1$.

d. Resistance grounded: Intentional insertion of resistance into the system grounding connection; $R_0 \geq 2X_0$.

e. High-resistance grounded: The insertion of nearly the highest permissible resistance into the grounding connection; $R_0 \leq X_{0c}/3$ where X_{0c} is the capacitive zero-sequence reactance.

f. Grounded for serving line-to-neutral loads: $Z \leq Z_1$ where Z is the system fault impedance.

2. Evaluate Disadvantages and Advantages

a. Solidly grounded: Provides for the highest level of fault current to permit maximum ability for overcurrent-protection for isolation of faulted circuit. Fault current may need to be limited if equipment ratings are to be met. Will trip on first fault (it is this factor that occasionally leads to use of ungrounded circuits). Provides greatest ability for protection against arcing faults. Provides maximum protection against system overvoltages because of lightning, switching surges, static, contact with another (high) voltage system, line-to-ground faults, resonant conditions, and restriking ground faults. Limits the difference of electric potential between all uninsulated conducting objects in a local area.

b. Effectively grounded: Permits the use of lower-rated (80 percent) surge arresters. Reduces fault current in comparison with solidly grounded circuits. The reactance limitations provide a fault-relaying current of at least 60 percent of the three-phase short-circuit value.

c. Reactance grounded: In order to limit the transient overvoltage, $X_0 \leq 10X_1$. This usually results in higher fault currents than resistance-grounded systems. Used to reduce zero-sequence fault current to generator fault-current rating (normally line-to-line rating).

d,e. Resistance grounded: At high resistance, extreme transient overvoltages are limited to 250 percent of normal. This system is intended to reduce fault damage, mechanical stresses, stray currents, and flash hazards. It requires sophisticated relaying. At the low-resistance end, this system allows fairly large fault current so it minimizes high-resistance grounding advantages. It is easier to relay.

f. Grounded line-to-neutral load: Used for single-phase loads.

Related Calculations: Some proponents of not grounding circuits suggest that the floating circuit provides a degree of safety because the first accidental contact with a live line would not present a shock hazard. However, the capacitance of lines may allow a dangerous level of current to be conducted through a person.

RECOMMENDED GROUND RESISTANCE FOR SOLIDLY GROUNDED SYSTEM

Determine the appropriate maximum ground resistance permitted.

Calculation Procedure:

1. Decide if System Is Residential-Commercial

Compare the projected system with Fig. 1. If loads are 120/240-V single phase, the projected system is classed as residential-commercial. A maximum ground resistance of 25 Ω is recommended.

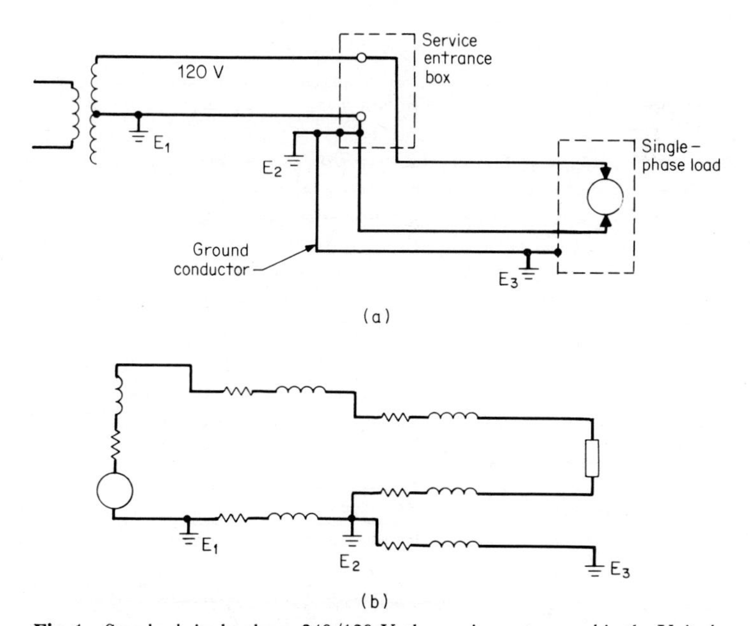

Fig. 1 Standard single-phase, 240/120-V, three-wire system used in the United States. (*a*) Circuit. (*b*) Equivalent circuit.

2. *Decide if System Is Light Industrial*

Compare the projected system with Fig. 2. If loads are primarily single phase on a three-phase system, the system is classed as light industrial. The maximum ground resistance recommended is 5 Ω.

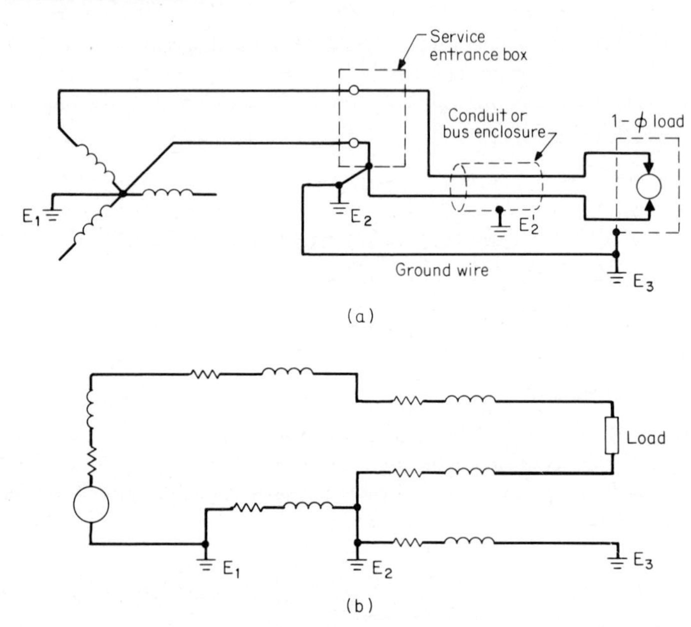

Fig. 2 Low-voltage, single-phase load connected to a three-phase, four-wire circuit. (*a*) Circuit. (*b*) Equivalent circuit.

3. *Decide if System Is Heavy Industrial or a Substation*

Compare the projected system with Fig. 3. If the loads are primarily three-phase as in the figure, then the maximum ground resistance recommended is 1 Ω.

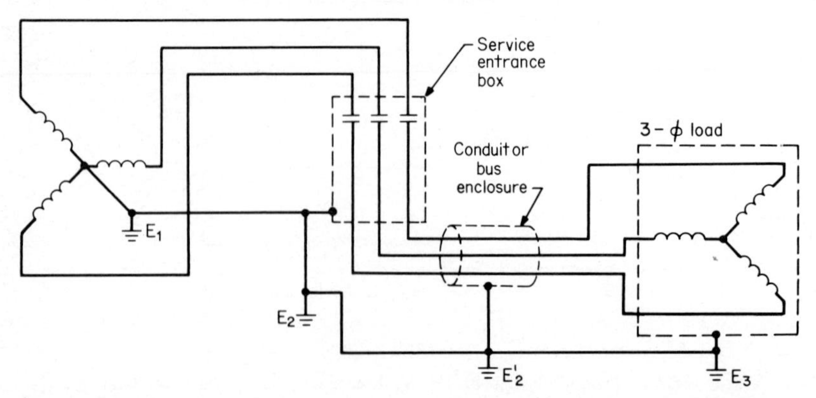

Fig. 3 480/277-V, four-wire, three-phase to three-phase load. E_1, E_2, E'_2, and E_3 are ground connections to earth.

4. Consider Lightning Protection

When a significant lightning threat is expected, use lightning arresters coupled with a maximum ground resistance of 1 Ω.

FORTUITOUS CONSTRUCTION GROUNDS

Underground metallic water pipe, well casings, metallic building frames, and concrete piers provide a resistance connection to ground. Decide if these features provide the minimum required resistance.

Calculation Procedure:

1. Consider Underground Piping

Experimental data indicate metallic underground water systems, metallic underground sewers, or underground metallic gas-pipe systems have a ground resistance of less than 3 Ω.

2. Consider Well Casings and Metallic Building Frames

These two types of systems have resistances to ground of less than 25 Ω.

3. Consider Nonmetallic Underground Construction

Wooden water pipes, plastic pipes, and nonconductive gaskets provide a circuit interruption. It is recommended that in all grounding construction, a ground-impedance measurement be made.

CONCRETE PIER GROUNDS

Determine the ground resistance of one or multiple concrete piers each formed of four rebars with spacer rings. Each rebar is 3 m (10 ft) long. The soil resistivity is 15,800 Ω·cm.

Calculation Procedure:

1. Calculate the Ground Resistance of One Pier

A reinforced-concrete pier of four rebars will have approximately one-half the resistance of a simple driven rod 1.59 cm (⅝ in) in diameter of the same length.

Use the earth resistance of a driven 3-m rod (from calculation of driven ground below), which is 64.8 Ω. One reinforced-concrete pier is one-half this value, or 32.4 Ω earth resistance.

2. Calculate the Ground Resistance for a Multiple Pier

In the case of multiple piers (arrangement of Fig. 4), or footings, divide the resistance of a single pier by half the number of outside piers. *Do not* include interior piers. For Fig. 4 we have eight exterior piers. Thus, the resistance is $32.4/(\frac{1}{2})(8) = 8.1\ \Omega$.

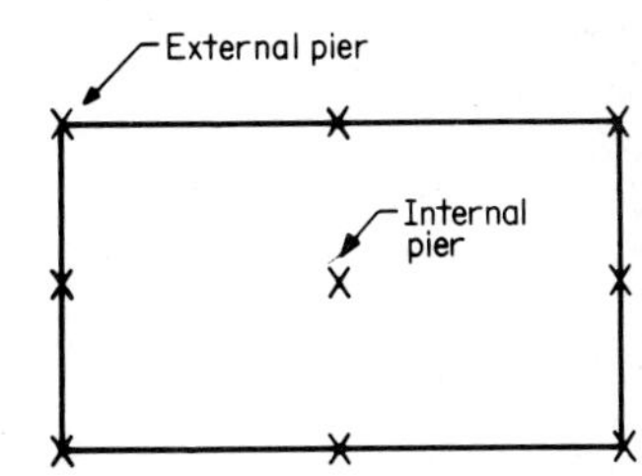

Fig. 4 Multiple concrete pier arrangement.

GROUND RESISTIVITY AND DESIGN OF DRIVEN GROUNDS

Select and design a ground rod for a commercial building in sandy loam of moderate sandiness in a climate with a 61-cm (2-ft) frost line and 1.27 m (50 in) of rain a year.

Calculation Procedure:

1. Choose First Calculation-Attempt Configuration

Table 1 shows some of the various shapes of manufactured ground rods. The most common is a 3-m (10-ft) cylindrical rod. National Electrical Code (**NEC®**) requires a minimum of 2.4-m (8-ft) driven length; a 3-m rod, therefore, is long enough to meet the

TABLE 1 Formulas for Calculation of Resistances to Ground*

Hemisphere, radius a	$R = \frac{\rho}{2\pi a}$
One ground rod, length L, radius a	$R = \frac{\rho}{2\pi L}\left(\ln\frac{4L}{a} - 1\right)$
Two ground rods, $s > L$; spacing s	$R = \frac{\rho}{4\pi L}\left(\ln\frac{4L}{a} - 1\right) + \frac{\rho}{4\pi s}\left(1 - \frac{L^2}{3s^8} + \frac{2L^4}{5s^4}\cdots\right)$
Two ground rods, $s < L$; spacing s	$R = \frac{\rho}{4\pi L}\left(\ln\frac{4L}{a} + \ln\frac{4L}{s} - 2 + \frac{s}{2L} - \frac{s^2}{16L^2} + \frac{s^4}{512L^4}\cdots\right)$
Buried horizontal wire, length $2L$, depth $s/2$	$R = \frac{\rho}{4\pi L}\left(\ln\frac{4L}{a} + \ln\frac{4L}{s} - 2 + \frac{s}{2L} - \frac{s^2}{16L^2} + \frac{s^4}{512L^4}\cdots\right)$
Right-angle turn of wire, length of arm L, depth $s/2$	$R = \frac{\rho}{4\pi L}\left(\ln\frac{2L}{a} + \ln\frac{2L}{s} - 0.2373 + 0.2146\frac{s}{L} + 0.1035\frac{s^4}{L^2} - 0.0424\frac{s^4}{L}\cdots\right)$
Three-point star, length of arm L, depth $s/2$	$R = \frac{\rho}{6\pi L}\left(\ln\frac{2L}{a} + \ln\frac{2L}{s} + 1.071 - 0.209\frac{s}{L} + 0.238\frac{s^3}{L^8} - 0.054\frac{s^4}{L^4}\cdots\right)$
Four-point star, length of arm L, depth $s/2$	$R = \frac{\rho}{8\pi L}\left(\ln\frac{2L}{a} + \ln\frac{2L}{s} + 2.912 - 1.071\frac{s}{L} + 0.645\frac{s^2}{L^8} - 0.145\frac{s^4}{L^4}\cdots\right)$
Six-point star, length of arm L, depth $s/2$	$R = \frac{\rho}{12\pi L}\left(\ln\frac{2L}{a} + \ln\frac{2L}{s} + 6.851 - 3.128\frac{s}{L} + 1.758\frac{s^2}{L^3} - 0.409\frac{s^4}{L^4}\cdots\right)$
Eight-point star, length of arm L, depth $s/2$	$R = \frac{\rho}{16\pi L}\left(\ln\frac{2L}{a} + \ln\frac{2L}{s} + 10.98 - 5.51\frac{s}{L} + 3.26\frac{s^3}{L^3} - 1.17\frac{s^4}{L^4}\cdots\right)$
Ring of wire, diameter of ring D, diameter of wire d, depth $s/2$	$R = \frac{\rho}{2\pi^2 D}\left(\ln\frac{8D}{d} + \ln\frac{4D}{s}\right)$
Buried horizontal strip, lenght $2L$, section a by b, depth $s/2$, $b < a/8$	$R = \frac{\rho}{4\pi L}\left[\ln\frac{4L}{a} + \frac{a^2 - \pi ab}{2(a+b)^2} + \ln\frac{4L}{s} - 1 + \frac{s}{2L} - \frac{s^2}{16L^2} + \frac{s^4}{512L^4}\cdots\right]$
Buried horizontal round plate radius a, depth $s/2$	$R = \frac{\rho}{8a} + \frac{\rho}{4\pi s}\left(1 - \frac{7}{12}\frac{a^2}{s^2} + \frac{33}{40}\frac{a^4}{s^4}\cdots\right)$
Buried vertical round plate radius a, depth $s/2$	$R = \frac{\rho}{8a} + \frac{\rho}{4\pi s}\left(1 + \frac{7}{24}\frac{a^2}{s^2} + \frac{99}{320} + \frac{a^4}{s^4} + \cdots\right)$

*Approximate formulas. Dimensions must be in centimeters to give resistance in Ω. ρ = resistivity of earth in Ω·cm.

Source: H. B. Wright, *AIEE,* vol. 55, 1936, pp. 1319–1328.

Code. As to diameter, **NEC®** requires 1.59-cm (⅝-in) minimum diameter for steel rods and 1.27-cm (½-in) minimum diameter for copper or copper-clad steel rods. Minimum practical diameters for driving limitations for 3-m rods are:

1.27 cm (½ in)—average soil
1.59 cm (⅝ in)—most soils
1.91 cm (¾ in)—very hard soils or more than 3-m driving depth

A practical selection, suitable for most soils and **NEC®** specifications, is a rod 1.59 cm (⅝ in) in diameter by 3 m (10 ft) long.

2. *Select Resistivity*

Table 2 indicates a variation of resistivity for sandy loam of 1020 to 135,000 Ω·cm with an average of 15,800 Ω·cm. To decide which to choose, examine the effect of moisture in Table 3. Because 15,800 Ω·cm lies in the 10 to 15 percent moisture range, our 1.27 m (50 in) per year indicates this will be a conservative estimate. If there is an indication the climate is seasonal with long drought periods, implying that the soil moisture may drop very low, one would want experimental data, or use a resistivity closer to the 135,000 Ω·cm end of the resistivity range.

TABLE 2 Resistivity of Different Soils

Soil	Resistivity, Ω·cm		
	Minimum	*Average*	*Maximum*
Ashes, cinders, brine, waste	590	2,370	7,000
Clay, shale, gumbo, loam	340	4,060	16,300
Same, with varying proportions of sand and gravel	1,020	15,800	135,000
Gravel, sand, stones with little clay or loam	59,000	94,000	458,000

The data indicate our design will be subject to possibly significant variations in soil temperature. Table 4 shows, as the temperature rises from 10°C (50°F) to 20°C (68°F),

TABLE 3 Effect of Moisture Content on Resistivity of Soil

Moisture content, percent by weight	Resistivity, Ω·cm	
	Top soil	*Sandy loam*
0	>10⁹	>10⁹
2.5	250,000	150,000
5	165,000	43,000
10	53,000	18,500
15	19,000	10,500
20	12,000	6,300
30	6,400	4,200

TABLE 4 Effect of Temperature on Resistivity of Sandy Loam, 15.2 Percent Moisture

Temperature		Resistivity, Ω·cm
°C	*°F*	
20	68	7,200
10	50	9,900
0 (water)	32	13,800
0 (ice)	32	30,000
−5	23	79,000
−15	14	330,000

a decrease of 27 percent in resistivity. Within the accuracy of our selection of the appropriate resistivity, this variation may be ignored. However, the variation from 20°C (68°F) to a temperature below freezing shows an increase in resistivity. To be conservative we will assume the frozen 61 cm (2 ft) of soil is completely insulating.

3. Compute *R* for Single Rod

From Table 1 for one ground rod of length L and radius a, we have $R = (\rho/2\pi L)[\ln(4L/a) - 1]$, where L is 297 cm (9.75 ft) [since we have a 3-m (10-ft) rod driven with 3 cm of the rod exposed] minus 61 cm (2 ft), equal to 236 cm (7.75 ft); a is 1.59 cm (⅝ in); and $\rho = 15{,}800\ \Omega\cdot\text{cm}$. Substituting values, we calculate R as 64.8 Ω. This is significantly greater than the value indicated under the "Recommended Ground Resistance for Solidly Grounded Commercial System" of 25 Ω maximum. Our first design is unsatisfactory; we therefore examine an alternative design.

4. Compute *R* for Two Rods, Close Spacing ($s < L$)

From Table 1 we have $R = (\rho/4\pi L)[\ln(4L/a) + \ln(4L/s) - 2 + s/2L - s^2/16L^2 + s^4/512L^4 \cdots]$, where $L = 236$ cm, $a = 1.59$ cm, $s = 100$ cm ($< L = 239$ cm), an arbitrary selection, and $\rho = 15{,}800\ \Omega\cdot\text{cm}$.

We calculate $R = 40.1\ \Omega$. Because this resistance still exceeds the recommended 25 Ω, we can consider a further variation.

5. Compute *R* for Two Rods, Wide Spacing ($s > L$)

From Table 1 we have $R = (\rho/4\pi L)[\ln(4L/a) - 1] + (\rho/4\pi s)(1 - L^2/3s^8 + 2L^4/5s^4 \cdots)$.

We select an arbitrary spacing of $s = 400$ cm (13⅛ ft), obtaining $R = 35.5\ \Omega$.

6. Decide if Two Rods at 400 cm Are Satisfactory

Recognizing the wide variability of sandy loam resistivity, one uses judgment in selecting the value of ρ. If the power system is a small commercial operation with a low-capacity transformer (and fault current will be moderate) or if the soil tends to be more loam than sand (4060 Ω·cm average resistivity as compared with 15,800 Ω·cm for sandy loam), then R for two widely spaced rods would be 9.12 Ω, which would be satisfactory. On the other hand, if the soil tends to be of high resistivity, or if the substation transformer is of large capacity, there would be a need to examine one of the lower-resistance systems, such as the "star" counterpoise shown in Table 1 or a ground grid (see next problem).

Related Calculations: Tables 2, 3, and 4 indicate the variability of the resistivity of earth as function of earth type, temperature, moisture, and chemical content. Resistivity is also sensitive to backfill compaction, earth pressure against the grounding metal, and the magnitude of fault current. It is common practice to assume for most ground calculations an earth resistivity of 10,000 Ω·cm, as compared with the 15,800 Ω·cm used in the preceding calculation.

Because R is directly proportional to resistivity, significant design variations from actual tested results can occur if a realistic resistivity is not selected. This can have severe consequences. For instance, a common lightning-stroke current of 1000 A through the 35.5-Ω ground resistance (see Step 5 above), would generate a 35,000-V transient on the electrical system. If the resistivity in the actual installation is 4 times our assumed value, a voltage transient of 142,000 V would result. For these reasons, it is prudent to test the finished installation as described later in this section.

GROUNDING WITH GROUND GRIDS

A grounding location consists of a thin overburden of soil over a rock substrate. The deep constructions from Table 1 are impractical. Design a horizontal-grid (four-mesh) grounding system for a light industrial application.

Calculation Procedure:

1. Define an Initial Design

Use a conductor radius r of 0.0064 m (¼ in), a conductor depth below earth surface, s, of 0.30 m (1 ft), and an overall grid width w of 6.1 m (20 ft). Use a four-mesh grid and a standard resistivity of 10,000 Ω·cm. (By way of illustration, Fig. 5 indicates a sketch of a nine-mesh grounding grid.)

2. Calculate Parameter for Abscissa of Fig. 6

a. $A = w^2 = 37.2\ \text{m}^2\ (400\ \text{ft}^2)$.

b. $2rs/A = 1.03 \times 10^{-4}$.

3. Read Ordinate from Fig. 6

Corresponding to an abscissa of 1.03×10^{-4}, the ordinate is 0.097 for a four-mesh grid.

4. Calculate R

Solving the equation $6.56\pi wR/\rho = 0.097$, find $R = 7.7\ \Omega$. This is close enough to the recommended 5 Ω for light industrial plant design.

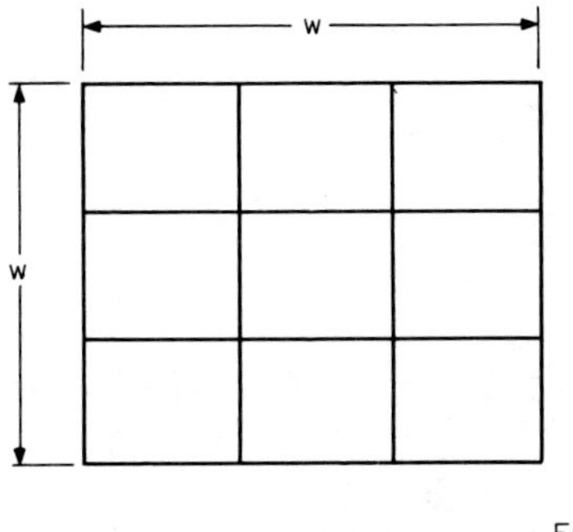

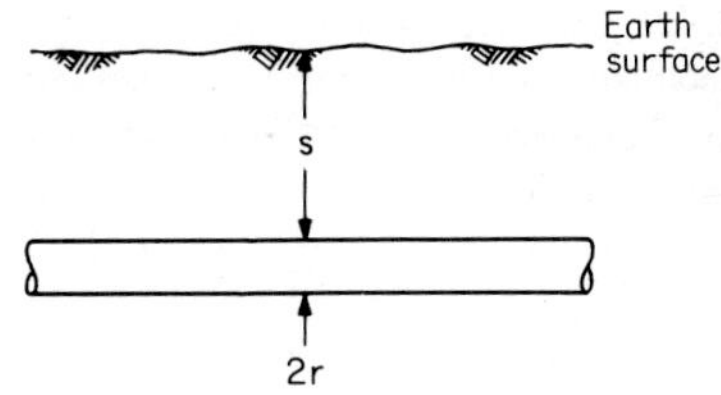

Fig. 5 Nine-mesh shallow grid; w = width of grid in meters, A = area of grid in square meters, r = radius of conductor in meters, ρ = resistivity of soil in Ω·cm, and s = depth of grid below earth surface in meters.

GROUNDING WITH A SIX-POINT STAR

The star horizontal-ground system of Table 1 is a possible alternative ground mat to the four-mesh grid designed in the previous section. Using the same length of ground con-

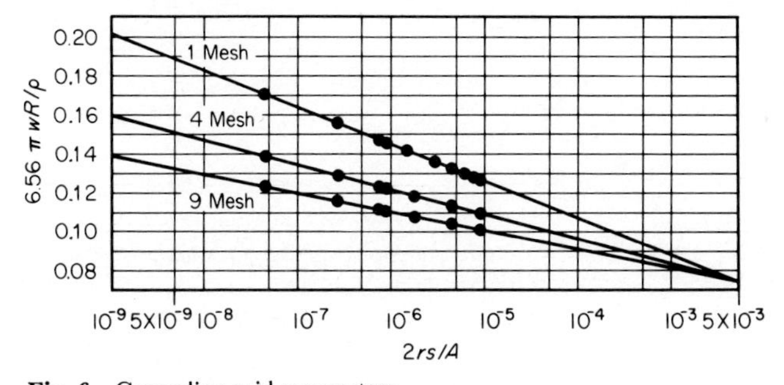

Fig. 6 Grounding grid parameters.

ductors and other dimensions as the four-mesh ground grid in the previous design, calculate the ground resistance of the star.

Calculation Procedure:

1. Calculate the Equivalent L of the Star Arm

The total ground conductor in the four-element ground mesh is 36.6 m (120 ft). Each of the six star arms is 36.6/6 = 6.1 m long.

2. Calculate R for the Star

Use $R = (\rho/12\pi L)[\ln(2L/a) + \ln(2L/s) + 6.851 - 3.128s/L + 1.758s^2/L^3 - - 0.49s^4/L^4 \cdots]$, where $\rho = 10{,}000\ \Omega\cdot\text{cm}$, $L = 610$ cm (20 ft) arm length, $s = 30$ cm (2 ft) depth, and $a = 1.59$ cm (⅝ in) diameter. Substituting values, find $R = 7.7\ \Omega$, the same value as in the mesh design in the preceding problem.

EFFECT OF GROUND-FAULT DISTANCE FROM GROUND POINT

The formulas in Table 1 are based on the resistance from a ground electrode of the given geometry to a hemisphere at an infinite radius. Calculate the variation in resistance for a hemispherical ground electrode of radius a as a function of distance to the outer hemispherical electrode, the distance being less than infinity.

Calculation Procedure:

1. Express Resistance between Two Hemispheres Using Ohm's Law

Ohm's law indicates $R = \rho L/A$. For the two hemispheres the incremental resistance dR between two spheres of area $A = 2\pi r^2$ separated by a distance dr is $dR = \rho(dr/2\pi r^2)$. Integration yields $R = (\rho/2\pi)(1/a - 1/r_2)$, where a is the radius of the inner sphere and r_2 is the radius of the outer sphere.

2. Calculate R as a Function of r_2

Define r_2 as na where $n = 1, 2, 3, \ldots, \infty$; hence, $R = (\rho/2\pi a)(1 - 1/n)$. Substituting for various values of $n = 1, 2, \ldots, \infty$, we obtain:

n	R
1	$0 \times \frac{\rho}{2\pi a}$
2	$\frac{1}{2} \times \frac{\rho}{2\pi a}$
5	$\frac{4}{5} \times \frac{\rho}{2\pi a}$
10	$\frac{9}{10} \times \frac{\rho}{2\pi a}$
20	$\frac{19}{20} \times \frac{\rho}{2\pi a}$
50	$\frac{49}{50} \times \frac{\rho}{2\pi a}$
100	$\frac{99}{100} \times \frac{\rho}{2\pi a}$

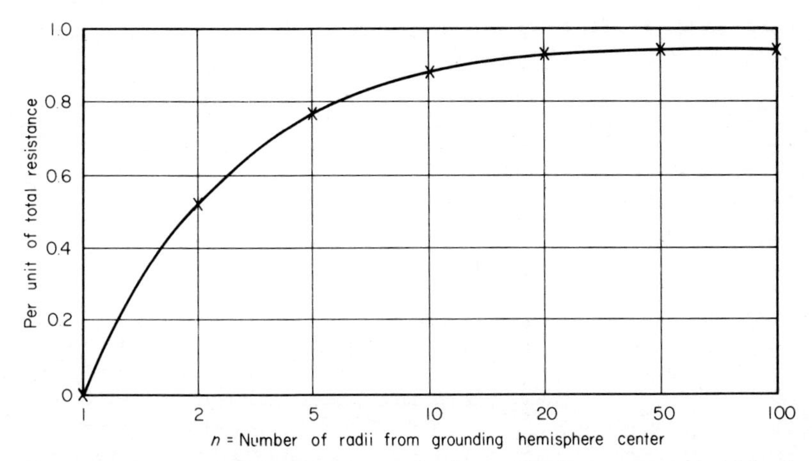

Fig. 7 Resistance between two hemispheres of radius a and radius na compared with hemisphere of radius a and radius infinity.

These results are plotted in Fig. 7. Note that 90 percent of the resistance is developed in the ground voltage drop over a distance of the first 10 radii.

EFFECT OF RESISTIVITY OF EARTH AND INDUCTANCE ON DEPTH OF GROUND CURRENT

Figure 8 indicates the distributed path of ground current between A and B. Accepted engineering practice represents the distributed current path as a single conductor of 30-

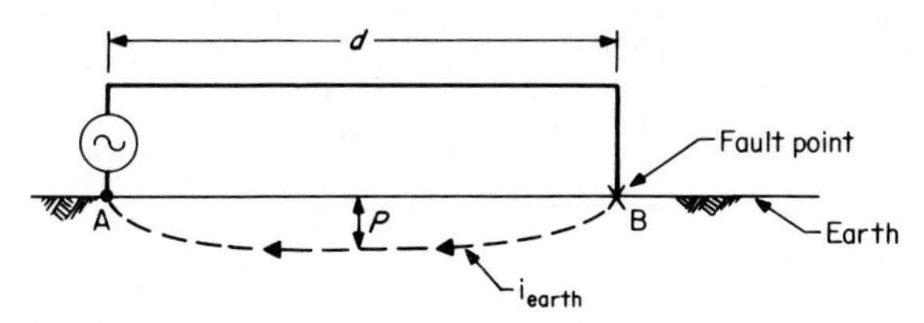

Fig. 8 Path of fault current. d = distance between two ground points. P = depth of equivalent conductor when $d \doteq \infty$.

cm (1-ft) radius with the same inductive and resistive voltage drop as the distributed current case. Calculate the distance P for resistivities of 340, 10,000, and 458,000 Ω·cm from Table 2.

Calculation Procedure:

1. ***Calculate*** **P for ρ of 340 Ω·cm**

$P = 8.5\sqrt{\rho}$ m $= 157$ m.

2. ***Calculate*** **P for ρ of 10,000 Ω·cm**

$P = 8.5\sqrt{\rho}$ m $= 850$ m.

3. ***Calculate*** **P for ρ of 458,000 Ω·cm**

$P = 8.5\sqrt{\rho}$ m $= 5750$ m.

IMPEDANCE OF LONG-DISTANCE FAULTS COMPARED WITH GROUNDING RESISTANCE

For a fault to earth, with resistivity of 10,000 Ω·cm, between points far apart with respect to penetration depth P (Fig. 8) of 850 m, calculate the fault impedance. The conductor is 1.27 cm (½ in) in radius with a resistance of 0.54 Ω/km. Compare it with the recommended grounding resistance.

Calculation Procedure:

1. Calculate Impedance

Impedance $\mathbf{Z} = r_c + 0.063 + j0.18 \log_{10} (P/\text{GMR})$, where $\mathbf{Z}$ is the complex impedance per kilometer, r_c is the conductor resistance per kilometer, 0.063 is the resistance per kilometer of the assumed 30-cm radius underground equivalent conductor (which, to engineering accuracy, is independent of earth resistivity), P is the penetration depth of the assumed earth conductor, and GMR is the geometric mean radius of the conductor in the same dimensions as P. $\text{GMR} = \sqrt{R_1R_2} = \sqrt{(0.0127)(0.30)} = 0.062$ m. Substituting in the expression for impedance, find $\mathbf{Z} = 0.6 + j0.74\ \Omega$.

2. Compare the Long-Line Impedance with Grounding Resistance

If the grounding resistance is kept to the recommended 1 Ω for heavy industrial applications, then the line impedance will be on the same order as the grounding impedance. Note the need for keeping the grounding resistance below 5 and 25 Ω for lighter-duty applications.

IMPEDANCE OF SHORT-DISTANCE FAULTS COMPARED WITH GROUNDING RESISTANCE

Use the standard formula for reactance of two parallel conductors to examine the inductive reactance X of a conductor and a metallic ground return. The radius of each is 1.27 cm.

Calculation Procedure:

1. Calculate X for Various Spacings

Use $X = 0.34 \log_{10}/\sqrt{D_{12}/\text{GMR}}\ \Omega/\text{km}$, where D_{12} is the distance between the center line of the two conductors and GMR is the geometric mean radius of the two conductors in the same dimensions as D_{12}. $\text{GMR} = \sqrt{(1.27\text{ cm})(1.27\text{ cm})} = 0.0127$ m. We obtain the following table.

D_{12} as multiple of 0.0127-m GMR	*D_{12}, m*	*Ω/km*
10	0.13	0.17
100	1.3	0.34
1000	13	0.51

2. Compare the Reactive Drop with Ground Resistance

Because even with the 30-m spacing between the normal conductor and metallic ground return, resistance per unit length is only 0.51 Ω/km, it is clear the ground resistance of the grounding element dominates the fault impedance for short line faults for all practical grounding resistances.

Related Calculations: Where great accuracy in impedance is desired, as in distance relaying design, a computer approach is required. Electric Power Research Institute Reports N-1605, vols. 1 and 2, describe the computer programming approach to solve the distributed-constant problem to the desired accuracy. In this calculation, orthogonal cubes are assumed. These cubes have front and back surfaces of equipotentials. The sides of the cubes are current sheets. By summing the currents into and out of the cube to zero, and solving for voltage, the interactive process can be made to converge to a solution of the unique gradients and equipotentials.

TESTING GROUND RESISTANCE

Test a ground electrode (station fence) for its ground resistance and calculate ground resistance.

Calculation Procedure:

1. *Run Test*

Set up instrumentation as shown in Fig. 9. At each distance shown in Fig. 10, measure V and I, and calculate static ground resistance.

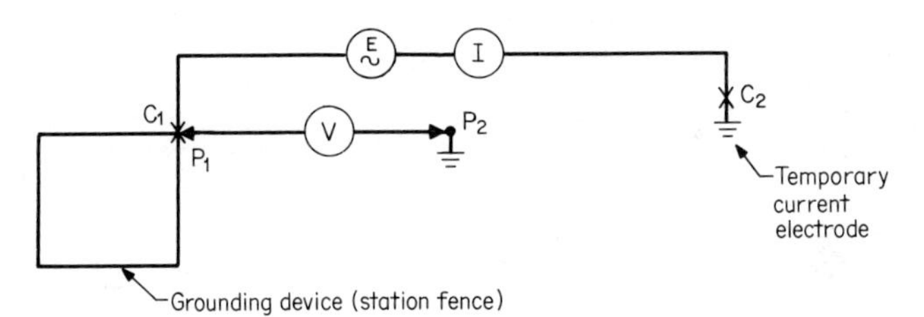

Fig. 9 Fall-of-potential three-electrode ground-resistance test arrangement. V is a voltmeter, E is voltage service, I is an ammeter, and P_2 is the potential electrode.

2. *Examine Fig. 10 for Asymptote*

If plot does not have an asymptote, then move C_2 to a greater distance and repeat. Use Table 5 to give the minimum distance to C_2. If an asymptote occurs, as in Fig. 10, the ground resistance is 1.6 Ω.

TESTING GROUND RESISTIVITY

Make a test to measure resistivity of soil using the test arrangement of Fig. 11.

Calculation Procedure:

1. *Set Up Test*

Place four equally spaced electrodes as in Fig. 11. If we are interested in the resistivity for use in the calculations of a standard 3-m (10-ft) ground rod, set b at 1½ m. Fix a at

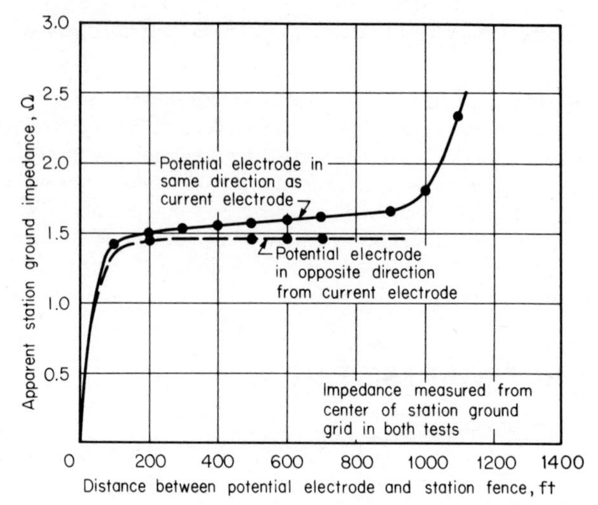

Fig. 10 Results of fall-of-potential test.

TABLE 5 Minimum Auxiliary Electrode C_2 vs. Grounding Element Maximum Dimension

Maximum grounding element dimension, m	*Minimum distance to C_2, m*
0.6	21
1.2	30
1.8	37.5
2.4	42
3.0	48
3.6	51
4.2	57
4.8	60
5.4	63
6.0	66
12.0	96
18.0	117
24.0	135
30.0	150
36.0	165
42.0	170
48.0	192
54.0	204
60.0	213

15¼ m (50 ft). Apply a voltage to the outer current electrodes to generate a current of 1 A.

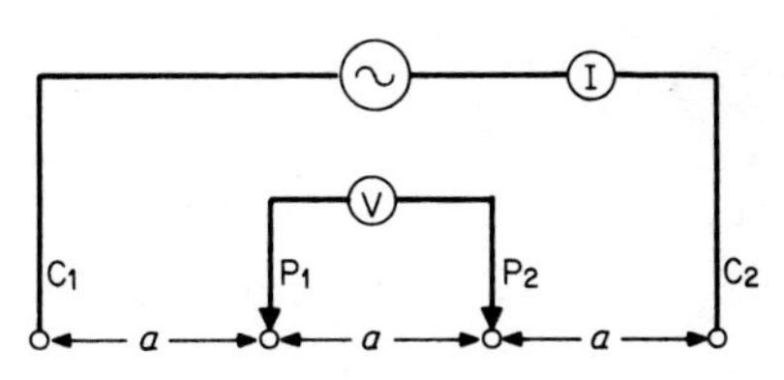

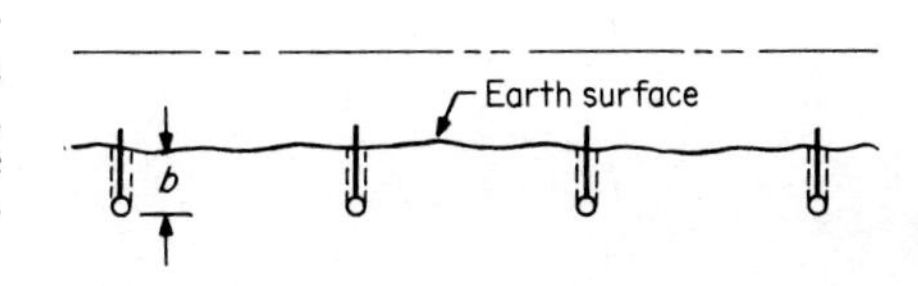

Fig. 11 Four-point resistivity test circuit. I is an ammeter, and V is a voltmeter; ρ = resistivity in Ω·cm, a = spacing in cm, b = depth in cm, and R = resistance in Ω. If $b \ll a$, a in cm, then $\rho = 12.6aR$ Ω·cm. *Note:* Usual practice for $b \ll a$ is to use rods for electrodes, not insulated probes.

2. *Make Measurements*

Measure the potential between points P_1 and P_2. Let us assume a reading of 2 V is measured.

3. *Calculate R*

$R = V/I = 2/1 = 2\ \Omega$.

4. *Calculate ρ*

Use $\rho = 12.6aR\ \Omega\cdot\text{cm} = (12.6)(a)(2)\ \Omega\cdot\text{cm}$, where $a = 1525$ cm. Hence, $\rho = 38{,}430\ \Omega\cdot\text{cm}$.

Section 15 POWER-SYSTEM PROTECTION

W. D. Connolly
Consulting Engineer

J. Gohari
Supervising Engineer

W. L. Hinman, P.E.*
Assistant Chief Engineer,
Gibbs & Hill, Inc.

The material contained in this section represents the authors' approach and does not necessarily represent Gibbs & Hill practice. The authors express their appreciation to the Gibbs & Hill editorial and Automated Transcribing Service staffs for their help in the preparation of this section.

REFERENCES Blackburn—*Applied Protective Relaying,* Westinghouse Electric; Fink and Beaty—*Standard Handbook for Electrical Engineers,* McGraw-Hill; Hicks—*Standard Handbook of Engineering Calculations,* McGraw-Hill; Horowitz—*Protective Relaying for Power Systems,*

*Now with ASEA Inc.

IEEE; Mason—*The Art and Science of Protective Relaying,* Wiley; Newcombe, et al.—*Protective Relay Application Guide,* Scolor Press; Wagner, et al.—*Electrical Transmission and Distribution Reference Book,* Westinghouse Electric; Warrington—*Protective Relays: Their Theory and Practice,* vol. 1, Wiley.

SETTING TRIP DEVICE ON CIRCUIT BREAKER

Determine the proper setting for a solid-state trip device on a circuit breaker protecting a 460-V, 70-hp, three-phase motor (Fig. 1). The solid-state sensor is rated 150 A, with "long-time" and "instantaneous" operating characteristics. The long-time overcurrent sensor pickup setting criterion is based on approximately 1.2 times motor full current. The instantaneous pickup setting is based on 10 times full-load current so that the sensor can override motor inrush current and current wave asymmetry.

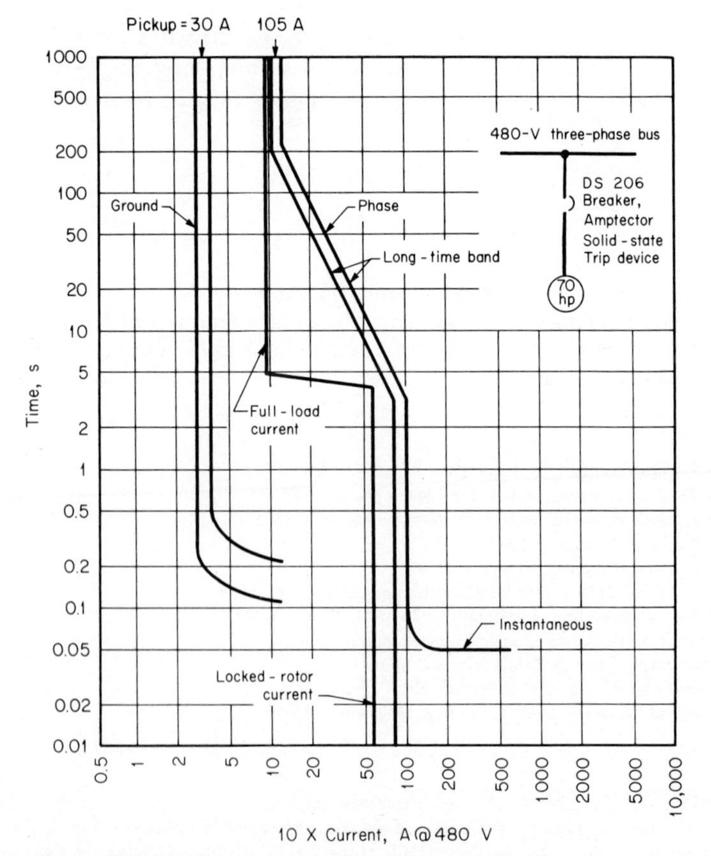

Fig. 1 Time-current characteristics of solid-state trip device.

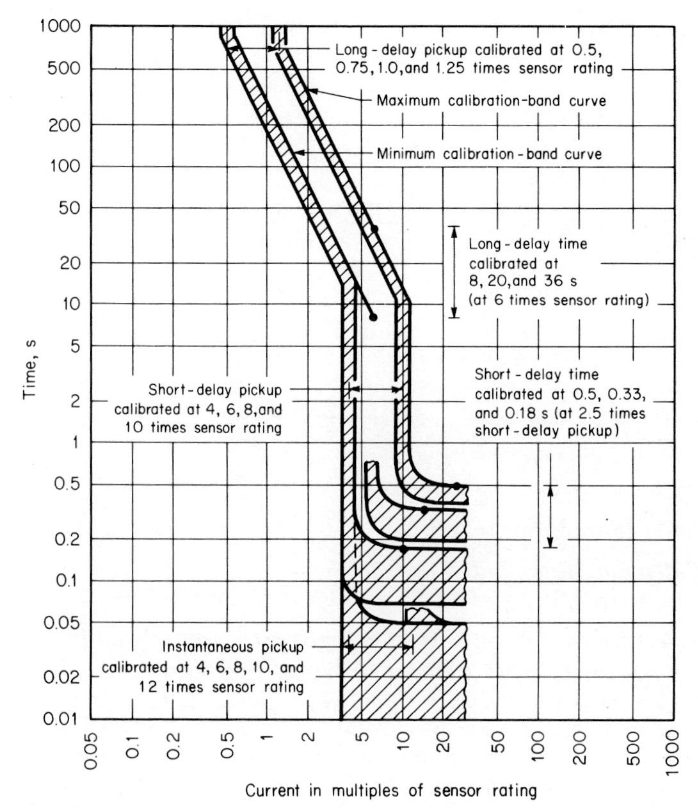

Fig. 2 Available adjustments of phase and ground sensor. (Westinghouse Electric)

Calculation Procedure:

1. Calculate the Motor Full-load Current

Assume hp = kVA for induction motors. Use $I_{FL} = \text{hp}/\sqrt{3}\text{kV}_{LL}$, where I_{FL} = full-load current and kV_{LL} = line-to-line voltage in kilovolts. Hence, $I_{FL} = 70/(\sqrt{3})(0.46) = 87.8$ A.

2. Select Sensor Settings

Figure 2 illustrates the available adjustments of the phase and ground sensor contained within the circuit-breaker assembly. In the phase unit:

a. For long delay setting, use $0.7 \times$ sensor rating $= (0.7)(150 \text{ A}) = 105$-A pickup.

b. Select time delay for long-time setting. Use 4-s delay at 6 times sensor rating.

c. Set instantaneous unit. Use 6 times sensor rating, or $(6)(150) = 900$ A.

For the ground unit:

a. Pickup current fixed by the sensor rating at $(0.2)(150) = 30$ A.

b. Select time band. Use minimum time setting; select 0.21-s band.

Related Calculations: The long-time pickup is set at 120 percent of motor full-load current. The instantaneous unit pickup is set at about 10 times motor full-load current. Figure 1 indicates the coordination of phase- and ground-sensor settings with the starting and running curves of the 70-hp motor.

SETTING TIME-OVERCURRENT RELAYS FOR 400-hp MOTOR

Determine correct setting for the time-overcurrent relays protecting a 400-hp, three-phase, 60-Hz, 4160-V induction motor. The current transformer ratio C.T. = 75/5. Protective-relay current ranges are:

1. Time-overcurrent unit taps: 2, 2.5, 3, 3.5, 4, 5, and 6 A.
2. High-dropout instantaneous unit: 4 to 8 A.
3. Standard instantaneous unit: 20 to 80 A.

Calculation Procedure:

1. Calculate Motor Full-Load Current

Use $I_{FL} = hp/\sqrt{3}kV_{LL} = 400/(\sqrt{3})(4.16) = 55.5$ A.

2. Select Relay Tap on Time-Overcurrent Unit to Provide Approximately 115 Percent Overload Protection

Try tap 5 and calculate the percent overload permitted; C.T. = 75/5 = 15/1. Use percent overload = [(C.T.)(relay tap)/(motor full-load current)](100 percent) = [(15)(5)/55.5](100 percent) = 135 percent overload. This 135 percent value is too high; try next-lower relay tap: [(15)(4)/55.5](100 percent) = 108 percent overload. Use tap 4.

3. Select Time-Dial Setting on Protective Relay

The time-dial setting must coordinate with the motor starting curve. Obtain relay characteristic curves from the manufacturer and compare relay curves to the motor starting curve; select time-dial setting 2.

4. Select Setting or High-Dropout Instantaneous-Trip (IT) Unit for Stalled Rotor Protection

The desired setting is approximately 160 percent of the motor full-load current: (1.6)(55.5 A) = 88.8 A. Use 90-A setting on high-dropout unit for ease of setting.

5. Calculate Percentage of Full-Load Current Permitted before Motor Trips

The percentage is: (90 A/55.5 A)(100 percent) = 162 percent of motor full-load current.

6. Select Setting for Standard IT Unit to Provide Short-Circuit Protection

Set IT unit at about 10 times motor full-load current to override motor inrush current. Symmetrical inrush current is 300 A (Fig. 3). Then $10I_{FL}$ = (10)(55.5 A) = 550 A. Divide 550 A by C.T. ratio: 550/15 = 36.7 A, secondary. Select 34-A setting to provide

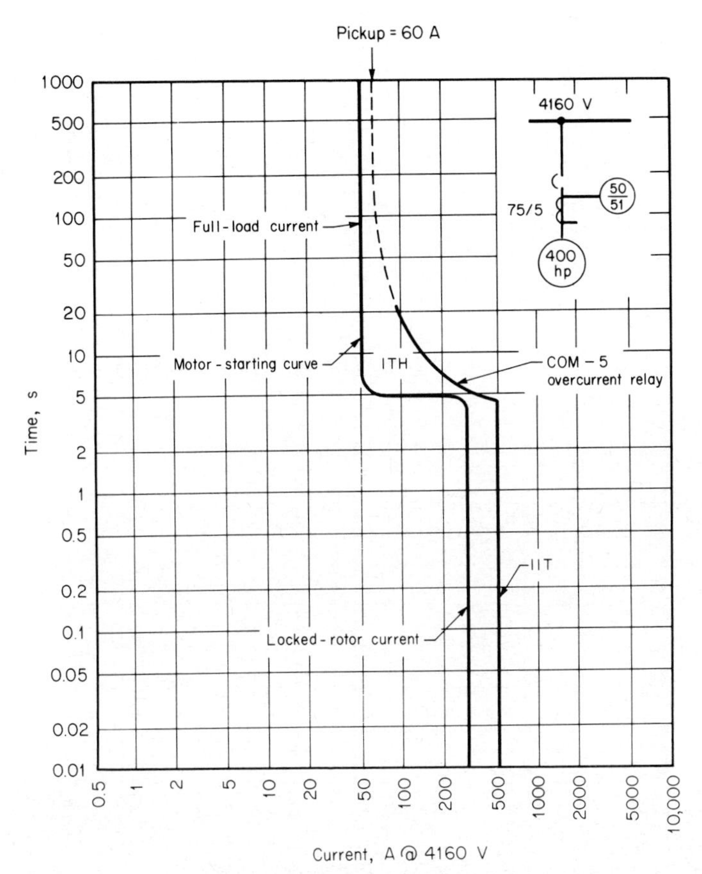

Fig. 3 Overcurrent time-current characteristic curves.

improved margin of protection, and still compensate for asymmetric inrush current: (34 A)(15/1) = 510 A; (510/55.5)(100 percent) = 918 percent of full load.

Related Calculations: The time-overcurrent unit is set to provide protection for current values above 108 percent overload and is set to alarm only. The high-dropout instantaneous unit is set at 162 percent of full load and is adjusted to trip for stalled rotor protection. The standard instantaneous-trip unit is set at approximately 9 times full-load current to provide short-circuit protection.

SETTING TIME-OVERCURRENT RELAYS FOR 7000-hp MOTOR

Determine correct setting for the time-overcurrent relays protecting a 7000-hp, three-phase, 60 Hz, 4160-V induction motor. Motor-starting and rotor thermal-limit curves are provided in Fig. 4. The current-transformer ratio is 1200/5. Relay 1 is rated: 2- to

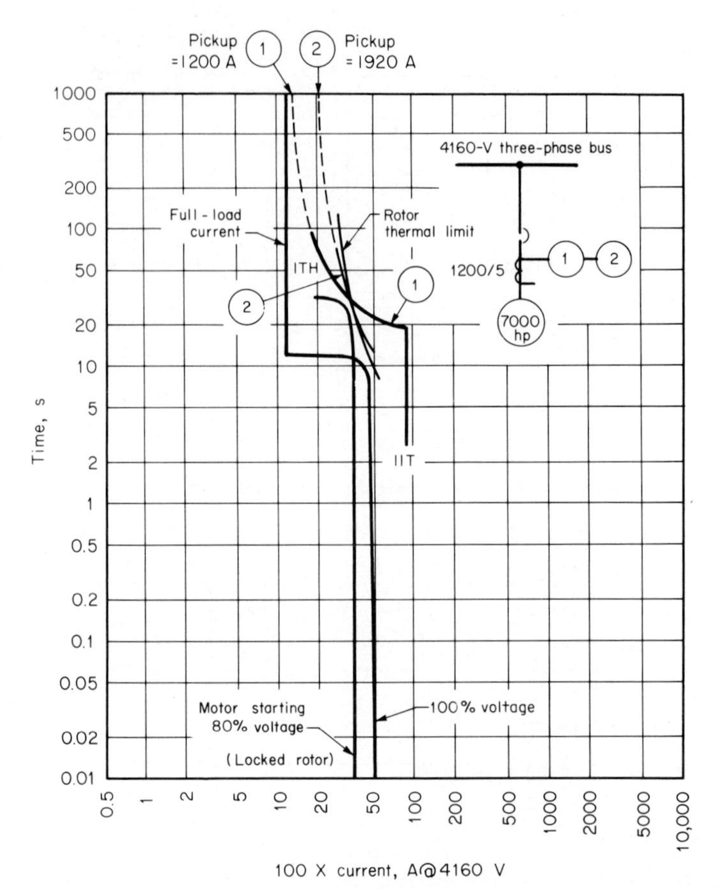

Fig. 4 Motor-starting and rotor thermal-limit curves.

6-A time-overcurrent unit with taps of 2, 2.5, 3, 3.5, 4, 5, and 6 A; 4- to 8-A high-dropout instantaneous unit; and a 20- to 80-A standard instantaneous unit.

Calculation Procedure:

1. Calculate Motor Full-Load Current

$I_{FL} = 7000/(\sqrt{3})(4.16) = 972$ A.

2. Set Relay 1

Select a tap on the time-overcurrent relay that provides approximately 120 percent overload protection. C.T. ratio is 1200/5 or 240/1. Since 120 percent of motor full-load current is $(1.2)(972 \text{ A}) = 1166$ A, the secondary current is $1166 \text{ A}/240 = 4.86$A.

Use relay tap 5. Tap number × C.T. ratio = primary current = $(5)(240) = 1200$ A. Overload protection is $(1200/972)(100 \text{ percent}) = 123$ percent.

Obtain relay characteristic curve from manufacturer. Select time-dial setting on relay 1 such that the relay will not trip during motor start-up; use time-dial setting 7. Figure

4 shows the curve for relay 1 plotted for 1200-A (primary) pickup with a time-dial setting of 7.

Select setting of the high-dropout instantaneous-trip unit, which is used for stalled motor protection. Desired setting is about 170 percent of motor full-load current: (1.7)(972) = 1652 A; 1652/240 = 6.88 A, secondary.

Use 7-A pickup for ease of setting. Calculate percent above full-load current: setting × C.T. ratio = primary current, or (7.0)(240) = 1680 A; (1680/972)(100 percent) = 173 percent full-load current.

3. Select Setting of Standard IT Unit for Short-Circuit Protection

Figure 4 shows the locked-rotor current to be 4800 A at 100 percent starting voltage. Set the IT unit above locked-rotor current to accommodate possible asymmetry of inrush current, but below the maximum three-phase fault current available at the 4160-V bus. The desired setting is approximately 1.7 times locked-rotor current: $(\sqrt{3})(4800)$ = 8314 A; 8314/240 = 34.6 A secondary. Use 34.5-A setting on the standard instantaneous unit.

4. Set Relay 2

Relay 2, intended to protect the motor from thermal damage during start-up, has a 1- to 12-A time-overcurrent range with taps of 1, 1.2, 1.5, 2.0, 2.5, 3.0, 3.5, 4, 5, 6, 7, 8, 10, and 12 A. It is desired to set relay 2 so that its characteristic operating curve will fall below the rotor thermal-limit curve and above the knee of the motor-starting curves as shown in Fig. 4. Position the operating curve of relay 2 over the rotor thermal-limit curve as plotted in the figure. Reading the pickup current required for relay 2, observe that approximately 2000-A primary current is required. Transfer primary current to relay secondary current and select the nearest tap on relay 2: 2000/240 = 8.33-A secondary current. Use tap 8.

5. Calculate Primary Current for Tap 8

The primary current is (8)(240) = 1920, which is the pickup setting for relay 2. By overlaying the relay 2 characteristic curves over the rotor thermal-limit curve in Fig. 4, select the time-dial setting for relay 2; use time-dial setting of 9.5. Plot relay 2 protective curve with motor startup and thermal-limit curves to ensure coordination.

SETTING THERMAL BREAKER

Determine the proper setting for a molded-case thermal breaker used to protect a 1-hp, 480-V, three-phase motor. Select proper motor thermal relay (heater) to be used with the motor starter. (*Note:* A molded-case thermal breaker protects for heavy short circuits above motor inrush current; a thermal relay provides protection for overload.)

Calculation Procedure:

1. Record Nameplate Data of Motor

Pertinent data are: full-load current = 1.95 A and locked-rotor current = 12 A.

2. Calculate 125 Percent Full-Load Current

The current is (1.25)(1.95) = 2.44 A.

3. Select Heater

The *minimum* heater rating is equal to, or greater than, the motor full-load current. Select a heater with an operating range of 2.16 A to 2.43 A. These heaters are designed

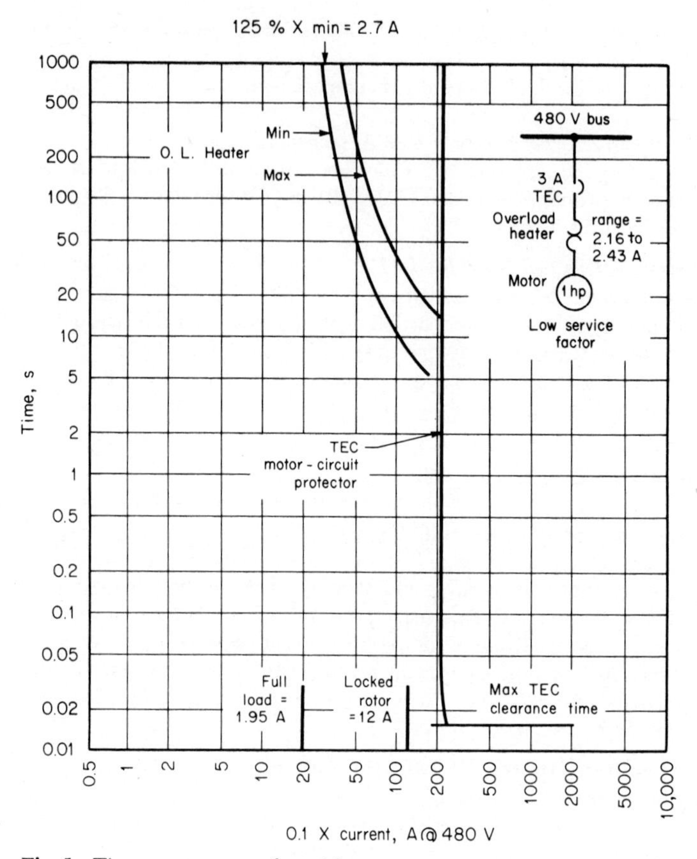

Fig. 5 Time-current curves for a 1-hp motor.

to trip at 125 percent of their *minimum* operating current value; 125 percent of 2.16 A is 2.7 A.

4. Calculate Percent Overload Protection Afforded to Motor

The overload protection is (2.7/1.95)(100 percent) = 138 percent. Hence, at full load, the motor will be tripped at 138 percent of full-load current if the current persists for the time values shown in Fig. 5.

5. Select Thermal Breaker

The rating of the breaker should at least equal the motor full-load current; select a 3.0-A breaker.

6. Set Breaker Instantaneous-Trip Unit

The unit should be set to $\sqrt{3}$ times motor locked-rotor current; therefore, setting = $(\sqrt{3})(12)$ = 20.8 A. Select nearest available setting on breaker: 23 A.

7. *Prepare Coordination Curve*

See Fig. 5.

APPLICATION OF DIFFERENTIAL RELAYS

Figure 6 illustrates the application of an overall differential relay to protect a 5500-hp, three-phase, 4160-V motor. Determine the setting and selection of relay.

Calculation Procedure:

1. Set Relay

No setting is required except for the time dial on the relay. The dial is set at position 1 to provide minimum speed of operation under internal fault conditions.

This type of relay is available either with a 10 or 25 percent minimum sensitivity. These figures refer to the percent of restraining current which must flow in the relay operating coil for tripping of the relay. The 10 percent relay with a pickup of 0.18 A is used for this application with current transformers of the C-400 class.

2. Use Current Sensors

Figure 6 also illustrates an alternate method of differential protection using a current sensor in each phase lead to the motor. Because the three-phase motor presents a balanced load, the differential relay can be set for minimum sensitivity of 2.0-A primary current. Select a 2.0-A pickup setting and an operating time of five cycles.

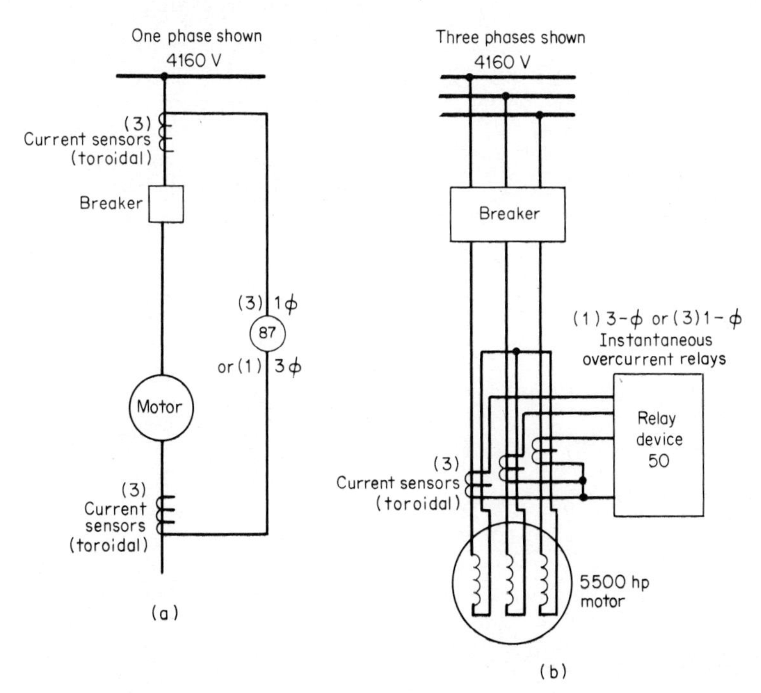

Fig. 6 Differential protection of 5500-hp motor with (*a*) differential and (*b*) instantaneous overcurrent relays.

3. *Provide Ground-Fault Protection*

For three-phase motors, ground protection is provided by one sensor enclosing all three phases and supplying one ground relay, as in Fig. 7. Set relay for 5.0-A pickup and a time delay of 0.1 s.

DIRECTIONAL-CONTROL SELECTION FOR OVERCURRENT RELAYS

Determine whether directional control units for phase instantaneous overcurrent relays (device 50) should be recommended for protection of a 34.5-kV subtransmission line. The overcurrent-relay pickup-setting should be 1.5 times maximum external-fault current. Assume that coverage of an additional 20 percent of the line would justify the cost of a directional unit. Primary line impedance is 10 Ω. Local- and remote-bus, three-phase, short-circuit apparent power ratings, with the line open, are 180 MVA and 390 MVA, respectively.

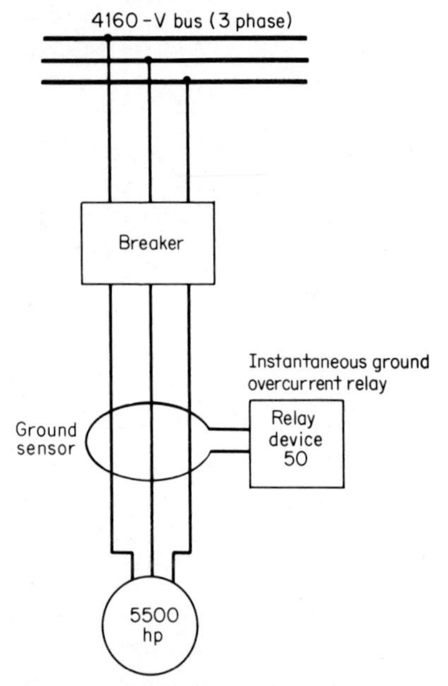

Fig. 7 Providing ground-fault protection.

Calculation Procedure:

1. *Calculate Source Impedances*

Because the bus short-circuit apparent power ratings are given with the line open, calculation of the source impedances is straightforward. For the local source, $Z_s = V_{LL}^2/VA = 34{,}500^2/(180 \times 10^6) = 1190/180 = 6.61\ \Omega$. For the remote source: $Z_u = 34{,}500^2/(390 \times 10^6) = 1190/390 = 3.05\ \Omega$.

2. *Calculate Relay Fault Currents*

To determine whether phase instantaneous-overcurrent relays are applicable at all, it is necessary to calculate the relay current for the following three-phase faults: (*a*) close-in reverse fault (F1), (*b*) close-in forward fault (F2), and (*c*) forward fault just beyond remote bus (F3). The relative magnitudes of I_{F1} and I_{F3} will further determine whether directional control is desirable. If the stronger source is behind the relay, then directional control is normally not recommended.

To visualize these faults, and to assist in writing the equations to determine fault currents, first draw a one-line diagram showing primary system impedances and fault locations, as in Fig. 8. The relay fault current, in primary amperes, is calculated by $I_{REL} = V_{LN}/Z_{TOT}$, where $V_{LN} = V_{LL}/\sqrt{3} = 34{,}500/\sqrt{3} = 19{,}920$ V. For I_{F1}, $Z_{TOT} = 10 + 3.05 = 13.05\ \Omega$. Hence, $I_{F1} = 19{,}920/13.05 = 1526$ A. For I_{F2}, $Z_{TOT} = 6.61\ \Omega$, and $I_{F2} = 19{,}920/6.61 = 3014$ A. For I_{F3}, $Z_{TOT} = 6.61 + 10 = 16.61\ \Omega$; $I_{F3} = 19{,}920/16.61 = 1199$ A.

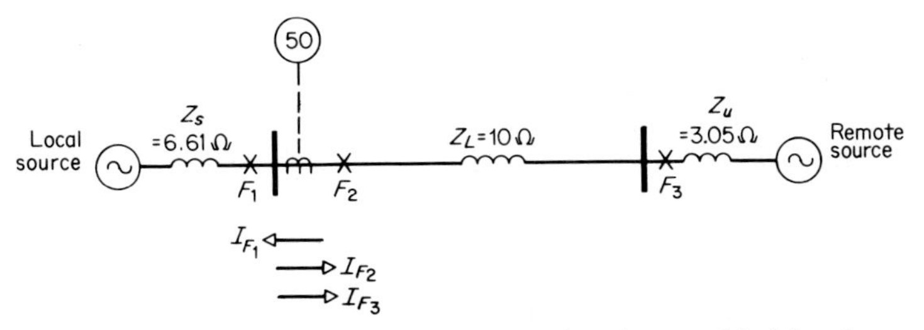

Fig. 8 One-line diagram showing primary system impedances and fault locations.

3. Determine Percent Coverage of Line without Directional Control

The nondirectional instantaneous overcurrent unit is a relatively inexpensive and simple unit. It provides high-speed tripping for close-in heavy faults. Lower-current faults, farther out in the line, are cleared by time-overcurrent relays.

The maximum external-fault current seen by relay 50 of Fig. 8 is 1526 A (I_{F1}). The pickup setting is, therefore, $I_{pu} = 1.5I_{F1} = (1.5)(1526) = 2289$ A. To determine the percent coverage of the line, use $I_{pu} = V_{LN}/[Z_s + (n/100)Z_L]$, where n is the percent coverage. Solving for n, we find $n = [100(V_{LN} - I_{pu}Z_s)]/I_{pu}Z_L = \{100[19{,}920 - (2289 \times 6.61)]\}/(2289)(10) = 20.9$ percent.

4. Determine Percent Coverage of Line with Directional Unit Control

With directional unit control, the maximum external fault current seen by relay 50 is 1199 A (I_{F3}). The pickup setting is therefore $I_{pu} = 1.5I_{F3} = (1.5)(1199\text{ A}) = 1799$ A. The percent coverage of the line is: $n = 100(V_{LN} - I_{pu}Z_s)/I_{pu}Z_L = (100)[19{,}920 - (1799)(6.61)]/(1799)(10) = 44.6$ percent.

5. Determine Whether a Directional Unit Is Justified

The increased coverage achieved by adding a directional unit is $44.6 - 20.9 - 23.7$ percent. Because the increased coverage exceeds 20 percent of the line, the cost of a directional unit is justified.

Related Calculations: The source impedances Z_s and Z_u have been assumed to be constant. In practice, these source impedances vary. Calculations can be made to determine the line coverage, in percent, provided by a nondirectional, instantaneous overcurrent relay on a radial line, considering the maximum and minimum expected values of equivalent source impedance Z_s. Once again, the setting of the overcurrent unit pickup should be based on 1.5 times the maximum external fault current seen by the relay.

LOADABILITY OF PHASE-DISTANCE RELAYS

The zone 2 phase-distance relay (device 21P-2) on a long 345-kV transmission line is set for 8.2 Ω, secondary impedance, with the zone 2 time set for 0.25 s. The relay maximum torque angle is 75°. The C.T. ratio is 2000/5 and the P.T. ratio is 345,000/115. Emergency load current can, for short periods, go well above 2000 A. Power factor during emergency loading will be no less than 0.85. How high can the 0.25-s peak load exist without causing the relay to trip the breaker? If the peak load results in undesired tripping, straight-line "blinder" relays may have to be added.

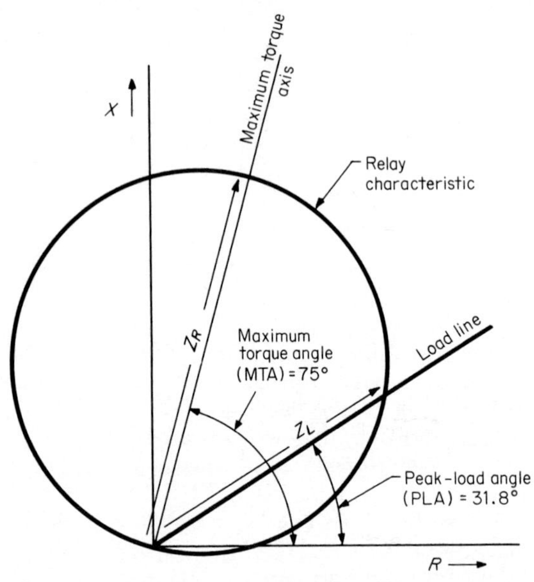

Fig. 9 An R-X diagram showing relay characteristic and load line.

Calculation Procedure:

1. Calculate Angle for Smallest Power Factor during Emergency Loading

The minimum power factor during emergency loading has been given as 0.85. Therefore, the angle θ that corresponds to this power factor is $\theta = \cos^{-1} 0.85 = 31.8°$.

2. Draw an R-X Diagram Showing Relay Characteristics and Load Line

The diagram, drawn in secondary ohms, is given on Fig. 9.

3. Determine Relay Reach at Peak Load Angle

Referring to Fig. 9, note that the peak-load-angle (PLA) relay reach (Z_L) may be expressed in terms of maximum-torque-angle (MTA) relay reach (Z_R): $Z_L = Z_R \cos(\text{MTA} - \text{PLA}) = 8.2 \cos(75° - 31.8°) = 5.98\ \Omega$.

4. Make Graphical Check of Z_L

The measured lengths of phasors Z_R and Z_L in Fig. 9 are $Z_R = 5.2$ cm and $Z_L = 3.8$ cm; therefore, $Z_L = (8.2)(3.8)/(5.2) = 6.0\ \Omega$.

5. Determine the Primary Current $I_{L(pri)}$

$I_{L(pri)} = V_{LN}/Z_{L(pri)}$, where $V_{LN} = V_{LL}/\sqrt{3} = 345{,}000/\sqrt{3} = 199{,}186$ V, and $Z_{L(pri)} = Z_{L(sec)}R_p/R_c$ where R_p = P.T. ratio = 345,000/115 = 3000 and R_c = C.T. ratio = 2000/5 = 400. Therefore, $Z_{L(pri)} = (5.98)(3000)/400 = 44.85\ \Omega$. Finally, $I_{L(pri)} = 199{,}186/44.85 = 4441$ A. (The 0.25-s peak load current must be less than 4441 A to prevent undesired breaker tripping.)

Related Calculations: Closely related to the loadability is the "arc coverage" of the distance relay. This is a measure of the amount of fault resistance that will be accommodated by the relay. The procedure is to select a fault location along the maximum-torque axis of the relay and then determine, by phasor addition, the maximum resistance phasor (horizontal on the R-X diagram) that can be added to the fault location to produce a resultant phasor that remains within the operating circle of the relay.

INFEED EFFECT ON GROUND-DISTANCE RELAY SETTING

Set an overreaching ground-distance relay to cover 125 percent of the apparent line-end impedance on a three-terminal 230-kV transmission line. Assume all line and source impedances are homogeneous (same ratio of zero-sequence impedance to positive-sequence impedance), with $p = 3.0$ (where p is defined as Z_0/Z_1). The positive-sequence line impedance is 0.4 Ω/km (0.65 Ω/mi). The line distances are indicated on Fig. 10. The positive-sequence source impedances are: Z_s (local) = 3.2 Ω and Z_t (third terminal) = 10.0 Ω. The remote-source impedance Z_u is not significant to this problem. The C.T. ratio is 1200/5.

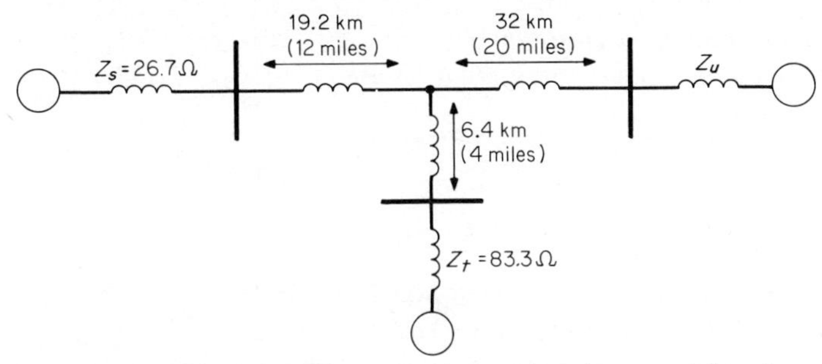

Fig. 10 One-line diagram illustrating sources, impedances and line distances. Source impedances are given in relay Ω. Remote source impedance Z_u is not significant to the problem.

Calculation Procedure:

1. Calculate the Source Impedances in Primary Ohms

Convert Z_s and Z_t to primary ohms, using $Z_{pri} = Z_{sec}(R_p/R_c)$, where R_p = P.T. ratio = 230,000/115 = 2000 and R_c = C.T. ratio = 1200/5 = 240; $R_p/R_c = 2000/240 = 8.33$. Therefore, $Z_{s(pri)} = (3.2)(8.33) = 26.7$ Ω and $Z_{t(pri)} = (10.0)(8.33) = 83.3$ Ω.

2. Calculate Line Positive-Sequence Primary Impedances

The per-kilometer positive-sequence line impedance Z_1 is given as 0.4 Ω/km. From the line lengths in Fig. 10, the line impedances may be calculated.

a. Line section from source Z_s to tap: $Z_1 = (0.4)(19.2) = 7.8$ Ω.

b. Line section from source Z_t to taps: $Z_1 = (0.4)(6.4) = 2.6$ Ω.

c. Line section from source Z_u to tap: $Z_1 = (0.4)(32) = 13$ Ω.

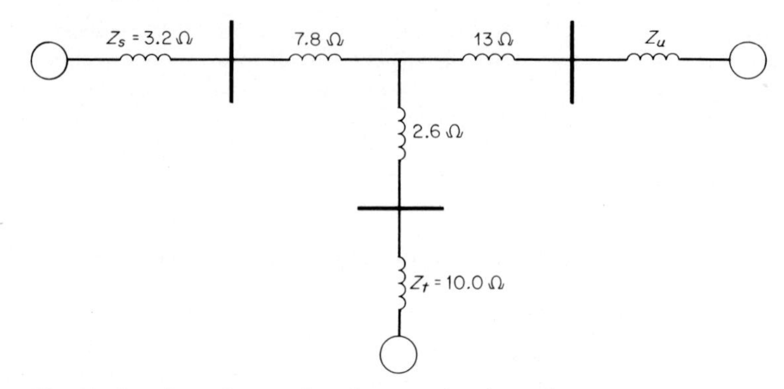

Fig. 11 Impedance diagram. Impedances are in primary Ω.

3. *Draw a One-Line Diagram Showing All Significant Impedance Values*

The significant impedance values, in primary ohms, have now all been calculated. As an aid in the calculations which follow, these impedances are given in Fig. 11.

4. *Lump the Series Impedances*

For ease of calculation, the source and line segment impedances are lumped together:

a. From equivalent sources Z_s to tap: $Z_{\text{I}} = 3.2 + 7.8 = 11.0\ \Omega$.

b. From equivalent source Z_t to tap: $Z_{\text{II}} = 10.0 + 2.6 = 12.6\ \Omega$.

c. From tap to remote bus (see Fig. 12): $Z_{\text{III}} = 13.0\ \Omega$.

5. *Determine Apparent Impedance Seen by Relay at Z_s Bus for a Ground Fault at the Remote Bus*

Because the impedances are homogeneous, the analysis may be performed in terms of positive-sequence impedances. The current that feeds this fault from the left-hand side (i.e., the three-terminal line) will flow entirely through impedance Z_{III}. This current, called I_F, will comprise currents from sources Z_s and Z_t in the following ratio: $I_{Zs} = I_F Z_{\text{II}}/(Z_{\text{I}} + Z_{\text{II}}) = I_F(12.6/23.6)$ and $I_{Zt} = I_F Z_{\text{I}}/(Z_{\text{I}} + Z_{\text{II}}) = I_F(11.0/23.6)$. There-

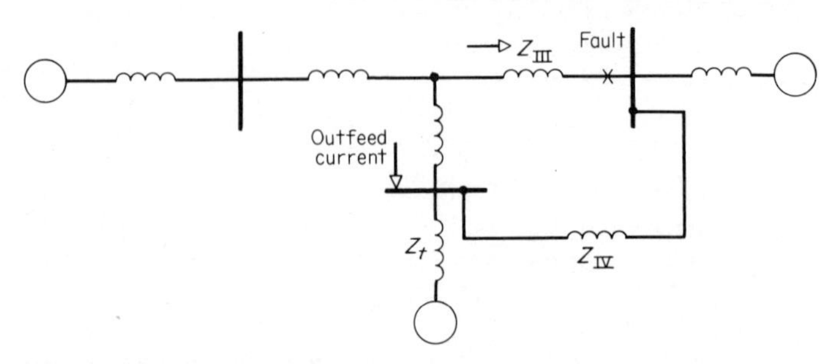

Fig. 12 Illustrating outfeed current.

fore, the apparent impedance seen by a relay at the Z_s bus is: $7.8 + Z_{III}I_F/I_{Zs} = 7.8 + 13I_F/I_F(12.6/23.6) = 7.8 + (13)(23.6/12.6) = 32.1\ \Omega$.

6. Calculate Relay Setting

As previously stated, the overreaching relay setting is to be 125 percent of the apparent line-end impedance. Therefore, the setting in primary ohms is $Z_R = (1.25)(32.1)\ \Omega = 40.1\ \Omega$.

Related Calculations: A related calculation is the determination of "outfeed" current on a three-terminal line. Outfeed is the current that flows out of the third terminal for an internal fault near the remote terminal. This condition is illustrated on Fig. 12. The outfeed condition is quite unusual, but can occur if the third terminal source is weak (Z_t large) and the parallel tie is strong (Z_{IV} small with respect to Z_{III}).

GENERATOR STATOR GROUND-RELAY SELECTION

Generator stator ground-fault protection is to be provided by an overvoltage relay connected across the distribution transformer secondary in parallel with the generator grounding resistor. The distribution transformer primary voltage may be rated line-to-line or line-to-neutral. The secondary may be rated 120 V or 240 V. The overvoltage relay, device 59, may be obtained in one of two ratings: 80 V or 200 V. The pickup of the relay is 7.5 percent of its continuous rating: 6 V or 15 V, respectively. The relay is to be used for alarming and must be capable of being energized continuously by the maximum-voltage ground fault.

Select the distribution transformer and relay combination that will detect ground faults over the maximum part of the stator winding. Determine what percent of the stator winding is protected.

Calculation Procedure:

1. Examine the Continuous Rating Capability of Each Transformer-Relay Combination

There are eight transformer-relay combinations: (*a*) line-line primary, 120-V secondary, 80-V relay; (*b*) line-line primary, 120-V secondary, 200-V relay; (*c*) line-line primary, 240-V secondary, 80-V relay; (*d*) line-line primary, 240-V relay; (*e*) line-line primary, 240-V secondary, 200-V relay; (*f*) line-neutral primary, 120-V secondary, 200-V relay; (*g*) line-neutral primary, 240-V secondary, 80-V relay; and (*h*) line-neutral primary, 240-V secondary, 200-V relay.

For a maximum-voltage ground fault (full neutral displacement), the developed transformer secondary voltage will be:

Combinations *a* and *b*: $120/\sqrt{3} = 69.3$ V

Combinations *c* and *d*: $240/\sqrt{3} = 138.6$ V

Combinations *e* and *f*: 120 V

Combinations *g* and *h*: 240 V

After these maximum voltages are compared with the relay ratings, it is clear that the following combinations are within the continuous relay ratings: (*a*) maximum voltage:

69.3 V, relay rating: 80 V; (*b*) maximum voltage: 69.3 V, relay rating: 200 V; (*d*) maximum voltage: 138.6 V, relay rating: 200 V; (*f*) maximum voltage: 120 V, relay rating: 200 V.

2. Determine Optimum Transformer-Relay Combination

The permissible combinations (for continuous rating) are now examined for maximum sensitivity:

a. Relay pickup: (80)(7.5 percent) = 6 V; stator-winding balance point: 6/69.3 = 8.6 percent.

b. Relay pickup: (200)(7.5 percent) = 15 V; stator-winding balance point: 15/69.3 = 21.6 percent.

d. Relay pickup: (200)(7.5 percent) = 15 V; stator-winding balance point: 15/138 = 10.8 percent.

f. Relay pickup: (200)(7.5 percent) = 15 V; stator-winding balance point: 15/120 = 12.5 percent.

3. Determine Stator-Winding Coverage for Optimum Combination

The minimum balance point (most sensitive relay protection) for a permissible combination (i.e., one that is within the relay rating for a maximum-voltage ground fault) is: combination (*a*) with a balance point of 8.6 percent. Therefore, the stator-winding coverage is 100 − 8.6 = 91.4 percent.

Related Calculations: The generator grounding resistor is connected across the secondary of the distribution transformer, in parallel with the generator stator ground relay. The resistor size is selected to provide enough resistor damping to eliminate any possibility of ferroresonance. Too little resistor current will not adequately suppress ferroresonance and too much resistor current will subject the machine to excessive damage during a ground fault. The optimum resistor current value has been determined to be equal to the capacitive current that flows during a ground fault.

To determine this value, the phase-to-ground capacitance must be calculated for the following circuit elements: the machine winding, the cabling, and the low-voltage-side winding of the step-up transformer. These values are then added to determine the total phase-to-ground capacitance. The resulting capacitive current during a ground fault must be balanced by an equal resistive current to provide optimum grounding.

GENERATOR-FIELD GROUND-RELAY SENSITIVITY

A generator-field ground relay is connected across the exciter terminals (Fig. 13). The exciter is rated 125 V dc. For purposes of this calculation, assume that the exciter is

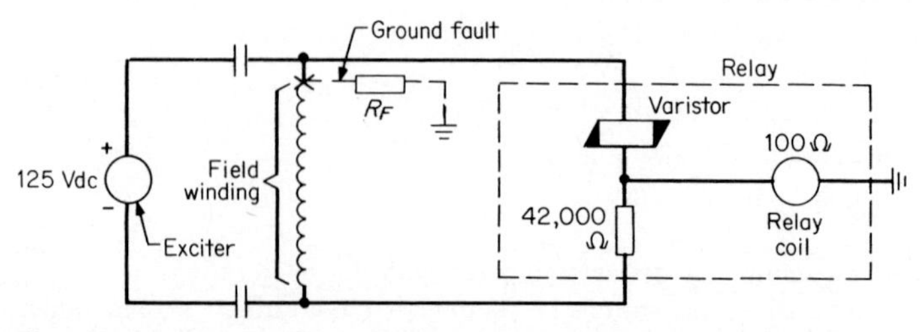

Fig. 13 Generator-field ground relay connected across exciter terminals.

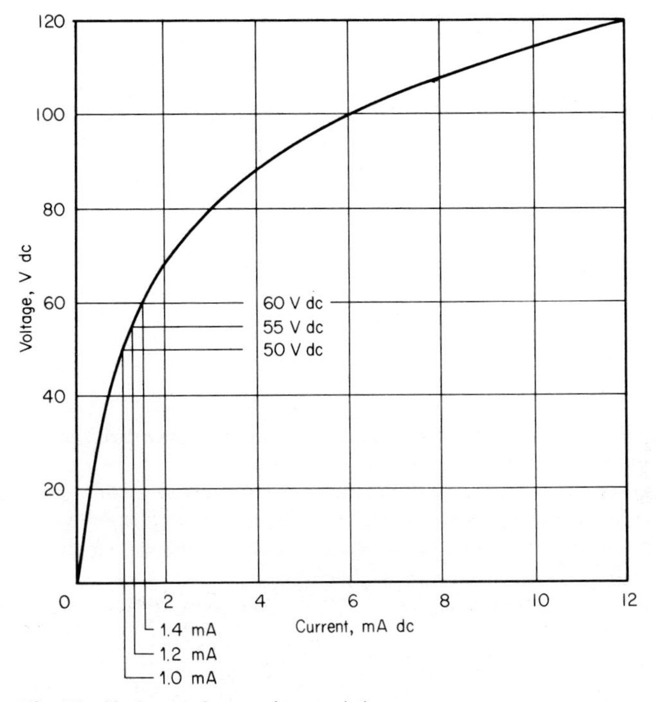

Fig. 14 Varistor resistance characteristic curve.

producing rated voltage. The relay circuit is connected to ground through a sensitive operating coil, with a relay pickup of 0.47 mA. As in any bridge-type circuit, a ground-fault location on the field winding exists for which no relay coil voltage is developed. This is called the null point. The purpose of the varistor (nonlinear resistor) is to shift the null point for any changes in exciter voltage. The fixed resistance values are given in Fig. 13 and the varistor resistance characteristic is shown on Fig. 14.

Assume that a ground fault occurs at the full positive-voltage point of the field winding, as indicated in Fig. 13. Further, assume that the fault has a fault resistance R_F associated with it. Determine the maximum value of fault resistance for which the relay will just pick up.

Calculation Procedure:

1. Redraw Relay Circuit

The first step is to redraw the circuit to permit simplification and allow reduction to a series-parallel equivalent (Fig. 15). Note that the ground fault completes the relay circuit through R_F to allow relay coil current to flow.

2. Solve for R_F

Because one of the circuit elements (the varistor) is nonlinear, the most straightforward way to determine the maximum value of R_F is to make assumptions with regard to the operating point on the varistor curve, and then make successive calculations of the relay current magnitude. Continue until the relay current is very near pickup (0.47 mA). At

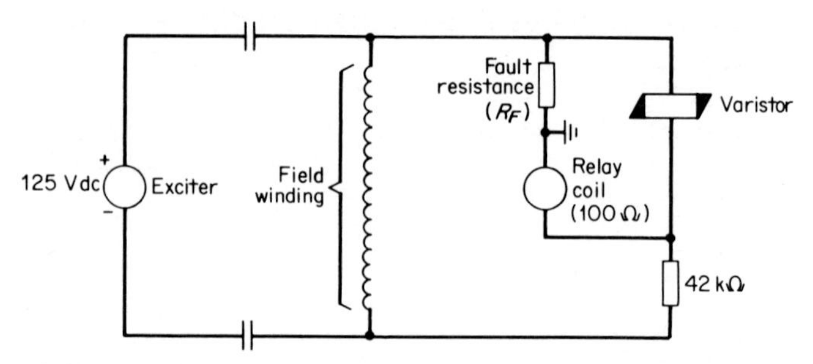

Fig. 15 Simplified circuit of generator-field ground relay. Ground fault is at positive terminal of field winding.

this point, the value of R_F required to obtain the proper voltage division will be the limiting value of the ground-fault resistance.

The first assumption of varistor operating voltage is entirely arbitrary. Assume that approximately one-half the exciter voltage (60 V dc) is the varistor voltage. Examination of Fig. 14 reveals that the varistor characteristic gives 1.4 mA at 60 V dc. This is equivalent to a varistor resistance $R_V = 42{,}900\ \Omega$.

To solve for R_F, write the expression for the equivalent resistance of R_F in parallel with the varistor: $R_{eq} = R_F R_V/(R_F + R_V) = 42{,}900 R_F/(R_F + 42{,}900)$. Note that in this calculation and the calculations which follow, the resistance of the relay coil (100 Ω) is ignored because it is very small in comparison with the fault resistance R_F.

The magnitude of R_{eq} may be determined by voltage division: $R_{eq}/42{,}000 = 60\ \text{V}/65\ \text{V}$ or $R_{eq} = (60/65)(42{,}000) = 38{,}769\ \Omega$; therefore, $38{,}769 = 42{,}900 R_F/(R_F + 42{,}900)$. Solving, we find $R_F = 402{,}612\ \Omega$.

The current through the relay coil is: $60/402{,}612 = 0.149$ mA. The relay current is well below the relay pickup current of 0.47 mA. Therefore, the varistor must be operated at a voltage that will give a significantly higher relay coil current.

Now assume a lower voltage, say 50 V, across the varistor. From Fig. 14, the varistor allows 1.0 mA at 50 V dc; therefore, $R_V = 50{,}000\ \Omega$. Following the same procedure as above, one obtains $R_F = 63{,}636\ \Omega$, and finds that the current through the relay coil is $50/63{,}636 = 0.79$ mA.

This amount of relay current is considerably above the relay pickup current, 0.47 mA. Therefore, the threshold point of the relay will be at a varistor voltage that is between the voltages used in the first two assumptions.

Assume a varistor operating voltage that is midway between the first and second assumed values, i.e., 55 V dc. For $V_V = 55$ V dc, the varistor current is 1.2 mA and $R_V = 45{,}833\ \Omega$. Then, $R_{eq} = 45{,}833 R_F/(R_F + 45{,}833)$. One finally obtains $R_F = 117{,}859\ \Omega$ and the relay current $= 0.47$ mA. Because the pickup current of the relay is 0.47 mA, the threshold operating voltage of the varistor is 55 V dc, and the limiting value of fault resistance R_F is 117,859 Ω.

Related Calculations: A similar step-by-step calculation procedure makes it possible to determine the ground-fault resistance sensitivity for faults at other points on the generator field winding. Similarly, the effect of exciter voltage variations on relay sensitivity may be determined.

CONVERSION OF WATT-VAR DIAGRAM TO *R-X* DIAGRAM

Assume that the impedance-element setting of a generator loss-of-field relay is to be based on the machine capability curves and the steady-state-stability limit curve. The machine characteristics are normally plotted on a watt-var diagram; the relay characteristic is shown on an *R-X* diagram. The machine capability curve at 208 × 10^3 Pa (30 psi) hydrogen gas pressure is shown in Fig. 16. Convert this curve to the equivalent *R-X* diagram curve.

Calculation Procedure:

1. Graphically Determine Several Power–Reactive-Power Coordinates on Capability Curve

Draw several straight-line intercepts from the origin of the power–reactive-power diagram. One of the intercepts should go through the point of discontinuity that joins the stator-limited region of the capability curve with the stator-end-iron-limited region. Only one intercept is required in the stator-limited region, because this curve is known to be a circle with its center at the origin, on both the power–reactive-power diagram and the *R-X* diagram. Intercepts at 30° intervals are suggested in the stator-end-iron-limited region. It is sufficient to measure the apparent-power magnitude at each intercept; these magnitudes are summarized in Table 1.

2. Convert Apparent-Power Values to Impedance Values

In order to obtain the impedance equivalent of the apparent-power value, use $Z_{pu} = 1/\text{kVA}$. In the conversion from the power–reactive-power diagram to an *R-X* diagram, the angle remains the same. Therefore, by converting apparent-power values to Z (i.e.,

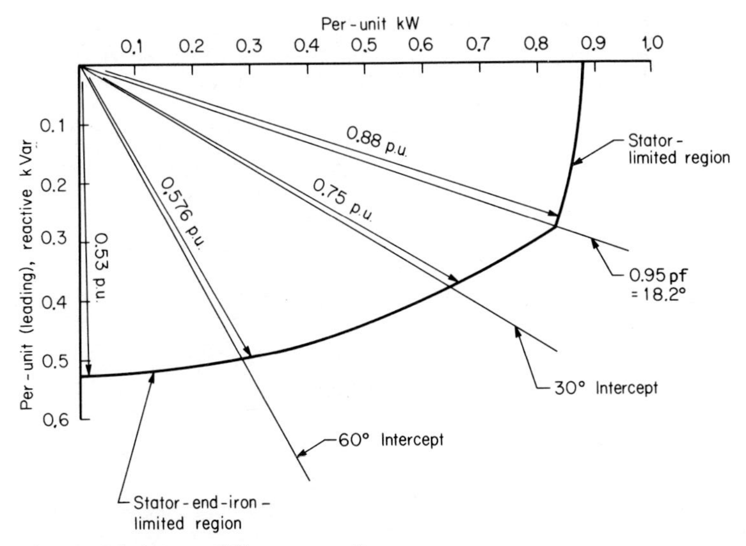

Fig. 16 Machine capability watt-var diagram.

TABLE 1 Apparent Power Measurements at Several Intercepts

Angle, measured from +kW axis, degrees	*Per-unit apparent power, kVA*
18.2 (= $\cos^{-1} 0.95$)	0.88
30	0.75
60	0.576
90	0.53

TABLE 2 Impedance Equivalents of kVA Values in Table 1

Angle, measured from +R axis, degrees	*Per-unit impedance, Z*
18.2	1.14
30	1.33
60	1.74
90	1.89

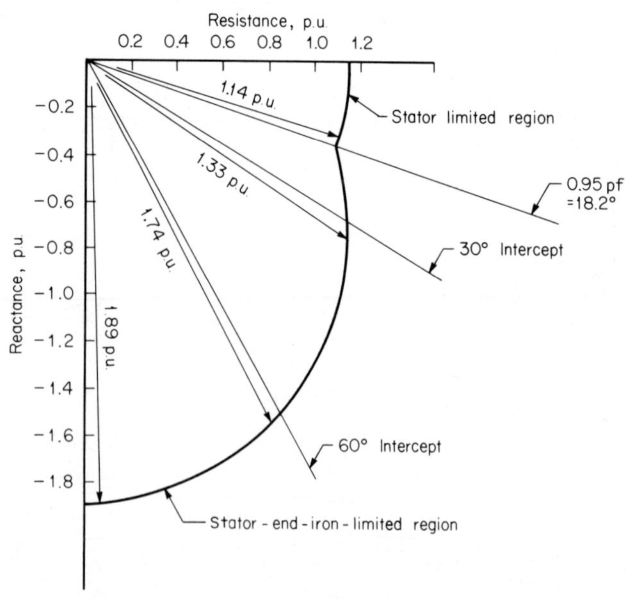

Fig. 17 Machine capability *R-X* diagram.

impedance) values, one readily accomplishes a conversion from a power–reactive-power diagram to an *R-X* diagram. Table 2 lists the impedance equivalents of the apparent-power values of Table 1.

3. *Plot Machine Capability Curve on R-X Diagram*

Using the values in Table 2, intercepts are drawn and the curve is plotted in Fig. 17.

Related Calculations: Similar calculations are performed to convert shunt capacitance and shunt reactor characteristics, normally given in MVA, to equivalent impedance values.

Section 16 POWER-SYSTEM STABILITY

Cyrus Cox
Professor of Electrical Engineering, South Dakota School of Mines and Technology

Norbert Podwoiski
Principal Engineer, Transmission Planning, Detroit Edison

REFERENCES Anderson and Fouad—*Power System Stability and Control,* Iowa State University Press; Byerly and Kimbark—*Stability of Large Electric Power Systems,* IEEE Press; Crary—*Power System Stability,* Wiley; Elgerd—*Electric Energy Systems Theory: An Introduction,* McGraw-Hill; IEEE—*Modern Concepts of Power System Dynamics,* IEEE Press; Kimbark—*Power System Stability,* Wiley; Meisel—*Principles of Electromechanical-Energy Conversion,* McGraw-Hill; Stagg and El-Abiad—*Computer Methods in Power Systems Analysis,* McGraw-Hill; Stevenson—*Elements of Power System Analysis,* McGraw-Hill; Westinghouse—*Electrical Transmission and Distribution Reference,* Westinghouse Electric Corp.

INTRODUCTION

Power-system stability calculations fall into three major categories: transient, steady-state, and dynamic stability. Power-system response to major disturbances, such as faults or loss of generation, is referred to as *transient stability*. Steady-state and dynamic stability include response to small disturbances, such as load changes. Steady-state stability neglects the effects of automatic voltage and frequency control, whereas dynamic stability considers some of the important effects of automatic control.

Analysis of power system stability is complex and nonlinear. As a consequence, final design decisions are generally based on detailed computer simulations. The simplified calculations in the procedures are only approximations. Nevertheless, simplified calculations do provide a starting point for, and a check of, computer simulations and an understanding of the factors that influence power-system stability. In addition, in preliminary design, detailed system data are generally not known and simplified calculations may be all that is required or feasible.

The major simplification in the calculations is that a generating facility can be represented reasonably well by a single machine connected through an external impedance to an infinite bus. Transient and steady-state stability calculations further assume that the machine can be represented as a constant internal voltage behind an impedance. Dynamic stability calculations assume that the machine equations can be linearized about an initial operating point for small perturbation analysis.

The first procedure presented is the analysis of a synchronous machine connected to an infinite bus. This method is a basic step in many of the procedures which follow. The remaining procedures are presented in an order which parallels the stability design process for a generating facility, i.e., in the order of transient, steady-state, and dynamic stability. In practice, the stability design process ends with detailed computer simulations to verify overall stability design.

SYNCHRONOUS MACHINE–INFINITE BUS ANALYSIS

A 1000-MVA, 26-kV synchronous machine is connected to an infinite bus through a generator-step-up (GSU) transformer and an external system reactance as shown in Fig. 1. Using the machine MVA rating as a common volt-ampere base, the machine transient reactance X'_d is 40 percent on a 26-kV base; the GSU transformer reactance is 10 percent on the 24.7/345-kV base; the external system reactance is 25 percent on a 345-kV base. The infinite-bus voltage is given on a per-unit basis referred to 345 kV. Compute the voltage behind transient reactance and construct the machine phasor diagram.

Calculation Procedure:

1. Compute Base-Matching Transformer per-Unit Tap Ratio

Because the GSU transformer-primary base voltage and the machine base voltage do not match, a base-matching ideal transformer is inserted between the machine and GSU

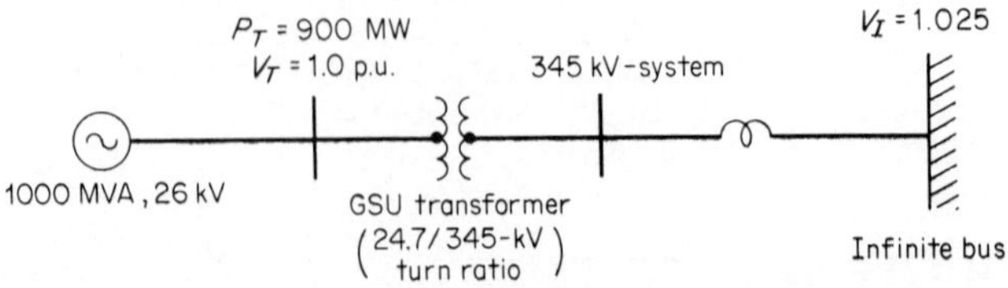

Fig. 1 Synchronous machine connected to an infinite bus through a GSU transformer and external system reactance.

transformer reactances. On the machine side, the tap ratio a = (GSU transformer primary kV base)/(machine kV base) = 24.7/26 kV = 0.95 per unit. The equivalent circuit of the base-matching ideal transformer is shown in Fig. 2.

2. *Sketch and Label the Equivalent Network*

To simplify the resultant equivalent network, reflect the GSU reactance, external system reactance, and the infinite-bus voltage to the machine side of the base-matching ideal transformer. Use $X_{eq} = a^2(X_{GSU} + X_s)$, where X_{eq} = equivalent reactance between the machine terminal and the infinite bus, a = ideal-transformer tap ratio on a per-unit basis, X_{GSU} = GSU transformer reactance, and X_s = external system reactance. Thus, in per-unit quantities, $X_{eq} = (0.95)^2(0.10 + 0.25) = 0.316\ \Omega$ per unit. The equivalent infinite-bus voltage $V_{Ieq} = aV_I = (0.95)(1.025) = 0.974$ per-unit V. The equivalent network of the power system of Fig. 1 is displayed in Fig. 3. The infinite bus is chosen as the reference bus with the angle set to 0°. The machine terminal is bus T.

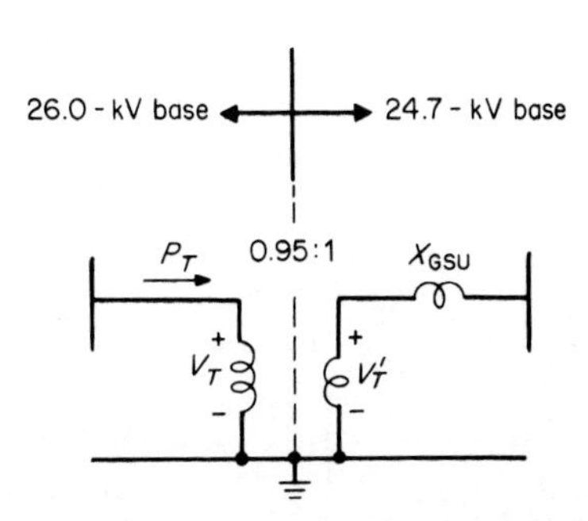

Fig. 2 Equivalent circuit of base-matching ideal transformer. The voltage relation for an ideal transformer is $V_T = (0.95/1.0)\ V'_T$.

3. *Compute the Machine-Terminal Voltage Angle*

Use the relation for the real-power flow between two buses of an ac network when resistances are neglected: $P_{ij} = V_iV_j \sin \delta_{ij}/X_{ij}$, where P_{ij} = real-power flow from bus i to j, V_i and V_j are the bus voltages, δ_{ij} = angle between bus i and j with bus j taken as the reference, and X_{ij} = equivalent reactance between buses i and j. For buses T and I in Fig. 3, the known values are, on a per-unit basis: $P_{ij} = P_T$ = 900 MW/1000 MVA = 0.90; $V_i = V_T = 1.0$; $V_j = V_{Ieq} = 0.974$; $X_{ij} = X_{eq} = 0.316$. Substituting these values in the power-flow relation and solving for the unknown machine-terminal voltage angle $\delta_{ij} = \delta_T$ yields: $\delta_T = \sin^{-1} [(0.90 \times 0.316)/(1.0)(0.974)] = 17.0°$.

4. *Compute the Machine Reactive Output*

Use the relation: $Q_{ij} = (V_i/X_{eq})(V_i - V_j \cos \delta_{ij})$, where Q_{ij} = the reactive power flow from bus i to j; the remainder of the quantities are as defined and computed in Step 3. Thus $Q_{ij} = Q_T = (1.0/0.316)(1.0 - 0.974 \cos 17.0°) = 0.220$ Mvar per unit or 220 Mvar.

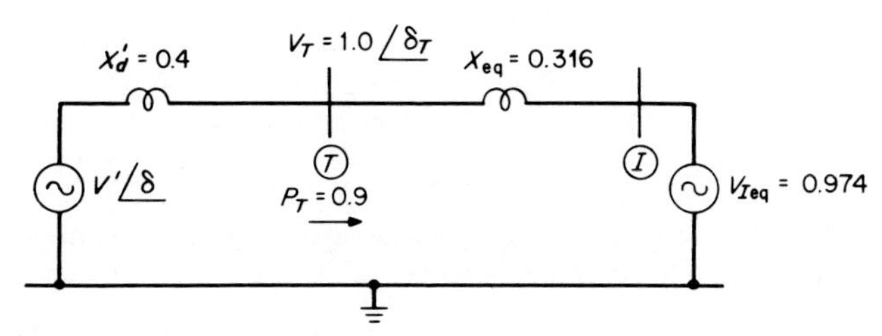

Fig. 3 Equivalent network of power system of Fig. 1.

5. *Compute the Machine-Terminal Current*

Use $\mathbf{I_T} = \mathbf{S_T^*}/\mathbf{V_T^*}$, where $\mathbf{I_T}$ = complex machine-terminal current, $\mathbf{S_T^*}$ = complex conjugate of the machine apparent power $(P_T - jQ_T)$, and $\mathbf{V_T^*}$ = complex conjugate of the machine-terminal voltage. Hence, $\mathbf{I_T} = (0.90 - j0.220)/1.0\underline{/-17.0°} = 0.926\underline{/-13.7°}/1.0\underline{/-17.0°} = 0.926\underline{/3.3°}$ per-unit A.

6. *Compute the Voltage behind Transient Reactance*

Use $\mathbf{V'} = \mathbf{V_T} + jX_d'\mathbf{I_T}$, where $\mathbf{V'}$ = complex voltage behind transient reactance; the other symbols are the same. Thus, $\mathbf{V'} = 1.0\underline{/-17.0°} + (0.926\underline{/3.3°})(0.40\underline{/90°}) = (0.935 + j0.662) = 1.15\underline{/35.2°}$.

7. *Construct the Machine Phasor Diagram*

The phasor diagram for this machine is provided in Fig. 4. The reference axis is chosen as the angle of the infinite bus voltage. The angle δ is the angle between $\mathbf{V'}$ and the

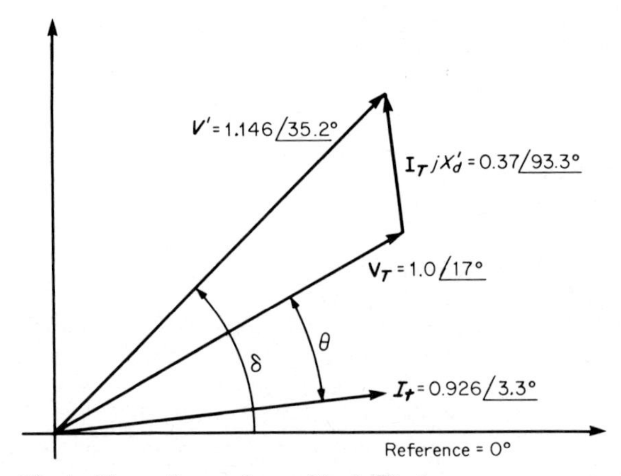

Fig. 4 Phasor diagram for machine in Fig. 1.

reference axis. The angle θ is the power-factor angle and is computed from the relation $\theta = \theta_V - \theta_I$, where θ_V = machine-terminal voltage angle and θ_I = machine-terminal current angle. Thus, $\theta = 17.0° - 3.3° = 13.7°$. The machine operating power factor $\text{pf} = \cos\theta = \cos 13.7° = 0.972$. Since the machine-terminal current angle lags the voltage angle $(\theta > 0)$, by convention this is a lagging power factor. If the current leads the voltage $(\theta < 0)$, then the power factor is leading.

Related Calculations: This general procedure is included because it is a basic step used in several other stability calculations. The equivalent system reactance used in the procedure is the Thevenin impedance at the plant's high-voltage bus. The system reactance can be estimated by the relation $X_s = 1.0/I_{sc}$, where I_{sc} = per-unit magnitude of the short-circuit contribution from the system for a three-phase fault at the plant's high-voltage bus. The infinite-bus voltage is the Thevenin open-circuit voltage at the plant's high-voltage bus. In general, this voltage is given as a range of values rather than one specific value.

SOLUTION OF THE SWING EQUATION

The equivalent network of a synchronous machine connected to an infinite bus is shown in Fig. 5. All quantities in Fig. 5 are given per-unit and the impedance values are expressed on a 1000-MVA base. The machine operates initially at synchronous speed (377 rad/s on a 60-Hz system). A three-phase fault is simulated by closing switch S1 at time $t = 0$. S1 connects the fault bus F to ground through a fault reactance X_f. Compute the machine angle and frequency for 0.1s after S1 is closed. Compute the machine angle and frequency with $X_f = 0$ and 0.15 per-unit Ω.

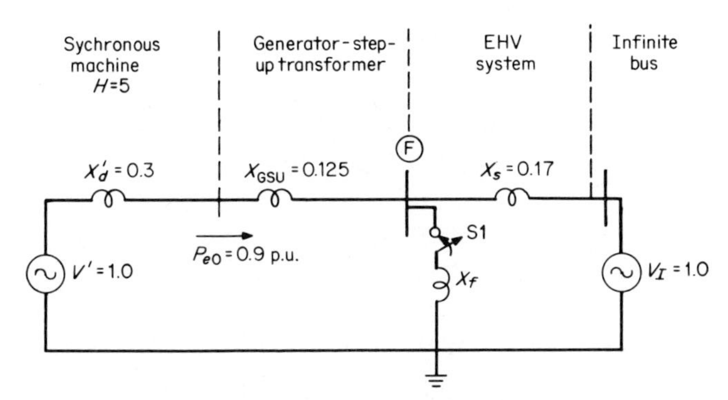

Fig. 5 Equivalent network of single machine connected to an infinite bus. All quantities are per-unit on common 1000-MVA base.

Calculation Procedure:

1. *Compute the Initial Internal Machine Angle*

Use $\delta_0 = \sin^{-1}(P_{e0}X_{\text{tot}}/V'V_I)$, where δ_0 = initial internal machine angle with respect to the infinite bus (for this machine δ_0 is the angle of the voltage V'), P_{e0} = initial machine electrical power output, V' = voltage behind transient reactance X'_d, V_I = infinite-bus voltage, and X_{tot} = total reactance between the voltages V' and V_I. Thus, $\delta_0 = \sin^{-1}[(0.9)(0.3 + 0.125 + 0.17)/(1.0)(1.0)] = 32.3°$.

2. *Determine Solution Method*

Computing synchronous-machine angle and frequency changes as a function of time requires the solution of the swing equation, $d\delta^2/dt^2 = (\omega_0/2H)(P_m - P_e)$, where δ = internal machine angle with respect to a synchronously rotating reference (infinite bus), ω_0 = synchronous speed in rad/s, H = per-unit inertia constant in s, t = time in s, P_m = per-unit machine mechanical shaft power, and P_e = per unit machine electric-power output. The term $(P_m - P_e)$ is referred to as the machine accelerating power and is represented by the symbol P_a.

If P_a can be assumed constant or expressed explicitly as a function of time, then the swing equation will have a direct analytical solution. If P_a varies as a function of δ, then numerical integration techniques are required to solve the swing equation.

Because in this calculation the solution of the swing equation is required for only a relatively short period of time (0.1 s), P_m can be assumed to remain constant. Thus, P_a will vary in a manner similar to P_e.

In the network of Fig. 5, for $X_f = 0$ the voltages at bus F and P_e during the fault are zero. Thus, $P_a = P_m =$ constant and therefore with $X_f = 0$ there is a direct analytical solution to the swing equation. For the case in which $X_f = 0.15$ (or $X_f \neq 0$), the voltages at bus F and P_e during the fault are greater than zero. Thus, $P_a = P_m - P_e$ = a function of δ. Therefore, with $X_f = 0.15$, numerical integration techniques are required for solution of the swing equation.

3. *Solve Swing Equation with* $X_f = 0$

Use the following procedure:

a. Compute the machine accelerating power during the fault. Use $P_a = P_m - P_e$; for this machine, $P_m = P_{e0} = 0.90$ per unit; $P_e = 0$. Thus, $P_a = 0.90$.

b. Compute the new machine angle at time $t = 0.1$ s. The solution of the swing equation with constant P_a is $\delta = \delta_0 + (\omega_0/4H)P_a t^2$, where the angles are expressed in radians and all other values are on a per-unit basis. Thus,

$$\delta = \frac{35.2°}{57.3°/\text{rad}} + \left(\frac{377}{4 \times 5}\right) 0.90\ (0.1)^2 = 0.783 \text{ rad, or } 44.9°$$

c. Compute the new machine frequency at time $t = 0.1$ s. The machine frequency is obtained from the relation $\omega = d\delta/dt + \omega_0$, where $d\delta/dt = (\omega_0 P_a t)/2H$. Thus, $\omega = [(377)(0.90)(0.1)/(2)(5)] + 377 = 380.4$ rad/s, or 60.5 Hz.

4. *Solve Swing Equation with* X_f **= 0.15 per-unit**

Use the following procedure.

a. Select a numerical-integration method and time step. There are many numerical-integration techniques for solving differential equations including Euler, modified Euler, and Runge-Kutta methods, etc. The Euler method is selected here. Solution by Euler's method requires expressing the second-order swing equation as two first-order differential equations. These are: $d\delta/dt = \omega(t) - \omega_0$ and $d\omega/dt = (\omega_0 P_a t)/2H$. Euler's method involves computing the rate of change of each variable at the beginning of a time step. Then, on the assumption that the rate of change of each variable remains constant over the time step, a new value for the variable is computed at the end of the step. The following general expression is used: $y(t + \Delta t) = y(t) + (dy/dt)\,\Delta t$, where y corresponds to δ or ω and Δt = time step; $y(t)$ and dy/dt are computed at the beginning of the time step. A time step of 1 cycle (0.0167 s) is selected.

b. From Fig. 6, determine the expression for the electrical power output during this time step. For this network, Case 3 is used where $X_G = X_d' + X_{GSU} = 0.3 + 0.125 = 0.425$ per-unit Ω, $X_s = 0.17$ Ω, and $X_f = 0.15$ Ω. Thus,

$$P_E = \left(\frac{(1.0)(1.0)}{0.424 + 0.17 + [(0.425)(0.17)/0.15]}\right) \sin\delta$$
$$= 9.930 \sin\delta$$

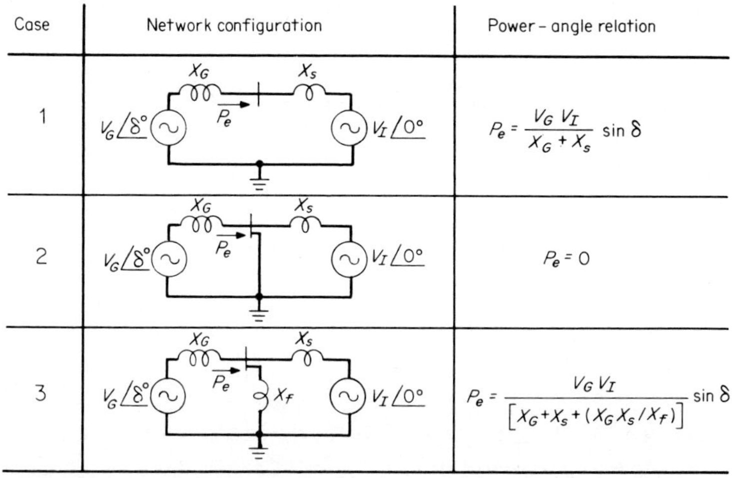

Case	Network configuration	Power – angle relation
1	$V_G \angle \delta°$, X_G, P_e, X_s, $V_I \angle 0°$	$P_e = \dfrac{V_G V_I}{X_G + X_s} \sin \delta$
2	$V_G \angle \delta°$, X_G, P_e, X_s, $V_I \angle 0°$	$P_e = 0$
3	$V_G \angle \delta°$, X_G, P_e, X_s, X_f, $V_I \angle 0°$	$P_e = \dfrac{V_G V_I}{[X_G + X_s + (X_G X_s / X_f)]} \sin \delta$

Fig. 6 Power-angle relations for general network configurations; resistances are neglected. $V_G \angle \delta$ = internal machine angle, X_G = machine reactance, X_s = system reactance, P_e = machine electrical power output, $V_I \angle 0°$ = infinite bus voltage, and X_f = fault reactance.

c. Compute $P_a(t)$ at the beginning of this time step ($t = 0$ s). Use $P_a(t) = P_m - P_e(t)$ for $t = 0$, where $P_m = P_{e0} = 0.90$ per unit; $P_e(0) = 0.930 \sin 35.2°$. Thus, $P_a(t = 0) = 0.90 - 0.930 \sin 35.2° = 0.36$ per unit.

d. Compute the rate of change in machine variables at the beginning of the time step ($t = 0$). The rate of change in the machine phase angle $d\delta/dt = \omega(t) - \omega_0$ for $t = 0$, where $\omega(0) = 377$ rad/s. Thus, $d\delta/dt = 377 - 377 = 0$ rad/s. The rate of change in machine frequency is $d\omega/dt = \omega_0 P_a(t)/2H$ for $t = 0$, where $P_a(0) = 0.36$ per unit as computed in *c*. Thus, $d\omega/dt = (377)(0.36)/(2)(5) = 13.68$ rad/s^2.

e. Compute the new machine variables at the end of the time step ($t = 0.167$ s). The new machine phase angle is $\delta(0.0167) = \delta(0) + (d\delta/dt)\,\Delta t$, where the angles are expressed in radians and Δt = time step = 0.167 s. Thus, $\delta(0.167) = 35.2°/(57.3°/\text{rad}) + (0)(0.167) = 0.614$ rad, or 35.2°. The new machine frequency is $\omega(0.167) = \omega(0) + (d\omega/dt)\,\Delta t = 377 + (13.7)(0.167) = 377.228$ rad/s, or 60.04 Hz.

f. Repeat *c*, *d*, and *e* for the desired number of time steps. Table 1 displays the remaining calculations for the 0.1-s solution time. The machine frequency at 0.1 s = 377.334 rad/s or 60.21 Hz; the corresponding machine angle is 38.3°.

Related Calculations: The general procedure described here is similar to algorithms used in sophisticated stability programs for computers. In summary, this general procedure consists of computing: (1) the initial conditions; (2) electric output of the machine(s); (3) rate of change of the machine phase angle, frequency, and other state variables; (4) new values for the state variables at the end of the time step. Steps (2) through (4) are repeated for the selected number of time steps.

TABLE 1 Computations for Solution of Swing Equation by Euler's Method

Time (sequential)	*Frequency*		*Angle*					$\frac{d\delta}{dt}\Delta t$	$\frac{d\omega}{dt}\Delta t$
	rad/s	*Hz*	*Rad*	*Degrees*	$P_a(t)$	$d\delta/dt$	$d\omega/dt$		
0	377	60	0.614	35.2	0.363	0	13.7	0	0.228
0.0167	377.288	60.04	0.614	35.2	0.363	0.228	13.7	0.0038	0.228
0.0334	377.456	60.07	0.617	35.4	0.361	0.456	13.6	0.0076	0.227
0.0501	377.683	60.1	0.624	35.8	0.355	0.683	13.4	0.0114	0.223
0.0668	377.906	60.15	0.637	36.5	0.347	0.906	13.1	0.015	0.218
0.0835	378.124	60.18	0.651	37.3	0.335	1.12	12.6	0.0187	0.210
0.100	378.334	60.21	0.668	38.3					

$P_a = 0.930 \sin\delta$, $d\delta/dt = \omega(t) - \omega_0$
$d\omega/dt = \omega_0 P_q(t)/2H$, $\Delta t = 0.0167$ s, $\delta(t + \Delta t) = \delta(t) + d\delta/dt\,\Delta t$
and $\omega(t + \Delta t) = \omega(t) + d\omega/dt\,\Delta t$.

CRITICAL FAULT-CLEARING TIME

Compute the critical fault-clearing (CFC) time for a zero-impedance three-phase fault on line 1 in the power system of Fig. 7*a*. The CFC time is the maximum time a fault can be sustained while transient stability is maintained.

Calculation Procedure:

1. Compute Machine Initial Conditions

Figure 7*b* displays the equivalent network of the power system in Fig. 7*a* with the three required machine initial conditions: $P_e = 0.90$ per unit = electrical power output, $V' = 1.0$ = voltage behind transient reactance, and $\delta_0 = 0.723$ rad (41.1°) = phase angle with respect to the infinite bus.

2. Compute Machine Power-Angle Relations

Use Fig. 6 to determine the machine power-angle relations before, during, and after the fault. Before the fault, breakers B1 and B2 are closed and switch S1 is open. Thus, the network is Case 1 of Fig. 6 where $X_G = X'_d + X_{\mathrm{GSU}} = 0.40 + 0.15 = 0.55$ and $X_s = X_{L1}X_{L2}/(X_{L1} + X_{L2}) = (0.4)(0.4)/(0.4 + 0.4) = 0.2$. Substituting for Case 1 the power-angle relation before the fault, we find $P_{eb} = [(1.0)(1.025)/(0.55 + 0.20)] \sin\delta = 1.36 \sin\delta$. During the fault, switch S1 closes and the network is in the form of Case 2. Thus, the power-angle relation during the fault is $P_{ed} = 0$. After the fault, breakers B1 and B2 open to clear the fault and the network is of the form of Case 1, where X_G = same as before the fault and $X_s = X_{L2} = 0.40$. Substituting for Case 1, we find the power-angle relation after the fault; $P_{ea} = [(1.0)(1.025)/(0.55 + 0.40)] \sin\delta = 1.08 \sin\delta$. The peak values for the power-angle relations occur when $\delta = \pi/2$ ($\delta = 90°$).

3. Compute the Maximum Allowable Machine Angle

Use $\delta_m = \pi - \sin^{-1}(P_e/P'_{ea})$, where δ_m = maximum allowable machine angle ($\pi/2 < \delta_m < \pi$) and P'_{ea} = peak value of the power-angle relation after the fault is

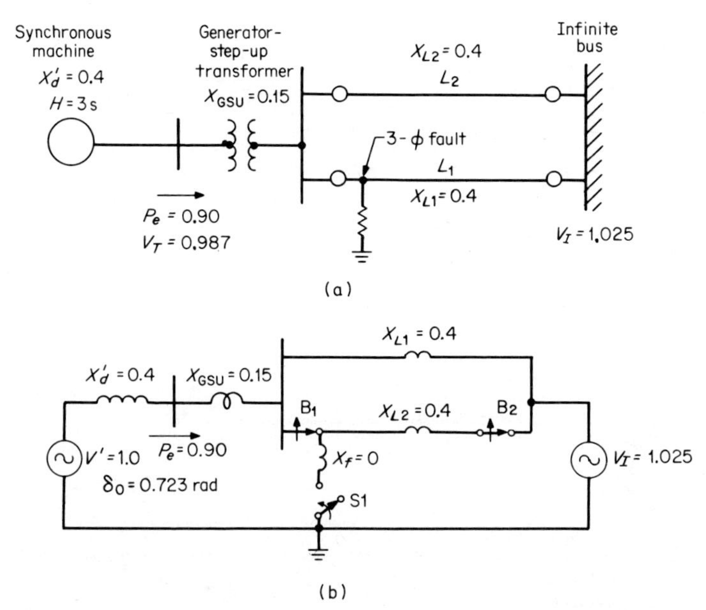

Fig. 7 Computing CFC time. (*a*) Typical power system. (*b*) Equivalent network.

cleared. Thus, $\delta_m = \pi - \sin^{-1}(0.9/1.08) = 3.14 - 0.985 = 2.16$ rad (124°). Figure 8 displays a plot of the three power-angle relations with the initial and maximum allowable angles indicated.

4. Compute Ratios of the Power-Angle Relations

Compute the ratio $r_1 = P'_{ed}/P'_{eb}$, where P'_{ed} = peak value of the power-angle relation during the fault and P'_{eb} = peak value of power-angle relation before the fault. Thus, $r_1 = 0/1.36 = 0$. Compute the ratio $r_2 = P'_{ea}/P'_{eb}$, where P'_{ea} = peak value of power-angle relation after the fault is cleared. Thus, $r_2 = 1.08/1.36 = 0.794$.

5. Compute the Critical Fault-Clearing Angle

The CFC angle is the maximum angle through which the machine phase angle can swing while maintaining stability. To compute the CFC angle use: $\delta_c = \cos^{-1}\{[1/(r_2 - r_1)][(P_e/P'_{eb})(\delta_m - \delta_0) + r_2 \cos \delta_m - r_1 \cos \delta_0]\}$, where the angles are expressed in radians and all symbols are as previously defined and computed. Thus, $\delta_c = \cos^{-1}\{[1/(0.794 - 0)][0.9/(1.36)(2.16 - 0.72) + 0.794 \cos 2.16]\} = \cos^{-1} 0.64 = 0.874$ rad (50.1°).

6. Compute the Critical Fault-Clearing Time

CFC time can be determined by solving the machine swing equation. The time for the machine to swing from the initial angle δ_0 to the CFC angle δ_c is the CFC time. For this machine the electric power output during the fault is constant ($P_{ed} = 0$); therefore, there is a direct analytical solution for the CFC time. However, in cases where P_{ed} varies as a function of the angle δ, then numerical integration techniques will be necessary to compute the CFC time. Thus, for this machine, the CFC time in seconds is $t_c = \sqrt{4(\delta_c - \delta_0)H/\omega_0 P_e}$, where the angles are expressed in radians, H = per-unit inertia constant in seconds, ω_0 = synchronous speed (377 rad/s for 60 Hz system), and P_e =

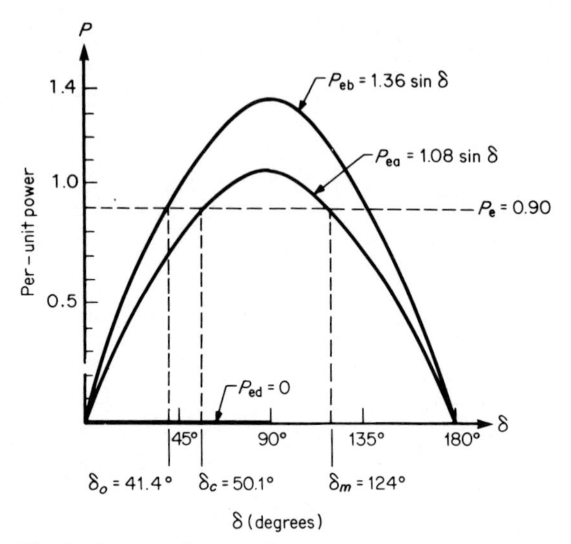

Fig. 8 Power-angle curves for computing critical fault-clearing time.

initial electrical power output. Hence, $t_c = \sqrt{(4)(0.874 - 0.720)(3)/(377)(0.9)} =$ 0.073 s, or 4.4 cycles.

Related Calculations: If the critical fault-clearing time is greater than the expected relay-breaker clearing times, then the plant is stable for the fault. The procedure presented here is based on the equal-area criterion for determining stability. A derivation for the relations used in the procedure can be found in several of the references listed at the beginning of the section.

SELECTING STABILITY DESIGN CRITERIA

Stability design criteria are to be selected for the generating facility shown in Fig. 9. The design basis for the plant is that the mean time to instability (MTTI—failure due to unstable operation) is greater than 500 yr.

Calculation Procedure:

1. Specify a Fault-Exposure Zone

A fault-exposure zone defines a boundary beyond which faults can be neglected in the selection of the stability criteria. For example, on long out-of-plant transmission lines it is typically unnecessary to consider line faults at the remote end. Specifying a fault-exposure zone is generally based on the fault frequency and/or the magnitude of the reduction in the plant's electric output during the fault.

For this plant, specify a fault-exposure zone of 50 km (31 mi); i.e., it will be assumed that the plant will maintain transient stability for faults beyond this distance. In addition, faults in the switchyard are considered outside the fault-exposure zone. This is done because switchyard-fault frequency is generally much less than line-fault frequency.

2. Compute Fault Frequency within the Exposure Zone

Use Table 2 to estimate the fault frequency per line. The values given are very general and based on composite data from numerous power industry sources. Line-fault fre-

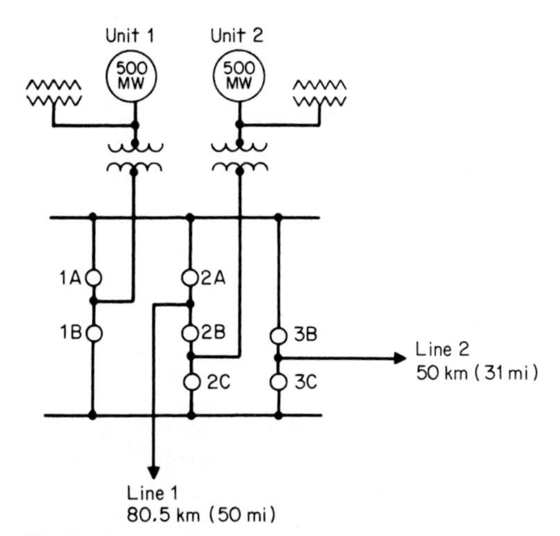

Fig. 9 A two-unit generating facility, switchyard, and out-of-plant transmission.

quency varies with factors such as line design, voltage level, soil conditions, air pollution, storm frequency, etc.

For this plant, use the typical line-fault frequency rates. Thus, the fault frequency within the exposure zone is $f =$ (number of lines)(fault frequency)(exposure zone/100 km), where the fault frequency is in faults/100 km·yr and exposure zone is in km. Then $f = (2)(1.55)(50/100) = 1.55$ faults/yr within the exposure zone.

TABLE 2 Typical Extra-High Voltage (EHV) Probability Data

Data description	*Optimistic*	*Typical*	*Pessimistic*
Line-fault frequency, faults/100 km·yr	0.62	1.55	3.1
Conditional probability that a relay-breaker fails to sense or open for a fault (3ϕ failure)*	0.0002	0.003	0.01

*Assumes nonindependent pole tripping. For independent pole tripping, probabilities are for a 1ϕ failure.

Use Table 3 to estimate the fault frequency on the basis of the number of phases involved. For this plant, use the composite values and assume that the unknown fault types consist entirely of single-phase faults. Thus, the three-phase fault frequency $f_{3\phi}$ = (percent of total faults which involve 3ϕ)(f) = (0.03)(1.55) = 0.465 − $3\ \phi$ faults/yr, or $1/f_{3\phi} \cong \text{MTBF}_{3\phi}$ = mean time between $3\ \phi$ faults = 1/0.0465 = 21.5 yr. Similar calculations are made for the other fault types and are tabulated below.

Fault type	*Faults per year*	*MTBF, years*
Three-phase	0.0465	21.5
Double-phase	0.140	7.17
Single-phase (+ unknown)	1.36	0.733

TABLE 3 Typical Composition of EHV Transmission-Line Faults

	Percent of total faults		
Type of fault	*765 kV*	*EHV composite*	*115 kV*
Phase to ground	99	80	70
Double-phase to ground	1	7	15
Three-phase to ground	0	3	4
Phase to phase	0	2	3
Unknown	0	8	8
Total	100	100	100

Note that in the above calculation $f \cong 1/\text{MTBF}$. The precise relation for the frequency is $f = 1/(\text{MTBF} + \text{MTTR})$, where MTTR = mean time to repair (i.e., time to clear the fault and restore the system to its original state). The term MTTR is required in the precise relation because a line cannot fail while it is out of service. However, for stability calculations MTTR ≪ MTBF and thus the approximation $f \cong 1/\text{MTBF}$ is valid.

3. *Compute Breaker-Failure Frequencies*

Only breaker failures in which both generating units in Fig. 9 remain connected to the transmission system are considered. Tripping a generating unit in the process of clearing a "stuck breaker" is sometimes used as a stability aid. For this plant and transmission system, the failure of breaker 2B for a fault on line 1 will result in tripping unit 2. Thus, only three breaker failures affect stability: 2A, 3B, and 3C.

From Table 2, the conditional probability that a relay/breaker will fail to sense or open for a line fault is $p_{\text{bf}} = 0.003$ (typical value for nonindependent pole tripping). The frequency of a three-phase fault plus a breaker failure is $f_{3\phi,\text{bf}} = f_{3\phi} p_{\text{bf}} B$, where B = number of breaker failures which impact stability. Thus, $f_{3\phi,\text{bf}} = (0.465)(0.003)(3) = 4.19 \times 10^{-4}$ three-phase faults plus breaker failures per year, or $1/f_{3\phi,\text{bf}} = 2390$ yr. Similar calculations are made for the other fault types and are tabulated below.

Fault type	*Fault plus breaker failures per year*	*Mean time between faults plus breaker failures, yr*
Three-phase	0.000419	2390
Double-phase	0.00126	793
Single-phase (+ unknown)	0.122	81.7

4. *Select the Minimum Criteria*

As a design basis, the plant must remain stable for at least those fault types for which the MTBF is less than 500 years. Thus, from the fault-type frequencies computed in Steps 2 and 3, only a three-phase or two-phase fault plus breaker failure could be eliminated to select the minimum criteria. A check is now made to ensure that neglecting three-phase and two-phase faults plus breaker failures is within the design basis.

5. *Compute the Frequency of Plant Instability*

Use $f_I = \Sigma$ frequencies of fault types eliminated in Step 4, where f_I = frequency of plant instability. Thus, $f_I = f_{3\phi,\text{bf}} + f_{2\phi,\text{bf}} = 0.000419 + 0.00126 = 0.00168$ occurrences of instability per yr, or $1/f_I \cong$ mean time to instability $= 595$ yr. Thus, neglecting three-phase and two-phase faults plus breaker failures in the criteria, the value is within the original design basis of MTTI > 500 yr. If the MTTI computed in this step was less than the design basis, then additional fault types would be added to the minimum criteria and the MTTI recomputed.

6. *Select the Stability Criteria*

The following stability criteria are selected:

a. Three-phase fault near the plant's high-voltage bus cleared normally.

b. Single-phase fault near the plant's high-voltage bus plus a breaker failure.

If the plant remains stable for the above two tests, the original design basis has been satisfied.

Related Calculations: Specifying a design basis, although not a step in the procedure, is critical in selecting stability criteria. No general guidelines can be given for specifying a design basis. However, most utilities within the United States specify some type of breaker failure test in their criteria.

In addition to neglecting low-probability fault types, low-probability operating conditions can also be neglected in the selection of stability criteria. As an example, suppose that for this plant it was determined that during leading power-factor operation, the plant could not remain stable for a single-phase fault plus a breaker failure. Further assume that the plant transition rate into leading power-factor operation $\lambda = 2/\text{yr}$ and the mean duration of each occurrence is $r = 8$ h. Use the following general procedure to compute the mean time to instability.

1. Sketch a series-parallel event diagram as shown in Fig. 10*a*. Any break in the continuity of the diagram causes instability. A break in continuity could be caused by Event 3, or Event 4, or the simultaneous occurrence of Events 1 and 2.
2. Reduce the series-parallel event diagram to a single equivalent event. By recursively applying the relations in Table 4, the diagram can be reduced to an equivalent MTTI. To reduce the parallel combination of Events 1 and 2 use $\lambda_{12} = \lambda_1\lambda_2(r_1 + r_2)/(1 + \lambda_1 r_1 + \lambda_2 r_2)$, where λ_{12} = equivalent transition rate for Events 1 and 2, λ_1 = transition rate for Event 1 in occurrences per year, λ_2 = transition rate for Event 2 in occurrences per year (note that for this case $f_2 \cong \lambda_2 = 0.122$ occurrences per year as computed in Step 3), r_1 = mean duration of Event 1 in years $= 8\ \text{h}/(8760\ \text{h/yr}) = 0.0009$ yr, r_2 = mean duration of Event 2 in years (the duration of a single-phase fault plus breaker failure $= 0$ yr). Thus, $\lambda_{12} = (2)(0.122)(0.0009)/[1 + (2)(0.0009)] = 0.0002$ transitions/yr. The diagram is now reduced to Fig. 10*b*. Note that the equivalent duration of Events 1 and 2 is $r_{12} \cong 0$ yr. The equivalent transition rate for the three series events is $\lambda_{\text{eq}} = \lambda_{12} + \lambda_3 + \lambda_4 = 0.0002 + 0.00126 + 0.000419 = 0.00188$ occurrences/yr. Since $r_{\text{eq}} \cong 0$ yr, we have $\lambda_{\text{eq}} \cong f_{\text{eq}}$ = frequency of instability f_I, or $1/f_I$ = MTTI $= 1/0.00188 = 532$ yr (Fig. 10*c*).

This general procedure can be used either to select a criterion or to determine the adequacy of final stability design.

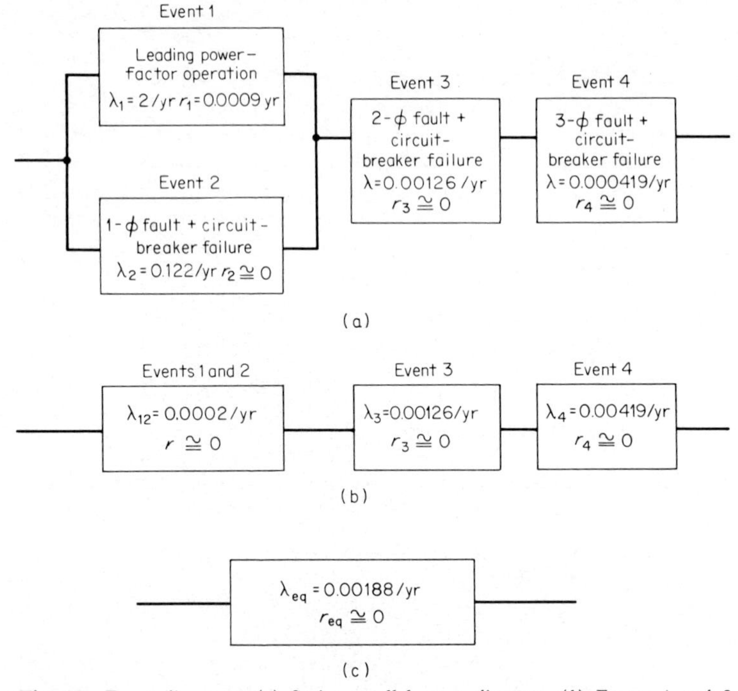

Fig. 10 Event diagrams. (*a*) Series-parallel event diagram. (*b*) Events 1 and 2 reduced to equivalent event. (*c*) All four events reduced to a single equivalent event.

TABLE 4 Equivalent Frequency-Duration Formulas for Occurrence of Events in Series or Parallel

Diagram / Variable	Series	Parallel
	λ_1, r_1 — λ_2, r_2	λ_1, r_1 ‖ λ_2, r_2
f_{12}^* = frequency, occurrences per year	1S $\dfrac{\lambda_1 + \lambda_2}{(1+\lambda_1 r_1)(1+\lambda_2 r_2)}$	1P $\dfrac{\lambda_1 \lambda_2 (r_1 + r_2)}{(1+\lambda_1 r_1)(1+\lambda_2 r_2)}$
r_{12} = mean time to repair, years	2S $\dfrac{\lambda_1 r_1 + \lambda_2 r_2 + (\lambda_1 r_1)(\lambda_2 r_2)}{\lambda_1 + \lambda_2}$	2P $\dfrac{r_1 r_2}{r_1 + r_2}$
λ_{12} = transition rate occurrences per year	3S $\lambda_1 + \lambda_2$	3P $\dfrac{\lambda_1 \lambda_2 (r_1 + r_2)}{(1 + \lambda_1 r_1 + \lambda_2 r_2)}$

*Although outage rate and frequency have same units of outage per year, they are not equivalent: λ is the reciprocal of mean time to failure (MTTF); f is the reciprocal of the sum of the MTTF and mean time to repair (MTTR). If MTTF $\gg$ MTTR, then $f \cong \lambda$.

TRANSIENT STABILITY AIDS

The generating facility of Fig. 9 cannot meet the stability design criteria. Provide a procedure to select a set of stability aids to meet the criteria. The number of transmission circuits is fixed.

Calculation Procedure:

1. Determine Critical Fault-Clearing Times

Compute or use computer simulation to determine the critical fault-clearing times for each contingency specified in the stability criteria.

2. Compute the Required Improvement in CFC Time

Use ΔCFC = relay-breaker clearing times − CFC time, where ΔCFC is the required improvement in CFC time. Table 5 displays the range of typical EHV (extra-high voltage) relay-breaker clearing times. Most components of the total clearing times

TABLE 5 Range of Typical EHV Relay-Breaker Clearing Times

	Time in cycles (60 Hz)		
Function	*Fast*	*Average*	*Slow*
Primary relay	0.25–0.5	1.0–1.5	2.0
Breaker clearing	1.0	3.0	3.0–5.0
Total normal clearing time	1.3–1.5	3.0–3.5	5.0–7.0
Breaker-failure detection	0.25–0.5	0.5–1.5	1.0–2.0
Coordination time	3.0	3.0–5.0	5.0–6.0
Auxiliary relay	0.25–0.5	0.5–1.0	1.0
Backup breaker clearing	1.0	2.0	3.0–5.0
Total backup clearing	5.75–6.5	9.0–13.0	15.0–20.0

are limited by the type of equipment used. The most common area to reduce ΔCFC is the time associated with the relay-coordination time for backup clearing. However, reduction in the margin time can result in erroneous backup clearing. The minimum achievable times displayed in Table 5 are associated with state-of-the-art equipment (1-cycle breakers and ultra-high-speed relaying).

3. Make Changes in Machine and GSU Transformer Parameters

A decrease in the machine transient reactance and/or an increase in the inertia constant can provide increases in CFC time. Table 6 displays typical ranges of these parameters

TABLE 6 Typical Range of Machine Transient Reactance and Inertia Constant for Modern Turbine-Generators

	Transient reactance, percent Ω		*Inertia constant, MWs/MVA*	
Type	*Low*	*High*	*Low*	*High*
Steam turbine, 1800 r/min	30	50	2.5	4.0
Steam turbine, 3600 r/min	20	40	1.75	3.5
Hydro-electric	20	35	2.5	6.0

TABLE 7 Typical Range of Impedance for Generator Step-Up Transformers

Nominal system voltage, kV	*Standard impedance in percent on GSU MVA base*	
	Minimum	*Maximum*
765	10	21
500	9	18
345	8	17
230	7.5	15
115–161	7.0	12
115	5.0	10

for modern turbine-generators. The most common way to achieve CFC increases in the generation system is to reduce the generator-step-up transformer (GSU) impedance. Table 7 displays the typical range of standard impedances for GSU transformers. GSU transformer impedances below the minimum standard can be obtained at a cost premium. Reductions in GSU transformer impedance may be limited by fault capabilities of the electrical system.

4. *Survey Transient Stability Aids*

Transient stability aids fall into three general categories, and the most commonly used aids are tabulated below.

Transient Stability Aids

Category	*Stability aid*
Machine control	High initial response excitation
	Turbine fast-valve control
Relay control	Independent pole tripping
	Selective pole tripping
Network control	Unit rejection schemes
	Series capacitors
	Braking resistors

Table 8 is a summary of typical improvements in CFC time and typical applications of the more commonly used stability aids. In general, stability aids based on machine control provide CFC increases of up to 2 cycles for delayed clearing and only marginal increases for normal clearing. Transmission relay control methods focus on CFC improvements associated with multiphase faults cleared in backup time. Network configuration control methods provide relatively large increases in CFC time for both primary and backup clearing.

5. *Select Potential Stability Aids*

In general, breaker-failure criteria have the most severe stability requirements. Thus, selection of stability aids is generally based on the CFC time improvements required to meet breaker-failure criteria. It should be noted that CFC time improvements associated with stability aids are not necessarily cumulative.

TABLE 8 Summary of Commonly Used Supplementary Transient Stability Aids

Stability aid	Maximum improvement in CFC time, cycles: Normal clearing	Maximum improvement in CFC time, cycles: Delayed clearing	Remarks
High initial response excitation	¼	2	Most modern excitation and voltage regulation systems have the capability for high initial response.
Independent pole tripping	N.A.	5	Reduces multiphase faults to single-phase faults for a breaker failure (delayed clearing). Increases relay costs.
Selective pole tripping	N.A.	5	Opens faulted phase only for single-phase faults. Generally only used at plants with one or two transmission lines. Increases relay cost and complexity.
Turbine fast-valve control	¼	2	Not applicable at hydro plants. Generally involves fast closing and opening of turbine intercept valves. Available on most steam turbines manufactured today.
Unit rejection schemes	Limited by amount generated that can safely be rejected		System must be capable of sustaining loss of the unit(s).
Series capacitors	Limited by amount of series compensation that can be added		May be required for steady-state power transfer. High cost; typically only economical for plants greater than 80 km (50 mi) from the load centers.
Braking resistors	Limited by the size of the resistor		High cost; typically only economical for plants greater than 80 km (50 mi) from the load centers.

6. Conduct Detailed Computer Simulations

Evaluation of the effects on stability of the stability aids outlined in this procedure is a nonlinear and complex analysis. Detailed computer simulation is the most effective method to determine the CFC time improvements associated with the stability aids.

7. Evaluate Potential Problems

Each transient-stability control aid presents potential problems which can be evaluated only through detailed computer analysis. Potential problems fall into three categories: (*a*) unique problems associated with a particular stability control method, (*b*) misoperation (i.e., operation when not required), (*c*) failure to operate when required or as expected. Table 9 briefly summarizes the potential problems associated with the stability aids and actions that can be taken to reduce the risk.

Related Calculations: Evaluation of stability controls is a complex and vast subject area. For more detailed discussions of stability controls see *Stability of Large Electric Power Systems* by Byerly and Kimbark, IEEE Press.

TABLE 9 Potential Problems Associated with Stability Controls

Stability aid	*Potential problems*	*Action to reduce risk*
High initial response excitation systems	Dynamic instability	Reduce response or add power system stabilizer (PSS)
	Overexcitation (misoperation)	Overexcitation relay protection
Independent pole tripping	Unbalanced generator operation (misoperation)	Generator negative-phase-sequence relay protection
Selective pole tripping	Sustain faults from energized phases	Add shunt reactive compensation
	Unbalanced generator operation (misoperation)	Generator negative-phase-sequence relay protection
Turbine fast valving	Unintentional generator trip	Maintain safety valves
Unit rejection schemes	Reduced generator reliability	Provide unit with capability to carry just the plant load (fast load runback)
Series capacitors	Subsynchronous resonance (SSR)	SSR filter
	Torque amplification	Static machine-frequency relay
	Self-excitation	Supplementary damping signals
Braking resistors	Instability (misoperation)	High-reliability relay schemes

STEADY-STATE STABILITY LIMITS

Compute the steady-state stability limit for a 595-MVA synchronous machine operating at 0.95 per-unit terminal voltage. Display the stability limits on the machine's reactive capability curve. The machine and system parameters are defined in Table 10.

Calculation Procedure:

1. Compute the Steady-State Stability Limit

The steady-state stability limits of a synchronous machine are described by a set of circles on the real- and reactive-power (P-Q) plane. Each circle corresponds to a different operating voltage. Use the following procedure to compute the center and radius of the circle corresponding to a 0.95 per-unit terminal voltage.

a. The center of the circle is at the point $P = 0$, $Q = [(1/X_e) - (1/X_d)]V_t^2/2$, where X_e = reactance between the machine terminals and infinite bus, X_d = machine synchronous reactance, and V_t = machine terminal voltage. Thus, $Q = [(1/0.4) - (1/1.8)](0.95)^2/2 = 0.877$ per unit. Hence the center of the circle is the point ($P = 0$, $Q = 0.877$).

b. The radius of the circle $r = [(1/X_e) + (1/X_d)]V_t^2/2$. Thus, $r = [(1/0.4) + (1/1.8)](0.95)^2/2 = 1.37$ per unit.

TABLE 10 Typical Machine and System Parameters* for a Unit Rated at 595 MVA, 3600 r/min, 22 kV, 0.90 pf

Direct-axis synchronous reactance	$X_d = 1.8\ \Omega$
Direct-axis transient reactance	$X'_d = 0.28\ \Omega$
Quadrature-axis synchronous reactance	$X_q = 1.75\ \Omega$
Inertia constant	H = 3.7 MWs/MVA
Direct-axis-transient open-circuit time constant	$T'_{do} = 4.1$ s
Equivalent reactance between machine terminals and infinite bus	$X_e = 0.4\ \Omega$

*Parameters expressed per unit on machine base.

2. Superimpose Limits on the Reactive Capability Curve

Figure 11 displays a typical reactive capability curve for a 595-MVA synchronous machine. The center of the circle is on the P axis ($P = 0$) with $Q = (0.877)(595\text{ MVA}) = 521$ MVAR. The radius $r = (1.37)(595\text{ MVA}) = 815$ MVA. A portion of this circle describing the steady-state stability limit is superimposed on the reactive capability curve. The points inside the circle are stable operating points in steady state. Note that the steady-state stability limit computed will limit the reactive capability of the machine in the leading power-factor range.

Related Calculations: A derivation of the relations used in the procedure can be found in "Underexcited Operation of Turbogenerators" by C. Adams and J. McLure, *AIEE Transactions*, vol. 67, part 1, 1948, pp. 521–528.

Transient and dynamic stability limits can also be superimposed on the machine's reactive capability curve.

DYNAMIC STABILITY LIMITS

Compute the range of the equivalent excitation system gain which will permit stable operation of the machine described in Table 10 at the operating point of 500 MW, 100 Mvar (leading), and 0.95 per-unit terminal volts. Represent the excitation system with the simplified linearized model in Fig. 12. Neglect the effect of machine damping and governor control.

Calculation Procedure:

1. Compute the Machine Terminal Current

Use $\mathbf{I_T} = \mathbf{S_T^*}/\mathbf{V_T^*}$, where $\mathbf{I_T}$ = complex machine terminal current, $\mathbf{S_T^*}$ = complex congujate of the machine apparent power, and $\mathbf{V_T^*}$ = complex conjugate of the machine terminal voltage. For this machine, let the terminal-voltage phase angle = 0°. Thus on a per-unit basis, $\mathbf{I_T} = [(500/595) - j(-100/595)]/0.95\angle 0° = (0.885 + j0.177) = 0.903\angle 11.3°$.

2. Compute the Infinite-Bus Voltage

Use $\mathbf{V_I} = \mathbf{V_T} - jX_e\mathbf{I_T}$, where $\mathbf{V_I}$ = complex infinite bus voltage, X_e = the external reactance from the machine terminals to the infinite bus, and the other symbols are as before. Thus, $\mathbf{V_I} = 0.95\angle 0° - [(0.40\angle 90°)(0.903\angle 11.3°)] = 1.08\angle -19.1°$.

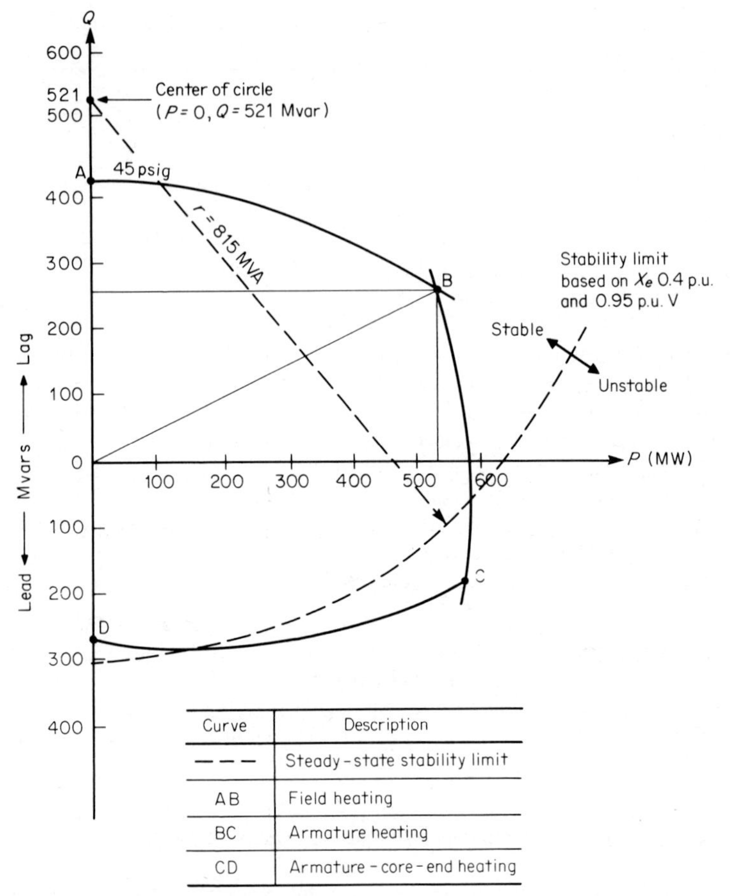

Fig. 11 Typical reactive capability curve of two-pole, 595-MVA, 3600-r/min, 22-kV, 0.90-pf synchronous machine.

3. Compute Remainder of Machine Conditions

The synchronous-machine model used in this determination of dynamic stability takes into account changes in field flux linkages and saliency. However, the model still neglects subtransient effects and saturation. Use the following procedure to compute the remaining machine conditions.

a. Compute the voltage behind quadrature axis (q axis) synchronous reactance X_q. Use $\mathbf{E_q} = \mathbf{V_T} + jX_q\mathbf{I_T}$, where $\mathbf{E_q}$ = complex voltage behind q-axis synchronous reactance. Thus, $\mathbf{E_q} = 0.95\underline{/0^\circ} + [(1.75\underline{/90^\circ})(0.903\underline{/11.3^\circ})] = 1.68\underline{/67.5^\circ}$.

b. Compute the machine internal phase angle with respect to the infinite bus (reference bus). Use $\delta = \theta_{eq} - \theta_I$, where δ = machine internal phase angle with respect to the infinite bus, θ_{eq} = phase angle of voltage $\mathbf{E_q}$, and θ_I = phase angle of infinite bus voltage = 19.1° as computed in Step 1. Thus, $\delta = 67.5^\circ - (-19.1^\circ) = 86.6^\circ$.

c. Compute the components of the machine terminal current along the direct and quadrature axes. Use $I_d = |\mathbf{I_T}| \sin(\delta - \alpha + \beta)$ and $I_q = |\mathbf{I_T}| \cos(\delta - \alpha + \beta)$,

Excitation System

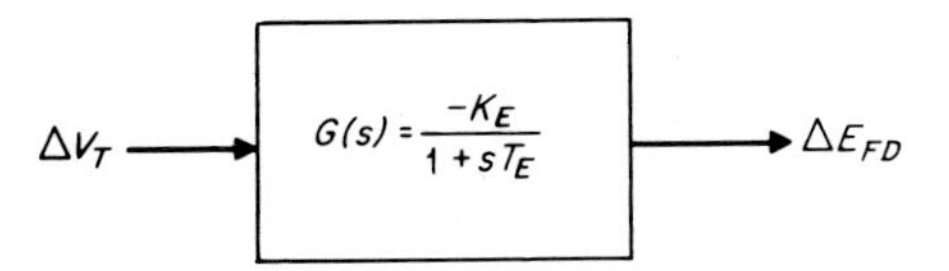

K_E = equivalent excitation system gain

T_E = equivalent excitation system time constant = 0.50 s

Fig. 12 Simplified linearized model of excitation-system response.

where I_d and I_q are the components of $\mathbf{I_T}$ along the direct and quadrature axes respectively, $|\mathbf{I_T}|$ = magnitude of the machine terminal current, δ = machine internal phase angle with respect to the infinite bus, α = phase angle of machine terminal current, and β = infinite-bus phase angle. For this machine $\delta = 86.6°$, $\alpha = 11.3°$, and $\beta = -19.1°$. Hence, $I_d = 0.903 \sin [86.6° - 11.3° + (-19.1°)] = 0.903 \sin 56.2° = 0.750$. $I_q = 0.903 \cos 56.2° = 0.502$.

d. Compute the voltage proportional to direct-axis (d-axis) flux linkages. Use $E'_q = |\mathbf{E_q}| - (X_q - X'_d)I_d$, where E'_q = voltage proportional to d-axis flux linkages; the other symbols are as previously defined. Thus, $E'_q = 1.68 - (1.75 - 0.28)(0.750) = 0.578$.

e. Compute the components of the machine terminal voltage along the d and q axes. Use $V_d = X_q I_q$ and $V_q = E'_q - X'_d I_d$, where V_d and V_q are the components of the machine terminal voltage along the d and q axes, respectively. Thus, $V_d = (1.75)(0.502) = 0.878$ and $V_q = 0.578 - (0.28)(0.750) = 0.368$.

4. *Compute the K Parameters*

Use the relations in Table 11 to compute the K parameters. The K parameters are functions of the machine loading and machine and system impedances. These parameters

TABLE 11 ***K*** **Parameters for Linearized Model of Synchronous Machine Connected to an Infinite Bus (Resistance Neglected)**

$$K_1 = \frac{X_q - X'_d}{X_1} I_q V_1 + \frac{E_q V_2}{X_3}$$

$$K_5 = \frac{X_q}{X_3}\frac{V_d V_2}{V_T} - \frac{X'_d}{X_1}\frac{V_q}{V_T} V_1$$

$K_2 = \frac{V_1}{X_1}$	$K_3 = \frac{X_1}{X_2}$
$K_4 = \frac{X_d - X'_d}{X_1} V_1$	$K_6 = \frac{X_e}{X_1}\frac{V_q}{V_T}$

$X_1 = X'_d + X_e$, $V_1 = V_I \sin \delta$, $X_2 = X_e + X_d$, $V_2 = V_I \cos \delta$, $X_3 = X_e + X_q$.

provide a convenient method for deriving the transfer functions or differential equations for the linearized d-q–axis model of a synchronous machine. The computed K parameters for this machine at the specified initial operating point are tabulated below.

K Parameters		
$K_1 = 1.2$	$K_2 = 1.59$	$K_3 = 0.309$
$K_4 = 2.41$	$K_5 = -0.122$	$K_6 = 0.225$

5. *Determine the Linearized Differential Equations*

Use the relations in Table 12 to determine the differential equations for the linearized model of the synchronous machine. The equations are expressed in the frequency domain ($s = d/dt$) and describe the response of the synchronous machine to small disturbances at a specific operating point. The excitation system response is described by the transfer function $G(s)$. From Fig. 12, $G(s) = -K_E/(1 + sT_E)$, where K_E = equivalent excitation system gain, T_E = equivalent excitation system time constant = 0.5 s, and $s = d/dt$. Governor control effects and machine damping are to be neglected. Thus, $H(s) = 0$ and $D = 0$. Substituting the computed K parameters for this machine and operating point, we find the differential equations for this machine are: $\Delta T_E = 1.21\ \Delta\delta + 1.59\ \Delta E_q'$, $\Delta E_q' = [0.309/(1 + 1.26s)]\ \Delta E_{FD} - 0.744\ \Delta\delta/(1 + 1.26s)$, $\Delta V_T = -0.122\ \Delta\delta + 0.225\ \Delta E_q'$, $\Delta E_{FD} = -K_e/(1 + 0.5s)\ \Delta V_T$, and $0.392s^2\ \Delta\delta = -\Delta T_E$.

The above equations describe the response of this synchronous machine to small disturbances at the operating point of 500 MW, 100 Mvar (leading), and 0.95 per-unit terminal volts.

6. *Select Method for Determining Stability*

There are numerous methods for evaluating the stability of the machine described by the set of linear differential equations determined in Step 5. The more commonly used methods used in synchronous-machine analysis are listed in Table 13.

7. *Specify the Range for the Equivalent Excitation Gain K_E*

Using any one of the methods in Table 13, give K_E a set of values and evaluate the stability of the machine. For this machine and initial operating point, K_E must be specified between approximately 0 and 115.

Related Calculations: Derivation of the K parameters and the linearized equations for a synchronous machine connected to an infinite bus can be found in the following references: W. G. Heffron and R. A. Phillips, "Effects of Modern Amplidyne Voltage Regulators," *AIEE Trans.,* August 1952; P. M. Anderson and A. A. Fouad, *Power System Control and Stability,* vol. 1, Iowa State University Press, 1977.

The general procedure described above can be used in a computer algorithm for determining dynamic-stability limits.

SELECTION OF AN UNDERFREQUENCY LOAD-SHEDDING SCHEME

An underfrequency load-shedding scheme is to be selected for a portion of a power system shown in Fig. 13. The scheme should protect this portion of the system from a total blackout in the event the two external power lines are lost. Load in the area varies 2 percent for each 1 percent change in frequency.

TABLE 12 Differential Equations for Linearized Model of a Synchronous Machine Connected to an Infinite Bus

Variable description	*Relation**
ΔT_E = electrical torque	$\Delta T_E = K_1\,\Delta\delta + K_2\,\Delta E'_q$
$\Delta E'_q$ = voltage proportional to field flux linkages	$\Delta E'_q = \dfrac{K_3\,\Delta E_{\mathrm{FD}}}{1 + sT'_{\mathrm{do}}K_3} - \dfrac{K_3K_4\,\Delta\delta}{1 + sT'_{\mathrm{do}}K_3}$
ΔV_T = terminal voltage	$\Delta V_T = K_5\,\Delta\delta + K_6\,\Delta E'_q$
ΔE_{FD} = field voltage	$\Delta E_{\mathrm{FD}} = G(s)\,\Delta V_T$
ΔT_M = mechanical torque	$\Delta T_M = H(s)s\,\Delta\delta$
$\Delta\delta$ = machine phase angle	$\dfrac{2H}{\omega_0}s^2\,\Delta\delta = \Delta T_M - \Delta T_E - ds\,\Delta\delta$

*K_i = K parameters, $G(s)$ = linearized transfer function for excitation system, $H(s)$ = linearized transfer function for governor control system, d = machine damping factor, H = machine inertia in MWs/MVA, and ω_0 = synchronous speed = 377 rad/s.

TABLE 13 Methods Commonly Used for Evaluating the Stability of the Linearized Synchronous-Machine Model

Method	*How determined*	*Major advantages*	*Major disadvantages*
Routh's criterion	Determine how many negative real roots in the charateristic polynomial: $a_n s^n\,\Delta\delta(s) + \cdots + a_1 s\,\Delta\delta(s) + a_0 = 0$	Easily applied for small-scale systems	Difficult to apply to *complex* or *large* systems forming characteristic polynomial
Root locus	Determine roots of the denominator of the transfer function: $\Delta T_E(s) = \dfrac{H(s)}{1 + GH(s)}\,\Delta\delta$	Provides actual root locations	Difficult to apply to *large* systems forming transfer function $1 + GH(s)$
Nyquist diagram	Plot frequency response of open-loop transfer function $GH(j\omega)$	Experimental frequency response data can be used	Difficult to apply to *large* system
Time response	Solve differential equations by numerical-integration techniques	Can be easily applied to *large* and *complex* systems	No information on *relative* stability
Eigenvalue analysis	Determine eigenvalues of state transition matrix $[A]$	Can be applied to *complex* and *large* systems; *relative* stability information	Formation of state transition matrix $[A]$ computation of the A-matrix eigenvalues

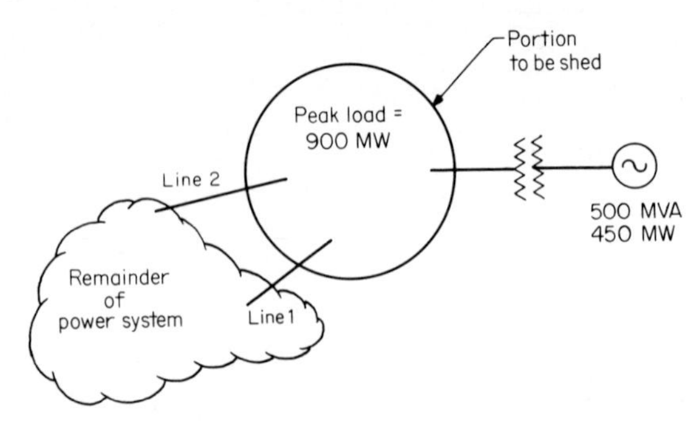

Fig. 13 Simplified diagram of shedding a portion of a power system.

Calculation Procedure:

1. *Select Maximum Generation Deficiency*

Selection of the maximum initial generation deficiency for which the load-shedding scheme should provide protection is an arbitrary decision. For the system of Fig. 13, an initial maximum generation deficiency of 450 MW is selected. This value is based on the difference between the peak load and the rated generator output.

2. *Compute Corresponding System Overload*

It is convenient to define a system overload as $\text{OL} = (L - P_m)/P_m$, where L = initial load and P_m = initial generator output. Thus, the maximum initial generation deficiency can be expressed in terms of a system overload as $\text{OL} = (900 - 450)/450 = 1.0$ per unit, or a 100 percent system overload.

3. *Select a Minimum Frequency*

The minimum frequency is the lowest allowable frequency that the system should settle to after all load shedding has occurred. In general, the minimum frequency should be above the frequency at which the generating unit will be separated from the system. Assume the generator trip frequency is 57 Hz and select a minimum frequency of 57.5 Hz.

4. *Compute the Maximum Amount of Load to be Shed*

The maximum load shed is that required to allow the system frequency to decay to the minimum frequency for the maximum system overload. Use $L_m = [\text{OL}/(\text{OL} + 1) - \alpha]/(1 - \alpha)$, where L_m = maximum load shed; OL = maximum initial system overload as selected in Step 1; $\alpha = d(1 - \omega_m/60)$, where d = system-load damping factor (given as 2); ω_m = minimum frequency. Thus, $2(1 - 57.5/60) = 0.833$ and $L_m = [1/(1 + 1) - 0.0833]/(1 - 0.0833) = 0.453$ per unit, or 45.3 percent of initial system load.

5. *Select Load-Shedding Scheme*

Selection of the load-shedding scheme involves specification of the number of load-shedding steps and the frequency set points. In general, the more load-shedding steps the better. However, as the number of steps increases the cost may also increase. In addition, too many steps may create relay coordination problems. Load should be shed gradually;

that is, each step should drop progressively more load. Frequency set points can be divided up in equal intervals from the maximum to minimum frequency set points. The table below displays the load-shedding scheme selected.

	Frequency set points	*Percent of initial load to be shed*
Step 1	59.5	5
Step 2	59.0	15
Step 3	58.5	25
Total		45

6. *Check Relay Coordination*

Relay coordination checks between adjacent steps are required to assure the minimum amount of load is shed for various initial system overloads. The reason for this check is that there may be a significant amount of time delay between the time the system frequency decays to a set point and the time the actual load shedding occurs. The time delays include relay pickup time, any intentional time delay, and breaker opening times. This time delay may cause the frequency to decay through two adjacent steps unecessarily. Use the following procedure to check relay coordination.

a. For load-shedding Steps 1 and 2, compute the initial system overload which will result in the frequency decaying to Step 2. Use $OL_2 = [L_d + \alpha/(1 - \alpha)]/(1 - L_d)$, where OL_2 = initial system overload for frequency to decay to Step 2; L_d = per-unit load that should be shed prior to Step 2; $\alpha = d(1 - \omega_2/60)$, where ω_2 is the relay set point for Step 2. Thus, $\alpha = (2)(1 - 59.0/60) = 0.033$ per unit and $OL_2 = [0.05 + 0.033/(1 - 0.033)]/(1 - 0.05) = 0.088$ per unit.

b. Compute the corresponding initial system load. Use $L_i = P_m(OL_2 + 1)$, where L_i is the initial system load which results in a 0.088 per-unit system overload. Thus, $L_i = 450(0.088 + 1) = 0.979$ per unit, or in MW, L_i = (per-unit load)(base MVA) = (0.979)(500) = 489 MW.

c. Conduct computer simulation. A digital computer simulation is made with an initial generation of 450 MW and an initial load of 489 MW. If relay coordination is adequate, load-shedding Step 2 should not shed any load for this initial system overload. If load-shedding Step 2 picks up and/or sheds load, then the relay set points can be moved apart, time delay may be reduced, or the load shed in each step may be revised.

d. Repeat steps *a* through *c* for the remaining adjacent load-shedding steps.

Related Calculations: This procedure is easily expanded for use with more than one generator. Digital computer simulations are required in the relay coordination steps to take into account the effects of automatic voltage regulation, automatic governor control, and load dependence on voltage and frequency.

Section 17 COGENERATION

Michael Nakhamkin, P.E., Ph.D.
Group Supervising Mechanical Engineer, Gibbs & Hill, Inc

REFERENCES Fink and Beaty—*Standard Handbook for Electrical Engineers,* McGraw-Hill; Giannuzzi, Horn, and Nakhamkin—"Optimization Design Considerations for Cogeneration Plant," ASME publ. 81-jPGC-GT-7; Baumeister—*Marks' Standard Handbook for Mechanical Engineers,* McGraw-Hill; Potter—*Power Plant Theory and Design,* Ronald Press.

INTRODUCTION

A cogeneration plant is a power plant that produces electrical power along with heat output, in the form of steam flow or water flow (through installation of steam-water heat exchangers), for industrial or residential consumption. The ratio of electric power to heat load varies, depending on the type of power plant. If the plant is located in an industrial complex, its main objective is often the supply of steam or hot water for industrial consumption. In this case, electric power output is considered as a by-product and is relatively small. The public utility companies have the inverse ratio of electric output to heat load, because heat output is considered as a by-product.

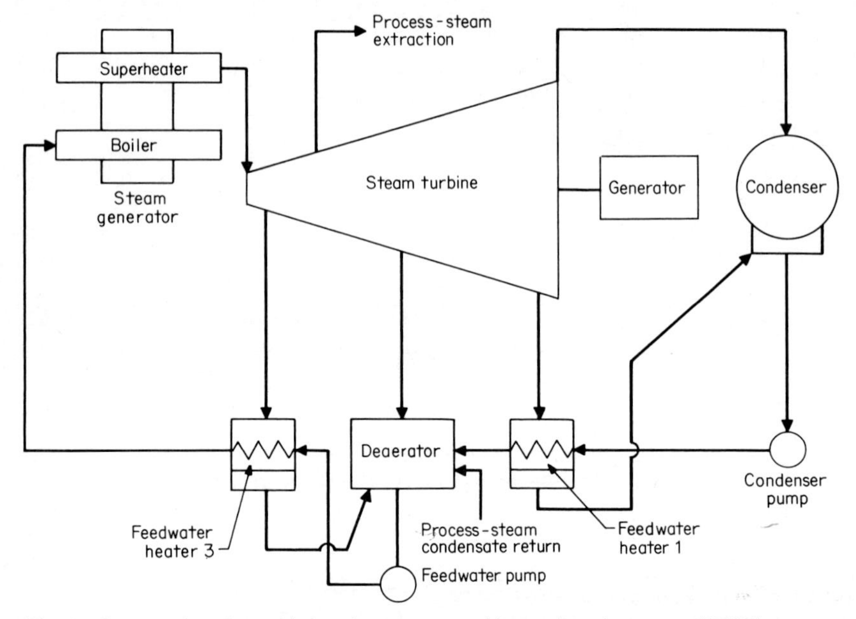

Fig. 1 Cogeneration plant cycle based on a steam turbine as the prime mover (STCP).

There are two conceptually different cogeneration plant types. One is the steam-turbine–based cogeneration plant (STCP) consisting of a steam turbine with the usual controlled steam extraction(s) for process steam supply (Fig. 1). The other is a gas-turbine–based cogeneration plant (GTCP) consisting of one or more gas turbines exhausting products of combustion through one or more heat-recovery steam generators (HRSGs) which produce steam for the heat supply (Fig. 2).

A cogeneration plant has the following major operational features: electric output in kWh, kJ; heat output in kJ (kcal, Btu); and heat rate in kJ/kWh (Btu/kWh). The cogeneration-plant heat rate requires special definition. For a conventional power plant, heat rate represents the heat consumption in kJ (kcal, Btu) per kilowatthour of electric output. However, this definition of heat rate is not applicable for cogeneration plants because it does not account for heat output.

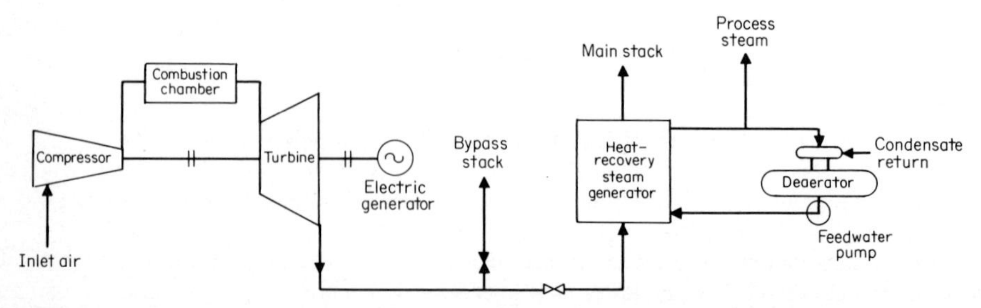

Fig. 2 Cogeneration-plant cycle based on a gas turbine as the prime mover (GTCP).

The heat rate calculations presented in this section are based on the following definition: heat rate $= (Q_1 - Q_2)/P$, where Q_1 = cogeneration plant heat input in kJ (Btu, kcal); Q_2 = conditional heat input with fuel to conventional steam generator to produce heat output equal to that produced by a cogeneration plant, in kJ (Btu, kcal); and P = electrical output in kWh. This heat rate definition assigns all the benefits from the combined power and steam generation to power production.

The thermodynamic efficiency of a cogeneration plant shall be evaluated using the expression: efficiency $= (P + H)/Q_1$, where P and H are the power and heat outputs of the cogeneration plant, respectively. They are expressed in the same heat units as Q_1.

POWER OUTPUT DEVELOPED BY TURBINE STAGES

Calculate the power output for the power plant of Fig. 3. The flows of steam through the various parts of the turbine are different as a result of steam extractions for the feedwater heating and process-steam supply.

Calculation Procedure:

1. *Divide Turbine into Sections*

To calculate the power output, the turbine is divided into sections (Fig. 4) which have constant steam flow and no heat addition or extraction.

2. *Calculate Total Power Output in Each Section*

The power output of each section is determined by multiplying the flow through that section by the enthalpy drop across the section (the difference between the steam enthalpy entering and that leaving the section). Hence, the section power output $= w_i \, \Delta H_i/3600$ where w_i = steam flow through section in kg/h (lb/h) and ΔH_i = enthalpy drop across section in kJ/kg (Btu/lb).

3. *Calculate Total Power Output*

Use:

$$\text{Total power output} = \sum_{i=1}^{n} w_i \, \Delta H_i/3600$$

where n is the number of sections. The calculated power output for each section of the turbine and the total power output of 96,000 kW is summarized in Table 1.

GENERATOR AND MECHANICAL LOSSES

Assume the power factor is 0.85 and the electric generator rating equals the 100 percent operating load (in kVA) plus 10 percent. Use a trial-and-error method in which the power output of the generator is assumed and the mechanical and generator losses are obtained from Figs. 5, 6, and 7, which give the respective losses as a function of the operating load in kVA. The generator power output plus the mechanical and generator

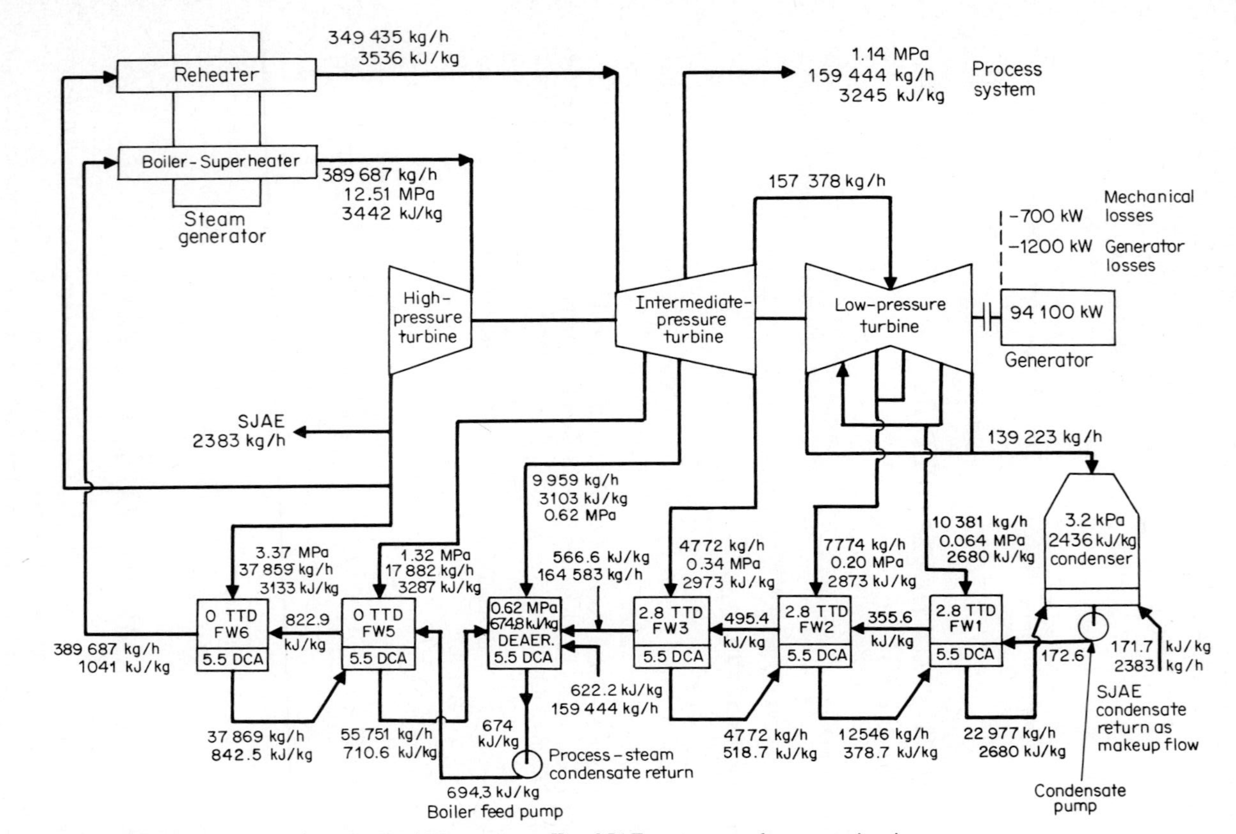

Fig. 3 STCP heat balance for a 94,000-kW generator. *Key:* SJAE = process-plant steam jet air ejector; P = MPa measured as absolute pressure; H = kJ/kg; W = kg/h.

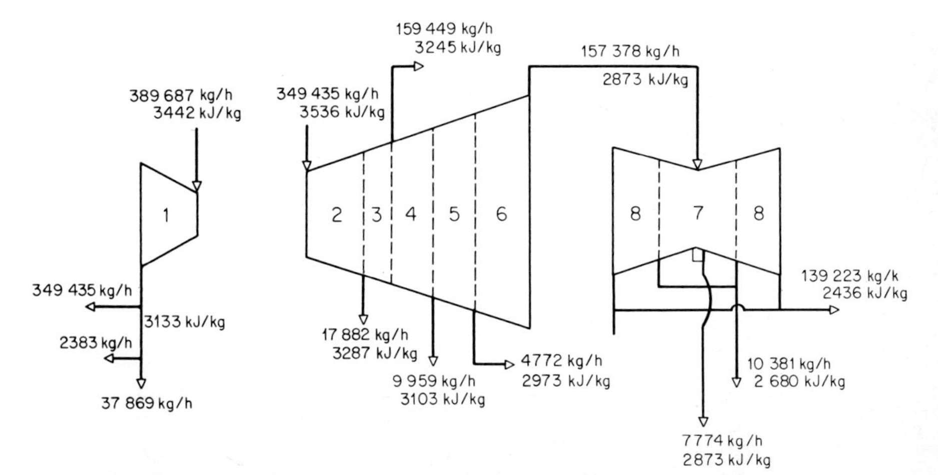

Fig. 4 Turbines in Fig. 3 divided into sections.

Flow in section 1 = 389 687		ΔH = 3442 − 3133 = 309
Flow in section 2 = 389 687 −	40 252 = 349 435	ΔH = 3536 − 3287 = 249
Flow in section 3 = 349 435 −	17 882 = 331 553	ΔH = 3287 − 3245 = 42
Flow in section 4 = 331 553 −	159 444 = 172 109	ΔH = 3245 − 3103 = 142
Flow in section 5 = 172 109 −	9 959 = 162 150	ΔH = 3103 − 2973 = 130
Flow in section 6 = 162 150 −	4 772 = 157 378	ΔH = 2973 − 2873 = 100
Flow in section 7 = 157 378 −	7 774 = 149 604	ΔH = 2873 − 2680 = 193
Flow in section 8 = 149 604 −	10 381 = 139 223	ΔH = 2680 − 2436 = 244

TABLE 1 Gross Power Output Calculations*

Steam turbine section number	*Steam flow through section, w_i, kg/h*	*Enthalpy drop across section, ΔH, kJ/kg*	*Power output, kW*
1	389,687	309	33,460
2	349,435	249	24,170
3	331,553	42	3,870
4	172,108	142	6,790
5	162,150	130	5,870
6	157,378	100	4,370
7	149,604	193	8,020
8	139,223	244	9,450
Sum of turbine-sections power outputs			96,000

*See Fig. 4.

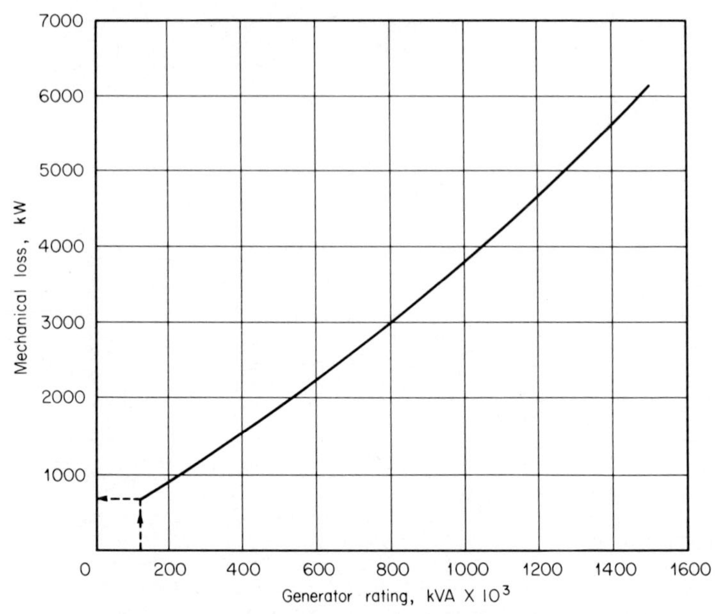

Fig. 5 Mechanical loss as a function of generator rating. If oil coolers are utilized in the condensate line, the recoverable loss = 0.85 mechanical loss.

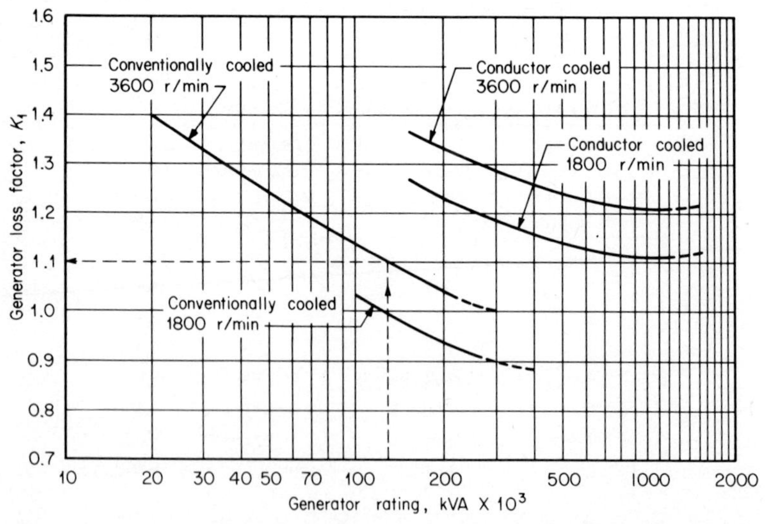

Fig. 6 Generator loss factor K_1 as a function of generator rating. Generator loss (at rated H_2 pressure = operating kVA $\left(\frac{K_i}{100}\right)$ K_2, K_1 from this figure, K_2 from Fig. 7. If hydrogen coolers are utilized in the condensate line, the recoverable loss at any kVA value = generator loss − 0.75 generator loss at the rated kVA value.

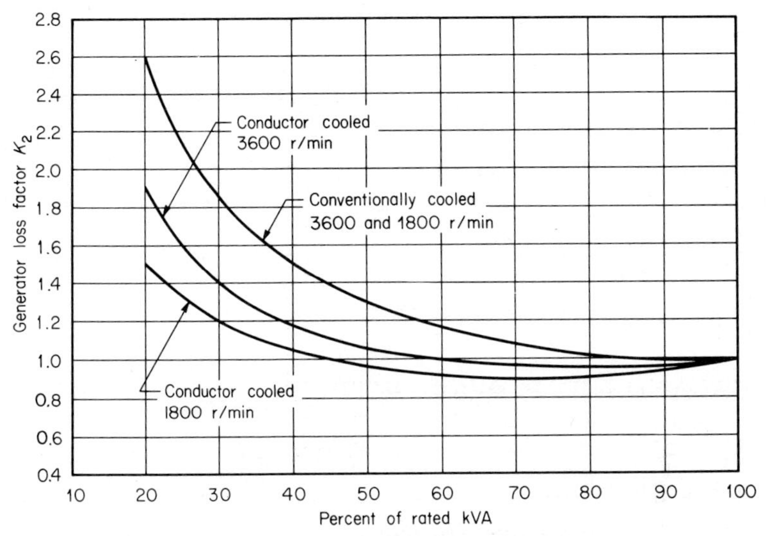

Fig. 7 Generator loss factor K_2 as a function of generator rating.

losses shall be equal to the sum of the turbine sections power outputs. Determine the generator and mechanical losses.

Calculation Procedure:

1. Assume Generator Power Output = 94,100 kW

Operating kVA = assumed power output/power factor = 94,100 kW/0.85 = 111,000 kVA. Generator rating = operating kVA × 1.1 = (111,000 kVA)(1.1) = 122,000 kVA.

From Fig. 5, the mechanical losses = 700 kW and the generator losses = (operating kVA)$(K_1)/100K_2$ = (111,000 kVA)(1.1)/(100)(1.0) = 1200 kW, where K_1 = 1.1 from Fig. 6 and K_2 = 1.0 from Fig. 7.

2. Check the Assumed Value of Generator Power Output

The generator power output (94,100 kW) + mechanical losses (700 kW) + generator losses (1200 kW) = 96,000 kW, which equals the sum of the power of the turbine sections (Table 1).

BOILER-FEED AND CONDENSATE PUMP POWER CONSUMPTION

Using the values given in Fig. 3, calculate the boiler-feed pump (BFP) and condensate pump (CP) power consumption.

Calculation Procedure:

1. Calculate Pump Power Consumption

Use: pump power consumption (kW) = enthalpy increase across the pump (kJ/kg) × mass flow (kg/s). Based on the values given in Fig. 4 for the boiler-feed pump, ΔH

= 694.3 − 674.8 = 19.5 kJ/kg and w = 9959 + 164,532 + 159,445 + 55,751 = 389,687 kg/h. Therefore, BFP = (19.5 kJ/kg)(389,687 kg/h)/(3600 kJ/kWh) = 2111 kW. For the condensate pump, ΔH = 172.6 − 171.7 = 0.9 kJ/kg and w = 139,223 + 22,977 + 2383 = 164,583 kg/h. Therefore, CP = (0.9 kJ/kg)(164,583 kg/h)/(3600 kJ/kWh) = 41 kW.

2. Determine Power Consumption of Electric Motors

In order to evaluate the power consumption of the electric motors of these pumps, assume a 90 percent motor efficiency for BFP and 85 percent for CP. Hence, BFP electric-motor power consumption = 2111 kW/0.9 = 2345 kW and CP electric-motor consumption = 41 kW/0.85 = 48.5 kW. The total CP and BFP motor power consumption is 2345 + 48.5 or approximately 2400 kW.

GROSS AND NET POWER OUTPUT

Calculate the gross and net power output for the power plant in Fig. 3.

Calculation Procedure:

1. Calculate Gross Power Output

Gross power output = sum of turbine-section power outputs − mechanical losses − generator losses = 96,000 − 700 − 1200 = 94,100 kW.

2. Calculate Net Power Output

Net power output = gross power ouptut − internal plant power consumption (for simplicity this is assumed to be only BFP and CP power consumption) = 94,100 kW − 2400 kW = 91,700 kW.

HEAT AND FUEL CONSUMPTION

Steam flow and enthalpy for characteristic cycle points for the cogeneration plant are given in Fig. 3. Assume the steam generator efficiency is 86 percent and no. 4 fuel oil having a heating value of 43,000 kJ/kg is used. Determine the heat and fuel consumption for the power plant.

Calculation Procedure:

The step-by-step calculations are summarized in Table 2.

HEAT RATE

The major parameters indicating the efficiency of a power plant are gross and net power-plant heat rates, which represent heat expenditure to produce 1 kWh of electrical energy. Using appropriate data in Table 2, determine the gross and net power-plant heat rates.

Calculation Procedure:

1. Calculate Gross Cogeneration Plant Heat Rate

Use: gross plant heat rate = $(Q_1 - Q_2)$/gross power output, where Q_1 = total heat added with fuel in cogeneration plant and Q_2 = heat in conventional steam gener-

TABLE 2 Heat and Fuel Consumption Calculations

Item no.	*Defined item*	*Source*	*Value*
1	Main steam flow	Heat balance, Fig. 3	389,687 kg/h
2	Main steam flow enthalpy	Heat balance, Fig. 3	3442 kJ/kg
3	Final feedwater enthalpy	Heat balance, Fig. 3	1041 kJ/kg
4	Enthalpy change across steam generator	Line (2) − Line (3)	2401 kJ/kg
5	Heat added to main flow	Line (1) × Line (4)	0.936×10^9 kJ/h
6	Hot reheat steam flow	Heat balance, Fig. 3	349,435 kg/h
7	Hot reheat enthalpy	Heat balance, Fig. 3	3536 kJ/kg
8	Cold reheat enthalpy	Heat balance, Fig. 3	3133 kJ/kg
9	Enthalpy change across reheater	Line (7) − Line (8)	403 kJ/kg
10	Reheater heat added	Line (6) × Line (9)	0.140×10^9 kJ/h
11	Total heat added in steam generator	Line (5) + Line (10)	1.076×10^9 kJ/h
12	Steam generator efficiency	Assumed	86%
13	Total heat added with fuel	$\frac{\text{Line (11)}}{\text{Line (12)}} \times 100$	1.25×10^9 kJ/h
14	No. 4 fuel-oil heating value	Assumed	43,000 kJ/kg
15	No. 4 fuel-oil consumption	Line (13): Line (14)	29,100 kg/h
16	Process-steam flow	Heat balance, Fig. 3	159,445 kg/h
17	Process-steam enthalpy	Heat balance, Fig. 3	3245 kJ/kg
18	Condensate return enthalpy	Heat balance, Fig. 3	622.2 kJ/kg
19	Enthalpy change across steam generator for process-steam flow	Line (17) − Line (18)	2622.8 kJ/kg
20	Heat input for process-steam supply	$\frac{\text{Line (16)} \times \text{Line (19)}}{\text{Line (12)}}$	0.49×10^9 kJ/h

ator. From Table 2, $Q_1 = 1.25 \times 10^9$ kJ/h and $Q_2 = 0.49 \times 10^9$ kJ/h. For gross power output = 94,100 kW, the gross plant heat rate = $(1.25 \times 10^9 - 0.49 \times 10^9)/94{,}100 = 8077$ kJ/kWh.

2. *Calculate Net Cogeneration-Plant Heat Rate*

Use: net plant heat rate = $(Q_1 - Q_2)$/(net power output). For net power output = 91,700 kW, the net plant heat rate = $(1.25 \times 10^9 - 0.49 \times 10^9)/91{,}700 = 8288$ kJ/kWh.

FEEDWATER-HEATER HEAT BALANCE

The heat balance of any heat exchanger is based on the law of conservation of energy; i.e., heat input minus heat losses is equal to heat output. A heat balance helps to determine any unknown flow or parameter if all other flows or parameters are known. Per-

TABLE 3 Feedwater Heater 5 Heat-Balance Calculations

Item	*Defined item*	*Source*	*Value*
1	Feedwater flow	Heat balance, Fig. 3	389,687 kg/h
2	Enthalpy of feedwater flow entering heater	Heat balance, Fig. 3	694.3 kJ/kg
3	Enthalpy of feedwater flow leaving heater	Heat balance, Fig. 3	822.9 kJ/kg
4	Enthalpy of extraction steam	Heat balance, Fig. 3	3287 kJ/kg
5	Enthalpy of drain flow from feedwater heater 6	Heat balance, Fig. 3	842.5 kJ/kg
6	Drain flow from feedwater heater 6	Heat balance, Fig. 3	37,869 kg/h
7	Enthalpy of drain flow from feedwater heater 5	Heat balance, Fig. 3	710.6 kJ/kg
8	Radiation losses in feedwater heater	Assumed	1.5%
9	Steam flow to feedwater heater 5	Feedwater-heater heat balance	17,883 kg/h

form heat-balance calculations to determine the required steam flow (assumed to be unknown) to high-pressure feedwater heater 5 (Fig. 3).

Calculation Procedure:

1. Write Heat Balance Equation

Denoting the unknown steam extraction flow as X and referring to Table 3, write the following heat balance for heater 5: $X \times$ Line (4) $\times$ [100 − Line (8)]/100 + Line (6) $\times$ Line (5) + [Line (1) $\times$ Line (2)] = Line (1) $\times$ Line (3) + [X + Line (6)] $\times$ Line (7), where the left-hand side of the equation represents heat flow into feedwater heater 5 and the right-hand side represents exiting heat flows.

2. Determine Steam Flow

The procedure is summarized in Table 3; X = 17,888 kg/h [Line (9)].

Gas-Turbine–Based Cogeneration Plant

A cogeneration-plant cycle with a gas turbine as the prime mover and the corresponding heat balance are shown in Fig. 8. The gas-turbine–based cogeneration plant consists of two major components: gas-turbine-generator(s) which produce electric power and heat-recovery steam generator(s) which produce process steam by recovering heat from the gas-turbine exhaust gases.

Gas-turbine performance data for a site temperature of 16°C, simple cycle application, and base load are assumed as follows: net generator output = 56,170 kW, net turbine heat rate HHV = 12,895 kJ/kWh, efficiency = 27.9 percent, airflow = 1,027,000 kg/h, turbine-exhaust temperature = 554°C, exhaust-gas flow = 1,043,200 kg/h, and light fuel oil HHV = 44,956 kJ/kg. (HHV ≡ high heating value of a fuel.)

The process-steam supply requirements are assumed as follows: process-steam flow = 159,445 kg/h, steam pressure from HRSG = 1.14 MPa (absolute pressure), steam

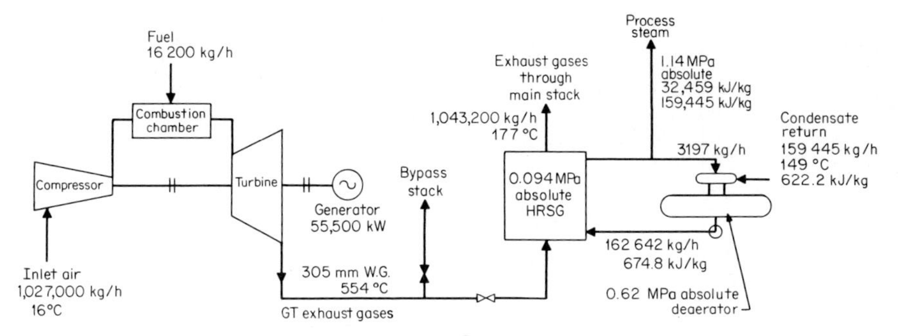

Fig. 8 GTCP heat balance.

enthalpy = 3245.9 kJ/kg, assumed pressure drop across HRSG = 305 mmH_2O gauge, condensate return temperature = 149°C, and condensate return enthalpy = 622.1 kJ/kg. Analyze the performance of the system.

GAS-TURBINE OUTPUT AND HEAT RATE IN CONGENERATION-PLANT MODE

A gas-turbine operating in the simple cycle mode (exhaust gases being vented to atmosphere and not into HRSG) has the following parameters at 16°C ambient air temperature: net power output = 56,170 kW and net turbine heat rate LHV = 12,985 kJ/kWh

A gas turbine operating in a cogeneration plant has an increase in back pressure because of the additional pressure drop in the flue gases across the HRSG, which is assumed to be 305 mmH_2O gauge.

Calculation Procedure:

1. Calculate Corrected Output and Heat Rate of Gas Turbine

From manufacturer's information the effect of the pressure drop across HRSG on the gas-turbine power output and heat rate is linear, and approximate recommendations are presented in Table 4.

Corrected output and heat rate of gas turbine in cogeneration-plant application area: power output = (56,170 kW)[1.00 − (0.004)(305 mmH_2O gauge)/(102 mmH_2O gauge)] = 55,500 kW and the heat rate = (12,985 kJ/kWh)[1.00 + (0.004)(305 mmH_2O gauge)/(102 mmH_2O gauge)] = 13,140 kJ/kWh.

TABLE 4 Effect of Pressure Losses on Gas Turbine Performance

	Effect on output	*Effect on heat rate*	*Increased exhaust temperature*
102 mmH_2O inlet	−1.4%	+0.4%	+1.1°C
102 mmH_2O exhaust	−0.4%	+0.4%	+1.1°C

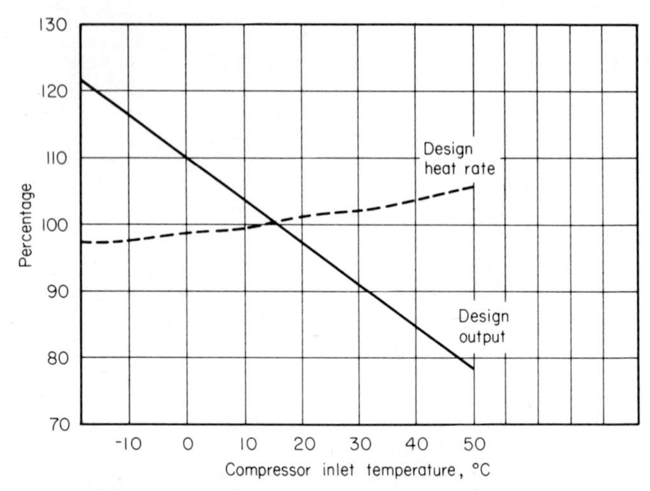

Fig. 9 Effect of compressor inlet air/temperature on gas turbine power output and heat rate. The fuel used was natural gas distillate oil.

2. *Calculate Compressor-Inlet Air-Temperature Effects on Plant Output and Gas-Turbine Heat Rate*

The influence of the compressor-inlet air-temperature on the gas-turbine power and heat rate is shown in Fig. 9. For 32°C ambient temperature (16°C is the design temperature), the correction factor for the heat rate is 1.025 and for the power output is 0.895. Therefore, the corrected power output = (55,496 kW)(0.895) = 49,670 kW and the corrected heat rate = (13,140 kJ/kWh)(1.025) = 13,470 kJ/kWh.

3. *Calculate Steam Productivity of HRSG Recovering Heat from Gas-turbine Exhaust Gases*

The calculation of steam productivity is based on the heat balance for HRSG. Heat transferred from gas turbine-generator exhaust gases less heat losses will be equal to the heat received by the HRSG medium (condensate steam). In calculation of HRSG steam productivity, the following is assumed; HRSG exhaust gas temperature is 177°C, which provides sufficient margin above the dew-point temperature of the products of combustion of no. 2 fuel oil (distillate) in a gas turbine. For conceptual calculations with acceptable accuracy, the specific heat of the exhaust gases can be obtained from air tables at average temperature, since the products of combustion in gas turbines have an air/fuel ratio much higher than theoretically required. Heat loss in an HRSG is assumed to be 2 percent. The HRSG heat-balance equation is: $W_gC_p(T_{g1} - T_{g2})(0.98) = W_{st}(H_1 - H_2)$, where W_g = exhaust-gas flow, 1,043,100 kg/h; T_{g1} and T_{g2} = temperatures of exhaust gases entering and leaving the HRSG, 554°C and 177°C, respectively; C_p = specific heat of air at average exhaust-gas temperature; and W_{st} = produced steam flow, kg/h. The average temperature = (554°C + 177°C)/2 = 366°C. From air tables, C_p = 1.063 kJ/kg°C and H_1 and H_2 = enthalpy of produced steam and entering HRSG feedwater, 3245.9 kJ/kg and 674.8 kJ/kg, respectively. From the solution of the heat-balance equation with the above data, the produced steam flow is W_{st} = 162,642 kg/h.

The process steam is the difference between the produced steam flow (162,642 kg/h) and the steam flow required by the deaerator (3197 kg/h, obtained from Fig. 8). Thus, process steam = 162,642 − 3197 = 159,445 kg/h.

4. Calculate Heat and Fuel Consumption

Heat consumption (kJ/h) = corrected gas-turbine heat rate (kJ/kWh) × corrected gas-turbine power (kW). Hence, (13,049 kJ/kWh)(55,496 kW) = 0.73×10^9 kJ/h.

Fuel consumption = (heat consumption: 0.73×10^9 kJ/h)/(no. 2 fuel oil HHV: 45,124 kJ/kg) = 16,200 kg/h.

5. Calculate Cogeneration-Plant Heat Rate

Cogeneration-plant heat rate = $(Q_1 - Q_2)$/net power output. Heat consumption $Q_1 = 0.73 \times 10^9$ kJ/h and heat input for process steam supply, $Q_2 = 0.49 \times 10^9$ kJ/h (from Table 2, line 20). Corrected gas-turbine power output = 55,496 kW. Therefore, cogeneration-plant heat rate = $(0.73 \times 10^9$ kJ/h $- 0.49 \times 10^9$ kJ/h)/(55,500 kW) = 4325 kJ/kWh.

COMPARATIVE ANALYSIS OF STCP AND GTCP

The selection of the most economical cogeneration-plant type—steam-turbine-based cogeneration plant vs. gas-turbine-based cogeneration plant—for special power and heat consumption requirements is the most important problem for the conceptual definition of a power plant. There are two approaches to an optimization of the cogeneration plant:

1. Comparative thermodynamic analysis of STCP and GTCP, which represents a comparison of the cost of a fuel and may be critical for the selection of the cycle in regions with high fuel cost and when the number of operating hours per year exceeds 6000.
2. Comparative economic analysis, which is based on the evaluation of the present-worth dollar values of the cogeneration plant capital and operating costs for two cogeneration-plant designs. This analysis requires, in addition to performance characteristics for both cogeneration plants, information relevant to the costs of equipment and other features which are not always available at the conceptual design phase.

Calculation Procedure:

Calculation procedures in this section represent a first approach for the conceptual selection of a cogeneration-plant cycle. Final selection is based upon more detailed calculations, along with other considerations such as water availability, environmental conditions, and operational personnel priorities, etc.

The conceptual analysis of the STCP and GTCP cycles, as well as a number of cogeneration plants' cycle optimizations, shows that the main criterion for efficiency comparison of both cogeneration plants is the ratio of required heat output to power output, Q/P. The following calculation procedure and results, presented as heat rate (HR) vs. Q/P curves for both cogeneration plant arrangements, will help the engineer to select a more efficient cogeneration plant cycle for special Q/P requirements.

Heat rate calculations for STCP and GTCP for various Q/P ratios are presented in Table 5. The calculations for STCP were done for three Q/P ratios and the respective heat balances presented in Figs. 3, 10, and 11. The heat rate calculations for GTCP were done for two Q/P ratios because of the evident linear character of this function. HR curves for both cogeneration plants are presented in Fig. 12.

TABLE 5 Heat Rate Calculations for Various STCP and GTCP Cycles

Item no.	Defined item	Source	STCP Case 1	STCP Case 2	STCP Case 3	GTCP Case 1	GTCP Case 2
1	Exporting steam	Assumed	159,445 kg/h	250,000 kg/h	300,000 kg/h	0	159,445 kg/h
2	Power output	Heat balances, Figs. 3, 10, 11	91,700 kW	72,630 kW	62,520 kW	55,500 kW	55,500 kW
3	Heat output	Line (1) × 2622.8, kJ/h	0.42×10^9 kJ/h	0.655×10^9 kJ/h	0.786×10^9 kJ/h	0	0.42×10^9 kJ/h
4	Heat output/power output ratio	Line (3)/Line (2), kJ/kWh	4,580 kJ/kWh	9,018 kJ/kWh	12,680 kJ/kWh	0	7636 kJ/kWh
		Line (1)/Line (2), kg/h per kilowatt	1.740 kg/h per kilowatt	3.40 kg/h per kilowatt	4.8 kg/h per kilowatt	0	2.9 kg/h per kilowatt
5	Heat input for process-steam supply*	Line (3)/0.86	0.49×10^9 kJ/h	0.76×10^9 kJ/h	0.91×10^9 kJ/h	0	0.49×10^9 kJ/h
6	Total heat input	Heat balances, Figs. 3, 10, 11	1.25×10^9 kJ/h	1.25×10^9 kJ/h	1.25×10^9 kJ/h	0.72×10^9 kJ/h	0.72×10^9 kJ/h
7	Heat rate	$\frac{\text{Line (6)} - \text{Line (5)}}{\text{Line (2)}}$	8353 kJ/kWh	6747 kJ/kWh	5406 kJ/kWh	13,050 kJ/kWh	4289 kJ/kWh

*In conventional steam generator with assumed efficiency 0.86.

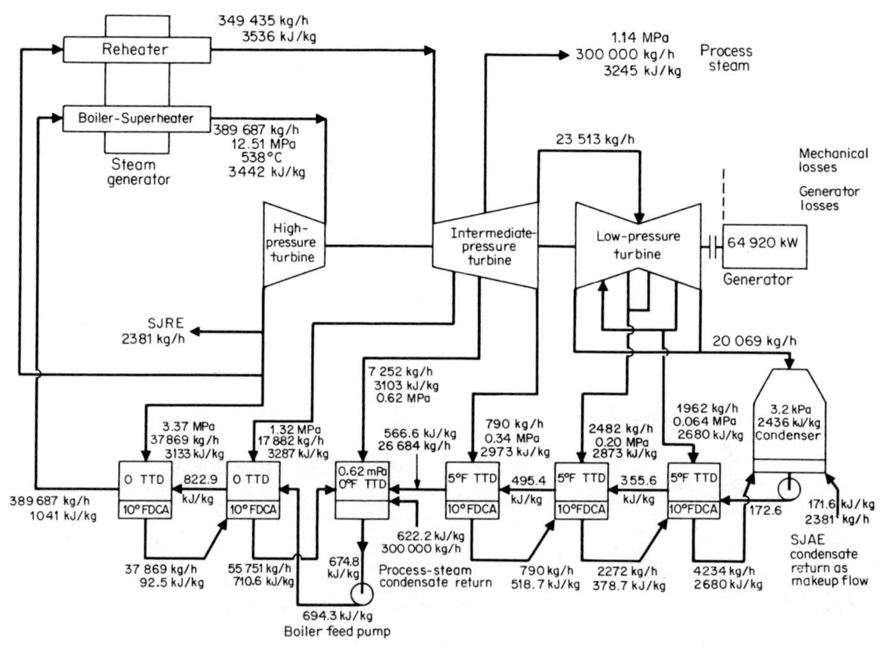

Fig. 10 Heat balance for 64,920-kW STCP. SJAE = process-plant steam jet air ejector.

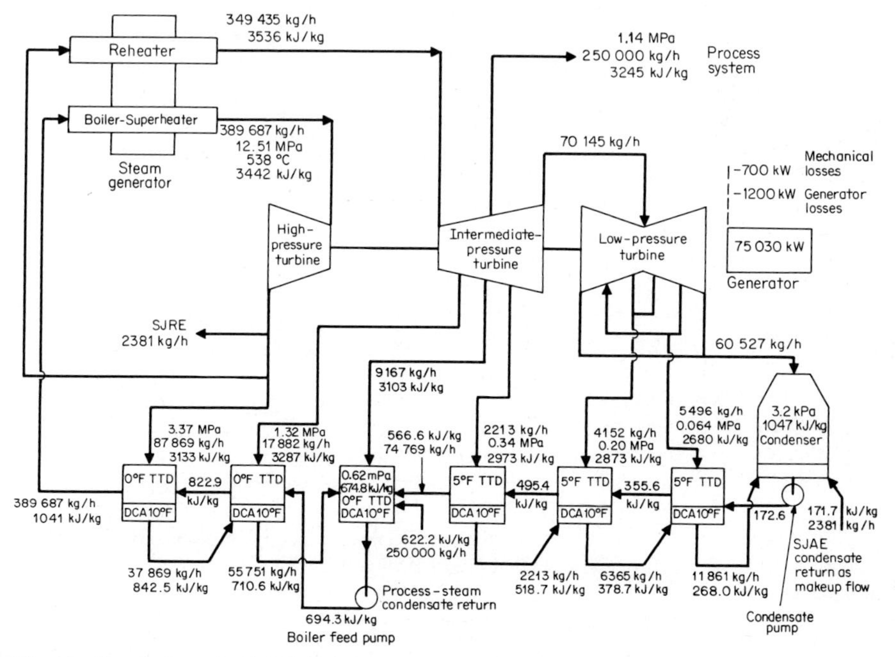

Fig. 11 Heat balance for 75,030-kW STCP. SJAE = process-plant steam jet air ejector.

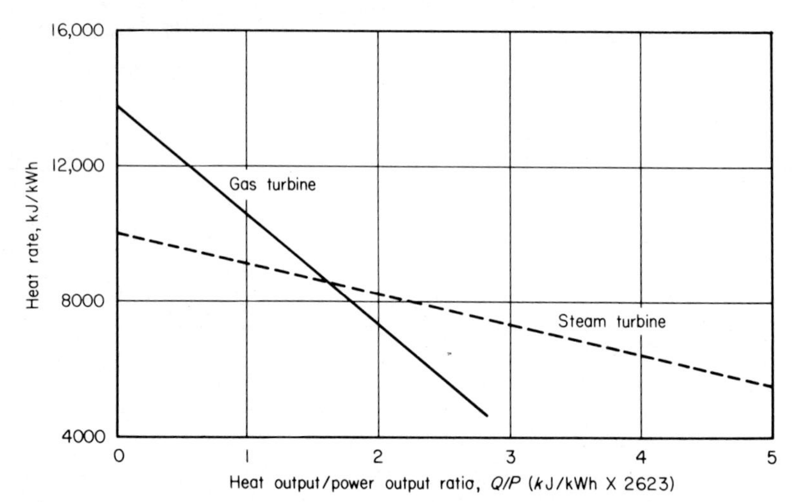

Fig. 12 Heat rate as a function of Q/P ratios for STCP and GTCP.

Analysis of the HR-vs.-Q/P curves shows that the break-even point for STCP and GTCP is at Q/P = 4200 kJ/kWh, or 1.6 kg/h per kilowatt. When the Q/P ratio is less than 4200 kJ/kWh, STCP is more economical (HR for STCP is less than for GTCP); where Q/P is higher than 4200 kJ/kWh, GTCP is more economical.

These results are based upon evaluation of selected STCP and GTCP cycle parameters. The calculations show that the conclusions reached with acceptable accuracy are applicable to various cycles; i.e., break-even points between STCP and GTCP exist and are located in the vicinity of the above Q/P ratio.

It is emphasized that these results can be used for preliminary conceptual selection of the cogeneration-plant arrangement in regions with high fuel costs, where capital costs of the equipment represent a small share of total evaluated costs. The final selection of the optimum cogeneration-plant arrangement for particular design and operational conditions is based on minimum present worth of total evaluated costs.

An economic analysis, based upon average equipment, fuel, and maintenance costs and current economic factors, shows that the break-even points for STCP and GTCP are located in the region of a smaller Q/P ratio. Thus it shortens even more the Q/P ratio's span (between zero and break-even Q/P), where the STCP is more economical than the GTCP. These results can be explained by the considerably lower installed costs of the GTCP which conpensate for the higher heat rate.

The number of expected operating hours per year for the congeneration plant is essential information in determining the applicable break-even Q/P ratio. The fewer annual operating hours, the lower the break-even Q/P ratio. Economic calculations show that if the number of operating hours per year is less than 2500, the break-even Q/P ratio is based only upon heat rate considerations.

The obtained results prove that for a considerable number of feasible power steam supply requirement ratios, the GTCP is more economical than the STCP. This is an important conclusion as it is generally assumed that because of its high heat rate, gas-turbine cogeneration cycles are inherently less economic.

Section 18 BATTERIES

Marco W. Migliaro, P.E.
Associate Consulting Engineer,
Ebasco Services Incorporated

REFERENCES ANSI/IEEE Standard 100-1977—*Standard Dictionary of Electrical and Electronics Terms;* ANSI/IEEE Standard 484-1981—*Recommended Practice for Installation Design, and Installation of Large Lead Storage Batteries for Generating Stations and Substations;* Hoxie—"Some Discharge Characteristics of Lead-Acid Batteries," *AIEE Trans. (Applic. and Industry)*, vol. 73, 1954; IEEE Standard 450-1980—*IEEE Recommended Practice for Maintenance, Testing, and Replacement of Large Lead Storage Batteries for Generating Stations and Substations,* IEEE Standard 485-1978—*IEEE Recommended Practice for Sizing Large Lead Storage Batteries for Generating Stations and Substations;* IEEE Standard 535-1979—*Standard for Qualification of Class 1E Lead Storage Batteries for Nuclear Power Generating Stations;* Fink and Beaty—*Standard Handbook for Electrical Engineers,* McGraw-Hill.

NUMBER OF CELLS FOR A 48-VOLT SYSTEM

Determine the number of lead-acid cells required of a battery for a nominal 48-V dc system (42-V dc minimum to 56-V dc maximum).

Calculation Procedure:

1. Compute Number of Cells

The nominal voltage of a lead-acid cell is 2.0-V dc; therefore, the number of cells = 48/2 = 24 cells.

2. Check the Minimum Voltage Limit

Minimum volts/cell = (min. volts)/(number of cells) = 42 V/24 cells = 1.75 V/cell. This is an accepted end-of-discharge voltage for a lead-acid cell.

3. Check Maximum Voltage Limit

Maximum volts/cell = (max. volts)/(number of cells) = 56 V/24 cells = 2.33 V/cell. This is an acceptable maximum voltage for a lead-acid cell. Therefore, select 24 cells of the lead-acid type.

Related Calculations: The lead-acid battery uses a highly reactive sponge lead for the negative electrode, lead dioxide as the active positive material, and sulfuric acid solution for the electrolyte. As the cell discharges, the active materials of both electrodes are converted into lead sulfate. The sulfuric acid electrolyte also takes part in the reaction, producing water. On charge, the reverse actions take place. The state of charge of the battery can be determined by measuring the specific gravity, which decreases on discharge and increases on charge. The discharge and charge reactions of the battery are:

$$Pb + PbO_2 + 2H_2SO_4 \underset{\text{charge}}{\overset{\text{discharge}}{\longleftrightarrow}} 2PbSO_4 + 2H_2O$$

At the end of the charge, electrolysis of water also occurs, producing hydrogen at the negative electrode and oxygen at the positive electrode.

General performance characteristics of the lead-acid battery are given in Fig. 1. Characteristics of commonly used batteries are summarized in Table 1.

The nominal voltage of the lead-acid cell is 2 V. The voltage on open circuit is a direct function of the specific gravity, ranging from 2.12 V for a cell with 1.28 specific gravity

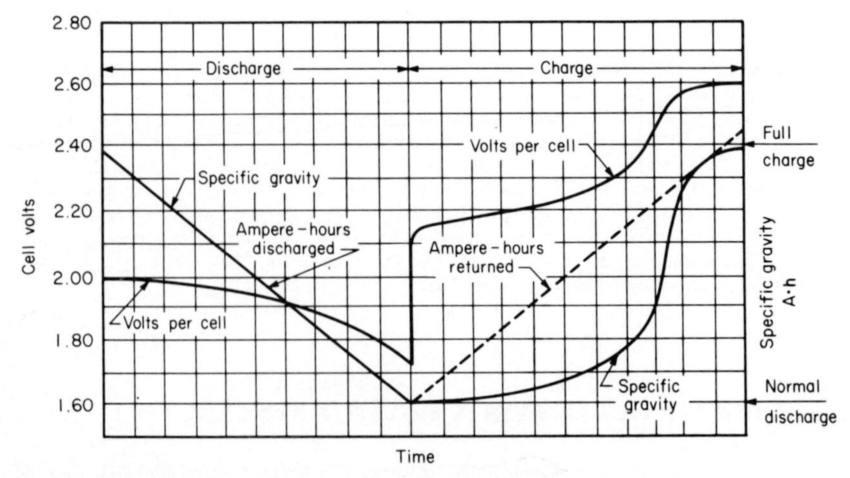

Fig. 1 Performance characteristics of lead-acid batteries. (From Fink and Beaty—*Standard Handbook for Electrical Engineers,* 11th ed., McGraw-Hill, p. 11–113.)

to 2.05 V at 1.21 specific gravity. Figure 2 presents typical discharge curves for the lead-acid cell. The end voltage is usually about 1.75 V but can be as low as 1.0 V at extremely high discharge rates, as in automotive starting service.

TABLE 1 Major Characteristics and Applications of Secondary Batteries*

System	*Characteristics*	*Applications*
Lead-acid:		
Automotive	Popular, low-cost secondary battery—moderate capacity, high-rate and low-temperature performance	Automobile starting, lighting, ignition (SLI); lawnmowers, tractors, marine, float service
Motive power	Designed for deep 6- to 9-h discharge, cycling service	Industrial trucks, materials handling; special types used for submarine power
Stationary	Designed for standby float service, long stand life	Emergency power—utilities, no-break systems
Sealed	Sealed, maintenance-free, low cost, good float capability	TV, portable tools, lights and appliances, radios and cassettes and tape players
Nickel-cadmium:		
Vented	Good high-rate, low-temperature capability; flat voltage, excellent cycle life	Aircraft batteries, industrial and emergency-power applications, communication equipment
Sealed	Good high-rate, low-temperature performance, excellent cycle life, maintenance-free	Photography, portable tools, appliances, standby power
Zinc–silver oxide	Highest energy density, good high-rate capability, low cycle life	Lightweight portable radio, TV, and communication equipment; torpedo propulsion, drones, submarines, and other military applications

*From Fink and Beaty—*Standard Handbook for Electrical Engineers,* 11th ed., McGraw-Hill, page 11–112.

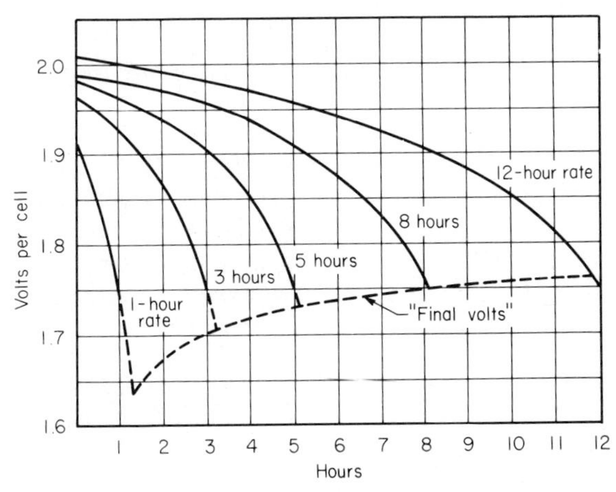

Fig. 2 Discharge curves of lead-acid batteries at different hour rates. (From Fink and Beaty—*Standard Handbook for Electrical Engineers,* 11th ed., McGraw-Hill, p. 11–114.)

NUMBER OF CELLS FOR 125-VOLT AND 250-VOLT SYSTEMS

Determine the number of lead-acid cells required of a battery for a nominal 125-V dc system (105 V dc minimum to 140 V dc maximum) and for a nominal 250-V dc system (210 V dc minimum to 280 V dc maximum).

Calculation Procedure:

1. Compute Number of Cells

If 1.75 V/cell is the minimum voltage of the 125-V dc system, the number of cells = min. volts/(min. volts/cell) = 105/1.75 = 60 cells.

2. Check Maximum Voltage

Using 2.33 V/cell, the maximum voltage = (number of cells)(max. volts/cell) = (60)(2.33) = 140 V dc. Therefore, for a 125-V dc system, select 60 cells of the lead-acid type.

3. Calculate Number of Cells for 250-V System

Number of cells = 210 V/(1.75/cell) = 120 cells.

4. Check Maximum Voltage

Max. voltage = (number of cells)(max. volts/cell) = (120)(2.33) = 280 V dc. Therefore, for a 250-V dc system, select 120 cells of the lead-acid type.

SELECTING NICKEL-CADMIUM CELLS

Select the number of nickel-cadmium (NiCd) cells required for a 125-V dc system with limits of 105 to 140 V dc. Assume minimum voltage per cell is 1.14 V dc for the NiCd cell.

Calculation Procedure:

1. Determine Number of Cells

Number of cells = min. volts/(min. V/cell) = 105/1.14 = 92.1 cells (use 92 cells).

2. Check Maximum Voltage per Cell

Max. volts/cell = max. volts/number of cells = 140/92 = 1.52 V/cell. This is an acceptable value for a NiCd cell.

Related Calculations: The active materials of charged nickel-cadmium cells are trivalent nickel oxide for the positive electrode and cadmium for the negative electrode. The alkaline electrolyte is a solution of potassium hydroxide. A simplified statement of the cell reaction is:

$$Cd + 2NiOOH + 2H_2O \underset{charge}{\overset{discharge}{\longleftrightarrow}} Cd(OH)_2 + 2Ni(OH)_2$$

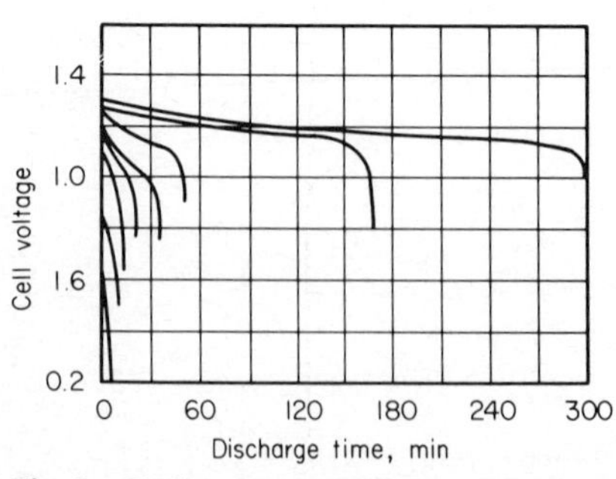

Fig. 3 Discharge curves of pocket plate NiCd cell at 25°C. (From Fink and Beaty—*Standard Handbook for Electrical Engineers,* 11th ed., McGraw-Hill, p. 11-119.)

During discharge (Fig. 3), the nickel oxide is reduced to the divalent state and the cadmium is oxidized. In addition, during charge (with the exception of sealed cells) hydrogen and oxygen are evolved by the positive and negative electrodes, respectively, as the cell reaches full charge. There is, however, little or no change in the bulk electrolyte concentration, and specific-gravity measurements give no indication of state of charge.

The nominal voltage of a NiCd cell is 1.2 V; the open-circuit voltage is 1.4 V.

LOAD PROFILES

Determine the worst-case load profile of a dc system for a generating station consisting of a 125-V dc nominal, stationary type lead-acid battery and a constant-voltage charger in full-float operation. The loads on the system and their load classification are as follows:

Load description	*Rating*	*Classification*
Emergency oil-pump motor	10 kW	Noncontinuous*
Controls	3 kW	Continuous
Two inverters (each 5 kW)	10 kW	Continuous
Emergency lighting	5 kW	Noncontinuous
Breaker tripping (20 at 5 A)	100 A	Momentary (1-s duration)†

*It will be assumed here that the equipment manufacturer requires the emergency oil pump to run continuously for 45 min following a unit trip.

†Occurs immediately following unit trip.

Calculation Procedure:

1. Determine Load Conditions for Battery

The first step in determining the worst-case load profile is to develop the conditions under which the battery is required to serve the dc system load. These conditions will vary according to the specific design criteria used for the plant. Load profiles will be determined for the following three conditions:

a. Supply of the emergency oil pump for 3 h (with charger supplying continuous load).

b. Supply of the dc system for 1 h upon charger failure.

c. Supply of the dc system for 1 h following a plant trip concurrent with loss of the auxiliary system's ac supply.

2. Develop Load List for Each Condition

A load list for each condition is required to determine the time duration of each load. Once this is done, the load profile may be plotted for each condition.

Condition a: The load list is as follows:

Load	*Current*	*Duration*
Emergency oil pump (inrush current)	250 A	0 to 1 min*
Emergency oil pump (full-load current)	80 A	1 to 180 min

*Although the inrush current lasts for a fraction of a second, it is customary to use a duration of 1 min for lead-acid batteries because the instantaneous voltage drop for the battery for the period of inrush is the same as the voltage drop after 1 min.

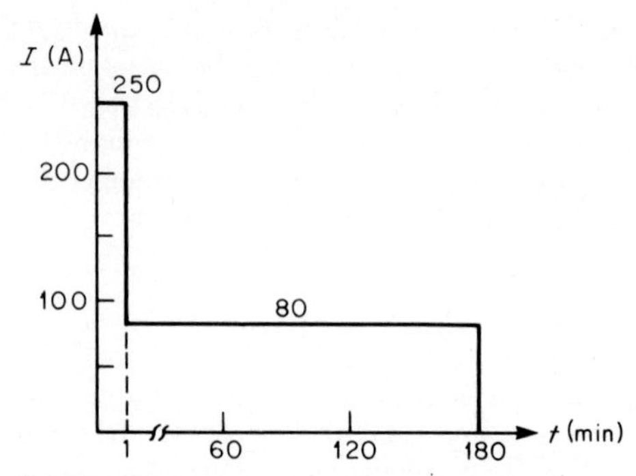

Fig. 4 An example of a load profile for a battery.

The load profile is plotted in Fig. 4.

Condition b: The load list is as follows:

Load	*Current*	*Duration*
Controls	24 A	0 to 60 min
Inverters (two)	80 A	0 to 60 min

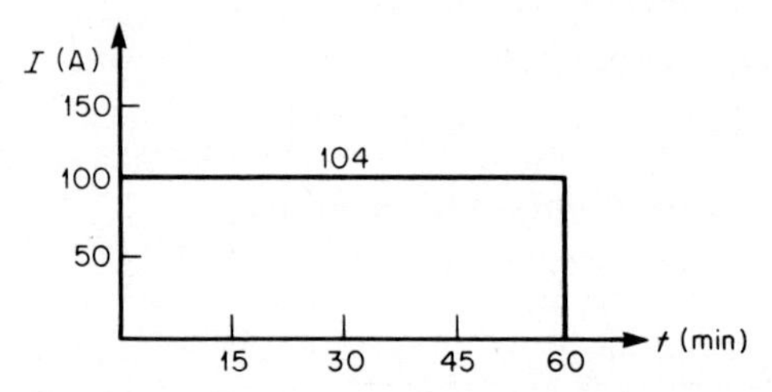

Fig. 5 Load profile for controls and two inverters.

The load profile is shown in Fig. 5.

Condition c: The load list is as follows:

Load	*Current*	*Duration*
Circuit-breaker tripping (20)	100 A	0 to 1 min*
Emergency oil pump (inrush)	250	0 to 1 min*
Emergency oil pump (full load)	80	1 to 45 min
Controls	24	0 to 60 min
Inverters (two)	80	0 to 60 min
Emergency lighting	40	0 to 60 min

*As stated previously for Condition *a*, the duration of one minute is used for lead-acid cells even though the load lasts for a fraction of a second. *Note:* It is assumed that breaker tripping and emergency oil pump inrush occur simultaneously.

The load profile is plotted in Fig. 6.

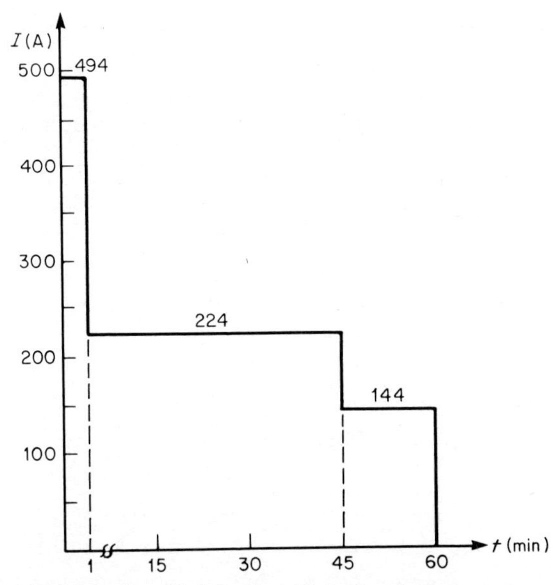

Fig. 6 Load profile for various kinds of equipment.

Related Calculations: Typically, a battery may have a number of load profiles (i.e., one for each set of specified conditions); for sizing the battery, however, the worst-case profile must be used. In some instances, the worst-case profile is apparent and may be used directly for battery sizing. In other cases, each profile should be used to calculate a battery size.

PROFILE FOR RANDOM LOAD

Determine the load profile of a lead-acid cell for the following load list which includes a random load (i.e., a load that may be imposed on the battery at any time during the duty cycle).

Load	*Current, A*	*Duration*
Circuit-breaker tripping (30)	150	0 to 1 s
Control	15	0 to 180 min
Fire-protection components	10	0 to 180 min
Emergency lighting	30	0 to 180 min
Sequence-of-events recorder	8	0 to 60 min
Oscillograph	17	0 to 1 min
	9	1 to 60 min
Emergency oil pump (inrush)	88	0 to 1 s
(full load)	25	1 s to 15 min (cycle repeats at the 60th and 120th minute)
Random load	45	1 min, occurring at any time from 0 to 180 min

Calculation Procedure:

1. Develop Method for Constructing Load Profile

The load profile can be constructed in a manner similar to that presented in the previous example, except that the random load must be considered separately. Because it is not known when the load occurs, the normal procedure is to develop a load profile without considering the random load. The battery is then sized on the basis of the profile and the effect of the random load is added to the portion of the load profile that is found to control the battery size.

2. Consider the First Minute

If the first minute of the load profile is examined separately, discrete loads may be identified to develop the 1-min profile, as shown in Figs. 7 and 8.

3. Construct Complete Load Profile

The complete load profile is shown in Fig. 9.

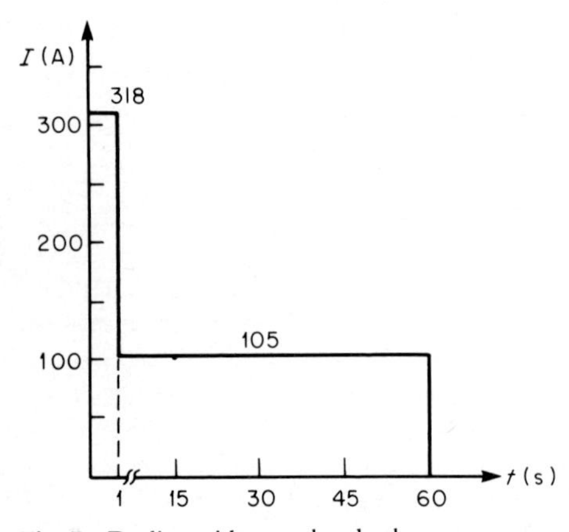

Fig. 7 Dealing with a random load.

Fig. 8 Load profile for the first minute.

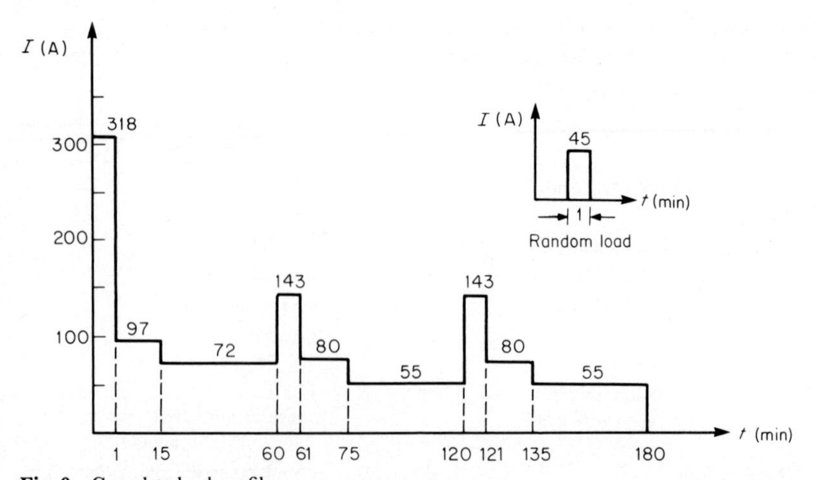

Fig. 9 Complete load profile.

Related Calculations: The load profile of Fig. 9 could be modified for use with NiCd cells by recognizing the availability of discharge rates under 1 min.

LOADS OCCURRING DURING FIRST MINUTE

Each of the following loads occur during the first minute; however, a discrete sequence of the loads cannot be established. Draw the load profile for the first minute, assuming a lead-acid cell is used.

Load	*Current, A*	*Duration*
Breaker closing	40	1 s
Motor inrush	110	1 s
Control	15	1 min
Miscellaneous	30	30 s

Calculation Procedure:

1. Develop Method for Drawing Load Profile

Because a discrete sequence cannot be established, the common practice is to assume that all of the loads occur simultaneously.

2. Draw Load Profile

The load profile for the first minute is drawn in Fig. 10.

Related Calculations: If NiCd cells are used whose discharge times are less than 1 min, some assumptions can be made. For example, one could assume all loads occur simultaneously for 1 s followed by control and miscellaneous loads for another 29 s followed by control load for 30 s. It can similarly be reasoned that the control load could occur for 30 s followed by control and miscellaneous loads for 1 s.

These possibilities are illustrated in Figs. 11 and 12. Each profile is then analyzed to determine which represents the worst case. The reader will find that determining which profiles represent the worst case will be intuitive, once a number of battery-sizing calculations have been performed.

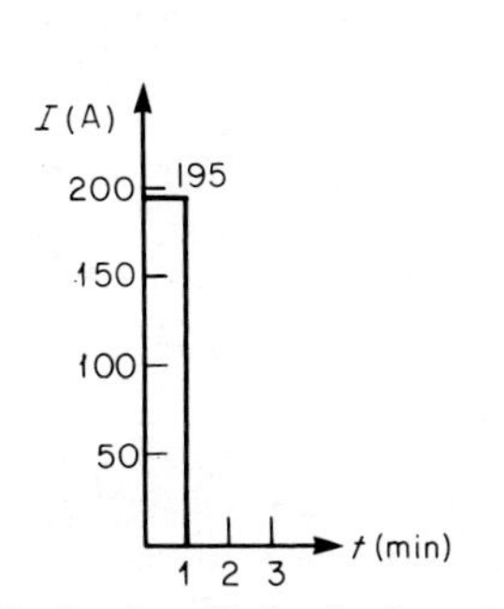

Fig. 10 Load profile for the first minute.

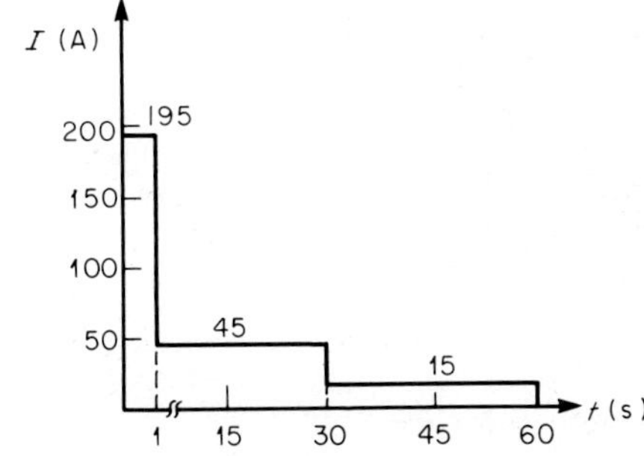

Fig. 11 A load profile to be analyzed for worst-case conditions.

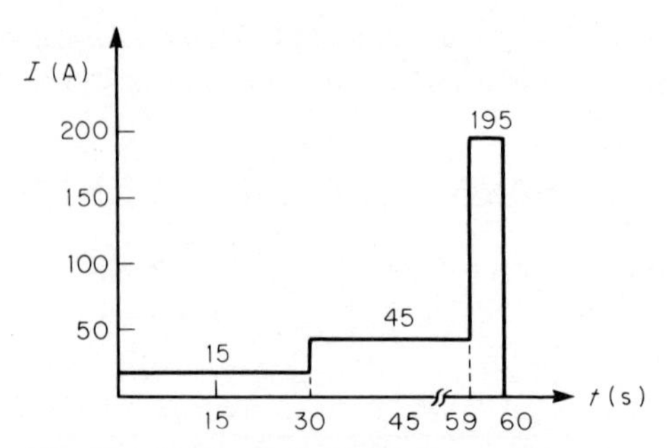

Fig. 12 Another load profile to be analyzed for worst-case conditions.

SIZING BATTERY FOR SINGLE-LOAD PROFILE

Given the load profile of Fig. 13, calculate the number of positive plates of lead-acid cell type X' required to supply the load. Assume: design margin = 10 percent, lowest electrolyte temperature is 10°C (50°F), 125-V dc system, 60 cells, 105-V dc minimum (i.e., 1.75-V/cell end-of-discharge voltage), and age factor = 25 percent.

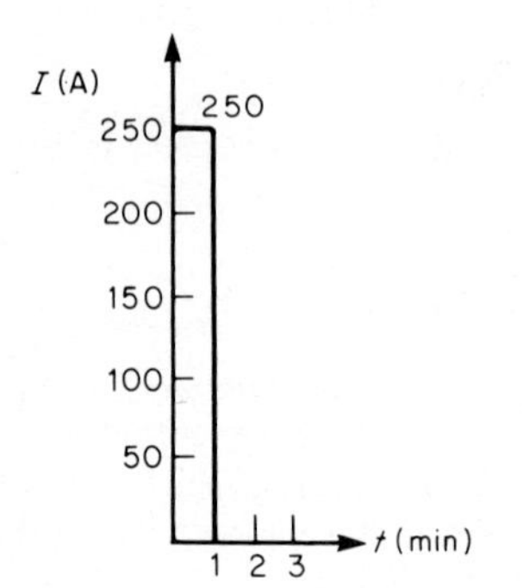

Fig. 13 Sizing battery for shown profile.

Calculation Procedure:

1. *Calculate Uncorrected Cell Size*

Use:*

Cell size (positive plates)

$$= \max_{S=1}^{S=N} \sum_{P=1}^{P=S} \frac{A_P - A_{(P-1)}}{R_T}$$

where S = section of load profile being analyzed, N = number of periods in the load profile, P = period being analyzed, A_P = amperes required for period P (note that A_0 = 0), T = time in minutes (seconds for NiCd cells) from the beginning of period P through the end of section S, and R_T = capacity rating factor representing the number of amperes that each positive plate can supply for T minutes at 25°C (77°F) to a specified end-of-discharge voltage.

Because there is only one section in the load profile, only one calculation need be made: number of positive plates (uncorrected) = $(A_1 - A_0)/R_T$. Because T = 1 min, R_T can be found for T = 1 min to 1.75 V/cell from manufacturer's data. Assume R_T = 75 A/positive plate; hence, number of positive plates (uncorrected) = (250 − 0)/75 = 3.333.

2. *Determine Required Size*

Required size = (uncorrected size)(temperature correction factor)(design margin)(age factor). From IEEE 485-1978, the temperature correction for 10°C (50°F) = 1.19.

*Formula from IEEE Standard 485-1978.

Therefore, required size = (3.33)(1.19)(1.10)(1.25) = 5.45 positive plates. Because it is impossible to get a fraction of a positive plate, it is normal practice to round the answer to the next higher whole number. Thus, for this example, six positive plates of cell type X' would be required.

Related Calculations: In determining the size of a battery for a specific application, adequate margin should be included for load growth of the dc system. Typically, a generating-station battery may be sized a number of times before it is purchased (e.g., conceptual sizing, followed by periodic resizing as load requirements are firmed up during the generating station design) followed by final sizing checks before it is placed in service. Each of these calculations require that a design margin be included; however, the margin will vary according to the type of calculation. For example, a design margin for a conceptual sizing might be 25 to 50 percent, but for the sizing calculation for purchasing the battery it might be only 10 to 20 percent. The design margin to be included is, therefore, dependent upon the specific battery installation.

As another example, consider a distribution substation with a single 138-kV line in and two 12-kV feeders out. If there were plans to expand the station to include an additional 138-kV line and four more 12-kV feeders within 5 yr, the battery would most likely be sized with enough design margin to carry the future load. However, if the expansion was not to take place for 15 years, the design margin might not include the future loads if, by economic analysis, it was determined to be more economical to size only for present load requirements and replace the battery with a larger one at the time the future load is added.

SIZING BATTERY FOR MULITPLE-LOAD PROFILE

Calculate the number of positive plates of a lead-acid cell type required to supply the load of Fig. 14. Assume: design margin = 10 percent, lowest electrolyte temperature is 21.1°C (70°F), age factor = 25 percent, 250-V dc system, 120 cells, 210 V dc minimum (i.e., 1.75 V/cell end-of-discharge voltage).

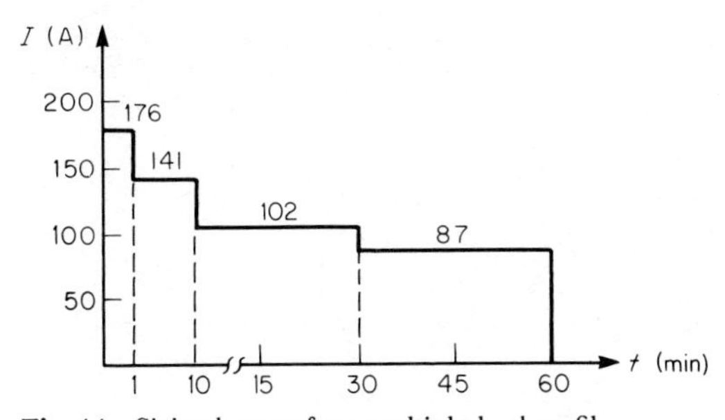

Fig. 14 Sizing battery for a multiple-load profile.

The load profile has four sections, each of which must be analyzed to determine which section controls the battery size, because the current in each period decreases with time. If the current in any period increased over that in the previous period, the section ending with the period just before the period of increased current would not have to be analyzed.

Calculation Procedure:

1. *Calculate Uncorrected Cell Size for Section 1 (Fig. 15)*

Assume $R_1 = 125$ A per positive plate ($T_1 = 1$ min). Hence, number of positive plates (uncorrected) $= (A_1 - A_0)/R_1 = (176 - 0)/125 = 1.41$ positive plates.

2. *Calculate Uncorrected Cell Size for Section 2 (Fig. 16)*

Assume $R_1 = 110$ A per positive plate ($T_1 = 10$ min) and $R_2 = 112$ A per positive plate ($T_2 = 9$ min). Number of positive plates (uncorrected) $= (A_1 - A_0)/R_1 + (A_2 - A_1)/R_2 = (176 - 0)/110 + (141 - 176)/112 = 1.6 - 0.31 = 1.29$ positive plates.

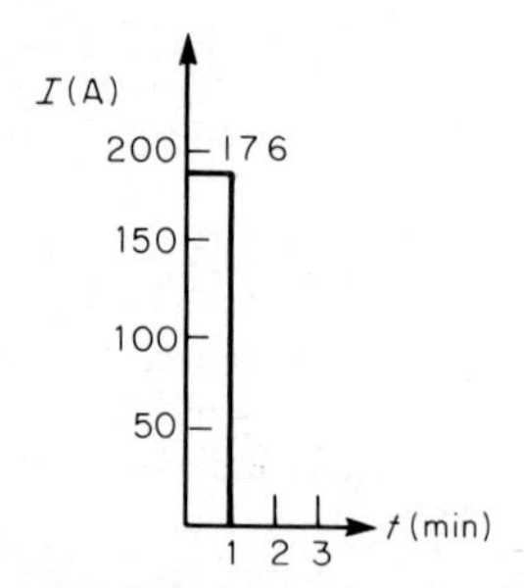

Fig. 15 Considering section 1 of a load profile.

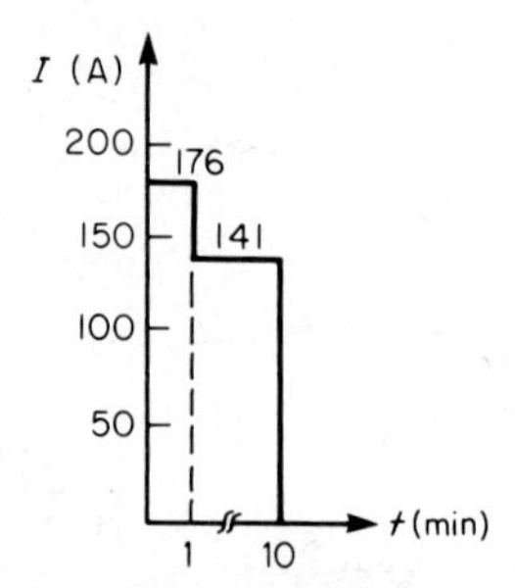

Fig. 16 Considering section 2 of a load profile.

3. *Calculate Uncorrected Cell Size for Section 3 (Fig. 17)*

Assume $R_1 = 93$ ($T_1 = 30$ min), $R_2 = 94$ ($T_2 = 29$ min), and $R_3 = 100$ ($T_3 = 20$ min) A per positive plate. Number of positive plates (uncorrected) $= (A_1 - A_0)/R_1 + (A_2 - A_1)/R_2 + (A_3 - A_2)/R_3 = (176 - 0)/93 + (141 - 176)/94 + (102 - 141)/100 = 1.89 - 0.37 - 0.39 = 1.13$ positive plates.

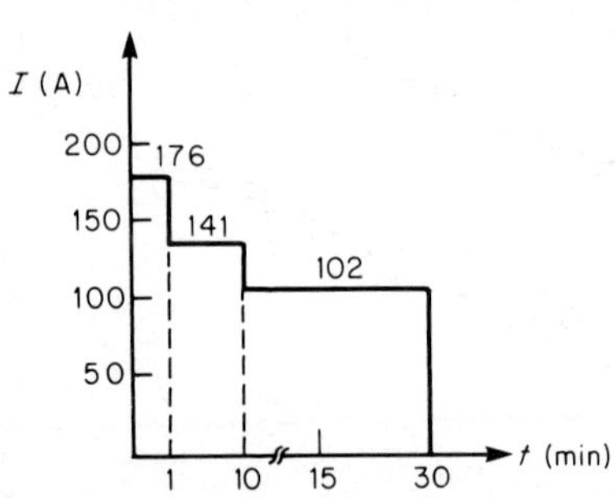

Fig. 17 Considering section 3 of a load profile.

4. *Calculate Uncorrected Cell Size for Section 4(Fig. 14)*

Assume $R_1 = 75$ ($T_1 = 60$ min), $R_2 = 76$ ($T_2 = 59$ min), $R_3 = 80$ ($T_3 = 50$ min), and $R_4 = 93$ ($T_4 = 30$ min) A per positive plate. Number of positive plates (uncorrected) $= (A_1 - A_0)/R_1 + (A_2 - A_1)/R_2 + (A_3 - A_2)/R_3 + (A_4 - A_3)/R_4 = (176 - 0)/75 + (141 - 176)/76 + (102 - 141)/80 + (87 - 102)/93 = 2.35 - 0.46 - 0.49 - 0.16 = 1.24$ positive plates.

5. *Determine Controlling Section*

Reviewing the positive plates required for each section, one finds that section 1 requires the most positive plates (i.e., 1.41) and is thus the controlling section.

6. *Determine Required Size*

With the correction factors and margin applied, required size = (max. uncorrected size)(temp. correction)(design margin)(age factor) = (1.41)(1.04)(1.10)(1.25) = 2.02 positive plates.

7. ***Select Cell***

Select a type Z' cell having three positive plates. This hypothetical cell would have a capacity of approximately 500 A·h at an 8-h rate, 25°C (77°F), to 1.75 V/cell, 1.210 specific-gravity electrolyte. Even though the cell requires only a fraction of a plate above 2, the selected cell is still rounded up to the next higher whole number.

Related Calculations: If, for example, the given load profile appears as shown in Fig. 18, period 2 does not have to be analyzed because it is exceeded by the current in period 3.

The effect of a random load is added to the controlling section of the battery size.

AMPERE-HOUR CAPACITY

Calculate the ampere-hour capacity of a lead-acid cell required to satisfy the load profile of Fig. 18. Assume: design margin = 10 percent, lowest electrolyte temperature is 26.7°C (80°F), age factor = 25 percent, 1.75 V/cell end-of-discharge voltage, 60 cells, 125-V dc system, 105 V dc minimum voltage.

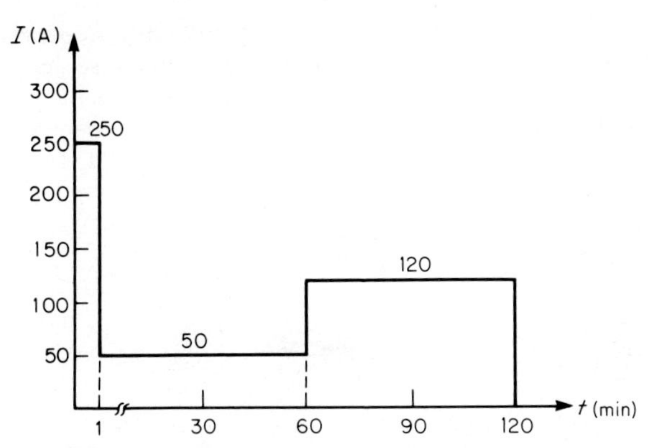

Fig. 18 Load profile where the current (50 A) in section 2 is exceeded by the current (120 A) in section 3.

Calculation Procedure:

1. ***Compute Uncorrected Ampere Hours for Section 1 (Fig. 19)***

Use:

$$\text{Cell size (A·h)} = \max_{S=1}^{S=N} \sum_{P=1}^{P=S} [A_P - A_{(P-1)}]K_T$$

where K_T = a capacity rating factor representing the ratio of rated ampere-hour capacity at a standard time rate, at 25°C (77°F), and to a standard end-of-discharge voltage, to the amperes which can be supplied by the cell for T minutes at 25°C (77°F), and to a given end-of-discharge voltage. (In the United States, the standard time rate is normally the 8-h rate; however, non-U.S. standards use 5- and 10-h standard rates as well.) The other terms have the same meanings as in the previous equations.

Assume $K_{T1} = 0.93$ ($T_1 = 1$ min); then, ampere-hours (uncorrected) $= (A_1 - A_0)K_{T1} = (250 - 0)0.93 = 232.5$ A·h.

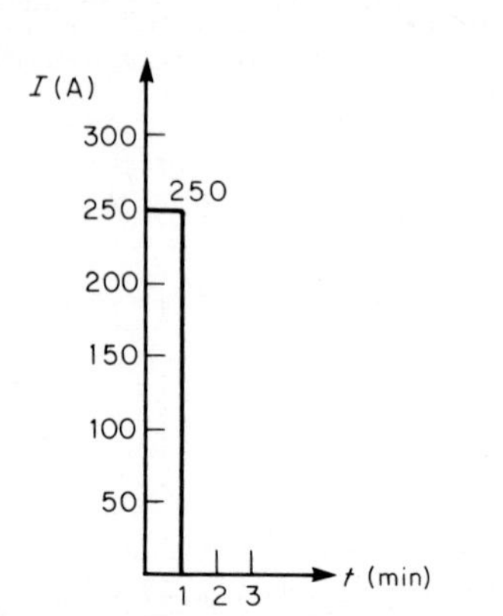

Fig. 19 Considering only section 1 of a load profile.

2. Compute Uncorrected Capacity in Ampere-Hours for Sections 1 and 3 (Fig. 18).
Assume $K_{T1} = 3$ ($T_1 = 120$ min), $K_{T2} = 2.95$ ($T_2 = 119$ min) and $K_{T3} = 2$ ($T_3 = 60$ min). Then, ampere-hours (uncorrected) $= (A_1 - A_0)K_{T1} + (A_2 - A_1)K_{T2} + (A_3 - A_2)K_{T3} = (250 - 0)3 + (50 - 250)2.95 + (120 - 50)2 = 750 - 590 + 140 = 300$ A.H.

3. Determine Controlling Section
Reviewing the capacity in ampere-hours required for each section, one finds section 3 requires the most capacity and is thus the controlling section.

4. Determine Required Size
Applying the correction factors and margin, one finds: required size = (max. uncorrected size)(temp. correction)(design margin)(age factor) = (300)(1.0)(1.10)(1.25) = 412.5 A·h. Therefore, select a standard type Z' cell with an ampere-hour rating at the 8-h rate greater than or equal to 412.5 A·h; e.g., a standard cell of 450 A·h might be selected.

Related Calculations: Common practice for temperatures above 25°C (77°F) is to use a factor of 1.0 rather than to take credit for the extra capacity available.

CHARGER SIZE

Determine the required charger output to recharge a lead-acid battery in 16 h while serving a dc system load of 20 A. The battery has the following characteristics: 500 A·h at the 8-h rate, 25°C (77°F), 1.210 specific-gravity electrolyte to 1.75 V/cell.

Calculation Procedure:

1. Compute Charger Current
The capacity of the required charger is calculated from the equation charger A $= (\text{AH})K/T + L$, where charger A = required charger output amperes, AH = ampere-hours removed from the battery, K = a constant to compensate for losses during the charge (normally 1.1 for lead-acid cells), T = desired recharge time in hours, and L = steady-state dc load to be served by the charger while the battery is being recharged.
Substitution of given values in the equation yields: charger A $= (500)(1.1)/16 + 20 = 54.38$ A.

2. Select Charger
Select the next standard-rated charger with an output that is greater than 54.38 A.

Related Calculations: Variables T and L are specific to the system under consideration and must be determined by the designer, although T is also a function of charging voltage. The higher the recharge voltage, the faster the recharge, within limits (consult the battery manufacturer).

Sometimes, however, the maximum dc system voltage may limit the recharge voltage. In these instances, a longer recharge time is required unless the system design allows the battery to be isolated from the system during recharge. (Some non-U.S. chargers have an internal dropping diode which regulates the dc bus voltage, while allowing the charger voltage to increase above the system limits for recharging the battery.)

RECHARGING NICKEL-CADMIUM BATTERIES

Calculate the required charger size to recharge a 340-A·h (at the 8-h rate) NiCd block battery having eight NC′ modules, using the two-rate charging method. It is desired to have 70 to 80 percent of the ampere-hour capacity (previously discharged) replaced within 10 h.

The battery is used on a 125-V dc system, maximum allowable voltage is 140 V dc, and there are 92 cells in the battery. If the high-rate charge can be accomplished in 10 h or less, how many additional hours will be required to replace the additional 20 to 30 percent of capacity? Also, assume the charger must carry 20 A steady-state dc system load.

Calculation Procedure:

1. Check Manufacturer's Data

From these data, it is determined that the battery can be recharged (first period) in 9 h if the recharge voltage is 1.50 V/cell or in 10 h at 1.55, 1.60, or 1.65 V/cell. In all cases, it is assumed that the charger output divided by the cell capacity during the first period is 0.1. All these values meet the specified criteria for recharge.

2. Check Maximum System Voltage

Maximum voltage per cell = max. system voltage/number of cells = 140/92 = 1.52 V/cell. From this result, the only acceptable recharge rate is 1.50 V/cell.

3. Calculate High-Rate Current

High-rate current = (0.1)(cell capacity) = (0.1)(340) = 34 A.

4. Determine Charger Size

Charger A = high-rate current + steady-state dc load = 34 + 20 = 54 A.

5. Select Charger

Select the next standard-rated charger with an output greater than 54 A. (From the manufacturer's data, an additional 170 h is required to restore the remaining 20 to 30 percent capacity.)

Related Calculations: NiCd batteries normally use two-rate charging. This type of charge accomplishes recharge over two time periods, each with its own charging rate. The battery manufacturer has data available to aid in the selection of the required charger. These data provide the charging times required to charge the battery assuming a specific voltage per cell recharge voltage and specific charger output as related to cell capacity during the first charge period.

For two-rate charging, the first period (or high-rate period) is the time required to replace 70 to 80 percent of the ampere-hour capacity discharged. The second period (or finish-rate period) is the time required to replace the remaining 20 to 30 percent of the discharged capacity.

Section 19 ECONOMIC METHODS

Bjorn M. Kaupang, P.E.
Manager, Generation Planning and Economics,
General Electric Company

REFERENCES AIEE—"Application of Probability Methods to Generation Capacity Problem," *AIEE Transactions on Power Apparatus and Systems,* February 1961; Billington, et al.—*Power System Reliability Calculations,* MIT Press; Day, et al.—"Optimizing Generation Planning," *Power Engineering,* July 1973; Dees, et al.—"The Effect of Load Growth Uncertainty on Generation System Expansion Planning," *Proceedings of the American Power Conference,* vol. 40, 1978; Galloway, et al.—"An Approach to Peak Load Economics," *AIEE Transactions,* part III, vol. 79, 1960; Galloway, et al.—"The Role of Pumped Storage in Generation Systems," *Proceedings of the American Power Conference,* vol. 6, 1964; Garver—"Effective Load-Carrying Capability of Generation Units," *IEEE Transactions on Power Apparatus and Systems,* 1966; Jeynes—*Profitability and Economic Choice,* Iowa State U. Press; Jordan, et al.—"The Impact of Load Factor on Economic Generation Patterns," *Proceedings of the American Power Conference,* vol. 38, 1976; Kirchmayer—*Economic Operation of Power Systems,* Wiley; Marsh—*Economics of Electric Utility Power Generation,* Oxford University Press; Marsh, et al.—"Combining Fossil Fueled High Efficiency, Nuclear Fueled, Pumped Hydro, and Peaking Gas Turbine Plants for Lower Total Costs," *Transactions of World Power Conference,* Lausanne Sectional Meeting, vol. 1, 1964; Park—*Cost Engineering Analysis,* Wiley.

COST OF MONEY

Find the weighted cost of money when given the bond interest rate i_B = 10 percent, the preferred stock interest rate i_p = 12 percent, and the return on common stock i_C = 15 percent. Also, the debt ratio (fraction of bonds) DR = 50 percent, the common-stock ratio (fraction of common stock) CR = 35 percent, and the preferred-stock ratio (fraction of preferred stock) PR = 15 percent.

Calculation Procedure:

1. Compute the Cost of Money

The weighted cost of money, i, is given by $i = i_B\text{DR} + i_P\text{PR} + i_C\text{CR}$. Substitution of given values in the equation yields $i = (10)(0.5) + (12)(0.15) + (15)(0.35) = 12.05$ percent.

Related Calculations: The basic sources of capital, other than capital surplus from operations, are the bond and stock markets for long-term capital needs and banking institutions for short-term (less than a year) borrowings. Bonds are the most common long-term debt instrument in the utility industry. The first-mortgage bond is the most senior and, therefore, has the first claim on a company's assets. This security is reflected in the relatively low interest rate for first-mortgage bonds.

Stocks could be issued as preferred, with a fixed dividend rate, or as common. In most companies, the only voting stock is the common stock. It also carries the highest risk and, therefore, the highest return.

Capitalization is an accounting term for total outstanding bonds and stock. The relationship between the bonds and stock making up the capitalization may be expressed as capitalization ratios which can be used for estimating the weighted cost of money.

LEVELIZED ANNUAL COST

Levelizing of nonuniform series of fixed and variable costs is often used in economic evaluation to compare the economic value of a cost series different in timing and magnitude. Find the levelized cost for the 5-yr cost series in Table 1. Assume a 10 percent discount rate (weighted cost of money) and a capital recovery factor CRF = 0.2638.

Calculation Procedure:

1. Determine Sum of Present Values of Each Annual Cost

Use $P = S/(1 + i)^N$ where S = future worth, P = present value, N = number of years, and i = interest rate. The calculated results are given in Table 1.

TABLE 1 Data for Calculating Levelized Costs

Year	*Annual cost, $*	*Present-value factor* $[1/(1+i)^N]$	*Present value P, $*
1	400	0.9091	363.64
2	600	0.8264	495.84
3	800	0.7513	601.04
4	1000	0.6830	683.00
5	1200	0.6209	745.08
Present value			2888.60

2. Compute Levelized Cost

The levelized annual cost is equal to the product of the present value and the CRF. Hence, the levelized annual cost = ($2888.60)(0.2638) = $762.01. A uniform series of $762.01 per year for 5 yr has the same present value as the actual cost series in the table.

SINKING FUND DEPRECIATION

Find the annual depreciation expense for a $1 investment with no salvage value after 5 yr. The discount rate is 10 percent and the sinking fund factor is 0.16380.

Calculation Procedure:

1. Find Accumulated Depreciation for Each Year

The accumulated depreciation at the end of year n may be calculated by

$$\sum_{1}^{n} d = \frac{(1+i)^n - 1}{(1+i)^N - 1}$$

For example, if $n = 2$, the accumulated depreciation is $[(1 + 0.1)^2 - 1]/[(1 + 0.1)^5 - 1] = 0.34398$. This and other values for n are tabulated in Table 2.

2. Find interest on last year's accumulated depreciation

Multiply the accumulated depreciation by 0.1. The resulting values are given in Table 2.

TABLE 2 Data for Sinking Fund Depreciation Calculation

Year n	*Sinking fund factor*	*Accumulated depreciation*	*Interest on last year's accumulated depreciation*	*Total annual depreciation expense d*
1	0.16380	0.16380	0	0.16380
2	0.16380	0.34398	0.01638	0.18018
3	0.16380	0.54218	0.03440	0.19820
4	0.16380	0.76020	0.05422	0.21802
5	0.16380	1.00000	0.07602	0.23982

3. *Find Annual Depreciation Expense*

This value may be found by adding the results from Steps 1 and 2 or by using the equation $d_n = i(i + 1)^{n-1}/[(1 + i)^N - 1]$. For example if $n = 2$, $d_n = 0.1(1.1)^{2-1}/(1.1^5 - 1) = 0.18018$. This and other values for n are tabulated in Table 2.

Related Calculations: Depreciation is a method of accounting to ensure recovery of the initial capital investment adjusted for a possible salvage value. The depreciation expense is typically based on a yearly basis, but the yearly amount does not necessarily reflect the actual loss of value of the investment.

There are several common methods of calculating annual depreciation expenses. All methods calculate an annual series of expenses which adds up to the initial capital expense less the salvage value at the end of useful life.

The most common method for calculating annual depreciation expense is the *straight-line method.* The annual expense is equal to the initial capital investment minus the end-of-life salvage value divided by the depreciation life in years. For example, an investment of $1 with no salvage value after 5 yr would have $1/5 = $0.20 per year in depreciation. The sinking fund depreciation method accumulates depreciation expenses more slowly than the straight-line method. The annual depreciation expense is equal to the sinking fund factor times the initial investment plus the interest charges on the previous accumulated depreciation.

SUM-OF-YEARS DIGIT DEPRECIATION

One common method used for a faster accumulation of depreciation expenses than the straight-line method is the sum-of-years digit method. The annual depreciation expense with this method is equal to the initial investment times the remaining life divided by the sum-of-years digits.

Determine the accumulated depreciation using the sum-of-years digit method for $1 investment over a 5-yr life.

Calculation Procedure:

1. *Find the Sum-of-Years Digit*

The sum-of-years digit = 1 + 2 + 3 + 4 + 5 = 15.

2. *Compute Annual Depreciation in Each Year*

Use $d_n = (N - n + 1)/[N(N + 1)/2]$ where d_n = annual depreciation. For example if $n = 2$, $d_n = (5 - 2 + 1)/[(5)(5 + 1)/2] = 0.26667$. This and other values for n are tabulated in Table 3.

TABLE 3 Data for Calculating Sum-of-Years Digit Depreciation

Year	*Remaining life*	*Remaining life sum-of-years digit*	*Accumulated depreciation*
1	5	0.33333	0.33333
2	4	0.26667	0.60000
3	3	0.20000	0.80000
4	2	0.13333	0.93333
5	1	0.06667	1.00000

3. Determine Accumulated Depreciation for Each Year

The accumulated depreciation for each year may be obtained by accumulating the annual values from Step 2. For example, if $n = 2$, the accumulated depreciation = $0.33333 + 0.26667 = 0.6000$. This and other values for n are tabulated in Table 3.

DECLINING-BALANCE METHOD

Another method which yields a faster depreciation than straight-line accumulation of depreciation expense is the declining-balance method. The rate of depreciation is applied to the remaining balance of initial investment minus accumulated depreciation. The rate is normally higher than the straight-line method. If the rate is double the straight-line method, we have a double declining-balance depreciation.

Find the annual depreciation expenses for a $1 investment depreciated at 40 percent for a 5-yr depreciation life.

Calculation Procedure:

1. Compute the Annual Depreciation

The annual depreciation may be computed by $d_n = r(1 - r)^{n-1}$ where r = rate of depreciation. For example, if $n = 2$, $d_n = 0.4(1 - 0.4)^{2-1} = 0.24$. This and other values for n are tabulated in Table 4.

2. Compute the Accumulated Depreciation

Accumulate the values found in Step 1. For example, if $n = 2$, the accumulated depreciation $= 0.40 + 0.24 = 0.64$. This and other values for n are tabulated in Table 4.

TABLE 4 Data for Declining-Balance Method Calculations

Year	*Remaining balance*	*Annual depreciation*	*Accumulated depreciation*
1	1.0	0.40	0.40
2	0.6	0.24	0.64
3	0.36	0.144	0.784
4	0.216	0.0864	0.8704
5	0.1296	0.0518	0.9222

Related Calculations: Because the accumulated depreciation does not reach 1.0 before n goes to infinity, it is common to transfer from the declining balance method to another depreciation method at some point during the depreciation life.

TRANSFER OF DEPRECIATION METHODS

Calculate the accumulated depreciation when a transfer is made to the straight-line depreciation method in the last 2 yr of the previous example.

Calculation Procedure:

1. Find Remaining Balance to Be Depreciated in Last 2 Years

From Table 4, the value is 0.216.

TABLE 5 Values of Annual and Accumulated Depreciation

Year	*Remaining balance*	*Annual depreciation*	*Accumulated depreciation*
1	1.0	0.40	0.4
2	0.6	0.24	0.64
3	0.36	0.144	0.784
4	0.216	0.108	0.8704
5	0.108	0.108	1.0

2. Find Annual Straight-Line Depreciation for Last 2 Years
The value is $0.216/2 = 0.108$.

3. Determine Accumulated Depreciation
The values are given in Table 5.

Related Calculations: The most common depreciation method is the straight-line method over the useful life of the equipment. The useful life should be based on knowledge of mechanical life as well as economic life. Technical improvements might make a device obsolete before its mechanical life is up.

The fast depreciation methods, like the sum-of-years digit and declining balance, are commonly used for calculating deductible expenses for income tax purposes. The effect is to delay, not to reduce, paying income taxes, thereby reducing external financing.

TAXES

Find the income tax and the resulting net income to the shareholders given the following information. The corporation has \$2500 in revenues and \$350 in annual operating expenses. The original investment was \$3000 and the annual interest expense is \$300. The depreciation life is 5 yr. The sum-of-years digit depreciation will be used for tax purposes and the straight-line method for book depreciation. The tax rate is 50 percent and the corporation is in its first year of operation.

Calculation Procedure:

1. Find Total Deductible Expenses
These values are given directly in the example except for depreciation. From $d_n = (N - n + 1)/[N(N + 1)/2]$ with $n = 1$, the first-year depreciation expense is \$1000. Total deductible expenses, therefore, for income tax purposes are: $350 + 300 + 1000 = \$1650$.

2. Compute Taxable Income
The revenues are given as \$2500 which gives $2500 - 1650 = \$850$ in taxable income.

3. Determine Income Tax
With a taxable income of 50 percent, the income tax is $(850)(0.50) = \$425$.

4. Calculate Income Statement for Shareholders
The only value not yet calculated is the book depreciation, which is $\$3000/5 = \$600/$ yr. The resulting income statement is provided in Table 6.

TABLE 6 Income Statement for Shareholders

Revenues	$2500	
Operating expenses		$ 350
Interest		300
Depreciation		600
Income taxes		425
Total expenses		$1675
Net income		$ 825

Related Calculations: For an industry or a utility, the common taxes are property, or ad valorem, taxes and income taxes. The ad valorem taxes are normally simple percentages of the assessed value of property and therefore are simple to calculate. Income taxes are normally levied by federal and most state governments by applying a fixed rate to a taxable income.

As a rule, taxable income is calculated differently from the income reported by the owners of the company. The major difference in the calculation comes from the treatment of depreciation. It is typical for a corporation to have two sets of books, one for income tax purposes and one for the shareholders.

INVESTMENT TAX CREDIT

This is a method used by government to encourage investment in new production facilities. The tax credit is normally calculated as a fixed percentage of new investment in a year and then subtracted directly from the income tax.

Find the effect of a 10 percent investment tax credit when applied to the previous income tax example.

Calculation Procedure:

1. Find Investment Tax Credit

The tax credit is ($3000)(0.10) = $300.

2. Compute Income Taxes

Table 7 shows the values from the income tax example modified by the investment tax credit.

3. Generate Income Statement to Shareholders

The new income statement is given in Table 8.

TABLE 7 Modified Tax Statement, Including Investment Tax Credit

Revenues	$2500	
Total deductions		$1650
Taxable income		850
Income tax		425
Investment tax credit, 10% of $3000		300
Income tax payable		$125

TABLE 8 New Income Statement for Shareholders

Revenues	$2500	
Operating expenses		$350
Interest		300
Depreciation		600
Income tax payable		125
Total expenses		1375
Net income		$1125

FIXED-CHARGE RATE

Assume that the tax rate is 50 percent, the return R multiplied by the weighted cost of money in equity is 0.045, accounting depreciation is 0.2, the levelized tax depreciation is 0.215, and the cost of money is 12 percent. Calculate the fixed-charge rate.

Calculation Procedure:

1. Find Income Tax Effect

Use $T = [t/(1 - t)](R_e - d_T - d_A)$ where t is the tax rate, R_e is the return R multiplied by the weighted cost of money in equity, d_T is the levelized tax depreciation based on sum-of-years digits, and d_A is the accounting depreciation. Substituting values yields $T = (0.5/0.5)(0.045 - 0.2150 - 0.2) = 0.03$.

2. Determine the Fixed-Charge Rate

The fixed-charge rate FCR is given by: $\text{FCR} = R + d_A + T + T_A + I$ where T_A is the ad valorem tax, R is the return needed to cover cost of money and depreciation, and I is the insurance. Assume $T_A + I = 0.025$; hence, $\text{FCR} = 0.0774 + 0.2 + 0.03 + 0.025 = 0.362$.

Related Calculations: Investment decisions often involve comparing annual operating expenses with the cost of capital needed for the investment and the revenues generated from a new project. For an electric utility, the revenues are regulated and it is therefore convenient to relate investment decisions to revenue requirements. The revenue requirements are equal to the annual expenses such as fuel costs and operations and maintenance (O&M) costs, plus the annual fixed charges on the investment. The fuel and operation and maintenance costs are obtained from estimates of production needs.

The assumed fixed charges are normally estimated by a levelized fixed-charge rate applied to the initial investment. The fixed charges should include a return to the shareholder, interest payments on debt, depreciation expenses, income tax effects, property taxes, and insurance.

In the first example, the cost of money was approximately 12 percent. For a 5-yr plant life and 12 percent cost of money, the annual charge necessary to recover the capital is the interest plus the sinking fund factor equal to $0.12 + 0.1574 = 0.2774$. The sinking fund could be regarded as a depreciation. It is common, however, to use straight-line depreciation where the depreciation rate is $1/5 = 0.2$. An excess of $0.2 - 0.1574 = 0.0426$ is thereby obtained. This gives a revenue requirement resulting from the cost of money and depreciation method equal to $0.12 - 0.0426 = 0.0774$.

The equation for the return R needed to cover the cost of money and depreciation is: $R = i - (d - \text{SFF})$ where i is the weighted cost of money, d is the depreciation used for accounting purposes, and SFF is the sinking fund factor.

Typical values for the fixed-charge rate in the private utility industry in 1980, for a 25- to 30-yr plant life, were 0.17 to 0.20. In government-sponsored utilities exempt from income taxes and, sometimes, from property taxes, and using only debt financing, the typical fixed-charge rate is 0.10 to 0.12.

In private, nonutility industries, the fixed-charge rate and the revenue-requirements methodology are seldom used. If calculated, however, the fixed-charge rate could vary from 0.20 to 0.40, depending on financing, project life, and management philosophy.

Rate of return, also called return on investment (ROI), is a more common measure in an industrial environment. This measure often refers to the discounted cash flow method of economic evaluation. Because most industries attempt to maximize the ROI, the approach is to estimate revenues and expenditures and then find the interest rate which, when used to discount the cash flows over the life of the project, will make the present-worth sum of the cash flows equal to the initial investment. An example of ROI and discounted cash flow analysis will be considered later.

REVENUE-REQUIREMENTS METHOD

Alternatives A and B have operational characteristics that allow both to perform the same service. The capital investment required is $50,000 and $48,000 for alternatives A and B, respectively. The useful life for both alternatives is 5 yr. Annual operating costs are given in Table 9. The weighted cost of money is 12 percent and the fixed-charge rate is 36.2 percent as calculated in the previous example. Make an economic choice between alternatives A and B.

Calculation Procedure:

1. Determine the Sum of Present Values of Revenue Requirements to Cover Annual Operating Costs

The calculated results are summarized in Table 10.

2. Find the Present Value of Annual Revenue Requirements for the Capital Investment

The levelized annual fixed charges are found by multiplying the investment by the fixed-charge rate. For alternative A, ($50,000)(0.362) = $18,100/yr; for alternative B, ($48,000)(0.362) = $17,376/yr.

TABLE 9 Annual Operating Costs of Alternatives A and B

	Annual operating costs	
Year	*Alternative A*	*Alternative B*
1	$6000	$6500
2	5800	6600
3	5600	6700
4	5400	6800
5	5200	7000

TABLE 10 Present Values of Revenue Requirement; i = 12 Percent

Year	*Alternative A*	*Alternative B*
1	$5,357	$5,804
2	4,624	5,261
3	3,986	4,769
4	3,432	4,322
5	2,951	3,972
Sum of present values	$20,350	$24,128

The present values are found by dividing the levelized annual fixed charges by the capital recovery factor, $\text{CRF} = i(1 + i)^N/[(1 + i)^N - 1]$ where i = interest rate and N = number of years. For 5 yr and i = 12 percent, $\text{CRF} = 0.12(1.12)^5/[(1.12)^5 - 1] = 0.2774$, and the present values of the annual revenue requirements for the capital investment are: alternative A, $18,100/0.2774 = $65,249 and alternative B, $17,376/0.2774 = $62,639.

3. *Find the Present Value of Total Revenue Requirements for the Two Alternatives*

This is the sum of the results obtained in Steps 1 and 2: alternative A, 20,350 + 65,249 = $85,599 and alternative B, 24,128 + 62,639 = $86,767. Alternative A, having the lowest present value of total revenue requirements, is the economic choice.

Related Calculations: The economic evaluation of engineering options is an important factor in equipment application. The capital cost of each alternative must be combined with its operating cost to develop a base for comparison. It is most likely that the capital costs differ for the alternatives, as do the operating costs.

The time period chosen for a study should include at least a major portion of the expected useful life of the equipment. In a utility, this would mean 15 to 20 yr, while study periods of only a few years could be appropriate for an industry.

Often, economic comparisons are made without considering the effects on the operation of the interconnected systems. This is acceptable if the alternatives have similar operational characteristics, i.e., equal annual energy production for generating units or operating hours for electric motors. If the effects on the interconnected system are uncertain, total system cost evaluations must be performed. The revenue-requirements method is typical of the utility industry while the discounted cash flow and payback-period methods are typical in industrial evaluation.

DISCOUNTED CASH-FLOW METHOD

Using the data in the previous example, choose between alternatives A and B by the discounted cash-flow method.

Calculation Procedure:

1. *Find Difference in Annual Cash Flows for Alternatives*

The annual cash flows are provided in Table 11 and include operating expenses, a tax depreciation based on a sum-of-years digit, and an ad valorem tax at a rate of 0.025.

TABLE 11 Annual Cash Flows for Alternatives A and B

	Year 1	*Year 2*	*Year 3*	*Year 4*	*Year 5*
Incremental investment	−2000				
Operating cost alternative B	6500	6600	6700	6800	7000
Operating cost alternative A	6000	5800	5600	5400	5200
Operating saving for incremental investment	500	800	1100	1400	1800
Tax depreciation	−667	−533	−400	−267	−133
Ad valorem tax	−50	−50	−50	−50	−50
Taxable income	−217	217	517	1083	1617
Tax at 50%	109	−109	−259	−542	−809
Net income	−108	108	258	541	808
Tax depreciation	667	533	400	267	133
Cash flow	559	641	658	808	941

2. *Find Discount Rate Using the Sum of Present Values Equal to the Investment Difference*

The discount rate is found by trial and error. Try $i = 15$ percent; this gives a present value PV equal to $559/1.15 + 641/1.15^2 + 658/1.15^3 + 808/1.15^4 + 941/1.15^5 = 2333$. For $i = 22$ percent, PV = 1964 is obtained. If $i = 21.22$ percent, PV = 2000.

3. *Make Economic Choice*

The result of the discounted cash-flow evaluation is a 21.22 percent rate of return on the incremental investment. If this rate equals, or exceeds, the hurdle rate established for this investment, alternative A is the economic choice.

DAILY-LOAD FACTOR

The daily-load factor is defined as the ratio of the load energy in the day to the energy represented by the daily peak demand multiplied by 24 h. Find the daily-load factor LF_D when the daily-load energy is 21 GWh and the daily peak demand is 1000 MW.

Calculation Procedure:

1. *Multiply the Peak Demand by 24 h*

The multiplication is $(1000)(24) = 24{,}000$ MWh = 24 GWh.

2. *Determine the Daily-Load Factor*

$$LF_D = 21 \text{ GWh}/24 \text{ GWh} = 0.875$$

Related Calculations: The shape of the daily-, seasonal-, and annual-load curves are important characteristics for operation and expansion of generation systems to meet the system load. Utilities record the chronological hourly loads on a continuous basis. A typical hourly load curve for a day is shown in Fig. 1.

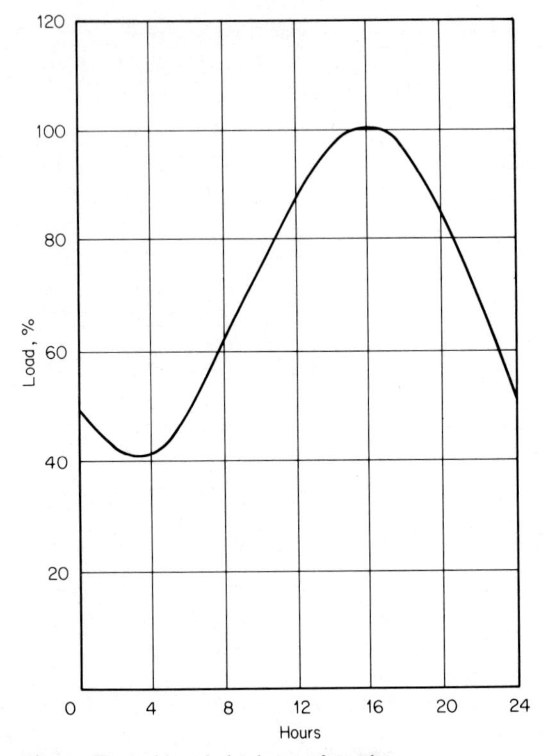

Fig. 1 Typical hourly load curve for a day.

ANNUAL-LOAD FACTOR

The annual-load factor is defined as the ratio of annual-load energy to the energy represented by the annual peak demand multiplied by 8760 h. It is possible to estimate the annual-load factor from the average daily-load factor by using typical daily and monthly daily peak-load variations.

Find the annual-load factor LF_A, when the average daily-load factor $LF_D = 0.875$, the ratio of average daily peak load to monthly peak load $R_{WM} = 0.85$, and the ratio of average monthly peak load to annual peak load $R_{MA} = 0.8$.

Calculation Procedure:

1. Compute LF_A

$$LF_A = LF_D R_{WM} R_{MA} = (0.875)(0.85)(0.8) = 0.595$$

Related Calculations: Load-duration curves are curves in which all the hourly loads in a time period (commonly a year) are arranged in descending order. This curve is not chronological. Because the area under the curve is the period-load energy, the curve is

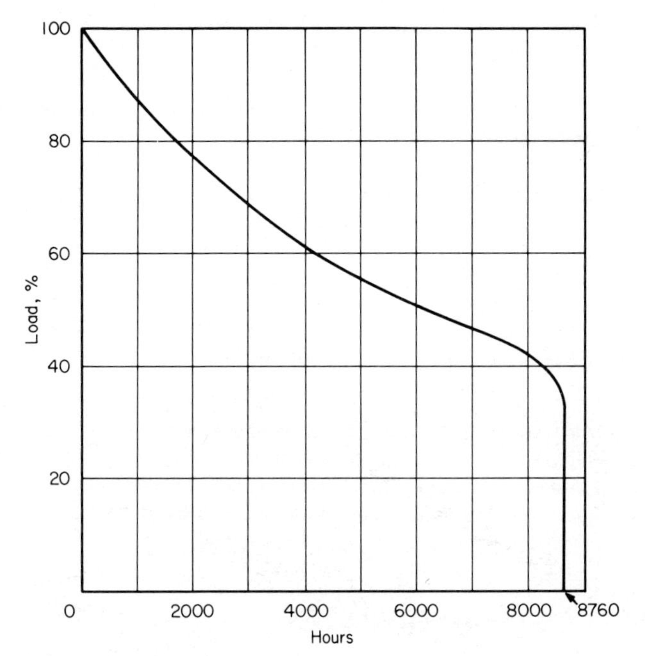

Fig. 2 Annual load-duration curve.

often useful in simplified utility economics calculations. A typical load-duration curve is provided in Fig. 2.

LOAD MANAGEMENT

The purpose of load management is to direct energy usage away from peak-load periods. Methods for load management include peak-sensitive rate structures and automatic control of power consumption to increase load diversity.

Find the annual-load factor LF_A when the average daily peak load is reduced 5 percent from 1000 MW to 950 MW. Assume $R_{WM} = 0.85$ and $R_{MA} = 0.8$.

Calculation Procedure:

1. Find Energy Associated with 950 MW for 24 h

The energy is $(950)(24) = 22{,}800$ MWh.

2. Find Resulting Annual-Load Factor with a Daily-Load Energy of 21,000 MWh.

$$LF_A = LF_D R_{WM} R_{MA} = (21{,}000/22{,}800)(0.85)(0.8) = 0.626$$

Related Calculations: The major problems with evaluating load-management devices is the difficulty in predicting the effect on peak demand and peak energy consumption before installation and actual operation of devices. The economic value of the reductions are also highly utility- and time-specific.

FORCED-OUTAGE RATE

Find the forced-outage rate FOR in percent for a generating unit which operated 6650 h in one year, with 350 h on forced outage and 1860 h on scheduled shut-down.

Calculation Procedure:

1. Compute FOR

Use FOR = FOH/(FOH + SH) where FOH = forced-outage hours and SH = service hours. Substituting values, find FOR = 350/(350 + 6650) = 0.05 = 5 percent.

Related Calculations: Total utility economic analysis includes the analysis of both the cost of reliability and the cost of production of all the generating units and the utility system loads they serve. The study time period should be long enough to adequately include at least the major economic effects of the alternatives studied. In some cases, this could mean as much as 20 yr.

SYSTEM ECONOMIC DISPATCH

Find the loading schedule for units A, B, and C given the data in Table 12. Linear interpolation is assumed between data points.

Calculations Procedure:

1. Find Breakeven Loading between Units

The values are calculated from the table values. For example, $17.55/MWh is found

TABLE 12 Data for Units A, B, and C

Data	*A*	*B*	*C*
Full load, MW	50	35	16
Heat rate, kJ/kWh	12,000	12,500	13,000
Btu/kWh	11,375	11,849	12,323
Fuel price, $/GJ	1.50	2.00	1.50
$/MBtu	1.58	2.11	1.58
Minimum load, MW	13	10	4
Incremental HR			
Minimum load, kJ/kWh	10,560	11,100	11,700
50% load, kJ/kWh	11,280	11,870	12,480
100% load, kJ/kWh	12,000	12,500	13,000
Incremental fuel cost			
Minimum load, $/MWh	15.84	22.4	17.55
50% load, $/MWh	16.92	23.74	18.72
100% load, $/MWh	18.00	25.00	19.50

TABLE 13 Loading Schedule for Units A, B, and C

		Unit loading, MW		
System load	*Fuel cost, $/MWh*	*A*	*B*	*C*
27	15.84	13	10	4
39	16.92	25	10	4
50	17.55	40	10	4
66	18.00	50	10	6
68	18.72	50	10	8
76	19.50/22.2	50	10	16
84	23.74	50	18	16
102	25.00	50	36	16

at a 40-MW loading on unit A and $18.00/MWh is found at a 6-MW loading level on unit C.

2. *Find Loading Schedule*

From the given data and Step 1, the results are tabulated in Table 13. The loading schedule shows all possible combinations of unit loadings listed in the order of increasing incremental fuel cost.

Related Calculations: The process of dispatching the committed generation is based on minimizing the cost of fuel. This results in an incremental generation loading, beyond minimum generation, in the order of increasing incremental fuel cost.

SYSTEM STORAGE

Find the cost saving realized by operating a storage device with conversion efficiency $\eta_S = 70$ percent on the utility system described by the daily load curve of Fig. 3. The storage reservoir has an additional storage capacity equivalent to 6 h charging at 50 MW as this day starts.

Calculation Procedure:

1. *Find Cost of 6-h Charging at 50 MW*

The values on the right-hand vertical axis in Fig. 3 are the variable fuel, operations, and maintenance costs in mills/kWh for the generating plants operating this day. (Cost of electricity is often expressed in mills/kWh or $/MWh). The cost proportion for 6-hr charging is estimated to be 60 percent at 12 mills/kWh, 40 percent at 15 mills/kWh. The cost of 6 h of charging, C_{CH}, then is $C_{CH} = (0.60)(12)(6)(50) + (0.40)(15)(6)(50) = \$3960$.

2. *Determine Cost of Stored Energy, in mills/kWh, Available for Discharging When Needed*

Because the energy available for discharge is less than the energy used for charging by the conversion efficiency of 70 percent, the unit cost of the energy for discharge, C_{DC}, is: $C_{DC} = 3960/(6)(50)(0.70) = 18.9$ mills/kWh.

3. *Find Cost of Existing Generation during the Daily Peak*

From Fig. 3, it is seen that existing generation would cost 35 mills/kWh.

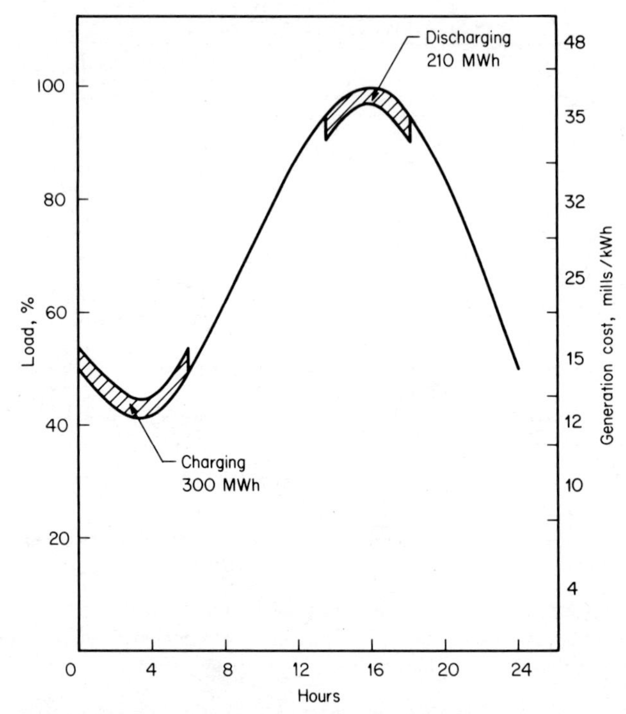

Fig. 3 Example of economic operation of utility system storage.

4. Find Cost Saving When Discharging Storage Device during Peak Period, Displacing Energy from Existing Generation

The energy available from the storage device is $(6)(50)(0.7) = 210$ MWh. The savings from operating the storage, S_{ST}, for 18.9 mills/kWh displacing energy at 35 mills/kWh is $S_{ST} = (210)(35 - 18.9) = \$3381$.

Related Calculations: These savings are energy savings only. Potential capacity savings in conventional generation are dependent on storage capacity and the cost of the storage system. These costs must be included for a total system economic analysis.

COST OF ELECTRICITY

Find the cost of electricity, C_E, from a steam turbine-generator plant having the following parameters:

Plant cost = \$1000/kW
Fixed charge rate = 20 percent
Fixed operation and maintenance costs = \$15/kW per year
Variable operation and maintenance costs = 8 mills/kWh
Heat rate = 10,000 kJ/kWh (9479 Btu/kWh)

Fuel cost = \$1.30/GJ (\$1.37/MBtu)

Capacity factor = 0.70 p.u.

Assumed inflation = 8 percent per year

Assumed discount rate = 12 percent per year

Study period = 30 yr

Calculation Procedure:

1. Find Plant Cost

Use the given values: annual plant cost = (1000)(0.20) = \$200/kW per year and annual operating hours = (8760)(0.70) = 6132 h/yr. Therefore, plant cost = $(200 \times 10^3)/6132$ = 32.6 mills/kWh.

2. Find Fixed Operation and Maintenance Costs

Fixed operation and maintenance costs = $(15 \times 10^3)/6132$ = 2.4 mills/kWh.

3. Compute Levelized Fuel Cost

The levelizing factor is found from $P = S/(1 + i)^N$ and CRF = 0.12414. Hence, fuel cost = $(10{,}000)(1.3)(2.06 \times 10^{-3})$ = 26.8 mills/kWh.

4. Find Levelized Variable Operation and Maintenance Costs

Operation and maintenance costs = (8)(2.06) = 16.5 mills/kWh.

5. Determine Cost of Electricity from Plant

The cost is found by adding the cost components found in Steps 1, 2, 3, and 4: cost of electricity = 32.6 + 2.4 + 26.8 + 16.5 = 78.3 mills/kWh.

Related Calculations: It should be noted that the result is a levelized value, and as such, is only used for comparison with other similar generation alternatives. This value is not directly comparable to the cost of electricity in any one year.

Life-cycle cost calculations may be made by estimating the changing operational and system economic conditions during the estimated lifetime of the unit. The life-cycle cost should be expressed in \$/kW per year and may include differences in unit rating and reliability, as well as the changes in unit capacity factor over the lifetime of the alternatives. Because the total lifetime must be included in the calculation, it is common to use levelized values for the variable components.

The effective capacities of the alternatives, as well as their capacity factors, may be different. It is therefore necessary to assume values for system replacement capacity and system replacement energy costs.

SCREENING CURVES

Screening curves (see Sec. 8) are useful for preliminary screening of alternatives with widely different characteristics. Screening curves also may be used for a preliminary evaluation of a new generation concept in comparison with conventional generation.

The screening curve is a plot of annual cost of generation expressed in \$/kW per year as a function of yearly hours of operation (capacity factor). These curves are straight lines with vertical intercepts at the fixed annual cost and with a slope determined by the variable fuel and operation and maintenance costs; levelized values should be used.

Develop screening curves for the three generation alternatives in Table 14.

TABLE 14 Developing Screening Curves for Three Generation Alternatives

Data	*A*	*B*	*C*
Plant cost, $/kW	900	450	200
Fixed O&M cost, $/kW per year	10	3	1
Fuel price, $/GJ	1.5	4	6
$/MBtu	1.58	4.22	6.33
Heat rate, kJ/kWh	10,000	9,000	13,000
Btu/kWh	9,477	8,530	12,321
Variable O&M cost, $/MWh	8	2	3
Levelizing factor, fuel and O&M	2.06	2.06	2.06
Fixed-charge rate, %	20	20	20

Calculation Procedure:

1. Find the Levelized Fixed Costs

The fixed costs are the plant cost and the fixed operation and maintenance costs in Table 14. The calculated results are given in Table 15.

TABLE 15 Levelized Fixed Costs

Plant	*Plant cost, $/kW per year*	*Levelized fixed O&M, $/kW per year*	*Total fixed costs, $/kW per year*
A	180	20.6	200.6
B	90	6.2	96.2
C	40	2.1	42.1

TABLE 16 Levelized Variable Costs

Plant	*Plant cost, $/kW per year*	*Levelized fixed O&M, $/kW per year*	*Total fixed costs, $/kW per year*
A	30.9	16.5	47.4
B	74.2	4.1	78.3
C	160.7	6.2	166.9

TABLE 17 Total Variable and Fixed Costs

Plant	*Total cost at 0 h/yr, $/kW per year*	*Total cost at 4000 h/yr, $/kW per year*
A	200.6	390.2
B	96.2	409.4
C	42.1	709.7

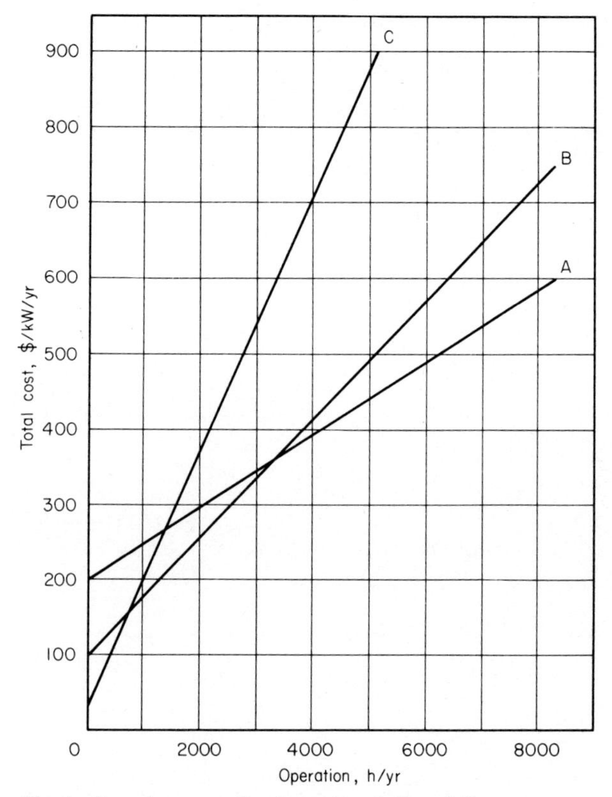

Fig. 4 Screening curves for alternatives A, B, and C.

2. *Determine the Levelized Variable Costs*

From the given values, the calculated values are provided in Table 16.

3. *Assuming 4000-h/yr Operation, Determine the Total Variable plus Fixed Costs*

The values in Step 2 are multiplied by 4000^{-3} to obtain cost in \$/kW per year. The results are summarized in Table 17.

4. *Plot the Screening Curves*

The resulting screening curves are plotted in Fig. 4.

Section 20 LIGHTING DESIGN

John P. Frier
Lighting Application Specialist,
General Electric Company

REFERENCES Frier and Frier—*Industrial Lighting Systems,* McGraw-Hill; Illuminating Engineering Society of North America—*IES Lighting Handbook,* Application Volume; Illuminating Engineering Society of North America—*IES Lighting Handbook,* Reference Volume.

AVERAGE ILLUMINATION LEVEL

A 3000-m^2 storage yard is lighted by two floodlights using 400-W high-pressure sodium lamps rated at 50,000 lumens (lm). The coefficient of utilization is 35 percent and the light loss factor is 80 percent. Calculate the average maintained illumination level.

Calculation Procedure:

1. Use the Lumen Method of Calculation

Because it relates the light produced by any light source to the response of the eye, the lumen output from any light source, regardless of spectral distribution, can be compared directly with the lumen output of any other source. The illumination level produced by a lighting system using a combination of incandescent and fluorescent lamps, for instance, would be the sum of the illumination produced by each system.

The lux (lx) is the International System (SI) unit of illuminance, or illumination: It is equal to the illumination of an area of one square meter produced by a uniform distribution over the surface of one lumen, or 1 lx = 1 lm/m^2. For calculation purposes, the expression is expanded to: $E = (N)(\text{LL})(\text{CU})(\text{LLF})/A$ where

E = the average maintained illumination, lx
N = number of lamps contributing light
LL = the lumen output of the lamp being used
CU = coefficient of utilization, which is the percentage of the lamp lumens falling within the lighted area
LLF = light-loss factor or maintenance factor, which is the product of the lamp lumen depreciation (LLD) and the luminaire dirt depreciation (LDD)
A = lighted area, m^2

2. *Perform the Calculation*

Substituting the given data in the equation for E, we obtain $E = (2)(50{,}000)(0.35)(0.80)/3000 = 9.3$ lx.

Related Calculations: There are two methods used for lighting calculations: the lumen method, which was employed above to calculate the average illumination level, and the point-by-point method, which is used to calculate the illuminance at a point. Both are based on the lumen, which is the integrated product of the radiant energy emitted by a light source in the visible portion of the electromagnetic spectrum (380 to 780 nm) and the photopic (direct) visual efficiency curve of the eye.

ILLUMINATION LEVEL AT A POINT

A floodlight emitting 25,000 candelas (cd) in the center of its beam is aimed at 60° to a point on the ground 20 m away. What is the resulting illumination at that point?

Calculation Procedure:

1. *Use the Point-by-Point Method of Calculation*

Point-by-point calculations are used to determine the illumination level at a point. This is useful in calculating the lighting uniformity and is also helpful in selecting the beam spread and maximum candlepower of the luminaire.

The illumination which any luminaire produces at a point is a function of its luminous intensity cd divided by the distance D, squared: $E = \text{cd}/D^2$. If the illumination strikes the surface at an angle, the area lighted is increased as a function of the cosine of the incident angle θ (Fig. 1). Hence, $E = (\text{cd})(\cos\theta)/D^2 = (25{,}000)(\cos 60°)/20^2 = 31.25$ lx.

2. *Use Another Formula for E*

A formula which is more convenient uses the mounting height MH instead of the diagonal distance D. The formula is $E = (\text{cd})(\cos^3\theta)/\text{MH}^2$. The mounting height is $(20)(\cos 60°) = 10$ m, so $E = (25{,}000)(\cos^3 60°)/10^2 = 31.25$ lx.

Factors to Consider: Calculations are affected by the light-loss factors and the recommended illumination levels for different activities. Major differences occur because of the types of luminaires and lamps that are normally used, and the manner in which the coefficient of utilization is derived.

Light-loss factor The light-loss factor, also called the maintenance factor, is used to increase the initial illumination level to compensate for the normal deterioration of the lighting system in use. The value can be calculated for the mean illumination level, which usually occurs at the midpoint of the luminaire's cleaning and relamping period. The LLF also can be calculated for the end of the relamping period, which is when the luminaire reaches its minimum output and the illumination level is at its lowest point. LLF values are supplied by the manufacturer.

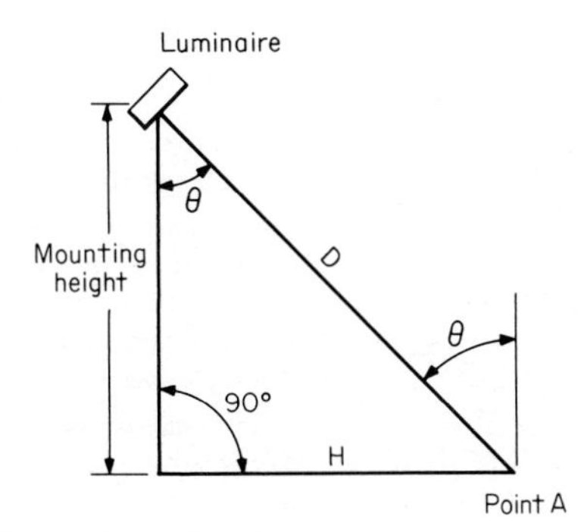

Fig. 1 When illumination strikes a surface at an angle, the area lighted is increased as a function of the cosine of the incident angle θ.

Levels of illumination Complete tables of recommended illumination levels are contained in the *IES Lighting Handbook* published by the Illuminating Engineering Society of North America (IES). The same tables, if adopted as American national standards, are also published by the American National Standards Institute (ANSI).

Interior lighting illumination recommendations (Table 1) consider the general area, the individual task being performed, and the age of the worker. Because of these variables, a range of illuminance values is suggested.

Light-source selection The cost of power to operate a lighting system is a major factor in selecting the type of lamp and luminaire. The lamp's maintained efficacy (lumens per watt, lm/W) and the luminaire coefficient of utilization and maintained efficiency are the key factors. Low-cost, inefficient systems can be justified only if the annual use is very low.

Table 2 provides overall comparison of commonly used lamp types, and Table 3 lists a few of the more popular types. Lamp improvements occur at frequent intervals, so recently published catalog material should be used.

Illuminance requirements can also be determined by measurements and analysis. Changes can also be made if it can be determined that improvements result in greater productivity or accuracy. Illuminance levels in adjacent areas should not vary by more than 3:1.

TABLE 1 Illuminance Recommended for Use in Selecting Values for Interior Lighting Design[1]

Category	*Range of illuminances[2] in lux (footcandles)*	*Type of activity*
A	20–30–50[3] (2–3–5)[3]	Public areas with dark surroundings
B	50–75–100[3] (5–7.5–10)[3]	Simple orientation for short temporary visits
C	100–150–200[3] (10–15–20)[3]	Working spaces where visual tasks are only occasionally performed
D	200–300–500[4] (20–30–50)[4]	Performance of visual tasks of high contrast or large size: e.g., reading printed material, typed originals, handwriting in ink and good xerography; rough bench and machine work; ordinary inspection; rough assembly
E	500–750–1000[4] (50–75–100)[4]	Performance of visual tasks of medium contrast or small size: e.g., reading medium-pencil handwriting, poorly printed or reproduced material; medium bench and machine work; difficult inspection; medium assembly
F	1000–1500–2000[4] (100–150–200)[4]	Performance of visual tasks of low contrast or very small size: e.g., reading handwriting in hard pencil on poor quality paper and very poorly reproduced material; highly difficult inspection
G	2000–3000–5000[5] (200–300–500)[5]	Performance of visual tasks of low contrast and very small size over a prolonged period: e.g., fine assembly; very difficult inspection; fine bench and machine work
H	5000–7500–10,000[5] (500–750–1000)[5]	Performance of very prolonged and exacting visual tasks: e.g., the most difficult inspection; extra fine bench and machine work; extra fine assembly
I	10,000–15,000–20,000[5] (1000–1500–2000)[5]	Performance of very special visual tasks of extremely low contrast and small size: e.g., surgical procedures

[1]Adapted from Table 1.2, *Guide on Interior Lighting,* Publication CIE No. 29 (TC4.1) 1975, Commission Internationale de l'Eclairage, Paris, France.

[2]Maintained in service.

[3]General lighting throughout room.

[4]Illuminance on task.

[5]Illuminance on task, obtained by a combination of general and local (supplementary) lighting.

Coefficient of utilization (CU) The coefficient of utilization is an important factor in the zonal-cavity method of lighting calculation. There are three major factors that influence the CU of an interior lighting system: the efficiency and photometric distribution of the luminaire, the relative shape of the room, and the reflectance of the room surfaces. These factors are combined in a coefficient of utilization table for each luminaire type. Table 4 shows sample tables for four commonly used luminaire types.

TABLE 2 Comparison of Commonly Used Lamp Types (Based on 400-W Sizes)

Lamp	*Initial lumens per watt (LPW)*	*Rated life, h*	*Lamp lumen depreciation (LLD), mean*	*CU†*	*Burning position*	*Minutes*		*Lamp cost*	*Color temperature*
						Warmup	*Hot restart*		
Incandescent	20	1,000 2,000 (quartz)	0.85	High	Any	0	0	Very low	3000 K
Mercury	55	24,000†	0.80*	Medium	Any	5–7	3–6	Low	3900 K
Fluorescent	80	18,000	0.85	Medium-low	Any	0	0	Low	4200 K
Metal halide	85–100	20,000 vert.	0.75–0.80	High	Any	2–4	10–15	Medium	4000 K
High-pressure sodium	125	24,000	0.90	High	Any	3–4	1	High	2100 K

*Based on 16,000 h.

†CU = Coefficient of utilization. Range indoor: high, 0.70+; medium, 0.50–0.70; low, less than 0.50. Range outdoor: high, 0.50+; medium, 0.40–0.50; low, 0.40.

TABLE 3 Characteristics of Some Popular Lamp Types

		Lamp lumen depreciation†			
Lamp type	*Initial lumens**	*Mean*	*End of relamping*	*Life, h‡*	*Luminaire line watts§*
Incandescent					
200 W	4,010			750	200
500 W	10,850			1,000	500
1000 W	23,740			1,000	1000
Fluorescent (low-energy types)					
48-in 40 W	3,050	0.88		20,000	91
96-in slimline	6,000	0.92		12,000	158
96-in high-output	9,100	0.87		12,000	253
96-in 1500 mA	12,300	0.78		10,000	420
Mercury (phosphor-coated, vertical operation)					
175 W	8,600	0.89	0.79	24,000+	210
250 W	12,100	0.86	0.75	24,000+	292
400 W	22,500	0.85	0.71	24,000+	453
1000 W	63,000	0.75	0.54	24,000+	1082
Metal halide (clear)					
175 W	17,500	0.75	0.84	15,000	210
250 W	20,500	0.83	0.69	10,000	300
400 W	40,000	0.75	0.64	20,000	465
1000 W	110,000	0.80	0.70	12,000	1090
High-pressure sodium (clear)					
100 W	9,500	0.90	0.73	24,000+	146
150 W	16,000	0.90	0.73	24,000+	199
250 W	27,500	0.90	0.73	24,000+	313
400 W	50,000	0.90	0.73	24,000+	476
1000 W	140,000	0.90	0.73	24,000+	1062

*Initial lumens can vary from lamp to lamp and with ballast type.

†Mean lumen depreciation is based on 40% life for fluorescent and metal halide lamps and 50% for other types. Life and lamp lumen depreciation for mercury lamps is based on 16,000 hours economic life.

‡Lamp life is affected by the number of hours burned per start and the type of ballast used.

§Fluorescent line watts are for two-lamp luminaires. Line watts for high-intensity discharge luminaires are for constant power or regulator ballasts.

For calculation purposes, the room is divided into three cavities, as shown in Fig. 2. The utilization of the lighting system is a function of the cavity ratio for each section: ceiling-cavity ratio $CCR = 5h_{CC}(L + W)/LW$, room-cavity ratio $RCR = 5h_{RC}(L + W)/LW$, and floor-cavity ratio $FCR = 5h_{FC}(L + W)/LW$, where h_{CC}, h_{RC}, and h_{FC} are as defined in Fig. 2.

The ceiling- and floor-cavity ratios are useful in adjusting the actual reflectance of the ceiling and floor surfaces to their effective reflectance based on the size and depth of the cavity. For shallow cavities (2 m or less), the actual surface reflectance can be used with little error. Correction tables are given in lighting handbooks.

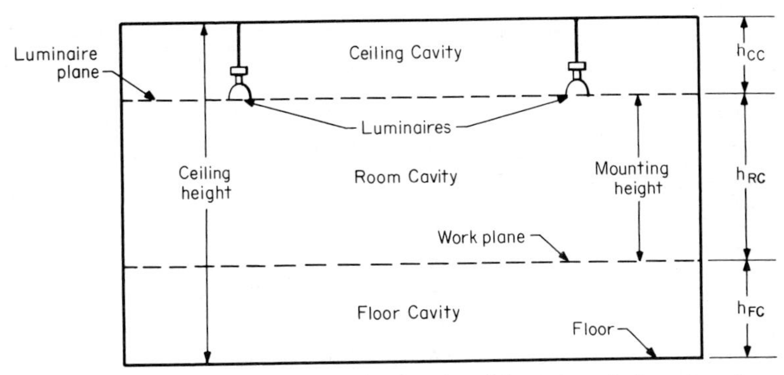

Fig. 2 Utilization of a lighting system is a function of the cavity ratio for each section. (Frier and Frier)

LIGHTING SYSTEM FOR AN INDOOR INDUSTRIAL AREA

A lighting system needs to be designed for a metalworking shop. The area of the shop is 12 m (40 ft) by 60 m (200 ft). (Conversion of meters to feet is approximate.) Total area is therefore 720 m^2 (8000 ft^2). The height of the room cavity, h_{RC}, is 4 m (13 ft). The height of the ceiling and floor cavities is 1 m (3 ft) each. In this facility, medium bench and machine work will be performed. Design an appropriate lighting system.

Calculation Procedure:

1. Find the Recommended Level of Illumination

The level recommended in Table 1 for medium bench and machine work (category E) is 1000 lx (100 fc) to allow older workers to see accurately.

2. Choose a Lamp Type and Lumen Rating

A high-pressure sodium (HPS) system is a good choice because of its high efficiency and long life. For machine shops or any manufacturing space requiring accurate seeing with a minimum of shadows, the maximum luminaire spacing should be close to the mounting height above the work plane. Luminaire spacing beyond 1½ times the mounting height will usually produce poor results in a manufacturing area.

As a rough approximation, half the initial lamp lumens are effective in producing the maintained illumination level. This can be used to calculate the maximum lamp lumens which each luminaire can have. The area per luminaire is equal to the spacing squared.

The maintained illumination lx (maintained) is lx (maintained) = ½(LL/area per luminaire). For a room cavity height of 4 m and a maintained illumination level of 1000 lx, the lamp lumens for a luminaire spacing equal to the mounting height would be: $1000 = \frac{1}{2}(LL/4^2)$. Solving, find LL = 32,000 lm. For a spacing equal to 1½ times the mounting height, LL = 72,000 lm. A 400-W high-pressure sodium lamp at 50,000 lm (Table 3) would be a good choice.

3. Select the Luminaire

When shiny metal surfaces are being worked on, it is desirable to have a luminaire with a refractor. The refractor spreads the light over a large area and prevents the lamp from being seen as a bright reflection on the surface of the work. In this problem, a high-intensity discharge (HID) luminaire which combines a reflector and refractor will be

TABLE 4 Coefficient of Utilization (Zonal-Cavity Method) for Four Commonly Used Luminaires

Typical luminaire	Typical distribution and percent lamp lumens	ρ_{CC}* →	80			70			50			30			10			0
		ρ_w† →	50	30	10	50	30	10	50	30	10	50	30	10	50	30	10	0
	Spacing criteria (SC)§	RCR‡ ↓	Coefficients of utilization for 20 percent effective floor cavity reflectance (ρ_{FC} = 20)															
Fluorescent unit with half prismatic lens, 4-lamp 2' wide; multiply by 1.10 for 2 lamp	SC = 1.4/1.2; 0% ↑; 60% ↓; 60°; ∥; ⊥	0	0.71	0.71	0.71	0.69	0.69	0.69	0.66	0.66	0.66	0.63	0.63	0.63	0.61	0.61	0.61	0.60
		1	0.65	0.63	0.61	0.63	0.62	0.60	0.61	0.59	0.58	0.59	0.57	0.56	0.57	0.56	0.55	0.54
		2	0.59	0.55	0.53	0.57	0.55	0.52	0.55	0.53	0.51	0.54	0.52	0.50	0.52	0.50	0.49	0.48
		3	0.53	0.49	0.46	0.52	0.49	0.46	0.50	0.47	0.45	0.49	0.46	0.44	0.47	0.45	0.43	0.42
		4	0.48	0.44	0.40	0.47	0.43	0.40	0.46	0.42	0.40	0.45	0.42	0.39	0.43	0.41	0.39	0.38
		5	0.43	0.39	0.35	0.43	0.38	0.35	0.42	0.38	0.35	0.40	0.37	0.34	0.39	0.36	0.34	0.33
		6	0.39	0.35	0.31	0.39	0.34	0.31	0.38	0.34	0.31	0.37	0.33	0.31	0.36	0.33	0.31	0.29
		7	0.36	0.31	0.28	0.35	0.31	0.28	0.34	0.30	0.27	0.33	0.30	0.27	0.33	0.30	0.27	0.26
		8	0.32	0.27	0.24	0.32	0.27	0.24	0.31	0.27	0.24	0.30	0.27	0.24	0.30	0.26	0.24	0.23
		9	0.29	0.24	0.21	0.29	0.24	0.21	0.28	0.24	0.21	0.27	0.24	0.21	0.27	0.23	0.21	0.20
		10	0.26	0.22	0.19	0.26	0.22	0.19	0.25	0.21	0.19	0.25	0.21	0.19	0.24	0.21	0.18	0.17
Porcelain-enameled reflector with 14° CW shielding	SC = 1.3; 13% ↑; 74% ↓	0	1.00	1.00	1.00	0.96	0.96	0.96	0.89	0.89	0.89	0.82	0.82	0.82	0.76	0.76	0.76	0.73
		1	0.88	0.85	0.82	0.85	0.82	0.79	0.79	0.77	0.74	0.73	0.72	0.70	0.68	0.67	0.66	0.63
		2	0.78	0.72	0.67	0.75	0.70	0.66	0.70	0.66	0.62	0.65	0.62	0.59	0.61	0.58	0.56	0.53
		3	0.69	0.62	0.57	0.66	0.60	0.56	0.62	0.57	0.53	0.58	0.54	0.51	0.54	0.51	0.48	0.46
		4	0.61	0.54	0.48	0.59	0.52	0.47	0.55	0.50	0.45	0.52	0.47	0.43	0.49	0.45	0.42	0.39
		5	0.54	0.46	0.41	0.52	0.45	0.40	0.49	0.43	0.39	0.46	0.41	0.37	0.43	0.39	0.36	0.33
		6	0.48	0.41	0.35	0.47	0.40	0.35	0.44	0.38	0.34	0.41	0.36	0.32	0.39	0.34	0.31	0.29
		7	0.43	0.36	0.31	0.42	0.35	0.30	0.40	0.34	0.29	0.37	0.32	0.28	0.35	0.31	0.27	0.25
		8	0.39	0.32	0.27	0.38	0.31	0.26	0.36	0.30	0.25	0.34	0.28	0.24	0.32	0.27	0.24	0.22
		9	0.35	0.28	0.23	0.34	0.27	0.23	0.32	0.26	0.22	0.30	0.25	0.21	0.28	0.24	0.20	0.19
		10	0.32	0.25	0.20	0.31	0.24	0.20	0.29	0.23	0.19	0.28	0.22	0.19	0.26	0.21	0.18	0.17

Wide-distribution ventilated reflector with clear HID lamp	SC = 1.5; $\frac{1}{2}$% ↑; $77\frac{1}{2}$% ↓	0	0.92	0.92	0.92	0.90	0.90	0.90	0.86	0.86	0.86	0.82	0.82	0.82	0.79	0.79	0.79	0.77
		1	0.85	0.82	0.80	0.83	0.81	0.79	0.79	0.78	0.76	0.76	0.75	0.74	0.74	0.72	0.71	0.70
		2	0.77	0.73	0.70	0.75	0.72	0.69	0.73	0.70	0.67	0.70	0.68	0.66	0.68	0.66	0.64	0.63
		3	0.70	0.65	0.61	0.68	0.64	0.60	0.66	0.62	0.59	0.64	0.61	0.58	0.62	0.59	0.57	0.56
		4	0.63	0.58	0.53	0.62	0.57	0.53	0.60	0.56	0.52	0.58	0.55	0.52	0.57	0.54	0.51	0.49
		5	0.57	0.51	0.47	0.56	0.51	0.47	0.55	0.50	0.46	0.53	0.49	0.46	0.52	0.48	0.45	0.44
		6	0.51	0.45	0.41	0.51	0.45	0.41	0.49	0.44	0.40	0.48	0.43	0.40	0.47	0.43	0.40	0.38
		7	0.46	0.40	0.35	0.45	0.39	0.35	0.44	0.39	0.35	0.43	0.38	0.35	0.42	0.38	0.34	0.33
		8	0.41	0.35	0.31	0.41	0.35	0.31	0.40	0.34	0.31	0.39	0.34	0.30	0.38	0.33	0.30	0.29
		9	0.37	0.31	0.27	0.37	0.31	0.27	0.36	0.30	0.27	0.35	0.30	0.27	0.34	0.30	0.26	0.25
		10	0.33	0.27	0.24	0.33	0.27	0.23	0.32	0.27	0.23	0.31	0.27	0.23	0.31	0.26	0.23	0.22
Low mounting-height reflector-refractor combination with clear HID lamp	SC = 1.8; 3% ↑; 73% ↓	0	0.88	0.86	0.84	0.86	0.84	0.82	0.81	0.80	0.78	0.77	0.76	0.75	0.73	0.73	0.72	0.70
		1	0.78	0.75	0.72	0.76	0.73	0.71	0.73	0.70	0.68	0.69	0.67	0.66	0.66	0.64	0.63	0.61
		2	0.69	0.64	0.61	0.67	0.63	0.59	0.64	0.60	0.57	0.61	0.58	0.56	0.58	0.56	0.54	0.52
		3	0.60	0.55	0.50	0.58	0.53	0.49	0.56	0.51	0.48	0.53	0.50	0.47	0.51	0.48	0.45	0.43
		4	0.52	0.46	0.41	0.50	0.45	0.40	0.48	0.44	0.40	0.46	0.42	0.39	0.44	0.41	0.38	0.36
		5	0.44	0.38	0.34	0.43	0.37	0.33	0.41	0.37	0.33	0.40	0.35	0.32	0.38	0.34	0.31	0.30
		6	0.37	0.32	0.28	0.36	0.31	0.27	0.35	0.31	0.27	0.34	0.30	0.27	0.33	0.29	0.26	0.25
		7	0.31	0.26	0.23	0.31	0.26	0.22	0.30	0.25	0.22	0.29	0.25	0.22	0.28	0.24	0.22	0.20
		8	0.26	0.22	0.19	0.26	0.22	0.19	0.25	0.21	0.18	0.24	0.21	0.18	0.23	0.20	0.18	0.17
		9	0.22	0.18	0.16	0.21	0.18	0.16	0.21	0.18	0.16	0.20	0.18	0.16	0.19	0.17	0.16	0.15
		10	0.18	0.16	0.15	0.18	0.16	0.15	0.17	0.16	0.15	0.17	0.15	0.14	0.16	0.15	0.14	0.14

*ρ_{CC} = percent effective ceiling-cavity reflectance.

†ρ_w = percent wall reflectance.

‡RCR = room-cavity ratio.

§SC = ratio of maximum luminaire spacing to mounting or ceiling height above work plane.

used. At a mounting height of 4 m, an enclosed industrial luminaire designed for low mounting heights is the best choice. Its wide beam produces complete overlap of the light from adjacent luminaires.

4. *Find the Coefficient of Utilization*

Table 4 gives the CU value for the chosen luminaire, which has a spacing criterion SC of 1.8. Assume a 30 percent ceiling-cavity reflectance, a 30 percent wall reflectance, and a 20 percent floor-cavity reflectance. The room-cavity ratio RCR is equal to $5h_{RC}(L + W)/LW = (5)(4)(60 + 12)/(60)(12) = 1440/720 = 2$. From the table, CU = 0.58 for this 400-W luminaire.

5. *Determine the Light-Loss Factors*

There are several light-loss factors which can cause the illumination level to depreciate in service. Of these, the two most commonly used are the lamp lumen depreciation and the luminaire dirt depreciation. The mean LDD value for a 400-W high-pressure sodium lamp is 0.9 (Table 3).

The LLD values vary considerably and can only be predicted accurately by experience with similar-type luminaires under similar service conditions. Suggested values are given in Table 5. If conditions within the plant are known to cause greater deterioration, it is more economical to clean the system at frequent intervals than to increase the initial illumination level to compensate for the loss.

TABLE 5 LDD Values for Some Luminaire Types

	Luminaire dirt depreciation (LDD)		
Luminaire type	*Light*	*Medium*	*Heavy*
Enclosed and filtered	0.97	0.93	0.88
Enclosed	0.94	0.86	0.77
Open and ventilated	0.94	0.84	0.74

For the enclosed and filtered 400-W luminaire, the light loss factor is $(0.9)(0.93) = 0.84$.

6. *Determine the Number of Luminaires N*

Use $N = EA/(\text{LL})(\text{CU})(\text{LLF})$ where E = maintained level of illumination, A = area of space to be lighted, LL = initial-rate lamp lumens, CU = coefficient of utilization, and LLF = light-loss factor. Hence, $N = (1000)(720)/(50{,}000)(0.58)(0.84) = 29.6$; use 30 luminaires.

7. *Determine the Final Luminaire Quantity and Spacing*

The average square spacing S of the luminaires can be determined by $S = \sqrt{A/N} = \sqrt{720/30} = 4.89$ m (16.1 ft).

The final luminaire quantity is frequently a compromise between the calculated quantity and the shape of the area being lighted. In this case, the length of the area is 5 times the width, so the number of luminaires per row should be approximately 5 times the number of rows. Thirty luminaires can be divided only into 2 rows of 15 or 3 rows of 10. Because the lighting uniformity and coverage will be better with three rows across the area, this would be the best solution.

There is no fixed rule on luminaire spacing. The final spacing, however, should be as close to the average square spacing as possible. The spacing between luminaires should also be twice the distance of the spacing from the walls.

Related Calculations: Check the luminaire spacing criterion to ensure that it exceeds the actual spacing-to-mounting-height ratio. The spacing-to-mounting height ratios are 1:1 and 1.5:1 for crosswise and lengthwise spacing, respectively. The luminaire spacing criterion is 1.8 so the uniformity should be excellent.

LIGHTING SYSTEM FOR AN OUTDOOR AREA

Determine the illumination level produced by a 400-W high-pressure sodium floodlight on an area 20 m wide and 40 m long. The floodlight is located on a 10-m pole midway along the 40-m side of the area (see Fig. 3). Also determine the illumination at points *A*, *B*, and *C*. The area is a construction site where excavation work will be done.

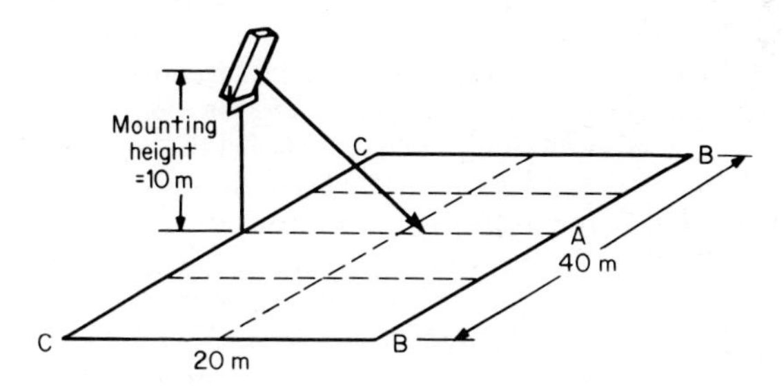

Fig. 3 Layout for the outdoor area-lighting design.

Calculation Procedure:

*1. **Determine the Luminaire Mounting Height (MH)***

If possible, the luminaire mounting height should be at least one-half the width of the area being lighted. Lower mounting heights will reduce utilization and uniformity.

*2. **Calculate the Aiming Angle***

Maximum utilization is obtained when the aiming line bisects the angle drawn between the near and far side of the lighted area. This will usually produce poor uniformity because there will be insufficient light at the far edge. The highest illumination level that can be produced away from the floodlight location occurs when the maximum floodlight candela value is directed at an angle of 54.7°, as in Fig. 4. This is approximately a 3-4-5 triangle, as shown. For an area which is twice as wide as the mounting height, this would be a logical aiming point. In this case, aim the floodlight 13 m across the 20-m-wide area.

*3. **Determine the Utilization Factor***

The most difficult part of area-lighting design is determining the location of the edges of the lighted area in relation to the floodlight aiming line. This is needed to determine the lumens which will fall inside the area. This can be greatly simplified if the area is divided into a grid having the same scale as the floodlight mounting height (Fig. 5).

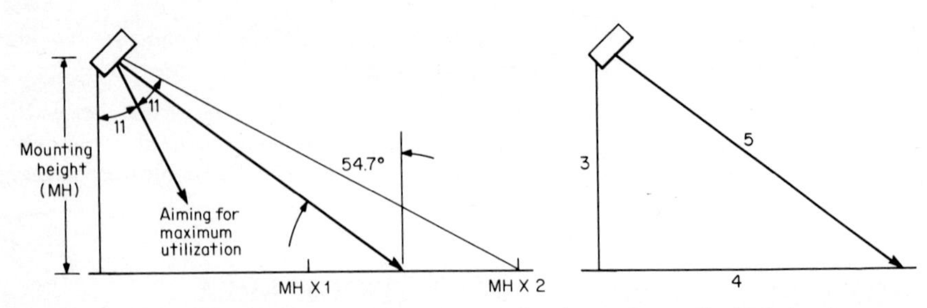

Fig. 4 The highest illumination level that can be produced away from a floodlight location occurs when the maximum floodlight candela value is directed at an angle of 54.7°. (Frier and Frier)

The sample area is 2 mounting heights (2 × 10 m = 20 m) wide and 4 mounting heights (4 × 10 m = 40 m) long. The floodlight is aimed 13/10 = 1.3 mounting heights across the lighted area. The luminaire is mounted in the center, so the area extends 2 mounting heights left and right of the aiming line. In all cases the grid must be drawn in line with and perpendicular to the aiming line, not the edges of the area being lighted. In this example they coincide, but this will not always be the case.

In order to determine the utilization, the dimensions of the lighted area must be translated into the beam dimensions of the floodlight. These are given in horizontal and vertical degrees. Figure 6 is used to convert from one to the other.

Plot the lighted area and aiming point on the chart. The area is 2 mounting heights wide and extends 2 mounting heights to each side of the aiming line. The aiming point is 1.3 MH across the area. This corresponds to a vertical angle of 53°.

The 53° aiming line corresponds to the 0.0 line on the photometric data for this floodlight, which is given in Fig. 7. The far edge corresponds to 63° or 10° above the aiming line. The lower edge is at the base of the luminaire or 53° below the aiming line. Locate these two lines on the lumen distribution chart in Fig. 7.

The right and left edges of the lighted area can be located on the lumen distribution curve by reference to the curved lines on the chart (Fig. 6). The far corner makes an angle of 42° with the luminaire. Plot the other points for the 10° increments on the photometric data below the aiming point. For instance, at 53° (0° on the photometric curve) the angle

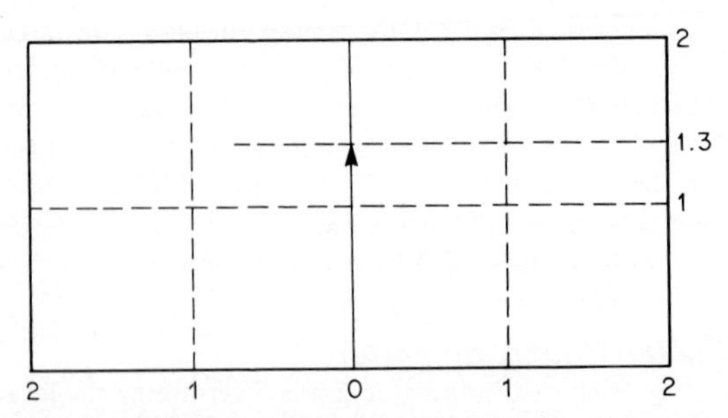

Fig. 5 Lighted area divided into a grid having the same scale as the floodlight mounting height.

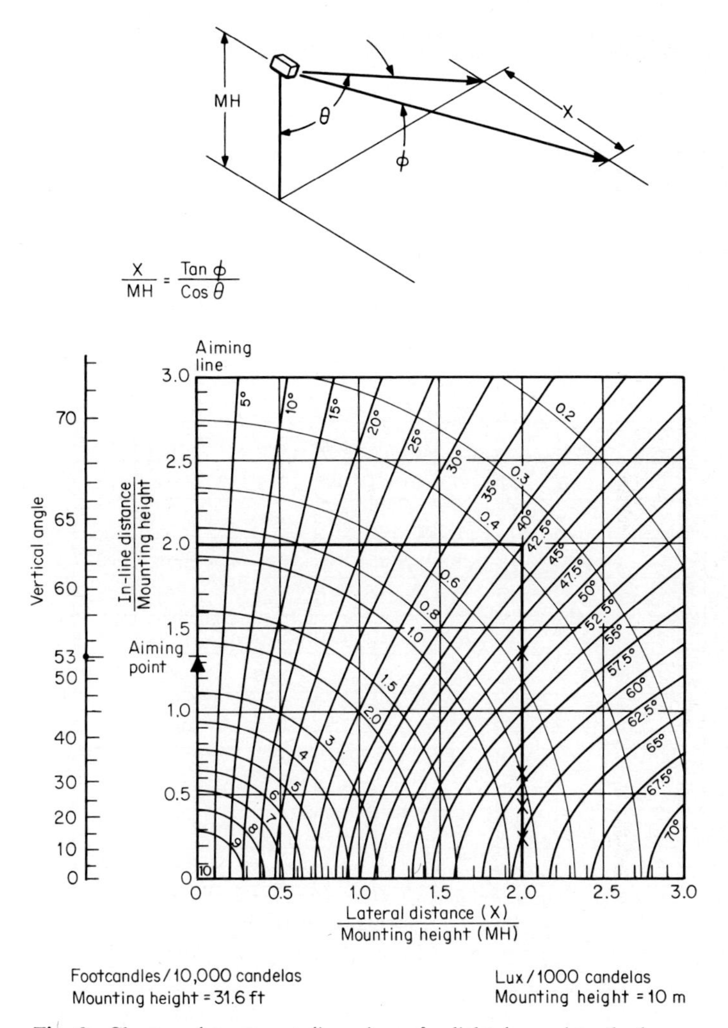

Fig. 6 Chart used to convert dimensions of a lighted area into the beam dimensions of a floodlight. It also can be used to find the level of illumination at a point. (Frier and Frier)

is 50°, and so forth. At the base of the pole the angle is 53°. A curved line drawn through all the points will show the location of the sides of the area on the photometric curve. In this case, only the right side was plotted. If the two sides are different, each would have to be plotted separately.

The total lumens which fall on the area are the sum of the lumens falling inside the area's boundary. To determine this, add lumen values shown in each block. Where the boundary cuts through the block, estimate the percentage of the value in the block that

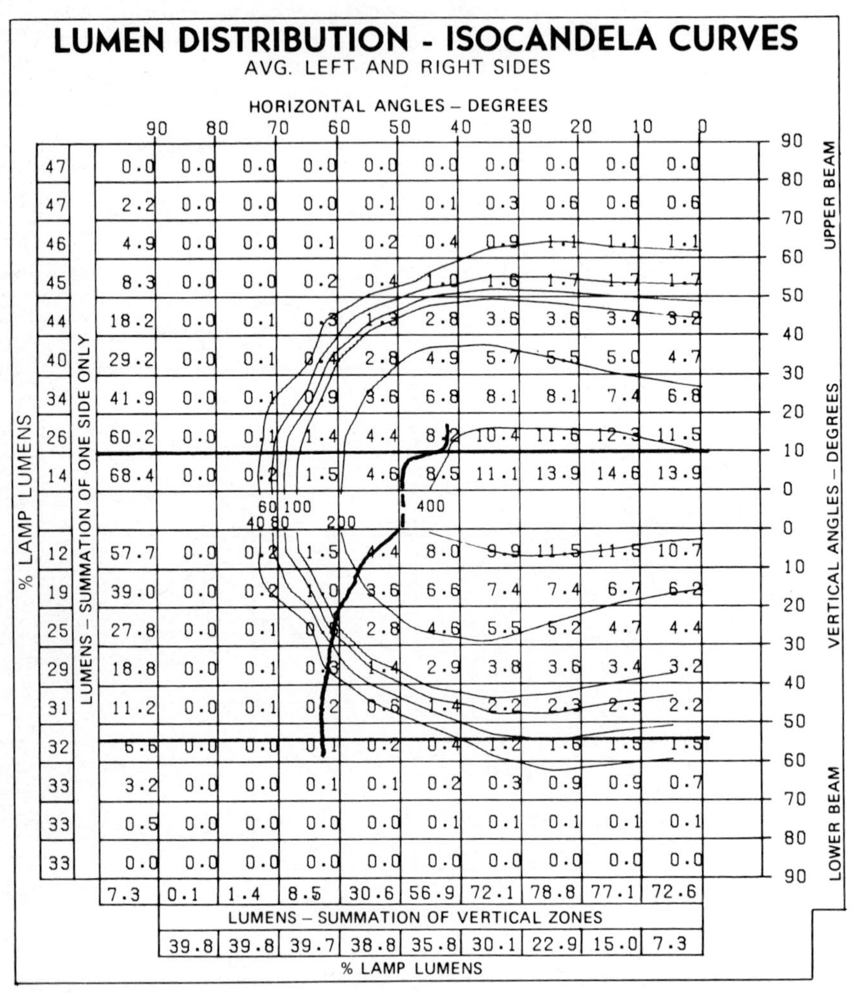

Fig. 7 Photometric data for a floodlight used in the outdoor area-lighting design. Plotted line indicates the right side of the lighted area. (General Electric Company, Lighting System Dept.)

falls inside the line. This would be proportional to the area. Table 6 provides the total lumens falling in the area of the sample problem.

4. *Calculate the Level of Illumination*

Average lux = (LL)(CU)(LLD)(LDD)/Area = (50,000)(0.4168)(0.9)(0.95)/(20)(40) = 22.3 lx. Table 7 shows that the minimum average recommended lux for this area is 20; this lighting system, therefore, will adequately light the construction site.

5. *Find Point-by-Point Illumination Values*

Refer to Fig. 6. The circular lines are *isolux* values based on 1000 cd. To determine the illumination at any point, the candela value of the floodlight at the same point must

TABLE 6 Total Lumens Falling on Outdoor Area-Lighting System

Vertical zone	*Horizontal angles*							*Total*
	0–10	*10–20*	*20–30*	*30–40*	*40–50*	*50–60*	*60–70*	
0–10	13.9	14.6	13.9	11.1	5.1*			58.6
0–10	10.7	11.5	11.5	9.9	8.0	2.2*		53.8
10–20	6.2	6.7	7.4	7.4	6.6	2.9		37.2
20–30	4.4	4.7	5.2	5.5	4.6	2.8	0.1*	27.3
30–40	3.2	3.4	3.6	3.8	2.9	1.4	0.1*	18.4
40–50	2.2	2.3	2.2	1.4	1.4	0.6	0.1*	11.1
50–60	0.5*	0.5*	0.5*	0.4*	0.1*			2.0
							Total right side	208.4
							Left side	208.4
							Total lumens	416.8
							Total lamp lumens 1000	
							Coefficient of utilization 41.68%	

*Estimated.

TABLE 7 Illumination Levels Recommended for Outdoor Lighting

General application	*Minimum average recommended lux*
Construction	
General	100
Excavation	20
Industrial roadways	
Adjacent to buildings	10
Not bordered by buildings	5
Industrial yard/material handling	50
Parking areas	
Industrial	10
Shopping centers	20–50
Commercial lots	20–50
Protective	
Entrances (active)	50
Building surrounds	10
Railroad yards	
All switch points	20
Body of yard	10
Shipyards	
General	50
Ways	100
Fabrication areas	300
Storage yards	
Active	200
Inactive	10

be known. The points can be located the same way as the area boundaries were determined. The mounting height used in the chart is 10 m. To calculate the level of illumination at any point, use the following relation: lux = [lux (from chart)](cd/1000)[(LL)/(LF)](LLF)(MHCF), where:

lux (from chart) = illumination in lx/1000 cd (Fig. 6)

cd = candela value taken from the photometric data isocandela curves (Fig. 7) at the same horizontal and vertical angles as indicated by the chart (must be corrected by dividing by 1000)

LL = lamp lumens

LF = lamp factor, which corrects the lamp lumens used in the photometric data to the rated lamp lumens used in the floodlight (in this case LF = 1000)

LLF = light loss factor = lamp lumen depreciation times luminaire dirt depreciation

MHCF = mounting-height correction factor, the ratio of the square of the mounting height used in the chart (Fig. 6) to the mounting height used in the problem (in this case, MHCF = 100/ MH^2 = 100/100 = 1)

Point *A* in Fig. 3 is along the aiming line. In Fig. 6 it is located at 63°, or 2MH. This point is between the 0.8-lx and 1.0-lx lines, so the value 0.9 can be given. Point *A* is 10° above the aiming point of 53°; in Fig. 7, this matches the isocandela curve marked 400. The candela value for this problem is, therefore, 400.

Substituting in the formula for point *A* yields lux = (0.9)(400/1000)(50,000/1000)(0.9)(0.95)(1) = 17.1 lx. Point *B* is located at a horizontal angle of 42° and the vertical angle is 10°. Substituting in formula, one obtains lux = (0.38)(400/1000)(50,000/1000)(0.9)(0.95)(1) = 7.22 lx.

Point *C* is outside the beam of the floodlight. The only contribution will be spill light from the floodlight. In practice, more than one floodlight can be used at each location to avoid the dark area along the near sideline.

ROADWAY LIGHTING SYSTEM

The information already known is that the street width is 20 m, the mounting height is 12 m, and the overhang of the luminaire is 2 m. The required average maintained level of illumination is 16 lx. It is necessary to determine the staggered spacing required to provide the specified illumination level, as well as the uniformity of illumination with staggered spacing.

Calculation Procedure:

1. Find the Level of Illumination

All the information needed to solve for the average illumination level and the minimum level is contained in the photometric data. Each roadway luminaire can be adjusted to produce a number of different light-distribution patterns. In addition, a luminaire can

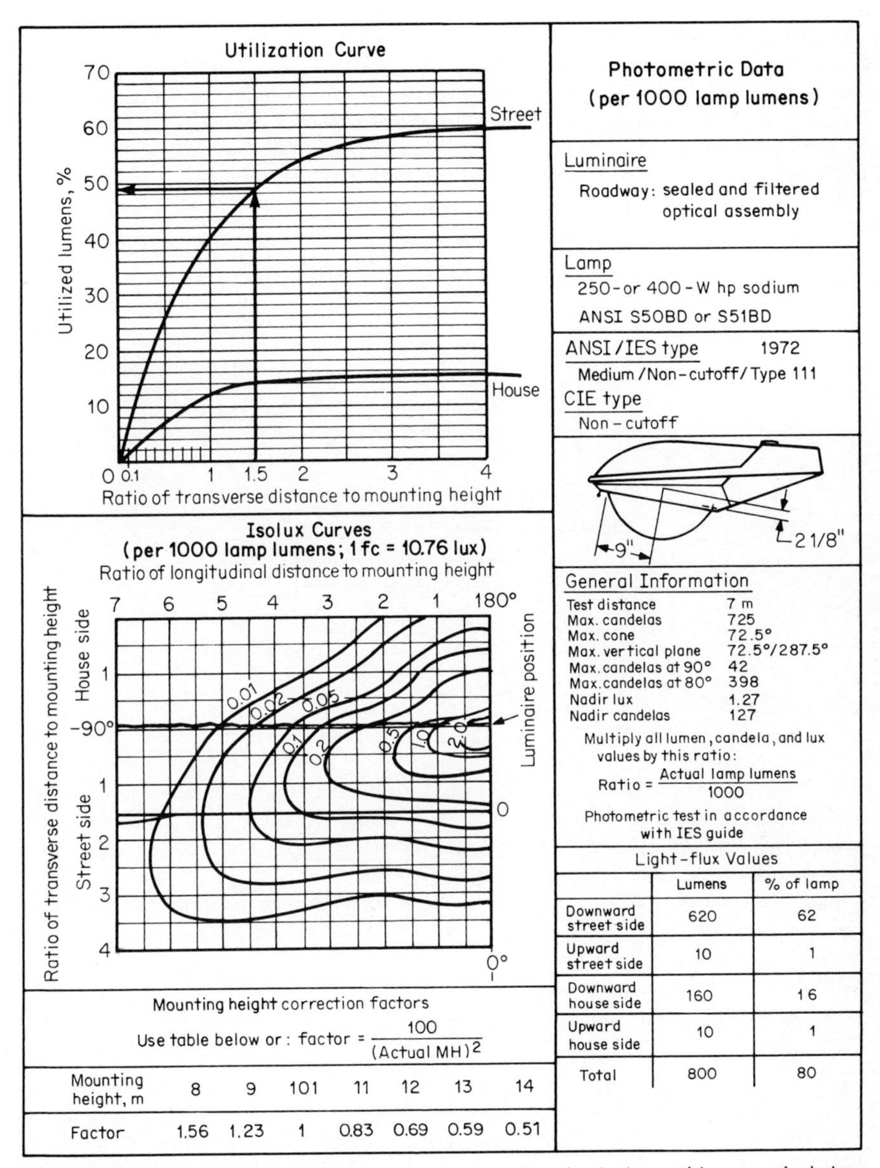

Light-flux Values	Lumens	% of lamp
Downward street side	620	62
Upward street side	10	1
Downward house side	160	16
Upward house side	10	1
Total	800	80

Mounting height, m	8	9	101	11	12	13	14
Factor	1.56	1.23	1	0.83	0.69	0.59	0.51

Fig. 8 Photometric data for a high-pressure sodium roadway luminaire used in a sample design. (General Electric Company, Lighting System Dept.)

usually use several different lamp wattages. It is necessary, therefore, to obtain from the manufacturer the photometric curves for a family of luminaires. Figure 8 provides photometric data for a 250- or 400-W high-pressure sodium luminaire. For this problem, the 250-W source will be used.

The equation for finding average illumination E is: $E = (\mathrm{LL})(\mathrm{CU})(\mathrm{LLF})/WS$

where W is the curb-to-curb street width and the other terms are as defined previously. Values are already given for E and W. For a 250-W HPS lamp, LL = 27,500 lm. Light-loss factors for roadway lighting are always determined for the minimum illumination in service. This occurs at the point where the luminaire is cleaned and relamped. For a 250-W HPS lamp, the LLD is 0.73 (Table 3). The suggested LDD for a sealed and filtered luminaire is 0.93 (Table 5). After the value for the CU is determined, it is then only necessary to solve the equation for average spacing S.

2. *Determine the Coefficient of Utilization*

Roadway luminaires are usually mounted over the roadway. They are leveled parallel to the roadway in an orientation perpendicular to the curb, as in Fig. 9. The area of the

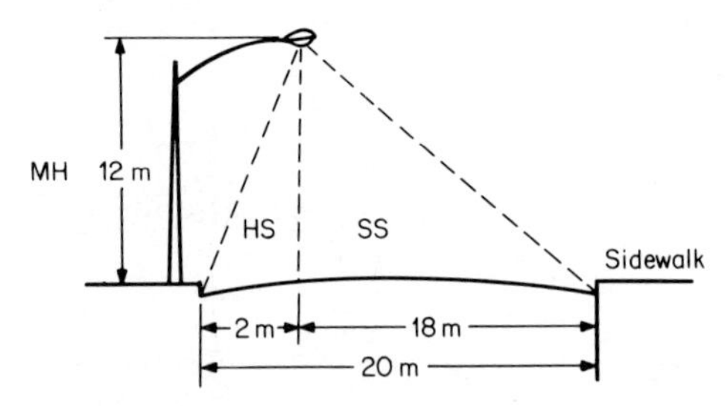

Fig. 9 Luminaire mounted over a roadway. (Based on Fig. 11-1, Frier and Frier)

road in front of the luminaire is called the street side (SS) and the area behind is called the curb, or house side (HS). The utilization of the luminaire is a function of the total lamp lumens which fall in each of the triangular sections.

Because the lumens falling on the roadway are the same if the angle subtended by the two curbs is the same, the utilization can be made a function of the ratio of the street side or house side transverse distance divided by the mounting height. The roadway is assumed to be continuous on either side of the luminaire.

The coefficient of utilization for each side can be read from the utilization curve in Fig. 8:

Ratio 1.35, street side, corresponds to CU	= 49%
Ratio 0.15, house side, corresponds to CU	= 2%
Total CU	= 51%

3. *Determine the Staggered Spacing Required*

Average staggered spacing can be determined by rewriting the basic equation and solving: $S = (\mathrm{LL})(\mathrm{CU})(\mathrm{LLD})(\mathrm{LDD})/EW = (27{,}500)(0.51)(0.73)(0.93)/(16)(20) = 9522/320 = 29.8$, or 30 m.

4. *Determine the Uniformity of Illumination*

The uniformity of illumination is usually expressed in terms of the ratio (average lux)/(minimum lux). The average was given as 16 lx at a spacing of 30 m, staggered.

The point of minimum illumination can be found by studying the isolux curves in Fig. 8 and taking contributions from all luminaires into account. Generally, the minimum value will be found along a line halfway between two consecutively spaced luminaires.

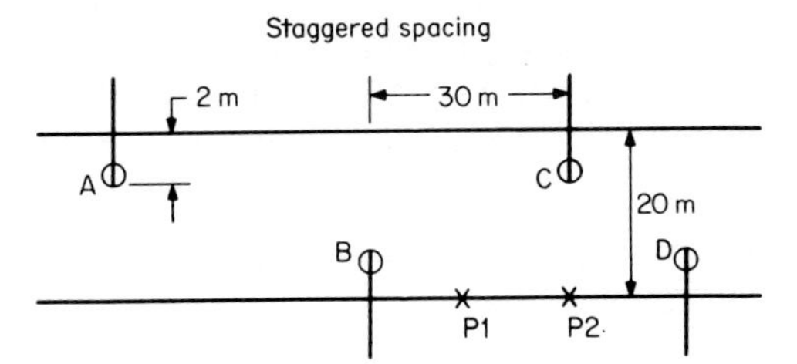

Fig. 10 Determining illumination at points *P*1 and *P*2.

However, this is not always the case, depending upon the geometry of the situation and the distribution pattern of the luminaire.

Related Calculations: All significant illumination contributions from luminaires should be determined for points *P*1 and *P*2, as illustrated in Fig. 10, in order to make sure the minimum illumination point is located. First, determine both transverse and longitudinal ratios of distance to mounting height relative to each of the luminaires. Using the ratios in Table 8 as coordinates, corresponding illumination values can be read from the isolux plot of Fig. 8. These values are tabulated as "illumination at test points" in Table 8.

TABLE 8 Values of Illumination at Test Points *P*1 and *P*2

Contributing luminaires	*Ratios for test points*				*Illumination at test points*	
	Transverse ratio		*Longitudinal ratio*			
	*P*1	*P*2	*P*1	*P*2	*P*1	*P*2
A	1.5	1.5	3.75	5	0.08	0.03
B	0.167	0.167	1.25	2.5	0.5	0.19
C	1.5	1.5	1.25	0	0.15	0.19
D	0.167	0.167	3.75	2.5	0.05	0.19
					Totals 0.78	0.60

The minimum illumination value of 0.6 lx occurs at point *P*2. This value is the initial value per 1000 lamp lumens. To convert to the actual maintained illumination level, use lux = [lx (min)](LL/1000)(LLD)(LDD)(MHCF). The mounting-height correction factor MHCF can be read from the chart in Fig. 8 as 0.69. Substituting values, find (0.6)(27,500/1000)(0.73)(0.93)(0.69) = 7.73 lx. The average to minimum illumination level is therefore 16/7.73 = 2.1:1. Because the recommended maximum ratio is 3:1, this spacing will produce better than minimum uniformity.

INDEX

ABOUT THE EDITORS

Arthur E. Seidman is Professor of Electrical Engineering, School of Engineering, Pratt Institute. He has many years of experience in industry, including a position as senior engineer at Sperry Gyroscope. He is the coauthor (with Milton Kaufman) of *Handbook of Electronics Calculations* and *Handbook for Electronics Engineering Technicians* (both McGraw-Hill).

Haroun Mahrous, D.Sc., P.E., is Professor of Electrical Engineering, School of Engineering, Pratt Institute. Dr. Mahrous has extensive industrial experience with IBM, Bell Telephone Laboratories, LILCO, Grumman, and the National Science Foundation.

Tyler G. Hicks, P.E., is Principal, International Engineering Associates. He has worked in both plant design and operation in a variety of industries and has taught at several engineering schools. Mr. Hicks is editor-in-chief of *Standard Handbook of Engineering Calculations* (McGraw-Hill) and author or coauthor of many other books.